Numerical Methods with Chemical Engineering Applications

Designed primarily for undergraduates, but also graduates and practitioners, this textbook integrates numerical methods and programming with applications from chemical engineering. Combining mathematical rigor with an informal writing style, it thoroughly introduces the theory underlying numerical methods, its translation into MATLAB programs, and its use for solving realistic problems. Specific topics covered include accuracy, convergence and numerical stability, as well as stiffness and ill-conditioning. MATLAB codes are developed from scratch, and their implementation is explained in detail, all while assuming limited programming knowledge. All scripts employed are downloadable, and built-in MATLAB functions are discussed and contextualized. Numerous examples and homework problems – from simple questions to extended case studies – accompany the text, allowing students to develop a deep appreciation for the range of real chemical engineering problems that can be solved using numerical methods. This is the ideal resource for a single-semester course on numerical methods, as well as other chemical engineering courses taught over multiple semesters.

Kevin D. Dorfman is Professor of Chemical Engineering and Materials Science at the University of Minnesota. He is the recipient of the Colburn Award of the AIChE, the DARPA Young Investigator Award, a Packard Fellowship, the NSF CAREER Award, and both the Dreyfus New Faculty and Teacher-Scholar Awards. He has co-authored over 100 papers.

Prodromos Daoutidis is Professor and Executive Officer in the Department of Chemical Engineering and Materials Science at the University of Minnesota. He is the recipient of the NSF CAREER Award, the PSE Model Based Innovation Prize, the Ted Peterson Award of the AIChE, the McKnight Land Grant Professorship, the Ray D. Johnson/Mayon Plastics Professorship, and the Shell Chair. He has been a Humphrey Institute Policy Fellow. He has co-authored two research monographs and over 220 papers.

Cambridge Series in Chemical Engineering

SERIES EDITOR
Arvind Varma, *Purdue University*

BOOKS IN SERIES

Numerical Methods with Chemical Engineering Applications

KEVIN D. DORFMAN AND PRODROMOS DAOUTIDIS

CAMBRIDGE
UNIVERSITY PRESS

University Printing House, Cambridge CB2 8BS, United Kingdom

One Liberty Plaza, 20th Floor, New York, NY 10006, USA

477 Williamstown Road, Port Melbourne, VIC 3207, Australia

314-321, 3rd Floor, Plot 3, Splendor Forum, Jasola District Centre, New Delhi - 110025, India

79 Anson Road, #06-04/06, Singapore 079906

Cambridge University Press is part of the University of Cambridge.

It furthers the University's mission by disseminating knowledge in the pursuit of education, learning and research at the highest international levels of excellence.

www.cambridge.org
Information on this title: www.cambridge.org/9781107135116
DOI: 10.1017/9781316471425

First published 2017

A catalogue record for this publication is available from the British Library

Library of Congress Cataloging in Publication data
Names: Dorfman, Kevin D., author. | Daoutidis, Prodromos, author.
Title: Numerical methods with chemical engineering
  applications / Kevin D. Dorfman and Prodromos Daoutidis.
Description: Cambridge, United Kingdom ; New York, NY :
  Cambridge University Press, [2017] | Series: Cambridge series
  in chemical engineering | Includes index.
Identifiers: LCCN 2016044117 | ISBN 9781107135116
Subjects: LCSH: Chemical engineering – Mathematics. | Numerical analysis.
Classification: LCC TP155.2.M36 D66 2017 | DDC 660.01/518–dc23
LC record available at https://lccn.loc.gov/2016044117

ISBN 978-1-107-13511-6 Hardback

Additional resources for this publication at www.cambridge.org/Dorfman

Contents

Illustrations

Tables

Programs

Preface

Why write yet another book on numerical methods? Having taught this course for a combined 10 years, we have continuously struggled to find a single textbook that works for chemical engineering students. There are outstanding books on the mathematical underpinnings of numerical methods, but the level of sophistication (theory–lemma) is often a turnoff for chemical engineering students and overkill at the introductory stage. There are also excellent generic books on numerical methods that include some engineering applications. We used one such book as a reference for the course, since many students do better with supplemental reading. The challenge with general books is that they do not incorporate sufficient chemical engineering examples, which leads to students wondering exactly how the textbook (which is often expensive) connects to the class. Moreover, with the explosion of online resources, it is quite easy to find explanations of standard numerical methods with the click of a mouse. (Whether those explanations are good or even correct is another issue entirely!) At the other end of the spectrum are chemical engineering textbooks solely devoted to applications with little discussion of the underlying numerical methods. These books provide a wealth of example problems, but they are not suitable for teaching students how to actually solve the problems.

Our goal in writing this texbook is to take a "Goldilocks" approach: not too rigorous, not too applied! At the same time, we want to firmly embed our discussion of numerical methods in the context of chemical engineering applications. The material presented in the book is based on the content of a course in numerical methods developed in our department more than 20 years ago. The book is intended for an undergraduate course in numerical methods, primarily for chemical engineering students. At Minnesota, this is a core course taught in the spring semester of the junior year. The students are taking this course concurrently with (i) Reaction Kinetics and Reactor Design and (ii) Mass Transfer and Separations. This location in the curriculum offers many opportunities to motivate the methods covered in the book and use them to solve realistic chemical engineering problems. For example, we have timed the numerical methods course and the reactors course so that, when we cover systems of ordinary differential equations in numerical methods, the reactors course covers systems of chemical reactions. The students have already taken Transport Phenomena and Thermodynamics in the fall, so these subjects are also covered in the book through examples of problems that could not be solved by hand in the preceding courses.

While we believe this to be the ideal location for the course, we provide sufficient background information for the different problems so that the students can solve them without necessarily having a previous course in the subject. As a result, the book can certainly be used early in the curriculum, for example in conjunction with Mass and Energy Balances, especially if the goal is to enhance the use of numerical methods in that class and to build a foundation in numerical methods that can be used later in the curriculum. The book would also fit well at the end of the chemical engineering curriculum, when the students have all of the scientific and engineering background but cannot yet solve "real" problems that lack closed-form solutions.

An emerging challenge that we are facing in our curriculum is a lack of programming experience. This is ironic, as it is hard to imagine that students have become any less reliant on computation in their education or their daily life. The problem is that our university (and others elsewhere) no longer requires entering engineering students to take an introductory programming course. Even when such a course is required, it rarely focuses on the aspects that are required for scientific computing. The logic for this evolution in the chemical engineering curriculum is two-fold. The first is commendable, namely creating a manageable course load for students so that they can graduate in a timely manner. Because of the recent expansion of the chemical engineering curriculum in areas such as biology, something needs to be removed to make space. The second is less defensible, namely that students will use packages such as MATLAB or Aspen to solve chemical engineering problems, so they no longer need to be taught basic programming. At a philosophical level, this argument is very troubling – it essentially argues that engineers do not need to understand how their tools work. At a more practical level, although programming environments (such as MATLAB) are prevalent nowadays, they are often used without a solid understanding of the underlying numerical methods. This "black box" approach can easily lead to erroneous analysis and interpretation of results. Knowledge of the underlying numerical methods is essential in order to be able to choose from a suite of options usually available, and use them correctly and efficiently. Numerical methods go hand-in-hand with programming. Moreover, with the discipline moving towards more fundamental, science-oriented research, the students' background in mathematics is often lacking, and so is their ability to appreciate the power of modern mathematics to provide solutions to cutting-edge problems across science and engineering. Our overall goal in this book is to address this problem with a single-semester, focused course that integrates basic programming skills and numerical methods to solve chemical engineering problems.

Organization of the Book

The book starts with a review of mathematical problems arising in chemical engineering, highlighting the (restrictive) assumptions which make them amenable to analytical solutions, and thus motivating the need for numerical methods. The problems covered in the rest of the text are (in order): linear algebraic equation systems, nonlinear algebraic equation systems, ordinary differential equation systems – initial value problems, ordinary differential equations – boundary value problems, classes of partial differential equations (initial-boundary and two-dimensional boundary value problems), and numerical integration. As such, the level of difficulty gradually increases, and the solution methods naturally build on those developed earlier. This arrangement of the material deviates from the standard approach in numerical methods in several substantial respects. For example, we have split up the material on numerical differentiation and integration, which often appear together early in a numerical methods course. The numerical differentiation appears in the context of boundary value problems, where it becomes clear why it is necessary, and the numerical integration is deferred to the end of the book, as its importance in chemical engineering is subordinate to solving nonlinear algebraic equations and differential equations.

For each class of problems, we start with a thorough derivation of the numerical approximation methods covered. We discuss in some length topics such as accuracy, convergence, and numerical stability properties of the derived methods. Topics such as ill-conditioning and stiffness are also covered as they are essential to be able to detect limitations inherent to the nature of the underlying physical/engineering problem we are trying to solve. We devote a separate chapter to dynamical systems, where we also provide an introduction to nonlinear dynamical phenomena.

We show how each numerical method can be programmed within MATLAB. We begin by assuming very limited programming experience, introducing elementary programming concepts in the first chapter. We then build on these ideas as we develop the necessary MATLAB codes from scratch, explaining in a step-by-step fashion their structure, function, and implementation. We then use (and modify) the codes that we built to solve chemical engineering problems. Importantly, this is *not* a book on using canned routines in MATLAB. Rather, we take advantage of MATLAB's easy debugging and interfacing with plot commands to make the course much simpler than would be the case if the students used a structured programming language like C. Indeed, the only canned numerical method that we use in the book is MATLAB's linear solver (the slash operator), and only after we have covered elementary methods for solving linear algebraic systems. We believe this to be a key feature of the book in that we essentially use the MATLAB environment to teach programming. Recognizing that students should also know how to use the canned routines as well, the end of each chapter provides a brief outline of MATLAB's built-in functions, but only after the students have learned the basics. All of the scripts appearing in the book are available as a downloadable resource.

In each chapter, in addition to short illustrative examples, we include detailed case studies that draw from core chemical engineering subjects (mass balances, thermodynamics, transport, kinetics, reactor design). We use these case studies to illustrate the value of numerical methods in solving realistic chemical engineering problems, but also to provide new perspective on problems that students have already seen (e.g. using continuation to generate phase diagrams). These examples are spaced out in such a way so that their frequency increases as the book proceeds. We have found this to be a useful way to organize a course using this book; as the material becomes more difficult, we slow down the pace of new numerical concepts and reinforce their applicability in chemical engineering.

Finally, numerous problems are included at the end of each chapter that address theoretical questions as well as application of the methods to chemical engineering problems. These problems have a wide range of difficulty and depth, ranging from simple short answer problems that test basic concepts to extended questions that we usually assign over several weeks for students to address in teams. A complete solution manual for instructors, including all of the codes required to solve the problems, is also available.

"Dorfman and Daoutidis's new book is a welcome addition to the undergraduate chemical engineering textbook literature. It provides a very attractive combination of programming, applied mathematics and chemical engineering applications, and is written in an accessible style. The incorporation of example MATLAB codes will be very helpful to students."

Michael D. Graham, University of Wisconsin-Madison

Acknowledgments

While we are the ones who put the pen to paper (or, more appropriate, fingers to keyboard) to write this book, its content is the result of several decades of development in the chemical engineering curriculum. The numerical methods course as part of the core chemical engineering curriculum was born out of the vision of H. Ted Davis more than 20 years ago. Numerous colleagues have contributed to the course over the years. We thank all of them, especially Bob Tranquillo, Jeff Derby, and Satish Kumar who have taught the course several times and helped shape its content. We are particularly indebted to Jeff Derby's contributions to early versions of several topics in the book, most notably in linear algebra and the discussion of the Lorenz attractor. We would also like to thank Andre Mkhoyan, Yiannis Kaznessis, and Matt Neurock for their contributions in recent teachings of the course as recitation instructors in the Minnesota team-teaching system. We have benefitted from the help of numerous excellent teaching assistants and undergraduates in preparing the problems and figures in this book. In particular, we thank Doug Tree, Sarit Dutta, Shawn Dodds, and Patrick Smadbeck for some of the most challenging problems in the book, and Yanal Oueis, Nate Lawdenski, Cody Palbicki, Albert Wu, Steven Cordes, Scott King, and Sarit Dutta for providing problems and figures, as well as finding many (but likely not all) typos in the book. While we have had ample assistance in preparing this text, we take responsibility for the accuracy of the final product.

This book would not have been finished without the support from the Department of Chemical Engineering and Materials Science under the leadership of Frank Bates and Dan Frisbie, who provided both the encouragement to undertake this project and financial resources to support many of the students acknowledged above. Likewise, we would have been stopped at the gates without the patience and loving support of our families.

1 Mathematical Modeling and Structured Programming

1.1 Elements of a Mathematical Model

At its heart, this book is about how to solve mathematical models of physical and engineering systems. In a mathematical model, we try to capture the essential chemistry and physics of a system. The resulting equations can be thought of as having the form

$$\mathcal{L}\mathbf{y} = \mathbf{f}(\mathbf{x}) \qquad (1.1.1)$$

where $\mathcal{L}$ is a mathematical operator, $\mathbf{y}$ is a vector of unknowns, $\mathbf{x}$ is a vector of known quantities, and $\mathbf{f}$ is a vector of (forcing) functions. We have written out Eq. (1.1.1) in a rather general form without worrying about the precise nature of the operator $\mathcal{L}$, and we will discuss some specific details shortly. For the moment, let's consider the general solution to the problem.

In principle, "all" we need to do is determine the inverse of the operator, $\mathcal{L}^{-1}$. If we apply this inverse operator to the left and right sides of Eq. (1.1.1), then we have

$$\mathcal{L}^{-1}\mathcal{L}\mathbf{y} = \mathcal{L}^{-1}\mathbf{f}(\mathbf{x}) \qquad (1.1.2)$$

By definition, the inverse operator "undoes" the operation $\mathcal{L}$, giving us back the unknowns that we are trying to compute,

$$\mathbf{y} = \mathcal{L}^{-1}\mathbf{f}(\mathbf{x}) \qquad (1.1.3)$$

Figure 1.1 provides a schematic illustration of the inverse operator – it acts on the vector of inputs to produce the outputs. In the end, the goal of this book is to see how computers can compute $\mathcal{L}^{-1}$ for different types of problems and then apply these techniques to examples in chemical engineering.

While Eq. (1.1.3) is a generic result, it is only useful if (i) we understand the meaning of the various quantities appearing in Eq. (1.1.1) and (ii) we can actually figure out what the inverse operator is and how to apply it. Let us consider a few simple examples.

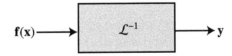

Figure 1.1 Idea behind a mathematical model. The inverse operator acts on the inputs to produce outputs.

Example 1.1   Explain the mathematical model

$$y + c = b \qquad (1.1.4)$$

in the context of Eq. (1.1.1).

Solution
This is a simple example you probably saw in your first algebra class, although probably without the mathematical formalism we are using here. The operator adds the constant c to the value of a scalar y,

$$\mathcal{L}y = y + c \qquad (1.1.5)$$

The forcing function is simply another constant, $f(x) = b$.

It should be apparent that the inverse operator, $\mathcal{L}^{-1}$, that solves Eq. (1.1.4) subtracts c,

$$\mathcal{L}^{-1}y = y - c \qquad (1.1.6)$$

Applying the inverse operator to both sides of the equation yields the solution to the algebraic problem in Eq. (1.1.4),

$$y = b - c \qquad (1.1.7)$$

Example 1.1 was a single, linear algebraic equation. In this case, the unknown vector **y** only contained one entry, which we denoted by the scalar y. Likewise, the forcing function contained a single scalar that we denoted as b. What if we have a system of algebraic equations?

Example 1.2   Explain the mathematical model

$$\mathbf{A}\mathbf{y} = \mathbf{b} \qquad (1.1.8)$$

in the context of Eq. (1.1.1) where **A** is a square matrix.

Solution
The operator for a system equation is a matrix of constant coefficients, **A**, that multiplies the unknowns **y**. The forcing function is a vector of constants, **b**.

You may recall Eq. (1.1.8) from your linear algebra classes in the form $\mathbf{A}\mathbf{x} = \mathbf{b}$, and we will use the latter form when we cover linear algebra. For the moment, we used the form in Eq. (1.1.8) to provide a direct correspondence with the notation in Eq. (1.1.1). Example 1.2 may also make it easier for you to see what we mean by the inverse operator $\mathcal{L}^{-1}$; for a system of linear algebraic systems, we need to compute the inverse of the matrix, $\mathbf{A}^{-1}$, and then left multiply it on both sides. If we apply this inverse operator to Eq. (1.1.8), we get

$$\mathbf{A}^{-1}\mathbf{A}\mathbf{y} = \mathbf{A}^{-1}\mathbf{b} \qquad (1.1.9)$$

Remembering that the product of a matrix and its inverse yields the identity matrix,

$$\mathbf{A}^{-1}\mathbf{A} = \mathbf{I} \qquad (1.1.10)$$

and that multiplying a vector by the identity matrix just returns that vector,

$$\mathbf{I}\mathbf{y} = \mathbf{y} \qquad (1.1.11)$$

we have the desired result

$$\mathbf{y} = \mathbf{A}^{-1}\mathbf{b} \qquad (1.1.12)$$

In addition to helping us explain the concept of an operator, the solution of Eq. (1.1.8) is the heart of almost all numerical methods and thus one of the most important results of this chapter. We will spend considerable time in Chapter 2 on methods to solve linear algebraic equations.

The concept of an operator is more general than the algebraic examples we have considered so far. For example, differentiation is also a mathematical operation,

$$\mathcal{L} = \frac{d}{dx} \qquad (1.1.13)$$

and its inverse is integration

$$\mathcal{L}^{-1} = \int \cdots dx \qquad (1.1.14)$$

Differential equations form an important part of chemical engineering modeling and the subject of a large portion of this book, from Chapters 4 to 7.

Example 1.3 Explain the mathematical model

$$\frac{dy}{dx} = y \qquad (1.1.15)$$

in the context of Eq. (1.1.1).

Solution
To have the correspondence with the definition of an operator, we should rewrite the equation as

$$\frac{dy}{dx} - y = 0 \qquad (1.1.16)$$

The operator of this equation is differentiation with respect to x and subtraction of the unknown function. The forcing function is zero. Equations where the forcing function is zero are called *homogeneous equations*.

The previous examples are linear problems, either linear algebraic equations or linear differential equations. While we will work with some linear problems in this book, for the most part to provide solvable problems to test out the numerical methods, many of the more interesting examples will be nonlinear problems. We can still use the formalism of an operator for nonlinear problems, but as the number and complexity of the equations increases, it becomes no longer possible to write out the operator in an explicit form. For example, consider the solution of the incompressible Navier–Stokes equations,

$$\nabla \cdot \mathbf{v} = 0 \tag{1.1.17}$$

$$\rho \left(\frac{\partial \mathbf{v}}{\partial t} + \mathbf{v} \cdot \nabla \mathbf{v} \right) = -\nabla p + \eta \nabla^2 \mathbf{v} \tag{1.1.18}$$

for the fluid velocity vector $\mathbf{v}$, density ρ, and pressure p. In the latter, ∇ is the gradient operator and η is the fluid viscosity. The system of equations described by Eqs. (1.1.17) and (1.1.18) is very challenging to solve. Indeed, it is not even obvious how to write these equations in the standard operator form of Eq. (1.1.1), let alone determine an inverse operator!

1.2   Some Analytically Solvable Chemical Engineering Models

One of our goals in this book is to teach you how to numerically solve chemical engineering problems that do not have closed-form, analytical solutions. In order to motivate the following chapters, let's take a moment to consider a few problems that you may have seen in other chemical engineering classes. One reason these problems appeared in your other chemical engineering classes is that they can actually be solved analytically in a closed form. We will briefly review the solution methods for these problems too, since we will need their analytical solutions later in the book to validate the numerical methods. We will also highlight small changes to the problem, such as which parameters are the knowns and which are the unknowns, or whether or not the system is isothermal, that make the problem unsolvable by analytical methods. In this way, we hope to make it clear why you need numerical methods and point the way forward to some of the examples you will see in later chapters. Moreover, as the problems get increasingly complicated, the analytical solution becomes rather difficult. As a result, you may want to solve these later problems using numerical methods too, if for no other reason than to check your analytical solution with a particular numerical example.

1.2.1   Vapor–Liquid Equilibrium

A two-component system reaches two-phase equilibrium if the chemical potentials of the species in each phase are equal. For an ideal vapor/liquid system, this thermodynamic condition leads to the system of equations

$$x_1 P_1^{sat}(T) = y_1 P \tag{1.2.1}$$

$$(1 - x_1) P_2^{sat}(T) = (1 - y_1) P \tag{1.2.2}$$

where x_1 and y_1 are the mole fractions of species 1 in the liquid and vapor phases, respectively, P is the system pressure, and P_i^{sat} is the saturation pressure of species i, which depends only on temperature. There are four variables and only two equations, so we need to specify two variables and then we can solve for the other two.

If we pick the temperature and pressure as the known parameters (i.e. the quantities in the forcing function of the mathematical model), then the equilibrium conditions reduce to a system of two linear algebraic equations. This system can easily be solved by adding together the two equations

$$x_1 = \frac{P - P_2^{sat}}{P_1^{sat} - P_2^{sat}} \tag{1.2.3}$$

and then using the equilibrium of species 1 to get the vapor composition,

$$y_1 = \frac{1 - P_2^{sat}/P}{1 - P_2^{sat}/P_1^{sat}} \tag{1.2.4}$$

This seems pretty easy! However, what happens if, instead of having two chemical species, we have a hundred species? For each species, we need to satisfy equilibrium

$$x_i P_i^{sat}(T) = y_i P, \quad i = 1, 2, \ldots, N \tag{1.2.5}$$

with the auxiliary equations

$$\sum_{i=1}^{N} x_i = 1 \tag{1.2.6}$$

and

$$\sum_{i=1}^{N} y_i = 1 \tag{1.2.7}$$

where N is the number of chemical species. (We will need to solve mass balance equations as well.) While each equation is linear in the mole fractions, solving a large system of equations by hand is not a trivial task. We will discuss the solution of large linear systems in Chapter 2, and revisit this particular example of multi-component flash in Section 2.7.

For linear systems, at least we know that we can find a solution (if one exists). For nonlinear problems, sometimes we can find a solution, sometimes not. For example, if we pick the liquid mole fraction and temperature as the known quantities, then we have a pair of nonlinear algebraic equations since y_1 and P multiply one another. However, this is still a solvable nonlinear problem because we can add the two equations to compute the system pressure,

$$P = x_1 P_1^{sat} + (1 - x_1) P_2^{sat} \tag{1.2.8}$$

and then use the equilibrium of species 1 to get the vapor composition,

$$y_1 = \frac{x_1 P_1^{sat}}{x_1 P_1^{sat} + (1 - x_1) P_2^{sat}} \tag{1.2.9}$$

If we do not know the temperature, we again end up with a nonlinear equation because the saturation pressures are in general nonlinear functions of temperature. The so-called Antoine equation is a common empirical relationship between saturation pressure and temperature,

$$\log_{10} P_i^{sat} = A_i + \frac{B_i}{T + C_i} \tag{1.2.10}$$

where A_i, B_i, and C_i are constants for a particular chemical moiety. Let's assume that we know the partial pressures of each species (i.e., y_1 and P). Then we need to solve the equations

$$x_1 10^{A_1 + \frac{B_1}{T + C_1}} = y_1 P \tag{1.2.11}$$

$$(1 - x_1) 10^{A_2 + \frac{B_2}{T + C_2}} = (1 - y_1) P \tag{1.2.12}$$

There is no nice trick to solve for the temperature and liquid phase composition. At this point, we are forced to resort to the numerical methods that we will teach you in Chapter 3.

The situation only becomes more complicated when we drop the assumption of an ideal liquid phase, which then requires an activity coefficient γ_i, or the assumption of an ideal vapor phase, which leads to computing the fugacity. In the former case, our system of equations becomes

$$x_1 \gamma_1(T, x_1) P_1^{sat}(T) = y_1 P \tag{1.2.13}$$

$$(1 - x_1) \gamma_2(T, x_1) P_2^{sat}(T) = (1 - y_1) P \tag{1.2.14}$$

The nonlinearities here can be very complicated, since the activity coefficients depend on both temperature and composition. Moreover, the dependence of the activity coefficient on these parameters is often exponential, which is a very nonlinear coupling between the unknowns. These challenging non-ideal equilibrium calculations will be the subject of a case study in Section 3.10.

1.2.2   Batch Reactor with a Single Reactant

In a batch reactor, the reactants are fed to a closed system at time $t = 0$ and undergo a reaction. If we have a single reactant, say species A, that is consumed by the reaction in a constant volume system, then the mass balance on this species is

$$\frac{dc_A}{dt} = -r(c_A) \tag{1.2.15}$$

where $r(c_A)$ is the rate of consumption of species A per unit volume and c_A is the concentration of species A. In our model, this reaction rate depends on the concentration of A but no other parameters, such as temperature. In order to determine the concentration of species A, we also need to specify its initial concentration,

$$c_A(t = 0) = c_0 \tag{1.2.16}$$

These types of problems are called initial value problems and form the basis for Chapters 4 and 5.

Equation (1.2.15) is a separable, first-order differential equation. As a result, we can write its solution as

$$\int_{c_0}^{c_A(t)} \frac{dy}{r(y)} = -\int_0^t dx \tag{1.2.17}$$

In the latter, we have used x and y as dummy variables of integration. The second integral is trivial, so we can formally write the solution as

$$\int_{c_0}^{c_A(t)} \frac{dy}{r(y)} = -t \tag{1.2.18}$$

If we can compute the integral on the left-hand side and then solve the resulting algebraic equation for $c_A(t)$, then we know the concentration of species A.

Example 1.4   Determine the concentration profile for a batch reactor with the consumption rate $r(y) = k_1 y$.

Solution
For a first-order reaction, Eq. (1.2.18) becomes

$$\int_{c_0}^{c_A(t)} \frac{dy}{y} = -k_1 t \tag{1.2.19}$$

The integral gives back natural logarithms,

$$\int_{c_0}^{c_A(t)} \frac{dy}{y} = \ln \frac{c_A(t)}{c_0} \tag{1.2.20}$$

We thus have

$$c_A(t) = c_0 \exp(-k_1 t) \tag{1.2.21}$$

The first-order kinetics in Example 1.4 corresponds to a linear differential equation. One of the nice aspects of separable differential equations is that we can also solve nonlinear differential equations, provided that we can still calculate the integral.

Example 1.5   Determine the concentration profile for a batch reactor with the consumption rate $r(y) = k_2 y^2$.

Solution
For a second-order reaction, Eq. (1.2.18) becomes

$$\int_{c_0}^{c_A(t)} \frac{dy}{y^2} = -k_2 t \tag{1.2.22}$$

The integral now gives inverse powers of concentration,

$$\int_{c_0}^{c_A(t)} \frac{dy}{y^2} = -\left(\frac{1}{c_A(t)} - \frac{1}{c_0}\right) \tag{1.2.23}$$

We can then solve for the concentration,

$$c_A(t) = \frac{c_0}{1 + c_0 k_2 t} \tag{1.2.24}$$

What do we do when we cannot compute the integral for a separable equation? Aside from trying to find someone smarter than us to do the math (which is always a good idea!) or seeing if the integral is in a book somewhere, the numerical methods in Chapters 4 and 8 will allow us to construct approximate numerical solutions to otherwise unsolvable problems.

1.2.3   Batch Reactor with Three Species

The single reactant problem, which appears in the very start of a class on chemical kinetics, is a relatively simple mathematical model. Let's consider now a somewhat more challenging problem with multiple chemical species. From a pencil-and-paper standpoint, even the linear problem we will solve here is reasonably involved and we might just want to solve the problem numerically anyway.

As a concrete example, consider the reversible reaction A $\rightleftharpoons$ B and the irreversible reaction B $\rightarrow$ C in a constant volume batch reactor. The conservation equations for these three species are

$$\frac{dc_A}{dt} = -k_1 c_A + k_{-1} c_B \tag{1.2.25}$$

$$\frac{dc_B}{dt} = k_1 c_A - k_{-1} c_B - k_2 c_B \tag{1.2.26}$$

$$\frac{dc_C}{dt} = k_2 c_B \tag{1.2.27}$$

subject to initial conditions for each species. In the latter equations, the rate constants k_1 and k_{-1} refer to the forward and reverse reaction rates for the first reaction, while the rate constant k_2 is the forward reaction rate constant for the second reaction.

The problem actually only requires simultaneously solving Eqs. (1.2.25) and (1.2.26). Once we know $c_B(t)$, then we can use the separability of Eq. (1.2.27) to get

$$c_C = k_2 \int_0^t c_B(x)dx + c_C(0) \tag{1.2.28}$$

where x is the dummy variable of integration.

To compute the other two concentration profiles, we first rewrite the problem for species A and B in matrix form,

$$\frac{d}{dt}\begin{bmatrix} c_A \\ c_B \end{bmatrix} = \begin{bmatrix} -k_1 & k_{-1} \\ k_1 & -(k_{-1} + k_2) \end{bmatrix}\begin{bmatrix} c_A \\ c_B \end{bmatrix} \tag{1.2.29}$$

The formal solution of this problem is, under some technical assumptions that we will not discuss here, expressed in vector form as

$$\mathbf{c} = \sum_i a_i \mathbf{e}_i e^{\lambda_i t} \tag{1.2.30}$$

where a_i are constants of integration to be determined from the initial conditions, λ_i are the eigenvalues of the coefficient matrix in Eq. (1.2.29), and $\mathbf{e}_i$ are the corresponding eigenvectors.

In order to solve the problem, we need to remember how to compute eigenvalues and eigenvectors. (This review will certainly help you out in Chapter 5, where eigenvalues play a key role in our analysis of systems of initial value problems.) Given a matrix $\mathbf{A}$, the eigenvalues are the solutions of

$$\det(\mathbf{A} - \lambda_i \mathbf{I}) = 0 \tag{1.2.31}$$

Once we know the eigenvalues, we can determine the eigenvectors as the non-zero vectors that satisfy

$$\mathbf{A}\mathbf{e}_i = \lambda_i \mathbf{e}_i \tag{1.2.32}$$

Example 1.6   Solve Eqs. (1.2.25)–(1.2.27) for the kinetic rate constants $k_1 = k_{-1} = 2$ and $k_2 = 3$ and the initial conditions $c_A(0) = 1$, $c_B(0) = 0$, and $c_C(0) = 0$.

Solution
For these kinetic constants, the coefficient matrix is

$$\begin{bmatrix} -k_1 & k_{-1} \\ k_1 & -(k_{-1} + k_2) \end{bmatrix} = \begin{bmatrix} -2 & 2 \\ 2 & -5 \end{bmatrix} \tag{1.2.33}$$

The eigenvalue equation is then

$$\det \begin{bmatrix} -2 - \lambda & 2 \\ 2 & -5 - \lambda \end{bmatrix} = 0 \tag{1.2.34}$$

Using the 2×2 determinant rule (see Appendix B), we have

$$\lambda^2 + 7\lambda + 6 = 0 \tag{1.2.35}$$

So the eigenvalues are $\lambda_1 = -1$ and $\lambda_2 = -6$.

We can then compute the corresponding eigenvectors. For λ_1, we need

$$\begin{bmatrix} -2 & 2 \\ 2 & -5 \end{bmatrix} \begin{bmatrix} e_1 \\ e_2 \end{bmatrix} = - \begin{bmatrix} e_1 \\ e_2 \end{bmatrix} \tag{1.2.36}$$

Using the first equation we have

$$-2e_1 + 2e_2 = -e_1 \tag{1.2.37}$$

If we pick $e_1 = 1$ then we have

$$-2 + 2e_2 = -1 \tag{1.2.38}$$

so we get $e_2 = 1/2$. In math classes, you are often told to make normalized eigenvectors of unit length, i.e. $e_1^2 + e_2^2 = 1$. For solving ODEs, this is not important because we will adjust the length of the vector to satisfy the initial condition through the calculation of the coefficients a_i in Eq. (1.2.30). Indeed, since we would prefer to avoid dealing with fractions now, let's just multiply the eigenvector by 2 to have

$$\mathbf{e}_1 = \begin{bmatrix} 2 \\ 1 \end{bmatrix} \tag{1.2.39}$$

Note that our choice automatically satisfies the second of Eq. (1.2.36),

$$2e_1 - 5e_2 = -e_2 \tag{1.2.40}$$

For λ_2, we need to find e_1 and e_2 such that

$$\begin{bmatrix} -2 & 2 \\ 2 & -5 \end{bmatrix} \begin{bmatrix} e_1 \\ e_2 \end{bmatrix} = -6 \begin{bmatrix} e_1 \\ e_2 \end{bmatrix} \tag{1.2.41}$$

Again using the first equation

$$-2e_1 + 2e_2 = -6e_1 \tag{1.2.42}$$

and the choice $e_1 = 1$ gives us $e_2 = -2$. So the second eigenvector is

$$\mathbf{e}_2 = \begin{bmatrix} 1 \\ -2 \end{bmatrix} \tag{1.2.43}$$

Using the eigenvalues and eigenvectors in the solution (1.2.30) gives us

$$\mathbf{c} = \begin{bmatrix} c_A \\ c_B \end{bmatrix} = a_1 \begin{bmatrix} 2 \\ 1 \end{bmatrix} e^{-t} + a_2 \begin{bmatrix} 1 \\ -2 \end{bmatrix} e^{-6t} \tag{1.2.44}$$

With the initial conditions $c_A(0) = 1$ and $c_B(0) = 0$, the coefficients are the solution to

$$\begin{bmatrix} 1 \\ 0 \end{bmatrix} = a_1 \begin{bmatrix} 2 \\ 1 \end{bmatrix} + a_2 \begin{bmatrix} 1 \\ -2 \end{bmatrix} \tag{1.2.45}$$

The solution to this problem is $a_1 = 2/5$ and $a_2 = 1/5$. As a result, the concentrations of species A and B are

$$\mathbf{c} = \begin{bmatrix} c_A \\ c_B \end{bmatrix} = \frac{2}{5} \begin{bmatrix} 2 \\ 1 \end{bmatrix} e^{-t} + \frac{1}{5} \begin{bmatrix} 1 \\ -2 \end{bmatrix} e^{-6t} \tag{1.2.46}$$

If you prefer to have this written in component form,

$$c_A = \frac{4}{5} e^{-t} + \frac{1}{5} e^{-6t} \tag{1.2.47}$$

$$c_B = \frac{2}{5} e^{-t} - \frac{2}{5} e^{-6t} \tag{1.2.48}$$

We can now go back and compute c_C. Since its concentration is zero at $t = 0$, we have

$$c_C = 3 \int_0^t \left(\frac{2}{5} e^{-x} - \frac{2}{5} e^{-6x} \right) dx \tag{1.2.49}$$

Evaluating the integrals gives

$$c_C = -\frac{6}{5}\left(e^{-t} - 1\right) + \frac{1}{5}\left(e^{-6t} - 1\right)$$

(1.2.50)

Grouping the terms we get

$$c_C = 1 - \frac{1}{5}\left(6e^{-t} - e^{-6t}\right)$$

(1.2.51)

While this example looks like a lot of work, it is actually a relatively "easy" chemical reaction problem since the reaction rates are all first order, and the end results are relatively benign. As we can see in Fig. 1.2, the startup of the reaction creates species B by the first reaction. However, the irreversible consumption of B to form C by the second reaction eventually leads to all of the A initially in the system to produce C. One interesting thing to note is that the eigenvalues, which appear in the exponential terms, are not the same as the original reaction rate constants. A nice way to think about this discrepancy is that the coupling between the two reactions leads to effective reaction rates that are different than the rates that we would observe for each reaction in isolation.

Things get a lot more complicated if we have nonlinear reactions. Nonlinear chemical systems can exhibit a fascinating array of behavior, including sustained oscillations (limit cycles) and the appearance/disappearance of steady states as system parameters change (bifurcations). While we will be able to use some of the concepts from linear systems to help in our analysis of nonlinear dynamic systems in Chapter 5, for the most part we will need to rely on numerical methods.

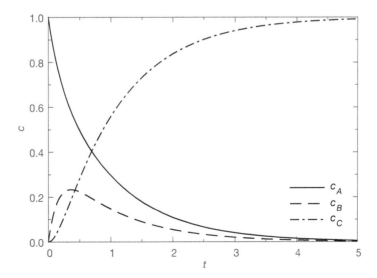

Figure 1.2 Concentration profile for Example 1.6.

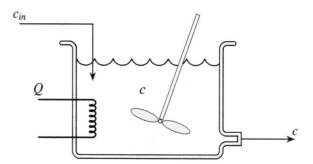

Figure 1.3 Isothermal continuous stirred tank reactor.

1.2.4   Continuous Stirred Tank Reactor

The continuous stirred tank reactor (CSTR) in Fig. 1.3 is one of the standard models for chemical reactors. In this problem, we consider a system maintained at a constant volume V and temperature T that takes in a volumetric flow rate q of a reactant at some feed concentration c_{in}. The reactor is well mixed and the reactant is consumed at a rate kc, where k is the rate constant and c is the concentration inside the tank at time t. We assume that there is some heat transfer Q such that the system remains isothermal. In order to maintain a constant volume, there must also be an outflow rate q that, for a well-mixed tank, also contains the reactant at the concentration $c(t)$. The differential equation governing the reactant concentration inside the tank at some time t is

$$V\frac{dc}{dt} = qc_{in} - qc - kcV \qquad (1.2.52)$$

where the first term represents the inflow, the second term is the outflow, and the last term is the consumption. To solve this problem, we also need to specify the initial concentration inside the tank, $c(t = 0) = c_0$.

Equations of the form of Eq. (1.2.52) are first-order, linear differential equations and can be solved using the integrating factor. We rewrite the equation in the form

$$V\frac{dc}{dt} + (q + kV)c = qc_{in} \qquad (1.2.53)$$

which is equivalent to

$$\frac{d}{dt}\left(ce^{(q/V+k)t}\right) = e^{(q/V+k)t}\frac{q}{V}c_{in} \qquad (1.2.54)$$

from the product rule for differentiation. We now have an exact differential on the left-hand side, which we can integrate to get

$$ce^{(q/V+k)t} = \int e^{(q/V+k)t}\frac{q}{V}c_{in}\,dt = a + \left(\frac{qc_{in}}{q + kV}\right)e^{(q/V+k)t} \qquad (1.2.55)$$

where a is a constant of integration. The general solution to the problem is then

$$c(t) = ae^{-(q/V+k)t} + \frac{qc_{in}}{q + kV} \qquad (1.2.56)$$

The constant of integration is determined by the initial condition,

$$c_0 = a + \frac{qc_{in}}{q + kV} \tag{1.2.57}$$

leading to the solution

$$c(t) = \left(c_0 - \frac{qc_{in}}{q + kV}\right) e^{-(q/V+k)t} + \frac{qc_{in}}{q + kV} \tag{1.2.58}$$

The CSTR model is one of the most deeply studied ones in reactor engineering. While the isothermal, first-order reaction problem has a simple solution, we can easily make the problem much more complicated with fairly innocuous changes in the problem statement. For example, if we allow for a second-order reaction, then

$$V\frac{dc}{dt} = qc_{in} - qc - kc^2V \tag{1.2.59}$$

In this case, there is no comparable solution method using an integrating factor. In Section 4.4, we will show you how to construct a numerical solution to this problem for an even harder situation where the inlet concentration is fluctuating in time. We will consider fluctuations again in Section 4.5, where we will look at how to control the level of the tank in the presence of inlet flow rate fluctuations. In both cases, we will benefit greatly from numerical methods to solve the problem.

The dynamics get even more interesting when we allow for a non-isothermal system (even with a first-order reaction). Here, we get a coupling between the concentration and temperature, through the reaction rate term that appears in both the material and energy balance equations. The reaction rate depends on the temperature, which itself depends on the heat released by the reaction. This is a classic problem (and one that all students at Minnesota are required to learn in the Amundson–Aris tradition) that we will discuss in great detail in Section 5.7. We also work out the numerical solution for a non-isothermal plug-flow reactor, the other standard model of chemical reactors, in Section 4.9.

1.2.5 Reaction–Diffusion in a Slab

The differential equations that we discussed so far are all examples of initial value problems, where we specify the state of the system at some time t and then integrate forward in time. Such problems are commonly encountered in chemical kinetics, which should be obvious from our choice of examples. The other class of ordinary differential equation (ODE) problems are boundary value problems, where we provide information at multiple points in the domain to solve the problem. Such problems arise most often in transport phenomena.

We will pay particular attention to second-order ODE boundary value problems, such as the steady state, one-dimensional reaction–diffusion equation

$$0 = D\frac{d^2c}{dx^2} - kc \tag{1.2.60}$$

where D is the molecular diffusivity of the species and k is a reaction rate constant. This problem needs to be solved subject to suitable boundary conditions, say at $x = 0$ and $x = L$.

Second-order, linear equations of the type of Eq. (1.2.60) can be solved using the characteristic polynomial. For the present problem, the characteristic polynomial is

$$Dr^2 - k = 0 \tag{1.2.61}$$

which has the real roots

$$r = \pm \sqrt{\frac{k}{D}} \tag{1.2.62}$$

The solution of the differential equation is then given as a combination of the exponentials of the roots of the characteristic polynomial,

$$c(x) = a \exp\left(\sqrt{\frac{kx^2}{D}}\right) + b \exp\left(-\sqrt{\frac{kx^2}{D}}\right) \tag{1.2.63}$$

In many problems, we can save ourselves some effort by working in terms of hyperbolic functions,

$$c(x) = a \cosh\left(\sqrt{\frac{kx^2}{D}}\right) + b \sinh\left(\sqrt{\frac{kx^2}{D}}\right) \tag{1.2.64}$$

where

$$\cosh x = \frac{e^x + e^{-x}}{2}, \quad \sinh x = \frac{e^x - e^{-x}}{2} \tag{1.2.65}$$

In either case, we use the boundary conditions to determine the value of the coefficients a and b.

Example 1.7   Solve Eq. (1.2.60) subject to boundary conditions $c(0) = 0$ and $c(L) = c_0$.

Solution
Let's use the solution in terms of hyperbolic functions. From the left boundary condition, we get $a = 0$. From the right boundary condition, we get

$$c_0 = b \sinh \phi \tag{1.2.66}$$

The dimensionless quantity

$$\phi = \sqrt{\frac{kL^2}{D}} \tag{1.2.67}$$

is known as the Thiele modulus, measuring the relative rates of reaction to diffusion in catalysis. We will encounter the Thiele modulus frequently in our reaction–diffusion problems in later chapters, where we will assume for notational simplicity that we are talking about catalytic systems. (For non-catalytic systems, this dimensionless number is called the Dämkohler number.) Putting everything together, the solution to the problem is

$$c(x) = c_0 \left[\frac{\sinh\left(\phi \frac{x}{L}\right)}{\sinh\phi} \right] \qquad (1.2.68)$$

Even relatively innocuous boundary value problems can lead to rather interesting behavior. For example, the characteristic polynomial can have complex roots $r = a + \iota b$ that lead to oscillatory solutions, or repeated roots that produce terms of the form xe^{rx}. The governing equations can also be inhomogeneous, for example if there was a zero-order reaction in addition to the first-order reaction in Eq. (1.2.60). The latter class of problems can be solved using methods such as variation of parameters or the method of undetermined coefficients. Things get more interesting for the linear problem when the coefficients are a function of position. We will see how to handle such problems in Section 6.4, when we consider the concentration profile in a packed bed.

In all of the cases we discussed above, the underlying problem is linear and analytical solution methods exist. The solution gets much more complicated when we introduce nonlinearities, such as a second-order reaction,

$$0 = D\frac{d^2c}{dx^2} - kc^2 \qquad (1.2.69)$$

You cannot use the characteristic polynomial to solve this equation anymore. We can get even more interesting behavior when we couple heat and mass transfer, similar to what we mentioned in the context of the CSTR, where the temperature distribution is nonlinearly coupled to the concentration field through the energy released or consumed by the reaction. We will show you how to solve these problems using numerical methods in Chapter 6, and consider the coupled heat and mass transfer problem in Section 6.6.

1.2.6 Unsteady Diffusion

Like boundary value problems, partial differential equations arise frequently in the context of transport phenomena and form the subject of Chapter 7. We are typically interested in the solution of unsteady problems giving rise to initial boundary value problems or steady state problems in multiple dimensions. For example, we might want to solve the unsteady diffusion equation

$$\frac{\partial c}{\partial t} = D\frac{\partial^2 c}{\partial x^2} \qquad (1.2.70)$$

subject to some initial concentration distribution and boundary conditions in x.

The linear problem in Eq. (1.2.70) can be solved using the method of separation of variables. The idea in separation of variables is to write

$$c(x,t) = f(x)g(t) \qquad (1.2.71)$$

If we substitute this into Eq. (1.2.70) we have

$$fg' = Df''g \qquad (1.2.72)$$

where the prime notation indicates differentiation. The latter equation can be separated to get

$$\frac{g'}{gD} = \frac{f''}{f} \qquad (1.2.73)$$

The left-hand side is a function only of t and the right-hand side is a function only of x. Since they are equal, they must be equal to some number

$$\frac{g'}{gD} = \frac{f''}{f} = -\lambda^2 \qquad (1.2.74)$$

There are an infinite number of constants λ (that are again called eigenvalues, with the corresponding functions f called eigenfunctions) that satisfy both equations. To figure out these eigenvalues, let's solve the f equation,

$$f'' + \lambda^2 f = 0 \qquad (1.2.75)$$

which is a second-order ODE similar to Eq. (1.2.60). The characteristic polynomial is

$$r^2 + \lambda^2 = 0 \qquad (1.2.76)$$

which has the roots $r = \pm \iota \lambda$. The solution is thus

$$f = a\cos(\lambda x) + b\sin(\lambda x) \qquad (1.2.77)$$

where we have already set the imaginary term obtained from the complex exponential to zero from the physics of the original problem. The boundary conditions determine the values of λ and which constants (a or b) are non-zero.

To figure out the time-dependent part g, we solve

$$\frac{g'}{g} = -D\lambda^2 \qquad (1.2.78)$$

We integrate to get

$$g = e^{-D\lambda^2 t} \qquad (1.2.79)$$

We have ignored the constant of integration here, since it is just prefactor of Eq. (1.2.79) that will be adsorbed into the coefficients in the next step.

Putting together the results for $f(x)$ in Eq. (1.2.77) and $g(t)$ in Eq. (1.2.79) with Eq. (1.2.71), one solution to the problem is

$$c(x,t) = [a\cos(\lambda x) + b\sin(\lambda x)]\,e^{-D\lambda^2 t} \qquad (1.2.80)$$

The complete solution of the problem is the sum of all linearly independent solutions

$$c(x,t) = \sum_{n=0}^{\infty} \left[a_n \cos(\lambda_n x) + b_n \sin(\lambda_n x) \right] e^{-D\lambda_n^2 t} \tag{1.2.81}$$

where a_n and b_n are coefficients corresponding to the eigenvalue λ_n. The solutions for negative values of n are not linearly independent due to the even/odd properties of the cosine and sine functions,

$$\cos(x) = \cos(-x) \tag{1.2.82}$$
$$\sin(x) = -\sin(-x) \tag{1.2.83}$$

The solution of Eq. (1.2.81) is a Fourier series, and the coefficients can be computed by expressing the initial condition $c(x,0)$ as a Fourier series,

$$c(x,0) = \sum_{n=0}^{\infty} a_n \cos(\lambda_n x) + b_n \sin(\lambda_n x) \tag{1.2.84}$$

and then comparing the terms in the latter series with terms in Eq. (1.2.81) at $t = 0$.

Example 1.8 Solve Eq. (1.2.70) subject to an initial condition $c(x,0) = 1$ and homogeneous boundary conditions at $x = 0$ and $x = L$.

Solution
We begin with analyzing the solution for $f(x)$ in Eq. (1.2.77). For this particular problem, the condition at $x = 0$ gives

$$f(0) = a = 0 \tag{1.2.85}$$

So we only need to consider the $\sin(\lambda x)$ terms. In order to satisfy the boundary condition at $x = L$,

$$f(L) = b \sin(\lambda L) = 0 \tag{1.2.86}$$

This would be satisfied if $b = 0$, but that is the trivial solution. It is also satisfied for

$$\lambda = \frac{n\pi}{L} \quad (n = 0, 1, 2, \ldots) \tag{1.2.87}$$

so that the solution for f is

$$f(x) = b \sin\left(\frac{n\pi x}{L} \right) \tag{1.2.88}$$

We can now use our results in Eq. (1.2.81),

$$c(x,t) = \sum_{n=1}^{\infty} c_n \sin\left(\frac{n\pi x}{L} \right) \exp\left(-\frac{n^2\pi^2 Dt}{L^2} \right) \tag{1.2.89}$$

where we removed the $n = 0$ term because it is zero for all x.

To determine the coefficients, we need to use the initial condition

$$c(x, 0) = 1 = \sum_{n=1}^{\infty} c_n \sin\left(\frac{n\pi x}{L}\right) \tag{1.2.90}$$

While the following calculation is standard, we will go through the details here in case you do not remember how to do it from your differential equations course. In doing so, let's assume x is dimensionless so that we can avoid having to carry through the factors of L in the analysis. We first multiply both sides by $\sin(m\pi x)$

$$\sin(m\pi x) = \sin(m\pi x) \sum_{n=1}^{\infty} c_n \sin(n\pi x) \tag{1.2.91}$$

Since m and n are independent variables, we can bring the $\sin(m\pi x)$ inside the summation,

$$\sin(m\pi x) = \sum_{n=1}^{\infty} c_n \sin(m\pi x) \sin(n\pi x) \tag{1.2.92}$$

We then integrate over the domain,

$$\int_0^1 \sin(m\pi x)dx = \int_0^1 \left[\sum_{n=1}^{\infty} c_n \sin(m\pi x) \sin(n\pi x)\right] dx \tag{1.2.93}$$

Let's evaluate the left-hand side first,

$$\int_0^1 \sin(m\pi x)dx = -\frac{1}{m\pi}[\cos(m\pi) - \cos(0)] \tag{1.2.94}$$

If m is even, then the answer is

$$\int_0^1 \sin(m\pi x)dx = 0 \quad (m = 2, 4, \ldots) \tag{1.2.95}$$

If m is odd, then we get

$$\int_0^1 \sin(m\pi x)dx = \frac{2}{m\pi} \quad (m = 1, 3, \ldots) \tag{1.2.96}$$

To evaluate the right-hand side, we can integrate each term in the sum individually,

$$\int_0^1 \left[\sum_{n=1}^{\infty} c_n \sin(m\pi x) \sin(n\pi x)\right] dx = \sum_{n=1}^{\infty} c_n \int_0^1 \sin(m\pi x) \sin(n\pi x)dx \tag{1.2.97}$$

We note that

$$\int_0^1 \sin(m\pi x) \sin(n\pi x)dx = 0 \quad (m \neq n) \tag{1.2.98}$$

Otherwise,

$$\int_0^1 \sin(m\pi x) \sin(n\pi x)dx = \frac{1}{2} \quad (m = n) \tag{1.2.99}$$

As a result,

$$\int_0^1 \left[\sum_{n=1}^{\infty} c_n \sin(m\pi x)\sin(n\pi x) \right] dx = \frac{c_m}{2} \qquad (1.2.100)$$

If we now use Eq. (1.2.96) and (1.2.100) in Eq. (1.2.93), we have

$$\frac{2}{m\pi} = \frac{c_m}{2} \quad (m = 1, 3, \ldots) \qquad (1.2.101)$$

Noting that the choice of index m or n is irrelevant, we have the constants

$$c_n = \frac{4}{n\pi} \quad (n = 1, 3, \ldots) \qquad (1.2.102)$$

The solution to the PDE is then

$$c(x,t) = \sum_{n=1,3,\ldots}^{\infty} \frac{4}{n\pi} \sin\left(\frac{n\pi x}{L}\right) \exp\left(-\frac{n^2\pi^2 Dt}{L^2}\right) \qquad (1.2.103)$$

There are a few things to keep in mind here. First, we ended up with a sine series because the boundary conditions were both on the value of c. If we change the boundary conditions, say to no-flux, then the series or its eigenvalues will need to change. Second, the boundary conditions need to be homogeneous to use separation of variables. If the boundary conditions are not homogeneous, you can often solve a transformed problem by subtracting a line that makes the new problem have homogeneous boundary conditions. For example, if $c(1) = 1$, then we would solve a problem for

$$\tilde{c} = c - (x/L) \qquad (1.2.104)$$

where the boundary conditions on $\tilde{c}$ are now homogeneous. Third, the terms in the expansion decay quickly. So as time increases, you need fewer terms in the series to get a good answer.

We will discuss the numerical solution of partial differential equations in Chapter 7. As you can see from the solution here, even solving linear PDEs by hand can be very challenging, and numerical solutions are sometimes just as useful as their analytical counterparts. Indeed, you still need numerical methods if you want to evaluate sums of the type appearing in Eq. (1.2.103). Moreover, there are problems in transport phenomena and other areas of science (notably in quantum mechanics) where the eigenvalue equations are not as simple as Eq. (1.2.86). We will see one such example in the context of reaction–diffusion in Section 3.5 where we need numerical methods just to compute the values of λ_n.

As was the case in many of our previous examples, there are also relatively small changes to the problem statement that can make the solution very challenging. For example, in Section 7.3, we will discuss unsteady diffusion through a membrane. The problem of the membrane separating two reservoirs is very similar to the one we solved here, which will allow us to make an "apples-to-apples" comparison of the difficulty and accuracy of the numerical solution versus the analytical one. We will also consider the case where the membrane connects an infinite reservoir to a finite tank, a seemingly small change in the problem statement that greatly benefits from a numerical solution.

In addition to unsteady diffusion problems, we will also look at how to solve two-dimensional reaction–diffusion problems in Chapter 7. In particular, Section 7.5 investigates how the concentration field is affected by changing from a first-order reaction, which can be solved analytically or numerically, to a second-order reaction, which can only be solved numerically.

1.3   Computer Math

The problems we have discussed in the previous section are (almost) all linear problems and thus constitute a very small subset of the problems that one encounters in chemical engineering. (This may seem surprising given that almost *every* equation you see in other chemical engineering classes has an analytical solution.) In order to solve "real" chemical engineering problems, we need numerical solutions. Before we get to the details of numerical methods, it is useful for us to briefly touch on a few key issues related to computer science and programming.

Nowadays, computers are an integral part of our everyday life. You can order pizzas, make an international phone call, watch a movie, or do your homework (God forbid!) on your computer. Indeed, in the case *Riley v. California*, Supreme Court Chief Justice Roberts wrote that mobile phones, which are essentially just handheld computers, "are now such a pervasive and insistent part of daily life that the proverbial visitor from Mars might conclude they were an important feature of human anatomy." This is certainly an appropriate view as an end-user of a computer – you treat the computer as a black box that is able to do certain tasks, and you do not worry about how those tasks are accomplished.

1.3.1   Three Key Operations

One of our key goals in this text is to open up this black box, at least in the context of performing scientific calculations. Indeed, figuring out how things work is the essence of engineering. In this respect, the things that a computer can do are quite limited – it can do arithmetic, store the results, and retrieve the results from the memory. For our development of the algorithms for numerical methods, it is useful to assume that computers can do only three higher-level things:

(1) *Solve a linear system of algebraic equations.* Computers certainly can add, subtract, multiply, and divide. As a result, they are ideally suited to solve the linear system

$$\mathbf{Ax} = \mathbf{b} \qquad (1.3.1)$$

provided, of course, that the system actually has a unique solution and that it is small enough to be solved in a reasonable amount of time. Even if Eq. (1.3.1) has a unique solution, it is not always trivial to find the solution. We will discuss a number of methods to solve this equation in Chapter 2, all of which represent a compromise between accuracy and speed. Ultimately, many of our numerical methods will reduce a more complicated problem, for example the solution of the nonlinear reaction–diffusion problem in Eq. (1.2.69), into a suitable linear equation that the computer can solve.

(2) *Evaluate a function.* You can almost always also evaluate a vector of functions of the form

$$\mathbf{y} = \mathbf{f}(\mathbf{x}) \qquad (1.3.2)$$

given that you know the inputs $\mathbf{x}$ and the functions in the vector $\mathbf{f}$. That being said, you still need to be concerned with the accuracy of your solution. For example, consider the set of functions

$$y_1 = x_1 + \frac{x_2}{x_3} \qquad (1.3.3)$$

$$y_2 = x_1 x_2^2 \qquad (1.3.4)$$

$$y_3 = \frac{x_1}{x_2 x_3} \qquad (1.3.5)$$

For many vectors $\mathbf{x} = [x_1, x_2, x_3]$, you can get a result for the vector $\mathbf{y} = [y_1, y_2, y_3]$. However, there are some possible problems that can arise. For example, if we use a vector $\mathbf{x} = [1, 1, 0]$, we get back an output of $\mathbf{y} = [\infty, 1, \infty]$. While the infinite results do not cause us any particular trouble in an analytical solution, they lead to numerical overflow in the computational solution. Even very large or very small numbers can cause significant problems that you would not notice in an analytical solution.

(3) *Iterative equations.* One of the most useful things that you can do on a computer is work with iterative equations. Let's consider the case where we want to know how some vector $\mathbf{x}$ evolves as a function of the number of times k that we apply some operation like Eq. (1.3.2). If we denote the value of this vector as $\mathbf{x}^{(k)}$, then the iterative equation is

$$\mathbf{x}^{(k+1)} = \mathbf{f}(\mathbf{x}^{(k)}) \qquad (1.3.6)$$

Provided that we know the initial value of the vector, $\mathbf{x}^{(0)}$, then repeated applications of Eq. (1.3.6) provide us with the value of $\mathbf{x}^{(k)}$ for a given value of k.

As we develop the subject of numerical methods in this book, it will become clear that every type of problem that we want to solve ultimately boils down to the three types of problems that we discussed here or some combination of them. Figure 1.4 provides you with a roadmap for the development of the different methods starting from a partial differential equation. In all cases, the numerical solution requires solving a linear system of algebraic equations, using an iterative equation, or both. The central importance of solving linear systems of algebraic equations is the reason why we will devote considerable time to this subject in Chapter 2, even if differential equations play a much larger role in the discipline. Indeed, a number of interesting applications of numerical methods

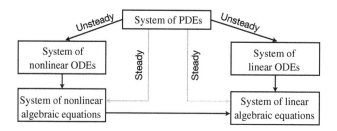

Figure 1.4 Connection between different types of numerical methods and the different types of problems that can be solved on a computer. Unsteady state PDEs are reduced to systems of ODEs, indicated by the black arrows, which then need to be reduced to systems of algebraic equations. Steady state PDEs are reduced directly to algebraic equations. In either case, if the resulting algebraic equations are nonlinear, the problem needs to be converted to solving (repeatedly) systems of linear algebraic equations.

in chemical engineering lie in dynamical systems, the subject of Chapter 5. These systems are described by differential equations, but you cannot understand the numerical solution of differential equations without first understanding the numerical solution of linear algebraic equations.

Since there are only a limited number of problems that can be solved, we always need to reduce a harder problem into one of the simpler problems described in this section. The cost for such a simplification is accuracy (and, sometimes, stability). We will spend quite some time in this text discussing both accuracy and stability. Even if you are using canned software routines to solve numerical problems, it is critical that you understand these issues. If not, you run the risk of GIGO: "garbage in, garbage out."

1.3.2   Words and Round-Off Error

Once we accept that computers can only do arithmetic (and store the results in the memory), we then need to think a bit about *how* computers go about doing arithmetic. At their heart, computers are electronic and have a finite precision. This leads to small errors when we do computer math. While the errors might be small for a single calculation, we often need to do millions of calculations to solve some of the problems we will encounter in this book. We might not be too worried about each small error, but they can add up and eventually lead to major errors in the final result.

In order to understand how computers do math, we need to touch on a few concepts in computer science. This description is by no means comprehensive, and whatever we might try to write today will surely be dated by the time you read this book. As a result, our goal here is to explain the general concept behind computer math, without worrying about the details of the actual implementation. For example, while computers work in base 2, we will do our work here in base 10 for simplicity.

Computers write numbers in the floating point format $m \times b^e$ where m is the mantissa, b is the base, and e is the exponent. The mantissa is constrained by

$$\frac{1}{b} \leq m \leq 1 \tag{1.3.7}$$

Figure 1.5 Illustration of 214.321 157 68 represented as a 16 bit word using base 10.

A simple example is to represent 157.68 in base 10. The mantissa is $m = 0.15678$ and the exponent is 3 such that $157.68 = 0.157\,68 \times 10^3$.

The fundamental unit of a number is a word. For the 16 bit word in Fig. 1.5, the entries are:

- entry 1 is the sign of the number
- entry 2 is the sign of the exponent
- entries 3–5 are the 3 bit exponent
- entries 6–16 are the 11 bit mantissa.

The structure of a word of any size (e.g., 16 bit) sets a limit on the precision (from the mantissa) and the size (from the exponent) of any numerical calculation. The exact structure of a word depends on the architecture of a computer and is really not interesting for the content of this book. The important thing for us to know in scientific computing is that there are a finite range and a finite quantity of numbers inside this range; if you get more deeply into scientific computing, then the details of the computer architecture become increasingly important.

Some numbers, such as integers, can be represented exactly; this is a very nice property of digital logic. Other numbers can be represented approximately and this introduces roundoff error. For example, suppose we have a 4 bit mantissa and I want to add 1.557 and 0.043 87. The first step is typically to make the exponents equal and then add the mantissa. So we would have here

$$0.1557 \times 10^1 + 0.004\,387 \times 10^1 = 0.160\,087 \times 10^1 \qquad (1.3.8)$$

However, since we only have four bits, the answer would either be 0.1600×10^1 or 0.1601×10^1, depending on how the computer does the rounding.

There is almost always some error in computer calculations due to "chopping" or round-off. The problem is small for one calculation. But what if you do the same calculation 10^8 times? It would take a fraction of a second but the answer would be way off. Even worse, consider a 4 bit mantissa and add 4000 to 0.0010:

$$0.4000 \times 10^4 + 0.000\,000\,10 \times 10^4 = 0.4000 \times 10^4 \qquad (1.3.9)$$

The addition does nothing! The take home message here is to be very careful when adding two numbers that are very different in magnitude or subtracting two numbers that are very close in magnitude.

1.4 Structured Programming

In order to get a computer to actually do some math for us, we need to write a set of instructions for a computer, known as a program. One of the most common comments we get at the start of this course at Minnesota surrounds programming, namely, why do chemical engineers need to learn programming? After all, if you wanted to be a programmer, you would have studied computer science. There are two answers to this question beyond the short-sighted ones that (i) students need to learn how to program to pass this class and (ii) programming is very helpful in other chemical engineering classes.

The first answer is that you need to have an elementary grasp of programming to understand numerical methods, and numerical methods *are* a key component of chemical engineering. It is true that you can solve many chemical engineering problems using canned software, and there are a number of "Numerical Methods for Chemical Engineers"-type books that are essentially handbooks for using such software. The text you are holding in your hands is not a handbook for using canned routines. Rather, we use MATLAB as a programming *environment* in order to teach you about numerical methods. This is not the most efficient way to write programs, but it is a very good way to learn how such programs work and probably sufficient for many of your immediate needs in computation. For example, in Chapter 4, we will discuss integration methods for initial value problems, such as the batch reactor problems appearing in the last section. The integration methods we will discuss are for the most part inferior to integration packages such as `ode45` that are integrated directly into MATLAB. So why should you learn the methods in this book? The answer is that engineers need to understand how things work, not just how to be an operator. Saying that the tools already exist so you do not need to understand how they work is tautology; this means that engineers are unnecessary because engineers exist! Once you know how the basic tools work, you can use more sophisticated tools with confidence. We will briefly discuss relevant built-in MATLAB functions at the end of each chapter, after you have sufficient background in the methodology to understand what theses black boxes do and where they might fail.

The second answer is more subtle. In a computer program, you give a set of instructions to the computer. The computer will do exactly what you say if it is possible, or it will tell you that what you ask is not possible. Even if you never write a program after you finish this book, learning how to give coherent instructions that can be carried out without any further clarifications from you is an invaluable skill for an engineer. At some point, almost every chemical engineer ends up in a supervisory role. Your ability to give concise, logical instructions in this context will be critical to your success. It is useful to keep this in mind as we proceed.

1.4.1   Learning to Program

There are generally two ways that one goes about learning to program. In the approach normally pursued in computer science, you learn about programming concepts (data structures, algorithms, logic) in an abstract way first and then apply the concepts to particular circumstances. This is no different than the traditional way of teaching chemical engineering, where you learn about abstract reactor concepts like CSTRs but then need to apply them to real reactors when you work in a plant. In computer science, there is also an emphasis on learning multiple programming languages since different languages are suited to different problems; it is illogical to use a fast but complicated language for scientific computing to write an app for your phone. The computer science approach is excellent training if your end goal is to become a computer programmer, since it provides a foundation that you can use for a wide range of problems later on.

Physical scientists and engineers often take a different approach to programming, learning only as much as they need to solve the problems of interest in a reasonable amount of time. Since as chemical engineers we are generally interested in generating results about a chemical process, this is an efficient approach; we do not waste time learning various programming concepts that do not come to bear on the problem at hand. The price we pay for this approach is evident in the programs themselves, which are often well below the optimum in terms of speed and memory management. If you ultimately want to be an expert in numerical methods for chemical engineering, you need to learn both the engineering/science approach *and* the computer science approach. However, if you just want to know enough about numerical methods to get the correct answer in a reasonable amount of time, which is our aim here, then the engineering/science approach is sufficient.

With this philosophy in mind, the present section introduces some basic logical concepts that you definitely need to know to understand the structure of later programs. We will do so in the context of MATLAB as a programming environment, and the subsequent programs in the book will be written in MATLAB. This choice is not because MATLAB is the best environment for scientific computing. In our opinion, MATLAB is the best environment for teaching scientific computing, and it is often a good starting point for developing more complicated codes that will eventually be ported to faster languages like C or Fortran. Each time we introduce a new program, we will spend some time discussing the structure of that program in the hope that your expertise will grow in a gradual way as we move forward. There is a bit of a learning curve at the start but, as we try to illustrate in Fig. 1.4, later numerical methods in the book build on the earlier ones. In this way, the programming component of the book gets easier as we move forward, since there are really only a few programming concepts that you need to learn.

At this stage in the process, we would like to outline how to set up **for** and **while** loops, implement logically operations with the **if-else** statements, and connect multiple programs together to create larger programs. The discussion below is by no means a complete tutorial on how to program in MATLAB. The easiest (and least expensive) way to learn how to do things in MATLAB is to search through the documentation for commands and search the web for snippets of code, in particular MATLAB's repository. If you prefer a more formal approach, there are a number of excellent books on programming in MATLAB that you can consult. We have also summarized the MATLAB programming features used in the book in Appendix A.

1.4.2 The for Loop

The **for** loop is one of the most important programming constructs. In a **for** loop, we specify in advance the number of times that we want to execute a particular piece of code, and then the computer repeatedly performs the specified list of calculations. This construct is one of the ways we go about evaluating iterative equations of the form of Eq. (1.3.6) – we specify the function evaluation and the upper limit of the counter k, and the program returns the output.

Example 1.9   Write a MATLAB function that determines the product

$$s = \prod_{j=j_0}^{n} j \tag{1.4.1}$$

Solution

In this program, we want to compute a product of a certain number of integers. For example, if we set $j_0 = 1$ and $n = 4$, then the product is

$$s = 1 \times 2 \times 3 \times 4 = 24 \tag{1.4.2}$$

A short MATLAB program to compute this product is:

```
1  function s = intro_for_loop(s,n)
2  % s = value of product
3  % n = number of steps
4  for j = s:n-1
5      s = s*(j+1);
6  end
```

Since this is our first MATLAB program, let's go through it line-by-line. The first line defines a function called `intro_for_loop` by starting with the word **function**. The function returns as its output the variable s and takes the two inputs in parenthesis, s and n. Note that there is no issue with the variable s being both an input and an output; the program takes in some initial value of the product in the variable s and keeps changing this variable until it has done n calculations.

Lines 2 and 3 start with % signs. These are comment lines that are not executed by the program. It is generally a good idea to provide comments in your code. While you are writing the code, it is easy to keep track of what you are doing. When you go back to the code six months later, this is rarely the case. In this book, we will add comment lines at the start of the code to define the variables inside the code and any dependencies (i.e., other functions that are required for the code to execute), but we will not provide extensive commenting elsewhere in the program. This is done simply to keep the functions within the page width of this book; our comment lines are often quite long and hard to fit on a page. Instead, we will provide the equivalent of comments in the text following each program, such as we are doing right now.

Lines 4–6 execute the **for** loop. The first line starts with the word **for** to tell MATLAB that we are starting a **for** loop, and then tells MATLAB the identity of the counter (j) and the range (for j from s to $n-1$ in increments of 1). The next line is the content of the **for** loop; our problem only requires a single calculation per iteration but we will see programs later on that have many lines of text inside the **for** block. Notice in line 5 that the variable s appears on both the right-hand side of the equation and the left-hand side of the equation. When you write any equation in a program, the result of the calculation on the right-hand side is stored in the memory location specified by the left-hand side. In this case, the operation is updating the value of s. The last line has only the word **end** to tell MATLAB that we are done with the **for** loop.

Let's see how the program in Example 1.9 works. If we type

```
>> intro_for_loop(1,2)
```

at the command line in MATLAB, we get back a response

```
ans =
     2
```

What happened here? We told MATLAB to start with an initial value $j_0 = 1$ and to compute up to the $n = 2$ term in the sum. Our loop just runs from $j = 1$ to $j = 1$, which means it executes once. As a result, we get back the result $s = 2 \times 1 = 2$. The example in Eq. (1.4.2) corresponds to

```
>> intro_for_loop(1,4)
```

which indeed returns the result $s = 24$.

There is a small problem with our program. Let's see what happens if we input

```
>> intro_for_loop(6,1)
```

This makes no sense in the problem statement, since we implicitly assume that $n > j_0$ in Eq. (1.4.1). The output from MATLAB is

```
ans =
     6
```

It seems that MATLAB decided that our request to run from $j = 6$ to $j = 1$ in increments of $+1$ was nonsense, so it just did not execute the **for** loop. This nonsense example illustrates the difference between the carefully crafted program of the computer scientist and the "just get the answer" program from the engineer. You need to be cautious when using a program of the latter style, since we are not putting in error checking to make sure the inputs are valid. Many of the programs in this book can be broken in a similar manner if they are not used as intended, but they still might give you output. As you might imagine, you need to be even more cautious when using "black box" programs, since they will generate output but do not let you see how this output was generated.

One important thing to know about functions is that the variables inside the function are "local" variables; they are only defined within the function. As a result, we need to send to the function all of the necessary information (in this case the starting index and the number of terms) and send as output the result of the calculation (in this case the product). Once the function has finished execution, all of the local variables are deleted from the memory. The use of local variables allows for the modular programming approach that we will introduce in Section 1.4.5, where multiple functions are put together to make one larger program. Since the variable names are only defined within a single function, there is no problem with having variables with different definitions in different functions. Likewise, the same variable can have different names in different functions. You can already see the concept in Example 1.9, where we renamed the variable j_0 in Eq. (1.4.1) as s in the program.

The price we pay for local variables is that we have to pass the information around between functions, which can get unwieldy when there is a lot of information to pass. An alternate approach is to use "global" variables in MATLAB that are defined in *all* functions. Using global variables can be an efficient way to program, but you have to be careful that you do not change their values inadvertently in the functions. Since the programs that we will write in this book are relatively simple and only pass a modicum of information, we will work exclusively with local variables to avoid these complications.

1.4.3   The while Loop

The **for** loop is great if you know how many times you want to execute a particular calculation. For example, when we study initial value problems, we will make steps forward in time until we reach the final time for the integration. These problems are ideally suited for using **for** loops. However, there will be other problems where we want to keep doing the calculation until we reach a desired degree of accuracy. In these cases, we do not know *a priori* how many iterations we need to reach the desired precision.

Iterative calculations with criteria for stopping the calculations take advantage of a different logical structure known as a **while** loop, illustrated in Fig. 1.6. In the **while** loop, we specify a particular set of criteria at the start. If the logical evaluation of these criteria is true, then we execute the loop and test the criteria again. Once we reach the point where the criteria are met, then we stop executing the loop. What happens if we never satisfy the criteria? If this happens, we get stuck in an infinite loop! This is

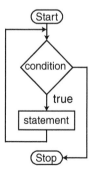

Figure 1.6 Flowchart illustrating the logic of a **while** loop.

the danger of the **while** loop, and we will show you in a moment how to get out of infinite loops. For the moment, let's see how to set up a while loop.

Example 1.10   Write a MATLAB program that computes the product in Eq. (1.4.1) until it exceeds a particular value s_{max}.

Solution
Since we want to compute enough terms to exceed some value s_{max}, this problem should be solved using a **while** loop.

```
1  function [s,k] = intro_while_loop(s,s_max)
2  % s_start = starting value
3  % s_max = stop after s exceeds this value
4  k = s; %counter
5  while s < s_max
6      s = s*(k+1);
7      k = k + 1;
8  end
```

Notice the difference between the first line of this program and the one we used for the **for** loop in Example 1.9. We still send the function the initial value j_0 as the variable s, but we have replaced the number of iterations n with the maximum value that we need to exceed, s_max. We are also returning *two* values from the function, the final value of the product, s, and the counter variable, k. These are returned in a 1×2 vector, which is the reason we used the square brackets for the output argument from the function.

Since the local variable k is undefined by the inputs, we define it in line 4 as the value $j_0 = s$. By default, MATLAB initializes all variables as a scalar with value of zero, so the program would still execute if we did not include line 4, but we would get the wrong answer! It is generally a good practice to initialize all variables before you use them, even if the programming language does not require it. In many programming languages, you are required to declare all variables at the outset so that the program creates space in the memory for them. While this is not required in MATLAB, you will find that some of the larger programs run *much* more efficiently if you initialize large vectors and matrices at the outset. We will come back to this point a bit later in the book in case you forget.

Lines 5–8 execute the **while** loop. Line 5 starts with the word **while**, which tells MATLAB that we are starting a block of text that executes if the criteria following the word **while** are true. In this case, we have a single criterion that the product has not yet exceeded smax. If this is true, we compute another term in the product on line 6 and update the counter on line 7. Line 8 has only the word **end** and tells MATLAB this is the end of the **while** loop. When MATLAB reaches the **end** command, it cycles back to the **while** line and tests the criterion again.

Let's see how this program operates, using the result in Eq. (1.4.2) as our guideline. If we type

```
>> intro_while_loop(1,20)
```

at the command line, we get back a response

```
ans =
    24
```

This is good, but we only got back the value of s from the calculation. If we want to get back the complete set of results, we need to type

```
>> [s,k] = intro_while_loop(1,20)
```

which gives us the reply

```
s =
    24
k =
     4
```

This is the same result that we had using the **for** loop with $j_0 = 1$ and $n = 4$.

1.4.4   If-else Statements

In addition to looping through repetitive calculations, the other logical construct that we need is the **if** statement illustrated in Fig. 1.7a. In an **if** statement, we provide some criteria. If these criteria are true, then MATLAB executes the entries in the **if** block. If the logical output of these criteria is false, then MATLAB does nothing. You can thus think of a **while** loop as continued repetition of an **if** block until the argument is false.

The **if** construct provides a single logical test, and we do the calculations in the **if** block if the logical test is true. You can easily think of cases where you want to

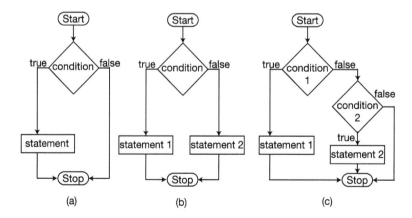

Figure 1.7 Flowchart illustrating the logic of (a) an **if** statement, (b) an **if-else** statement, and (c) an **if-elseif** statement.

do one thing if the statement is true and something else if the statement is false. You could construct such logic out of two separate **if** statements that test the true and false cases separately, but this logic is more readily implemented as the **if-else** statement illustrated in Fig. 1.7b. You can get even more complicated logic with more cases, which can use nested **if** statements or the **if-elseif** statement in Fig. 1.7c. If the logic gets really complicated, there is a **switch** structure that can clean up your code. Keeping with our theme of only worrying about the things we need, we will defer to discussion of these more complicated logical structures to later in the book.

While we have been talking in generalities so far, it is easiest to see how to implement an **if** statement in the context of a particular example.

Example 1.11   Write a MATLAB program that computes n terms in the sum

$$s = \sum_{j=1}^{n} |\sin j| \qquad (1.4.3)$$

Solution
Let's see how to compute this sum using **if** statements:

```
1  function s = intro_for_loop_if(n)
2  % n = number of steps
3  s = 0;
4  for j = 1:n
5      v = sin(j);
6      if v > 0
7          s = s + v;
8      else
9          s = s - v;
10     end
11 end
```

This program only takes in as an input the number of terms and returns the value of the sum as the output. In line 3, we initialize the value of the sum as $s = 0$. Lines 4–11 are a **for** loop that adds a new term to the previous value of the sum. In line 5, we find the next term $\sin j$ and store it as the variable v. We then execute an **if-else** statement to add the absolute value of $\sin j$. If the value of v is positive, then we just add it to the old value of s. If not, then we subtract the value of v such that we are adding its absolute value to the sum.

One of the most important uses of the **if** statement will be avoiding infinite loops in **while** statements. One way to avoid infinite loops is to modify **while** loops with a simple **if** statement that stops the loop if we exceed some maximum number of iterations. For example, we can take the program from Example 1.10 and add a bit of text to avoid getting stuck in an infinite loop:

```
 1  function [s,k] = intro_while_loop_if(s,s_max)
 2  % s_start = starting value
 3  % s_max = stop after s exceeds this value
 4  k = s; %counter
 5  while s < s_max
 6      s = s*(k+1);
 7      k = k + 1;
 8      if k > 100
 9          fprintf('Exceed max iterations!\n')
10          s = 0;
11          break
12      end
13  end
```

Lines 8–12 of this modified program handle the possibility of an infinite loop. In line 8, we see if the counter k has exceeded some maximum number of terms. If so, we do three things. First, we write a warning message to the screen using the fprintf command. Second, we set the value of s to a dummy variable so that we do not return a value that we think is meaningful. Third, the word **break** on line 11 tells MATLAB to stop the **while** loop.

We could have also stopped the **while** loop by setting s to a value that exceeds s_max. The danger with this second approach is that the value of s returned by the program could be interpreted as a meaningful value rather than garbage from breaking out of the loop. A way to get around this problem is to also send back a flag:

```
 1  function [s,k,flag] = intro_while_loop_flag(s,s_max)
 2  % s_start = starting value
 3  % s_max = stop after s exceeds this value
 4  % flag = 0 if k > kmax, 1 otherwise
 5  k = s; %counter
 6  while s < s_max
 7      s = s*(k+1);
 8      k = k + 1;
 9      if k > 100
10          fprintf('Exceed max iterations!\n')
11          s = 0;
12          flag = 0;
13          break
14      end
15  end
16  flag = 1;
```

In this modified program, we have added a third output called flag that is zero if the program quits before reaching the answer and 1 if the program exits successfully. This is a good programming practice, since another function can read the value of flag and know whether s and k are garbage.

Notice that none of the approaches we discussed here solves the intrinsic problem of getting stuck in an infinite loop – if you write a program where the **while** loop will not ever finish, you need to fix the program! The most common problem is a failure to update the variables for the criteria. In the latter example, if we never change the value of s, then the loop will never stop running. There are more subtle issues that can arise, and it is often hard to tell whether your program will get stuck in a **while** loop. As a result, we recommend that you include one of these approaches any time you use a **while** loop. The extra time required to evaluate the **if** statement is small compared to the frustration of staring at the screen while your program seems to do nothing.

Students sometimes like to use **if** statements to use **for** loops as **while** loops, and vice versa. For example, you could use an **if** statement to make a **while** loop become equivalent to a **for** loop by writing something of the form

```
1  k = 1;
2  while k <= n
3      do stuff here
4      k = k + 1;
5  end
```

Along the same lines, you could also turn a **for** loop into a questionable version of a **while** loop by inserting an **if** statement that breaks out of the **for** loop if some criteria are met:

```
1  for i = 1:n
2      do some calculations to get new value of x
3      if x < tol
4          break
5      end
6  end
```

In general, these hacks to the **for** and **while** constructs are unnecessary (and potentially ruinous for the code if done incorrectly), so you should use **for** loops when you know the number of iterations and **while** loops otherwise.

1.4.5 Modular Programming

The other crucial concept we need to introduce is the idea of a hierarchy of functions. A program is generally organized into a main program and subroutines. In MATLAB, the best way to organize your program is the following

```
1  function out = main(void)
2  . . . .
3
4  function out = subroutine_one(x)
5  . . . .
6
7
8  function out = subroutine_two(x)
9  . . .
10
11 etc.
```

The name of the main program is `main` and it takes no input and returns the value `out`. The `out` and `void` parts are optional – your main function does not need to return anything, and it does not need to take anything. The other functions written below the first function in the file usually take in arguments and return something. In MATLAB, these subfunctions can be contained inside the same file as the main function, or they can be separate files in your working directory (or the path for MATLAB). We will use the latter approach here, breaking up our programs into a number of smaller files. If all of these files are in the same directory, then the program will execute.

Let's see how this works in the context of Example 1.11. Instead of nesting the **if-else** inside the **for** loop, we could instead write a program that returns the absolute value of a number,

```
1  function x = absolute_value(x)
2  if x < 0
3      x = -x;
4  end
```

This is a very simple program; it just changes the sign of `x` if it is negative. We now want to call this program from another program. To do this, we modified the program in Example 1.11 to be

```
1  function s = intro_for_loop_abs(s,n)
2  % n = number of steps
3  s = 0;
4  for i = 1:n
5      s = s + absolute_value(sin(i));
6  end
```

As you can see, line 5 of this program calls the function `absolute_value(x)` to compute the absolute value of the function. If we place the program `absolute_value` in the same directory as `intro_for_loop_abs`, the latter program will call the former and return the desired answer. The function `absolute_value` could also be in the same file as `intro_for_loop_abs` so long as `absolute_value` appears *after* `intro_for_loop_abs`. In this ordering, `intro_for_loop_abs` is the main function and `absolute_value` is a subfunction.

If you have used MATLAB before, you may be wondering why we did not write line 5 of the program as `s = s + abs(sin(j))`, since MATLAB has the built-in function called `abs` that computes the absolute value. This is indeed a good idea, and we will use many of MATLAB's built-in functions as we proceed through this text. However, we will need to draw a balance between using built-in functions in MATLAB (for convenience) and writing our own functions (for learning). As a rule of thumb, we will use our own functions when we are learning a new concept, such as the **if-else** statement here, but take advantage of some of MATLAB's built-in functionality as we move forward.

1.5 Further Reading

The best reference for learning MATLAB is the online documentation. You can access this documentation by typing >> doc at the command line. If you follow it up with a command, like >> doc linspace then it will search the documentation for that command. There are also a large number of good reference books dedicated specifically to writing programs in MATLAB or understanding various functionalities. A few good options are listed here:

- R. K. Johnson, *Elements of MATLAB Style*, Cambridge University Press, 2011.
- S. Attaway, *MATLAB: A Practical Introduction to Programming and Problem Solving*, Butterworth-Heineman, 2009
- S. J. Chapman, *MATLAB Programming with Applications for Engineers*, Cengage Learning, 2013
- B. H. Hahn and D. T. Valentine, *Essential MATLAB for Engineers and Scientists*, Academic Press, 2010.

All of these books have similar content and provide relatively detailed references for performing basic functions in MATLAB. You may also want to consider the text

- S. C. Chapra, *Applied Numerical Methods with MATLAB for Engineers and Scientists*, McGraw Hill, 2012.

which has more detail on the numerical methods themselves along with sample programs written in MATLAB.

Problems

1.1 Identify the operator in the following equation:

$$\frac{\partial y}{\partial t} + \alpha \frac{\partial^2 y}{\partial x^2} = -ky + x^3$$

1.2 Is the operator of the following equation linear?

$$\frac{d^3 y}{dx^3} + x^2 y = \sin(\pi x)$$

1.3 What are the operator, unknown(s), forcing function and input(s) for

$$az^2 \frac{dy}{dt} + y = 0$$

1.4 What are the operator, unknown(s), forcing function and input(s) for

$$T^2 = 5x$$

1.5 What are the operator, unknown(s), forcing function and input(s) for

$$\frac{\partial \mathbf{f}}{\partial t} - \frac{\partial^2 \mathbf{f}}{\partial x^2} - 6\mathbf{f} + \mathbf{z} = 0$$

where $\mathbf{z}$ is a constant.

1.6   Solve the following ordinary differential equation

$$y' = 3xy, \quad y(3) = 1$$

1.7   Solve the following ordinary differential equation

$$\frac{dy}{dx} = \ln x, \quad y(1) = 2$$

1.8   Solve the following ordinary differential equation

$$y' = 3xy + x \quad y(3) = 1$$

1.9   Solve the following ordinary differential equation

$$\frac{dy}{dx} = y - 2, \quad y(0) = 2$$

1.10   Solve the following ordinary differential equation

$$y'' + 3y' + 2y = 0, \quad y(0) = 1, \quad y(2) = 2$$

1.11   Solve the following ordinary differential equation

$$y'' = -2y, \quad y(0) = 2, \quad y'(0) = 0$$

1.12   Solve the following ordinary differential equation

$$\frac{d^2y}{dx^2} - 2\frac{dy}{dx} = 0, \quad y(0) = 2, \quad \frac{dy}{dx}(0) = 1$$

1.13   Solve the convection–diffusion–reaction problem

$$v\frac{dc}{dx} = D\frac{d^2c}{dx^2} - kc$$

subject to no-flux at $x = 0$,

$$D\frac{dc}{dx} = 0$$

and a concentration $c = c_0$ at $x = L$. Before solving this problem, you should convert it into dimensionless form, wherein you can fix the Péclet number as $vL/D = 1$ and the Dämkohler number as $kL^2/D = 2$.

1.14   Solve the following ordinary differential equation

$$\frac{d^2y}{dx^2} = 4y + 1, \quad y(0) = 1, \quad \frac{dy}{dx}(0) = 0$$

1.15   Solve the following ordinary differential equation for the function $y(x)$

$$y'' + y = (1 - \pi^2)\sin \pi x$$

subject to $y(0) = 0$ and $y(1) = 1$.

1.16 Solve the system of ordinary differential equations

$$\frac{dx}{dt} = 3x + 2y$$

$$\frac{dy}{dt} = 4x + 2y$$

subject to $x(0) = 0$ and $y(0) = 1$.

1.17 Consider the reversible reaction A $\rightleftharpoons$ B with a forward rate constant of 3 and a reverse rate constant of 2. If we start with 1 mole of A and 1 mole of B, compute the number of moles of each species as a function of time. Confirm that you solution agrees with thermodynamic equilibrium in the limit $t \to \infty$.

1.18 Determine the trajectory of the kinetic system

$$\frac{dA}{dt} = 2A - 3B$$
$$\frac{dB}{dt} = -4A + B$$

with initial conditions $A(0) = 1$ and $B(0) = 2$. Does the system reach steady state?

1.19 Consider a slab of thickness $2W$ that is initially at a temperature T_H. The slab is in contact with a cold reservoir at some temperature $T_C < T_H$. The slab has a thermal conductivity k and the heat transfer coefficient h is infinitely large. Starting from the unsteady heat equation, find an appropriate set of dimensionless variables so that you can convert this problem into the form

$$\frac{\partial T}{\partial t} = \frac{\partial^2 T}{\partial x^2}$$

subject to the initial condition $T(x, 0) = 1$ and the boundary conditions $T(0, t) = 0$ and $T(1, t) = 0$.

1.20 Consider the partial differential equation

$$\frac{\partial y}{\partial t} = \frac{\partial^2 y}{\partial x^2}$$

subject to $\partial y / \partial x = 0$ at $x = 0$ and $y = 0$ at $x = 1$. When you solve this problem using separation of variables, what are the eigenvalues?

1.21 Solve the unsteady diffusion equation

$$\frac{\partial c}{\partial t} = \frac{\partial^2 c}{\partial x^2}$$

subject to boundary conditions $c(0, t) = 1$ and $\partial c / \partial x = 0$ at $x = 1$. The initial condition is given by the piecewise function

$$c(x, 0) = \begin{cases} 1 & (0 < x < \epsilon) \\ 0 & (\epsilon < x < 1) \end{cases}$$

You should also provide the physical interpretation of this problem in transport phenomena terms: What do the boundary conditions and initial conditions mean?

1.22   Solve

$$\frac{\partial T}{\partial t} = \frac{\partial^2 T}{\partial x^2}$$

subject to the initial condition $T(x, 0) = 1$ and the boundary conditions $T(0, t) = 0$ and $T(1, t) = 0$. Report your answer in terms of the non-zero Fourier modes.

1.23   Consider a box of length $L = 1$ that is initially in contact with a heat reservoir on the left at $T = 1$ and a heat reservoir on the right with $T = 0$. When this system reaches equilibrium, it will have a temperature distribution $T(x) = 1 - x$. When the system has equilibrated, the box is disconnected from the heat reservoir and replaced with insulating boundary conditions, $\partial T/\partial x = 0$ at both $x = 0$ and $x = 1$. Determine the temperature distribution inside this box, $T(x, t)$ for $t > 0$ where $t = 0$ is the time when the boundary conditions change. Since this problem is already stated in a dimensionless form, you can take the heat equation to have the form

$$\frac{\partial T}{\partial t} = \frac{\partial^2 T}{\partial x^2}$$

1.24   Use separation of variables to solve the unsteady-diffusion equation

$$\frac{\partial c}{\partial t} = \frac{\partial^2 c}{\partial x^2}$$

subject to no flux at the left boundary

$$\frac{\partial c}{\partial x} = 0 \quad \text{at } x = 0$$

and zero concentration $c(1, t) = 0$ on the right boundary. The initial condition is $c(x, 0) = 1$.

1.25   Use separation of variables to obtain an eigenfunction solution of the form

$$c(x, t) = \sum_{n=0}^{\infty} (c_n \sin \lambda_n x + d_n \cos \lambda_n x) \, e^{-\lambda_n^2 t}$$

for the unsteady-diffusion equation

$$\frac{\partial c}{\partial t} = \frac{\partial^2 c}{\partial x^2}$$

subject to the constant concentration boundary condition $c(0, t) = 0$ and the reaction–diffusion boundary condition

$$c - \frac{\partial c}{\partial x} = 0 \quad \text{at } x = 1$$

and the initial condition $c(x, 0) = 1$.

1.26   Express the 103493.234 as a 16 bit word in base 10.

1.27 Compute the following sums using base 10 and the indicated mantissa:

(a) 12.4235 + 0.2134 with a 3 bit mantissa
(b) 1.029439 + 4.3284 with a 2 bit mantissa
(c) 4.532938 + 0.004938 with a 5 bit mantissa

1.28 Determine the output of the sum 10.3 + 0.442 using 3 bits of precision.

1.29 Use the rules of operator precedence to compute the output for the following calculations:

```
1   (2-4)^-2+5
2   8+3*2^(4-2)
3   9-3^3+(2-3)^5
4   8/4*3+2*3/6
5   8/4*(3+2)*3/6
```

1.30 Determine the output of the following calculation:

```
1   x = 0;
2   for i = 1:2:5
3       x = x+1;
4   end
5   x
```

1.31 Determine the output of the following calculation:

```
1   A = zeros(3,3);
2   for i = 1:3
3       for j = 1:3
4           A(i,j) = i-j;
5       end
6   end
7   A
```

1.32 What are the entries of the matrix produced by the following code:

```
1   for i = 1:3
2       for j = 1:i
3           A(i,j) = i*j;
4       end
5   end
```

1.33 Determine the output of the following calculation:

```
1   stop = 0;
2   j = 2;
3   while j < 9
4       j = j^2;
5   end
6   j
```

1.34 What is the value of k at the end of this function?

```
1  function k = problem1_34
2  stop_loop = 2;
3  k = 0;
4  while stop_loop > 1
5      k = k+1;
6      stop_loop = stop_loop - 0.25;
7  end
```

1.35 Determine the output of the following calculation:

```
1   k = 5;
2   if k < 3
3       q = 3;
4       if k > 2
5           q = 4;
6       elseif k < 1
7           q = 5;
8       else
9           q = 6;
10      end
11  elseif k == 4
12      q = 2;
13  else
14      if k ~= 7
15          q = 1;
16      else
17          q = 0;
18      end
19  end
20  q
```

1.36 What is the output of the following MATLAB code?

```
1   function out = problem1_36
2
3   for i = 4:-1:1
4       x(i) = i^2;
5   end
6
7   counter = 3;
8   stop = 0;
9
10  while stop == 0
11      if x(counter) > 5
12          out = x(counter-1)-1;
13          stop = stop + 1;
14      else
15          out = 1;
16      end
17  end
```

Computer Problems

1.37 The goal of your problem is to write an m-file that generates a table in a text file with the trace of $\mathbf{A} \times \mathbf{A}^\dagger$ for a matrix $\mathbf{A}$ with the entries

$$A_{ij} = \frac{i+j}{ij}.$$

You must use `for` loops to (i) create the matrix $\mathbf{A}$, (ii) construct the transpose $\mathbf{A}^\dagger$, (iii) compute the product $\mathbf{A} \times \mathbf{A}^\dagger$, and (iv) compute the trace of $\mathbf{A} \times \mathbf{A}^\dagger$. Do this calculation for all 9 combinations of matrix dimensions 10, 100, and 1000 (i.e., where $\mathbf{A}$ is 10×10, 10×100, 10×1000, 100×10, etc.). For the output, create a 9×3 matrix where the first column is the number of rows in $\mathbf{A}$, the second column is the number of columns in $\mathbf{A}$, and the third column is the trace of $\mathbf{A} \times \mathbf{A}^\dagger$ for these conditions. Your program should write this matrix to a text file using the MATLAB routine `dlmwrite`. You should only have to run your program once to generate the entire output file.

1.38 In this problem, you are going to look into the error due to the addition of small numbers and the subtraction of very similar numbers. You should start with the large number $x = 10^{10}$. We want to look at different ways to add 1 to this number and then subtract x from the result. If the calculation is perfect, the final result should be 1. If there are round-off errors or chopping errors, you will not get 1.

Define a small number ϵ. In this problem, you will consider the values $\epsilon = 0.1$, $0.01, \ldots, 10^{-10}$. Your program should perform the following calculation for each value of ϵ,

$$R = \left(x + \sum_{j=1}^{\epsilon^{-1}} \epsilon \right) - x$$

where R is the final result of the calculation. To compute the term in brackets, compute the value of $x + \epsilon$, then add ϵ to that answer to get the value of $x + 2\epsilon$, add ϵ to that answer to get $x + 3\epsilon$, and so on so that computing the term in brackets involves the addition of very different sized numbers. In other words, do not compute the summation and then add it to x. You should not store the value of every step in the calculation – for $\epsilon = 10^{-10}$ this would be a huge memory cost and your program would take a very long time to run. This entire program should execute in less than a minute on a reasonable computer if you write it well. Your program should make a semilogx plot the value of R as a function of ϵ.

1.39 In numerical methods, there are often many ways to make a calculation that give the same answer but have vastly different efficiencies. In this problem, we are going to consider a common calculation in molecular dynamics simulations, which requires computing the term

$$LJ = \frac{1}{r^6} + \frac{1}{r^{12}}$$

(If you are familiar with this topic, we are going to compute the Lennard–Jones potential without any prefactors.) Let us consider two particles, one located at $(1, 0, 3)$ and another particle located at $(2, -2, -1)$ and the following two methods for computing the LJ potential between these two particles. The input to the calculation is the positions of the two particles.

Method 1

(1)   Calculate $r = \sqrt{\Delta x^2 + \Delta y^2 + \Delta z^2}$, where Δx is the distance $x_1 - x_2$ and so forth.
(2)   Calculate LJ by plugging the value of r into the formula above.

Method 2

(1)   Calculate $r^2 = \Delta x^2 + \Delta y^2 + \Delta z^2$.
(2)   Calculate $1/r^2$.
(3)   Calculate $1/r^6 = 1/r^2 \times 1/r^2 \times 1/r^2$.
(4)   Calculate $1/r^{12} = 1/r^6 \times 1/r^6$.
(5)   Calculate LJ by adding the results from the previous two steps.

Write a MATLAB program that determines the amount of time required to do Method 1 and Method 2 if you do the calculation 10, 100, ..., 10^8 times. Look at how the calculation time increases with the number of times you need to do it – in simulations, you would make this calculation an enormous number of times. For these calculations, you should use the `tic` and `toc` commands in MATLAB to get the time. Your program should automatically produce a log-log plot of the time required versus the number of calculations.

1.40   It is possible to obtain the value of π through the infinite sum

$$\pi = \sum_{k=1}^{\infty} \frac{(-1)^{k+1}}{0.5k - 0.25}.$$

In this problem, you will explore the convergence of this summation for a finite number of terms N,

$$\pi(N) = \sum_{k=1}^{N} \frac{(-1)^{k+1}}{0.5k - 0.25}.$$

Write a MATLAB function that computes the value of this sum up to a desired value of the residual

$$\epsilon = \frac{|\pi(N) - \pi(N-1)|}{\pi(N-1)}.$$

The latter expression is only valid for $N > 1$. Your program should compute the value of $\pi(N)$ and the number of terms N required to reach the first term where the residual is less than $\epsilon = 10^{-1}, 10^{-2}, \ldots, 10^{-8}$. Make one semilog plot of the eight outputs for the value of the sum versus ϵ and a log–log plot of the eight outputs for the number of terms needed to compute the sum for a given value of ϵ. Your program should also output the value of $\pi(N)$ and the value of N for $\epsilon = 10^{-8}$.

1.41   Consider the sum

$$S = \sum_{n=1}^{k} \frac{1}{n(n+1)}$$

Write a MATLAB function file that computes the sum and makes a semilog-x plot of the value of this sum as a function of k for $k = 1, 2, \ldots, 1000$. Your numerical solution should suggest the value of this sum for $k = \infty$. Derive the value of the infinite sum.

1.42 Consider the function

$$c(x, t) = \sum_{n=0}^{\infty} \frac{4(-1)^n}{(2n+1)\pi} \cos\left[\frac{(2n+1)\pi x}{2}\right] \exp\left[-\left(\frac{(2n+1)\pi}{2}\right)^2 t\right]$$

for the concentration profile as a function of time. In practice, we need to truncate the sum with a finite number k terms,

$$c(x, t) = \sum_{n=0}^{k} \frac{4(-1)^n}{(2n+1)\pi} \cos\left[\frac{(2n+1)\pi x}{2}\right] \exp\left[-\left(\frac{(2n+1)\pi}{2}\right)^2 t\right]$$

(a) Some functions are very difficult to represent as a Fourier series. The initial condition for this problem is one such function. Write a MATLAB program that automatically plots the value of $c(x, 0)$ for different values of k. It is convenient to use the `linspace` function in MATLAB to generate values of x for this problem. You should choose values of k that illustrate the difficulty in making the plot. To help get you started with modular programming, your program should include a subfunction `out=getc(x,t,k)` that returns the value of $c(x, t)$ for some position x and time t using k terms in the sum. Explain what happens to the behavior of the plot as a function of k. Also explain why the difference between the sum and the initial condition is different at $x = 0$ and $x = 1$.

(b) Now that we know how to plot the hardest solution, modify your program from part (a) to plot the concentration profile as a function of time. Pick a number of times for the plot that illustrate both the short-time and long-time behavior of the solution – in other words, it should be clear how the concentration profile is evolving as a function of time but the plot needs to be readable! Explain why your plot satisfies the boundary conditions and what happens at the steady state for the problem.

1.43 This problem involves analyzing the solution of

$$\frac{\partial T}{\partial t} = \frac{\partial^2 T}{\partial x^2}$$

subject to the initial condition $T(x, 0) = 1$ and the boundary conditions $T(0, t) = 0$ and $T(1, t) = 0$. The Fourier series for the temperature is

$$T(x, t) = \sum_{k=1}^{\infty} \frac{4}{(2k-1)\pi} \sin[(2k-1)\pi x] \exp[-(2k-1)^2 \pi^2 t]$$

(a) Write a program that prints to the screen a formatted table of the values of the temperature at $t = 0.1$ for $x = 0.1, 0.25, 0.5, 0.75,$ and 0.9 using 100 non-zero Fourier modes. Your program should use a subfunction that takes in the value of x, t, and the number of Fourier modes and returns the value of the temperature at this point. You will use this subfunction in the rest of the problem.

(b) We now want to analyze the accuracy of the solution as a function of the number of Fourier modes using 201 evenly spaced grid points between $[0,1]$. Let us define the "exact" solution as the value of the temperature using $k = 100$ Fourier modes. Write a program that computes the error in the sum using the definition

$$\epsilon \equiv \frac{||T_k - T_{100}||}{||T_{100}||}$$

for $k = 1, 2, \ldots, 25$. You should use the subfunction from part (a) in this calculation. Your program should automatically generate a semilog-y plot of ϵ versus k. The quantity $|| \ldots ||$ is a vector norm.

(c) In transport, you used graphical plots of the temperature profile to solve heat transfer problems. Here, we want to generate one such plot for our problem. Write a program that automatically generates a plot of the temperature over the domain $x \in [0.5, 1]$ for the times $t = 0.001, 0.005, 0.025, 0.05, 0.1$, and 0.2. (Since the problem is symmetric, there is no need to worry about the other side of the slab.)

1.44 For the following piecewise function:

$$y(x) = \begin{cases} -x & \text{for } -1 \leq x \leq 0 \\ 0 & \text{for } 0 \leq x \leq 1 \end{cases}$$

(a) Express $y(x)$ as a Fourier series,

$$y(x) = \frac{a_0}{2} + \sum_{n=1}^{\infty} a_n \cos n\pi x + b_n \sin n\pi x$$

(b) Write a MATLAB program that computes the value of the Fourier series for $n = 10$, $n = 100$ and $n = 1000$ using a grid in x that is linearly space from $[-1, 1]$ using 101 points. The program should automatically plot these three functions to the screen.

(c) Write a MATLAB program that determines the value of the error,

$$E = ||y_n - y^*||$$

where y_n is the approximation from the Fourier series using n terms and y^* is the exact value of the function, for $n = 1, 2, \ldots, 1000$. You can use the MATLAB function $\texttt{norm(x)}$ to compute the norm. The program should automatically create a semilog-x plot of this error as a function of the number of terms n. Explain why the error does not go to zero for $n \gg 1$. The quantity $|| \ldots ||$ is a vector norm.

2   Linear Algebraic Equations

2.1   Definition of a Linear System of Algebraic Equations

In this chapter, we will discuss a myriad of methods for solving a system of linear algebraic equations on the computer. The general form of a system of linear algebraic equations is

$$
\begin{aligned}
a_{1,1}x_1 + a_{1,2}x_2 + \ldots + a_{1,n}x_n &= b_1 \\
a_{2,1}x_1 + a_{2,2}x_2 + \ldots + a_{2,n}x_n &= b_2 \\
&\vdots \\
a_{n,1}x_1 + a_{n,2}x_2 + \ldots + a_{n,n}x_n &= b_n
\end{aligned}
\tag{2.1.1}
$$

where we are trying to solve for the various unknowns x_i. As shown in the above description, we will assume that the number of equations is equal to the number of unknowns. In the context of our previous definition of mathematical models, let's write this equation system in matrix form

$$
\mathbf{A}\mathbf{x} = \mathbf{b}
\tag{2.1.2}
$$

The coefficient matrix

$$
\mathbf{A} =
\begin{bmatrix}
a_{1,1} & a_{1,2} & \cdots & a_{1,n} \\
a_{2,1} & a_{2,2} & \cdots & a_{2,n} \\
\vdots & \vdots & & \vdots \\
a_{n,1} & a_{n,2} & \cdots & a_{n,n}
\end{bmatrix}
\tag{2.1.3}
$$

is the operator for this problem, the unknown vector is

$$
\mathbf{x} =
\begin{bmatrix}
x_1 \\
x_2 \\
\vdots \\
x_n
\end{bmatrix}
\tag{2.1.4}
$$

and the forcing function is the constant vector

$$\mathbf{b} = \begin{bmatrix} b_1 \\ b_2 \\ \vdots \\ b_n \end{bmatrix} \tag{2.1.5}$$

You may wonder why we are devoting so much time in this book to the "simple" problem of solving this system of equations. In particular, you may remember solving numerous mass and energy balance problems that ultimately reduced to a system of linear algebraic equations and hardly felt the need to resort to a computer to get an answer. This is because the problems you were trying to solve are too small, not because linear algebraic systems are easy to solve! Consider the two problems in Fig. 2.1. In most introductory chemical engineering classes, the problems are similar to Fig. 2.1a. These are small problems, and the unknowns are selected such that the equations are only weakly coupled. In these small problems, you can proceed easily through the equations, solving for one variable at a time and then proceeding to the next equation. In the worst case, the problems are designed so that you have to solve two or three simultaneous equations that are readily handled by substitution. Now consider the more complicated, highly coupled process in Fig. 2.1b. Do you think you could quickly solve for all of the unknowns by hand? If you answered yes, then you are much faster than us! For mere mortals, the solution of large linear algebraic problems is challenging, and we could easily spend an entire semester solely focusing on this single topic.

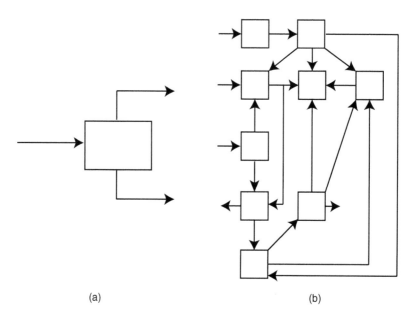

(a)   (b)

Figure 2.1 Comparison between (a) a simple mass balance problem that is easily solved by hand and (b) a more realistic mass balance problem that would drive you nuts to solve by hand!

The other reason for focusing so intensely on the solution of linear algebraic equations is that most of the numerical methods that we will discuss later in the book ultimately boil down to solving linear algebraic equations. For example, you might be tempted to argue that solving transport phenomena problems is more important (as an undergraduate) than mass balances. As a result, it would seem like we should spend most of our time on partial differential equations and leave the linear algebra problems for the birds. For example, unsteady diffusion problems require solving a partial differential equation. We will see in Chapter 7 how to convert linear partial differential equations into systems of linear algebraic equations. Linear algebra is the core of numerical methods, and easily deserving of the attention that we will pay to it.

2.2 Does a Solution Exist?

When confronted with a linear algebraic equation system, there are two important questions to ask:

(1) Does a solution exist?
(2) If so, is it unique?

With this pair of yes/no questions, there are three possible cases: (i) infinitely many solutions; (ii) a unique solution; and (iii) no solution. Note that these are the only possibilities for linear systems. As we will see in Chapter 3, things get much more complicated when we have systems of nonlinear equations.

Before we get down to a more sophisticated analysis, let's analyze some 2×2 systems of equations, since these have simple geometric interpretations that you have likely seen elsewhere. We simply need to plot the two equations in the (x_1, x_2) plane and see whether the two lines, corresponding to the two equations, intersect. For example, consider

$$x_1 + x_2 = 1$$
$$2x_1 + 2x_2 = 2 \tag{2.2.1}$$

As we can see in Fig. 2.2a, these are the same line – the second equation is just double the first equation. There are thus infinite intersections and thus infinitely many solutions.

Now consider instead the system

$$x_1 + x_2 = 1$$
$$x_1 + 2x_2 = 2 \tag{2.2.2}$$

These two lines are plotted in the (x_1, x_2) plane in Fig. 2.2b. The two lines intersect at $x_1 = 0$ and $x_2 = 1$, which is the unique solution to the problem.

To illustrate the third case, we can look at

$$x_1 + x_2 = 1$$
$$x_1 + x_2 = \frac{1}{2} \tag{2.2.3}$$

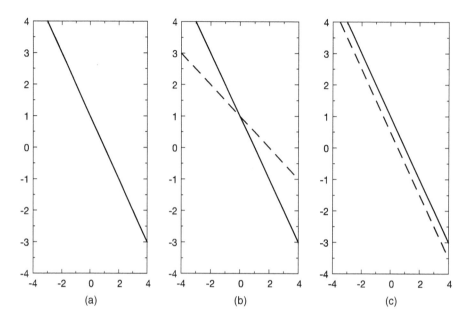

Figure 2.2 Plots of the three possibilities for a system of linear equations: (a) infinitely many solutions, (b) a unique solution, and (c) no solution.

The plot in Fig. 2.2c shows that these are two parallel lines. Parallel lines can never intersect, so there is no solution to this system of equations.

The geometrical arguments that we are making in two dimensions still apply in the higher-dimensional spaces that are required for larger systems of equations. For an $n \times n$ system, each equation yields a hyperplane in n-dimensional space, which is just the generalization of a plane in three-dimensional space. While this generalization is nice in an aesthetic sense, it is useless in terms of making figures and looking for the intersections between all of these planes. If we want to determine whether there is a solution for larger systems, we need a more robust mathematical framework. This can be done in the context of determinants, which are discussed in Appendix B.

2.3   Conditioning

Now that we know whether a solution exists or not, let's ask ourselves whether we can get a decent solution before we worry about how to go about computing it. In particular, we want to know if the solution **x** is very sensitive to small changes in the coefficients of either **A** or **b**. If so, then round-off errors can become a serious problem. Such systems are known as "ill-conditioned" systems, and they are very difficult to solve by hand and even worse to solve on a computer.

This is not just an issue of mathematical accuracy. In the context of any physical problem, and particularly in engineering problems, there is always some uncertainty in the values of the parameters. For example, consider a mass balance problem of the type in Fig. 2.1a. Each of the flow rates entering the system would have to be measured using a flow meter, and their composition would need to be obtained by chemical analysis. While textbooks normally present these data as infinitely accurate, that is never the case in practice. If the process you are trying to analyze is ill-conditioned, then you could end up with answers from the mass balance solution that seem to change in mysterious ways due to very small changes in the inputs. So understanding whether the problem you are analyzing is well-conditioned is clearly an issue in engineering practice as well as in numerical methods.

2.3.1 Vector Spaces

In order to understand the problem arising in an ill-conditioned problem, we first need to recall the idea of a vector space. We can write $\mathbf{Ax} = \mathbf{b}$ as

$$x_1 \begin{bmatrix} a_{1,1} \\ a_{2,1} \\ \vdots \\ a_{n,1} \end{bmatrix} + x_2 \begin{bmatrix} a_{1,2} \\ a_{2,2} \\ \vdots \\ a_{n,2} \end{bmatrix} + \cdots + x_n \begin{bmatrix} a_{1,n} \\ a_{2,n} \\ \vdots \\ a_{n,n} \end{bmatrix} = \begin{bmatrix} b_1 \\ b_2 \\ \vdots \\ b_n \end{bmatrix} \qquad (2.3.1)$$

In an n-dimensional vector space, any vector can be represented in terms of n linearly independent vectors which form a basis vector set. In order to have a unique solution to $\mathbf{Ax} = \mathbf{b}$, the columns of $\mathbf{A}$ must be a basis vector set, i.e. they should span the space of n-dimensional vectors. This is the case if the columns of $\mathbf{A}$ are linearly independent, which is equivalent to saying that $\det \mathbf{A} \neq 0$. If $\det \mathbf{A} = 0$ and the columns are linearly dependent, then we either have no solution or infinitely many solutions.

To see this in a more concrete way, consider the case

$$\mathbf{A} = \begin{bmatrix} 1 & 2 \\ 0 & 0 \end{bmatrix} \qquad (2.3.2)$$

Using the 2×2 formula for the determinant in Eq. (B.5), it is clear that $\det \mathbf{A} = 0$. Likewise, if we compare the columns of $\mathbf{A}$, we see that the second column is just twice the first column.

Now let's see what happens if we want to find a vector $\mathbf{x}$ that, when left multiplied by $\mathbf{A}$, leads to a vector

$$\mathbf{b} = \begin{bmatrix} 1 \\ 1 \end{bmatrix} \qquad (2.3.3)$$

Thinking about this linear algebra problem in terms of vectors, we are trying to find x_1 and x_2 such that

$$x_1 \begin{bmatrix} 1 \\ 0 \end{bmatrix} + x_2 \begin{bmatrix} 2 \\ 0 \end{bmatrix} = \begin{bmatrix} 1 \\ 1 \end{bmatrix} \qquad (2.3.4)$$

This is impossible because, as we see in Fig. 2.3a, the vector $\mathbf{b}$ is not in the vector space spanned by the columns of $\mathbf{A}$. So this is a case where there is no solution.

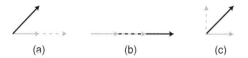

(a)   (b)   (c)

Figure 2.3 Interpretation of the solution of a linear system as the linear combination of basis vectors. The solid black vector is the vector **b** and the gray vectors are the columns of **A**. Case (a) is Eq. (2.3.4) and case (b) is Eq. (2.3.6). In both cases, the basis vectors do not span the two-dimensional vector space leading to (a) no solution or (b) infinitely many solutions depending on **b**. Case (c) is the case where the column vectors of **A** are linearly independent. As a result, they span the entire space and can be combined to produce any other vector. This is the case of a unique solution for any **b**.

A case where we get infinitely many solutions is

$$\mathbf{b} = \begin{bmatrix} 3 \\ 0 \end{bmatrix} \tag{2.3.5}$$

We now are trying to figure out x_1 and x_2 such that we can get

$$x_1 \begin{bmatrix} 1 \\ 0 \end{bmatrix} + x_2 \begin{bmatrix} 2 \\ 0 \end{bmatrix} = \begin{bmatrix} 3 \\ 0 \end{bmatrix} \tag{2.3.6}$$

There are infinitely many choices because, as we now see in Fig. 2.3, **b** is now in the vector space spanned by the columns of **A** but the columns of **A** are not linearly independent.

The problem we have studied above is a case where $\det \mathbf{A} = 0$. In two dimensions, if you have two non-parallel lines (i.e., $\det \mathbf{A} \neq 0$), then we see in Fig. 2.3c that these lines can be rescaled and added to produce any other line in the two-dimensional space. As a result, there is a unique solution in this case.

While we have focused here on two-dimensional problems for simplicity, the arguments outlined above apply in any higher dimensional space.

The concept of a vector space is not only important for determining the number of solutions, if any, but also for understanding the sensitivity of those solutions to perturbations in the matrix or the forcing function. To make our arguments more concrete, let's think about what happens if we have a matrix

$$\mathbf{A} = \begin{bmatrix} 1.000 & 1.000 \\ 1.000 & 1.001 \end{bmatrix} \tag{2.3.7}$$

The basis vectors for this system (the columns of **A**) are almost parallel, as illustrated in Fig. 2.4. Consider the constant vector

$$\mathbf{b} = \begin{bmatrix} 2 \\ 2 \end{bmatrix} \tag{2.3.8}$$

As we illustrate graphically in Fig. 2.4, the solution to $\mathbf{Ax} = \mathbf{b}$ requires finding a linear combination of the two basis vectors represented by the columns of **A** to get a new vector **b**. When we do this, we get

(a)   (b)

Figure 2.4 Geometrical comparison of (a) an ill-conditioned system and (b) a well-conditioned system of linear algebraic equations. In both cases, the vector **b** (in black) is the same. In an ill-conditioned system, small changes in **b** can greatly affect the value of **x** because the basis vectors are almost parallel.

$$\mathbf{x} = \begin{bmatrix} 2 \\ 0 \end{bmatrix} \tag{2.3.9}$$

For a slightly different constant vector

$$\mathbf{b} = \begin{bmatrix} 2 \\ 2.001 \end{bmatrix} \tag{2.3.10}$$

the solution that we get is totally different,

$$\mathbf{x} = \begin{bmatrix} 1 \\ 1 \end{bmatrix} \tag{2.3.11}$$

Figure 2.4 shows why the solution is so sensitive to the values in **b** – since the two column basis vectors are almost parallel, we need to use a very different mix of the two vectors to get a slightly different final vector **b**. This extreme sensitivity is a big problem for numerical calculations, because there will always be round-off error.

Obviously, we do not want to make figures like Fig. 2.4 every time we have a new matrix in order to analyze the vector space of **A**. Moreover, while it is useful to explain the concept of conditioning in the context of a 2×2 system of equations, we need to be able to analyze the conditioning of large problems. Indeed, if we only had to solve 2×2 systems of equations, there would be no need for this book!

2.3.2   Vector Norms

In order to quantify how ill-conditioned a system is, we need to introduce the idea of vector and matrix norms. A norm is defined as a non-negative, real-valued function that provides a measure of the size ("length") of a vector or matrix. While we are introducing the concept of norms at this stage to analyze the condition of a linear algebra problem, we will make extensive use of norms later in this chapter (for the convergence of iterative methods) and elsewhere in the book for even more complicated problems.

Let's assume that we have a vector

$$\mathbf{x} = \begin{bmatrix} x_1 \\ x_2 \\ \vdots \\ x_n \end{bmatrix} \tag{2.3.12}$$

We will consider three norms, the L_1, L_2, and L_∞ norms. The L_1 norm is defined as

$$||\mathbf{x}||_1 \equiv \sum_{i=1}^{n} |x_i| \tag{2.3.13}$$

and simply corresponds to adding up the absolute value of each entry in the vector. The L_2 (Euclidean) norm is

$$||\mathbf{x}||_2 \equiv \sqrt{\sum_{i=1}^{n} x_i^2} \tag{2.3.14}$$

When you think about the size of a vector, you usually think about the Euclidean norm. Finally, the L_∞ norm is defined as the entry in the vector with the largest absolute value.

$$||\mathbf{x}||_\infty \equiv \max_{1 \le i \le n} |x_i| \tag{2.3.15}$$

This is sometimes called the "maximum magnitude" or "uniform vector" norm.

2.3.3   Matrix Norms

In order to analyze the conditioning of a linear system, we also need to have equivalent definitions for the size of a matrix

$$\mathbf{A} = \begin{bmatrix} a_{1,1} & a_{1,2} & \cdots & a_{1,n} \\ a_{2,1} & a_{2,2} & \cdots & a_{2,n} \\ \vdots & \vdots & & \vdots \\ a_{n,1} & a_{n,2} & \cdots & a_{n,n} \end{bmatrix} \tag{2.3.16}$$

In what follows, we are going to write the equations for the square $n \times n$ matrix in Eq. (2.3.16), since these matrices will be the focus of our later analysis. It is easy to generalize the definitions of the norm to non-square matrices.

The L_1 norm is defined as

$$||\mathbf{A}||_1 \equiv \max_{1 \le j \le n} \sum_{i=1}^{n} |a_{i,j}| \tag{2.3.17}$$

In other words, you add up the absolute values of the entries in each column, and then pick the column that has the largest sum. This is called the "column-sum" norm. The name can be a bit confusing, since the sum in Eq. (2.3.17) is over the rows, so we will always call this quantity the 1-norm.

While we only have the Euclidean norm for a vector, there are a pair of 2-norms that we can use for matrices. The L_2 norm is defined by

$$||\mathbf{A}||_2 \equiv \sqrt{\lambda_{\max}} \qquad (2.3.18)$$

The quantity $\lambda_{\max}$ is the largest eigenvalue of the matrix $\mathbf{A}^\dagger\mathbf{A}$, where $\mathbf{A}^\dagger$ is the adjoint (or just the transpose, for real matrices) of $\mathbf{A}$. The quantity in Eq. (2.3.18) is also called the spectral norm. You can imagine that the spectral norm is expensive to calculate, since it requires the matrix multiplication and then solving an eigenvalue problem. If we want to use a simpler 2-norm, we can also use the natural extension of the Euclidean norm for a vector to a matrix,

$$||\mathbf{A}||_e \equiv \sqrt{\sum_{i=1}^{n}\sum_{j=1}^{n} a_{i,j}^2} \qquad (2.3.19)$$

This is also called the Frobenius norm.

The L_∞ norm is defined as

$$||\mathbf{A}||_\infty \equiv \max_{1\le i\le n} \sum_{j=1}^{n} |a_{i,j}| \qquad (2.3.20)$$

This quantity is similar to the L_1 norm, except we now sum up the absolute values of the entries in a given row and then pick the row that has the largest sum. If the L_1 norm is called the column-sum norm, then it makes sense that the L_∞ norm is called the "row-sum" norm. For matrices, the L_1 and L_∞ norms are used frequently because they are easier to compute.

2.3.4 Condition Number

Let us now see how we can use the idea of matrix and vector norms to assess the conditioning of the system $\mathbf{Ax} = \mathbf{b}$. We will begin with the problem we encountered in Fig. 2.4 where a small change in the vector $\mathbf{b}$ leads to a big change in the solution $\mathbf{x}$. In general, how much of a change might we expect to see for a given matrix $\mathbf{A}$?

To answer this question, let $\tilde{\mathbf{x}}$ denote the numerical solution and let $\mathbf{x}$ be the exact solution. We can then define a residual vector

$$\mathbf{R} = \mathbf{b} - \mathbf{A}\tilde{\mathbf{x}} \qquad (2.3.21)$$

that measures the error in the numerical solution. The idea behind a residual, which we will see extensively in other chapters, is that the residual vanishes if the numerical solution is the same as the exact solution, $\tilde{\mathbf{x}} = \mathbf{x}$. We can thus write

$$\mathbf{0} = \mathbf{b} - \mathbf{Ax} \qquad (2.3.22)$$

If we subtract Eq. (2.3.22) from Eq. (2.3.21) then we get

$$\mathbf{R} = \mathbf{A}(\mathbf{x} - \tilde{\mathbf{x}}) \qquad (2.3.23)$$

If we want to know how far we are from the solution to the equation, we need to solve Eq. (2.3.23),

$$\mathbf{x} - \tilde{\mathbf{x}} = \mathbf{A}^{-1}\mathbf{R} \qquad (2.3.24)$$

While Eq. (2.3.24) provides some information, it is not particularly easy to analyze. This is where the idea of a norm becomes important. A particularly important property of norms is that the norm of the product of two matrices is less than or equal to the product of the norms of the individual matrices. From this property, it follows that

$$||\mathbf{A}^{-1}\mathbf{R}|| \leq ||\mathbf{A}^{-1}||\,||\mathbf{R}|| \qquad (2.3.25)$$

so

$$||\mathbf{x} - \tilde{\mathbf{x}}|| \leq ||\mathbf{A}^{-1}||\,||\mathbf{R}|| \qquad (2.3.26)$$

If the residual is "small," then $||\mathbf{A}^{-1}||$ would need to be large in order for us to still be far from the exact solution. To a first approximation, $||\mathbf{A}^{-1}||$ provides a measure of the conditioning of a system.

We can be much more quantitative about conditioning. Let's go back to the problem we outlined at the outset, where we perturb $\mathbf{b}$ by some small quantity $\Delta\mathbf{b}$. In this case the solution $\mathbf{x}$ is also perturbed by some amount $\Delta\mathbf{x}$,

$$\mathbf{A}(\mathbf{x} + \Delta\mathbf{x}) = \mathbf{b} + \Delta\mathbf{b} \qquad (2.3.27)$$

Since we know that

$$\mathbf{A}\mathbf{x} = \mathbf{b} \qquad (2.3.28)$$

then

$$\mathbf{A}\Delta\mathbf{x} = \Delta\mathbf{b} \qquad (2.3.29)$$

As a result, the perturbation to the solution caused by the perturbation in the forcing vector is

$$\Delta\mathbf{x} = \mathbf{A}^{-1}\Delta\mathbf{b} \qquad (2.3.30)$$

We are now in a position to relate the change in $\mathbf{b}$ to the change in $\mathbf{x}$, since this is the problem that we run into in ill-conditioned systems. Using the same property of the norm of products of matrices as before, we get

$$||\Delta\mathbf{x}|| \leq ||\mathbf{A}^{-1}||\,||\Delta\mathbf{b}|| \qquad (2.3.31)$$

and

$$||\mathbf{b}|| \leq ||\mathbf{A}||\,||\mathbf{x}|| \qquad (2.3.32)$$

Once we convert all of the quantities in terms of norms, we can use normal algebraic manipulation for scalars. Inverting Eq. (2.3.32)

$$\frac{1}{||\mathbf{b}||} \geq \frac{1}{||\mathbf{A}||\,||\mathbf{x}||} \qquad (2.3.33)$$

and solving for $||\Delta \mathbf{b}||$ from Eq. (2.3.31) we get

$$||\Delta \mathbf{b}|| \geq \frac{||\Delta \mathbf{x}||}{||\mathbf{A}^{-1}||} \qquad (2.3.34)$$

If we combine these two results, we have

$$\frac{||\Delta \mathbf{b}||}{||\mathbf{b}||} \geq \frac{||\Delta \mathbf{x}||}{|\mathbf{x}|} \left(\frac{1}{||\mathbf{A}|| \, ||\mathbf{A}^{-1}||} \right) \qquad (2.3.35)$$

The usual way to express this result is

$$\frac{||\Delta \mathbf{x}||}{||\mathbf{x}||} \leq \text{cond}(\mathbf{A}) \frac{||\Delta \mathbf{b}||}{||\mathbf{b}||} \qquad (2.3.36)$$

where the condition number is defined as

$$\text{cond}(\mathbf{A}) \equiv ||\mathbf{A}|| \, ||\mathbf{A}^{-1}|| \qquad (2.3.37)$$

The idea is that we now know how much we would expect our solution to change due to a small perturbation in the forcing function – the condition number serves as the "amplifying" factor. If the condition number is large, then a small change in the forcing function (e.g., just due to round-off error) can give rise to a large change in the solution.

Keep in mind that Eq. (2.3.36) sets an upper bound on the amount that the solution can change. There is no reason why the solution must have the maximum change. Likewise, the particular choice of the vector and matrix norms will change the values in this equation. So it is best to think of the condition number in Eq. (2.3.37) as something like a dimensionless number from transport phenomena – if this number is large, then you can run into problems.

In the latter exercise, we showed what happens to the solution when there is a small change in the constant vector. What happens when there is a small change in the matrix? Following the same logic as before, we would expect that

$$(\mathbf{A} + \Delta \mathbf{A})(\mathbf{x} + \Delta \mathbf{x}) = \mathbf{b} \qquad (2.3.38)$$

Since we know $\mathbf{A}\mathbf{x} = \mathbf{b}$, we can simplify the latter equation to

$$\Delta \mathbf{A}(\mathbf{x} + \Delta \mathbf{x}) = -\mathbf{A} \Delta \mathbf{x} \qquad (2.3.39)$$

Solving for $\Delta \mathbf{x}$ gives

$$-\Delta \mathbf{x} = \mathbf{A}^{-1} \Delta \mathbf{A}(\mathbf{x} + \Delta \mathbf{x}) \qquad (2.3.40)$$

If we use again the proprety of a norm of a product of matrices we have

$$||\Delta \mathbf{x}|| \leq ||\mathbf{A}^{-1}|| \, ||\Delta \mathbf{A}|| \, ||(\mathbf{x} + \Delta \mathbf{x})|| \qquad (2.3.41)$$

which can be rearranged to give

$$\frac{||\Delta \mathbf{x}||}{||(\mathbf{x} + \Delta \mathbf{x})||} \leq ||\mathbf{A}^{-1}|| \, ||\Delta \mathbf{A}|| \qquad (2.3.42)$$

If we multiply the right-hand side by $||\mathbf{A}||/||\mathbf{A}||$, then we get

$$\frac{||\Delta \mathbf{x}||}{||(\mathbf{x} + \Delta \mathbf{x})||} \leq \text{cond}(\mathbf{A}) \frac{||\Delta \mathbf{A}||}{||\mathbf{A}||} \tag{2.3.43}$$

We see again that the condition number relates the change in the solution relative to the perturbation of the entries in the matrix.

Example 2.1   What is the condition number for the matrix

$$\mathbf{A} = \begin{bmatrix} 2 & 1 \\ 1 & -2 \end{bmatrix} \tag{2.3.44}$$

which has the inverse

$$\mathbf{A}^{-1} = \frac{1}{5} \begin{bmatrix} 2 & 1 \\ 1 & -2 \end{bmatrix} \tag{2.3.45}$$

Solution
If we use the Euclidean norm for the condition number, we have

$$||\mathbf{A}|| = \sqrt{4 + 1 + 1 + 4} = \sqrt{10} \tag{2.3.46}$$

and

$$||\mathbf{A}^{-1}|| = \sqrt{(2/5)^2 + (1/5)^2 + (1/5)^2 + (2/5)^2} = \frac{\sqrt{10}}{5} \tag{2.3.47}$$

The condition number is then

$$\text{cond}(\mathbf{A}) = \sqrt{10} \frac{\sqrt{10}}{5} = 2 \tag{2.3.48}$$

so the matrix is well conditioned.

2.4   Cramer's Rule

Now that we know if we are going to run into a problem solving a linear algebra problem, let's get down to the business of actually solving these equations. One way to solve a linear system is through determinants, in a method known as Cramer's rule. As we will see, this approach is only useful for small systems, but it is the most elegant mathematical approach. This should teach you probably the most important lesson you can learn in numerical methods – sometimes the most beautiful mathematical method is a terrible numerical method.

In Cramer's rule, the solution of $\mathbf{Ax} = \mathbf{b}$ is given by

$$x_i = \frac{\det \mathbf{A}_i}{\det \mathbf{A}} \tag{2.4.1}$$

where $\mathbf{A}_i$ is the same as $\mathbf{A}$ except that the ith column is replaced with $\mathbf{b}$,

$$\mathbf{A}_i = \begin{bmatrix} a_{11} & \cdots & a_{1,i-1} & b_1 & a_{1,i+1} & \cdots & a_{1n} \\ a_{21} & \cdots & a_{2,i-1} & b_2 & a_{2,i+1} & \cdots & a_{2n} \\ \vdots & & \vdots & \vdots & \vdots & & \vdots \\ a_{n1} & \cdots & a_{n,i-1} & b_n & a_{n,i+1} & \cdots & a_{nn} \end{bmatrix} \tag{2.4.2}$$

Cramer's rule gives you yet another way to see how the determinant of $\mathbf{A}$ tells you if a solution exists and is unique. If $\det \mathbf{A} = 0$, then there is no solution or infinitely many solutions and Eq. (2.4.1) yields nonsense.

Applying Cramer's rule to a 2×2 system yields a nice formula that is worth remembering. We know that

$$\det \mathbf{A} = a_{11}a_{22} - a_{12}a_{21} \tag{2.4.3}$$

We then compute

$$\det \mathbf{A}_1 = \det \begin{bmatrix} b_1 & a_{12} \\ b_2 & a_{22} \end{bmatrix}$$
$$= b_1 a_{22} - b_2 a_{12} \tag{2.4.4}$$

and

$$\det \mathbf{A}_2 = \det \begin{bmatrix} a_{11} & b_1 \\ a_{21} & b_2 \end{bmatrix}$$
$$= a_{11}b_2 - a_{21}b_1 \tag{2.4.5}$$

If we use these results with Cramer's rule (2.4.1), we get

$$x_1 = \frac{b_1 a_{22} - b_2 a_{12}}{a_{11}a_{22} - a_{12}a_{21}} \tag{2.4.6}$$
$$x_2 = \frac{a_{11}b_2 - a_{21}b_1}{a_{11}a_{22} - a_{12}a_{21}} \tag{2.4.7}$$

Cramer's rule is best for small systems because the number of calculations required scales like $n!$ for $n \gg 1$. How big is $n!$? For a 100×100 matrix, the number of operations is approximately $100! \approx 10^{158}$. If you had an exoflop computer with 10^{18} operations per second, it would take 10^{140} seconds to compute the determinant. This is 10^{132} years. MATLAB can solve the whole problem in 200 milliseconds, which is mostly the computer overhead, so there must be a faster approach.

Example 2.2   Solve the system of equations

$$x_1 + 2x_2 - x_3 = 5$$
$$-x_1 + 3x_2 + x_3 = -2$$
$$x_1 - x_2 + x_3 = 3 \tag{2.4.8}$$

using Cramer's rule.

Solution

The original matrix is

$$\mathbf{A} = \begin{bmatrix} 1 & 2 & -1 \\ -1 & 3 & 1 \\ 1 & -1 & 1 \end{bmatrix} \tag{2.4.9}$$

and $\det \mathbf{A} = 10$. To compute x_1, we need

$$\det \begin{bmatrix} 5 & 2 & -1 \\ -2 & 3 & 1 \\ 3 & -1 & 1 \end{bmatrix} = 37 \tag{2.4.10}$$

so $x_1 = 3.7$. To compute x_2, we need

$$\det \begin{bmatrix} 1 & 5 & -1 \\ -1 & -2 & 1 \\ 1 & 3 & 1 \end{bmatrix} = 6 \tag{2.4.11}$$

so $x_2 = 0.6$. Finally, to compute x_3, we need

$$\det \begin{bmatrix} 1 & 2 & 5 \\ -1 & 3 & -2 \\ 1 & -1 & 3 \end{bmatrix} = -1 \tag{2.4.12}$$

so $x_3 = -0.1$.

2.5   Gauss Elimination

Cramer's rule is an excellent example of a very elegant mathematical method being totally useless in practice. Let's now start to talk about methods that actually work for large problems, beginning with Gauss elimination. It is extremely likely that you have seen Gauss elimination in a class in linear algebra and actually remember how to use it. If you already know how to do Gauss elimination, why do we start the numerical methods part of our book with such an elementary topic? There are three good reasons. First, Gauss elimination is an example of a *direct* method for solving linear algebraic systems and forms the basis for some of the more advanced direct methods that we will cover in this chapter. Second, even if you feel that you can solve a system of linear algebraic equations using Gauss elimination by pencil and paper, implementing the algorithm on a computer provides you with a straightforward example to understand how to translate your understanding to a computer. Indeed, many of the tricks and intuition that you might use to solve a problem by hand are generally not done by the computer – the computer does what you tell it to do, and nothing more. Third, we will use Gauss elimination as a canonical example for estimating the computational cost of a numerical method.

The basic premise in Gauss elimination is to transform the original system of equations (2.1.1) into one of the form

$$a_{1,1}x_1 + a_{1,2}x_2 + \cdots + a_{1,n-1}x_{n-1} + a_{1,n}x_n = b_1$$
$$a_{2,2}x_2 + \cdots + a_{2,n-1}x_{n-1} + a_{2,n}x_n = b_2$$
$$\vdots \qquad\qquad \vdots$$
$$a_{i,i}x_i + \cdots + a_{i,n-1}x_{n-1} + a_{i,n}x_n = b_i$$
$$\vdots \qquad\qquad \vdots$$
$$a_{n-1,n-1}x_{n-1} + a_{n-1,n}x_n = b_{n-1}$$
$$a_{n,n}x_n = b_n \qquad (2.5.1)$$

Equation (2.1.2) can be written in matrix form as

$$\mathbf{U}\mathbf{x} = \mathbf{b} \qquad (2.5.2)$$

where $\mathbf{U}$ is an upper triangular matrix,

$$
\mathbf{U} = \begin{bmatrix}
a_{1,1} & a_{1,2} & \cdots & \cdots & \cdots & \cdots & a_{1,n-1} & a_{1,n} \\
0 & a_{2,2} & \cdots & \cdots & \cdots & \cdots & a_{2,n-1} & a_{2,n} \\
\vdots & \vdots & \vdots & \vdots & \vdots & \vdots & \vdots & \vdots \\
0 & 0 & \cdots & 0 & a_{i,i} & \cdots & a_{i,n-1} & a_{i,n} \\
\vdots & \vdots & \vdots & \vdots & \vdots & \vdots & \vdots & \vdots \\
0 & 0 & \cdots & \cdots & \cdots & 0 & a_{n-1,n-1} & a_{n-1,n} \\
0 & 0 & \cdots & \cdots & \cdots & \cdots & 0 & a_{n,n}
\end{bmatrix} \qquad (2.5.3)
$$

Provided that all of the diagonal entries in $\mathbf{U}$ are non-zero, we can then solve this system by backward substitution (starting from the last equation and working backwards) to get

$$x_n = \frac{b_n}{a_{n,n}}$$
$$x_{n-1} = \frac{b_{n-1} - a_{n-1,n}x_n}{a_{n-1,n-1}}$$
$$\vdots \quad \vdots$$
$$x_1 = \frac{b_1 - a_{1,n}x_n - a_{1,n-1}x_{n-1} - \cdots - a_{1,2}x_2}{a_{1,1}} \qquad (2.5.4)$$

These solutions can be written in compact form as

$$x_i = \frac{1}{a_{i,i}}\left(b_i - \sum_{j=i+1}^{n} a_{i,j}x_j\right) \qquad (2.5.5)$$

provided that $a_{i,i} \neq 0$, for $i = n - 1, \cdots, 1$. We will see shortly that this form can be easily programmed in a computer. Note that this formula does not make sense for $i = n$, since the sum would go from $n + 1$ to n. So we need to compute x_n first as the "seed" to start the back substitution algorithm.

While the upper triangular form in Eq. (2.5.2) is the standard approach for Gauss elimination, there is no reason why we had to convert to an upper triangular matrix. Indeed, we could have worked with a lower triangular matrix system

$$\mathbf{Lx} = \mathbf{b} \tag{2.5.6}$$

where $\mathbf{L}$ is

$$\mathbf{L} = \begin{bmatrix} a_{1,1} & 0 & 0 & \cdots & 0 \\ a_{2,1} & a_{2,2} & 0 & \cdots & 0 \\ \vdots & \vdots & \vdots & \vdots & \vdots \\ a_{n,1} & a_{n,2} & & \cdots & a_{n,n} \end{bmatrix} \tag{2.5.7}$$

Such a system can be solved by forward substitution (starting from the first equation and solving sequentially),

$$x_1 = \frac{b_1}{a_{1,1}} \tag{2.5.8}$$

$$x_2 = \frac{b_2 - a_{2,1}x_1}{a_{2,2}} \tag{2.5.9}$$

$$\vdots \quad \vdots$$

$$x_n = \frac{b_n - a_{n,n-1}x_{n-1} - a_{n,n-2}x_{n-2} - \cdots - a_{n,1}x_1}{a_{n,n}} \tag{2.5.10}$$

2.5.1   Algorithm

Gauss elimination refers to the systematic transformation of a system of linear equations into an upper triangular form, and its subsequent solution using backward substitution. It thus consists of two steps:

(1) Forward elimination to convert $\mathbf{Ax} = \mathbf{b}$ to $\mathbf{Ux} = \mathbf{b}'$.
(2) Backward substitution to compute $\mathbf{x}$ from $\mathbf{Ux} = \mathbf{b}'$.

We now describe in a step-by-step fashion the elimination procedure, in a way that will lead naturally to its computer implementation. The original system we need to solve is

$$\begin{aligned} a_{1,1}x_1 + a_{1,2}x_2 + \cdots + a_{1,n}x_n &= b_1 \\ a_{2,1}x_1 + a_{2,2}x_2 + \cdots + a_{2,n}x_n &= b_2 \\ \vdots \qquad\qquad\qquad \vdots \\ a_{n,1}x_1 + a_{n,2}x_2 + \cdots + a_{n,n}x_n &= b_n \end{aligned} \tag{2.5.11}$$

Let k be the column number whose elements are to be eliminated and i be the row number where elimination takes place.

Step 1: $k = 1$. Eliminate all coefficients in the first column below $a_{1,1}$. For the ith row, multiply row $k = 1$ by $-a_{i,1}/a_{1,1}$ and then add to row i for $i = 2, \ldots, n$. Following this step, row 2 now becomes

$$\left(a_{2,2} - \frac{a_{2,1}}{a_{1,1}}a_{1,2}\right)x_2 + \cdots + \left(a_{2,n} - \frac{a_{2,1}}{a_{1,1}}a_{1,n}\right)x_n = b_2 - \frac{a_{2,1}}{a_{1,1}}b_1 \qquad (2.5.12)$$

which in shorthand notation can be written as

$$a_{2,2}^{(2)}x_2 + a_{2,3}^{(2)}x_3 + \cdots + a_{2,n-1}^{(2)}x_{n-1} + a_{2,n}^{(2)}x_n = b_2^{(2)} \qquad (2.5.13)$$

with the superscript 2 indicating a coefficient that has been updated once. (We will use the superscript 1 to indicate the original value.) Repeating for the other rows gives us

$$a_{1,1}^{(1)}x_1 + a_{1,2}^{(1)}x_2 + a_{1,3}^{(1)}x_3 + \cdots + a_{1,n-1}^{(1)}x_{n-1} + a_{1,n}^{(1)}x_n = b_1^{(1)}$$

$$a_{2,2}^{(2)}x_2 + a_{2,3}^{(2)}x_3 + \cdots + a_{2,n-1}^{(2)}x_{n-1} + a_{2,n}^{(2)}x_n = b_2^{(2)}$$

$$a_{3,2}^{(2)}x_2 + a_{3,3}^{(2)}x_3 + \cdots + a_{3,n-1}^{(2)}x_{n-1} + a_{3,n}^{(2)}x_n = b_3^{(2)}$$

$$\vdots \qquad\qquad \vdots$$

$$a_{n,2}^{(2)}x_2 + a_{n,3}^{(2)}x_3 + \cdots + a_{n,n-1}^{(2)}x_{n-1} + a_{n,n}^{(2)}x_n = b_n^{(2)} \qquad (2.5.14)$$

The definition of these new coefficients is convenient for the computer implementation of the method, since in each iteration we will be storing only the current entries of the matrix **A**.

Step 2: $k = 2$. Eliminate all coefficients in the second column below $a_{2,2}^{(2)}$. To do this, multiply the current row $k = 2$ by $-a_{i,2}^{(2)}/a_{2,2}^{(2)}$ and add to row i for $i = 3, \ldots, n$. The resulting system now looks like

$$a_{1,1}^{(1)}x_1 + a_{1,2}^{(1)}x_2 + a_{1,3}^{(1)}x_3 + \cdots + a_{1,n-1}^{(1)}x_{n-1} + a_{1,n}^{(1)}x_n = b_1^{(1)}$$

$$a_{2,2}^{(2)}x_2 + a_{2,3}^{(2)}x_3 + \cdots + a_{2,n-1}^{(2)}x_{n-1} + a_{2,n}^{(2)}x_n = b_2^{(2)}$$

$$a_{3,3}^{(3)}x_3 + \cdots + a_{3,n-1}^{(3)}x_{n-1} + a_{3,n}^{(3)}x_n = b_3^{(3)}$$

$$\vdots \qquad\qquad \vdots$$

$$a_{n,3}^{(3)}x_3 + \cdots + a_{n,n-1}^{(3)}x_{n-1} + a_{n,n}^{(3)}x_n = b_n^{(3)} \qquad (2.5.15)$$

Steps 3, 4, $\ldots$, $n-1$. Repeating this procedure for columns $k = 3, \ldots, n-1$, we finally obtain the system

$$a_{1,1}^{(1)}x_1 + a_{1,2}^{(1)}x_2 + a_{1,3}^{(1)}x_3 + \cdots + a_{1,n-1}^{(1)}x_{n-1} + a_{1,n}^{(1)}x_n = b_1^{(1)}$$

$$a_{2,2}^{(2)}x_2 + a_{2,3}^{(2)}x_3 + \cdots + a_{2,n-1}^{(2)}x_{n-1} + a_{2,n}^{(2)}x_n = b_2^{(2)}$$

$$a_{3,3}^{(3)}x_3 + \cdots + a_{3,n-1}^{(3)}x_{n-1} + a_{3,n}^{(3)}x_n = b_3^{(3)}$$

$$\vdots \qquad\qquad \vdots$$

$$a_{n-1,n-1}^{(n-1)}x_{n-1} + a_{n-1,n}^{(n-1)}x_n = b_{n-1}^{(n-1)}$$

$$a_{n,n}^{(n)}x_n = b_n^{(n)} \qquad (2.5.16)$$

which is in the desired upper triangular form. We can then solve with back substitution to obtain the solutions

$$x_i = \frac{1}{a_{i,i}^{(i)}} \left(b_i^{(i)} - \sum_{j=i+1}^{n} a_{i,j}^{(i)} x_j \right) \qquad (2.5.17)$$

provided that $a_{i,i} \neq 0$, for $i = n-1, \cdots, 1$, where $x_n = b_n^{(n)}/a_{n,n}^{(n)}$.

For the purpose of the computer implementation of the method, it makes little sense to distinguish between the $a_{i,j}$ and b_i coefficients. It is instead natural to work with the $n \times (n+1)$ augmented matrix,

$$\begin{bmatrix} a_{1,1} & a_{1,2} & \cdots & a_{1,n} & b_1 \\ a_{2,1} & a_{2,2} & \cdots & a_{2,n} & b_2 \\ \vdots & \vdots & & \vdots & \vdots \\ a_{n,1} & a_{n,2} & \cdots & a_{n,n} & b_n \end{bmatrix} \qquad (2.5.18)$$

and perform the forward elimination steps on all the matrix rows and columns. The operations are then implemented in three nested loops, whereby:

(1) For each column to perform forward elimination: $k = 1, \ldots, n-1$.
(2) For each row where elimination is to take place: $i = k+1, \ldots, n$.
(3) For each column position within the ith row: $j = k+1, \ldots, n+1$.

Note that j starts from $j = k+1$ and not k, since we already know that in the transformed matrix this term is zero. By not updating all the entries which fall below the main diagonal, these entries still have the values of the original entries of **A**, rather than **0**. However, these entries are never used in backward substitution, so not touching them makes no difference and saves us time.

Example 2.3   Solve the following system of equations by Gauss elimination:

$$\begin{bmatrix} 1 & 1 & 1 & 1 \\ 2 & 3 & 1 & 5 \\ -1 & 1 & -5 & 3 \\ 3 & 1 & 7 & -2 \end{bmatrix} \begin{bmatrix} x_1 \\ x_2 \\ x_3 \\ x_4 \end{bmatrix} = \begin{bmatrix} 10 \\ 31 \\ -2 \\ 18 \end{bmatrix} \qquad (2.5.19)$$

Solution

The augmented matrix is

$$
\begin{bmatrix}
1 & 1 & 1 & 1 & 10 \\
2 & 3 & 1 & 5 & 31 \\
-1 & 1 & -5 & 3 & -2 \\
3 & 1 & 7 & -2 & 18
\end{bmatrix}
=
\begin{bmatrix}
a_{1,j}^{(1)} & \cdots & b_1^{(1)} \\
a_{2,j}^{(1)} & \cdots & b_2^{(1)} \\
a_{3,j}^{(1)} & \cdots & b_3^{(1)} \\
a_{4,j}^{(1)} & \cdots & b_4^{(1)}
\end{bmatrix}
\qquad (2.5.20)
$$

The right-hand side corresponds to the nomenclature described above. After the first step of elimination, we have

$$
\begin{bmatrix}
1 & 1 & 1 & 1 & 10 \\
0 & 1 & -1 & 3 & 11 \\
0 & 2 & -4 & 4 & 8 \\
0 & -2 & 4 & -5 & -12
\end{bmatrix}
=
\begin{bmatrix}
a_{1,j}^{(1)} & \cdots & b_1^{(1)} \\
a_{2,j}^{(2)} & \cdots & b_2^{(2)} \\
a_{3,j}^{(2)} & \cdots & b_3^{(2)} \\
a_{4,j}^{(2)} & \cdots & b_4^{(2)}
\end{bmatrix}
\qquad (2.5.21)
$$

After the next step,

$$
\begin{bmatrix}
1 & 1 & 1 & 1 & 10 \\
0 & 1 & -1 & 3 & 11 \\
0 & 0 & -2 & -2 & -14 \\
0 & 0 & 2 & 1 & 10
\end{bmatrix}
=
\begin{bmatrix}
a_{1,j}^{(1)} & \cdots & b_1^{(1)} \\
a_{2,j}^{(2)} & \cdots & b_2^{(2)} \\
a_{3,j}^{(3)} & \cdots & b_3^{(3)} \\
a_{4,j}^{(3)} & \cdots & b_4^{(3)}
\end{bmatrix}
\qquad (2.5.22)
$$

After the last step,

$$
\begin{bmatrix}
1 & 1 & 1 & 1 & 10 \\
0 & 1 & -1 & 3 & 11 \\
0 & 0 & -2 & -2 & -14 \\
0 & 0 & 0 & -1 & -4
\end{bmatrix}
=
\begin{bmatrix}
a_{1,j}^{(1)} & \cdots & b_1^{(1)} \\
a_{2,j}^{(2)} & \cdots & b_2^{(2)} \\
a_{3,j}^{(3)} & \cdots & b_3^{(3)} \\
a_{4,j}^{(4)} & & b_4^{(4)}
\end{bmatrix}
\qquad (2.5.23)
$$

At the end of the process, we now have the form of Eq. (2.5.2),

$$
\mathbf{Ux} =
\begin{bmatrix}
1 & 1 & 1 & 1 \\
0 & 1 & -1 & 3 \\
0 & 0 & -2 & -2 \\
0 & 0 & 0 & -1
\end{bmatrix}
\begin{bmatrix}
x_1 \\ x_2 \\ x_3 \\ x_4
\end{bmatrix}
=
\begin{bmatrix}
10 \\ 11 \\ -14 \\ -4
\end{bmatrix}
= \mathbf{b}'
\qquad (2.5.24)
$$

We then apply the back substitution formula (2.5.17) to get

$$
x_4 = \frac{-4 - 0}{-1} = 4
$$

$$
x_3 = \frac{-14 - [(-2)(4)]}{-2} = 3
$$

$$
x_2 = \frac{11 - [(3)(4) - 3(1)]}{1} = 2
$$

$$
x_1 = \frac{10 - [(4)(1) + (3)(1) + (2)(1)]}{1} = 1
\qquad (2.5.25)
$$

PROGRAM 2.1   *We are now ready to look at our first implementation of a numerical method in MATLAB. The program below, called* `ngaussel.m`*, implements the "naive" Gauss elimination that we have discussed so far. (We will see shortly what exactly is naive about this algorithm.) The program takes as its inputs an n × n matrix* **A** *and an n × 1 vector b and returns an n × 1 vector* **x** *as the solution.*

```
 1   function x = linear_ngaussel(A,b)
 2   %  A = n x n matrix
 3   %  b = column vector, n x 1
 4   n=length(b);
 5   x=zeros(n,1);
 6   % Perform the forward elimination
 7   for k=1:n-1
 8       for i=k+1:n
 9           m=A(i,k)/A(k,k);
10           for j=k+1:n
11               A(i,j)=A(i,j)-m*A(k,j);
12           end
13           b(i)=b(i)-m*b(k);
14       end
15   end
16   % Perform the back substitution
17   x(n)=b(n)/A(n,n);
18   for i=n-1:-1:1
19       S=b(i);
20       for j=i+1:n
21           S=S-A(i,j)*x(j);
22       end
23       x(i)=S/A(i,i);
24   end
```

You will notice that the program implements Gauss elimination using two steps. In the first step, we do the forward elimination to create the upper triangular matrix **U** *via a series of three nested loops. As noted earlier, we do not overwrite all of the entries in* **A**, *just the ones that are eventually a part of* **U**. *This program does not use an augmented matrix, but rather keeps the constant vector* **b** *as a separate vector. At the conclusion of the forward elimination step, the matrix is transformed to an upper triangular form. In the second step, we do the backward substitution to compute the solution* **x**. *There are two things to notice about the back substitution step. First, the counter in line 18 goes backwards from $i = n - 1$ to $i = 1$ by adding the counter variable -1 between the initial and final value. Second, the value of* `x(i)` *is updated using a temporary variable* `S`, *which we use to represent the current value of the summation in Eq. (2.5.17).*

2.5.2   Computational Cost

The "cost" of performing our calculation is measured in two different kinds of currency – memory and time. Memory is simply the amount of space we need to use in the computer to solve our problem. This was once a severe restriction, and a variety of algorithmic tricks were used to fit larger and larger problems onto a computer. Yet, in virtually all of today's high-speed computers and workstations, memory is cheap and installed in large amounts. So memory limitations are increasingly rare; however, it is still useful to understand how and when an algorithm may be limited by memory considerations. The other "harder" currency of computational cost is simply wall-clock time. Even with the fastest machines of today, we are usually limited to solving problems within some reasonable amount of time.

Let's consider these two costs for Gaussian elimination. Estimating the memory requirements is quite easy. If we want to store the entire $n \times n$ matrix, $\mathbf{A}$, and the right-hand-side vector, $\mathbf{b}$, we need $(n + 1)n$ words of storage. Estimating wall-clock time is a bit more difficult, but we will argue that the solution time will be proportional to the number of mathematical operations we need to perform algorithm. Although this does not actually give us a time, we can measure the number of floating point operations per second (typically abbreviated as FLOPS) for our computer. Multiplying the number of operations by the FLOPS gives an estimate of the wall-clock time needed for the solution. So, for Gaussian elimination, in the forward elimination we perform $(n - k)(n - k)$ multiplications and additions during step k. In total, we perform

$$\sum_{k=1}^{n-1}(n - k)(n - k) = \frac{1}{3}n(n^2 - 1) \tag{2.5.26}$$

operations (where we've loosely defined an operation to be one multiplication plus one addition). The back substitution requires only $\frac{1}{2}n(n - 1)$ operations. Now of course, we are only really interested in the limitations we face for big systems, i.e., for $n \gg 1$. So all we really need to remember is that Gaussian elimination for a full matrix requires $\sim n^2$ words and $\sim n^3/3$ operations (the back substitution cost becomes significantly less than the forward elimination).

Gauss elimination is much better than Cramer's rule because of this scaling. We know that Cramer's rule scales with $n!$. Let's say we have a matrix with a big number of entries $n_1 = N \gg 1$ and we want to increase it to a new size $n_2 = \alpha N$, where $\alpha > 1$. For Gauss elimination, the increase in the amount of computational time would be

$$\frac{t_2}{t_1} \sim \left(\frac{\alpha N}{N}\right)^3 \sim \alpha^3 \tag{2.5.27}$$

For Cramer's rule, we have

$$\frac{t_2}{t_1} \sim \frac{(\alpha N)!}{N!} \tag{2.5.28}$$

We can estimate the factorial quantities using Stirling's approximation for the factorial,

$$\lim_{M \gg 1} M! \approx (2\pi M)^{1/2} \left(\frac{M}{e}\right)^M \tag{2.5.29}$$

If we use Eq. (2.5.29) in Eq. (2.5.28), we get

$$\frac{t_2}{t_1} \approx \frac{(2\pi \alpha N)^{1/2} (\alpha N/e)^{\alpha N}}{(2\pi N)^{1/2} (N/e)^N} \tag{2.5.30}$$

which becomes

$$\frac{t_2}{t_1} \approx \alpha^{\alpha N + 1/2} (N/e)^{N(\alpha - 1)} \qquad (2.5.31)$$

A convenient way to express this is in a logarithmic form,

$$\ln\left(\frac{t_2}{t_1}\right) \approx (\alpha N + 1/2)\ln \alpha + N(\alpha - 1)(\ln N - 1) \qquad (2.5.32)$$

This result depends on both the original matrix size, N, and the amount it increases.

Let's compare going from a 100×100 matrix to 1000×1000. For Gauss elimination, this corresponds to an increase in time of 10^3 or 1000 times longer. For Cramer's rule, we would get

$$\ln\left(\frac{t_2}{t_1}\right) \approx (1000 + 1/2)\ln 10 + 100(9)(\ln 100 - 1) \qquad (2.5.33)$$

which gives a final result

$$\frac{t_2}{t_1} \approx e^{5548} \approx 10^{2409} \qquad (2.5.34)$$

This is obviously a huge number!

Although Gauss elimination is better than Cramer's rule, still computational cost scales geometrically with our problem size. We'll examine ways to alleviate these costs in some of the following sections.

2.6   Pivoting

2.6.1   A Simple Example

One obvious case where Gauss elimination fails is if one of the coefficients on the diagonal vanishes, i.e., $a_{i,i} = 0$. Let's explore this further and discuss how it can be addressed. For example, consider the problem

$$x_1 + x_2 + x_3 = 1 \qquad (2.6.1)$$
$$x_1 + x_2 + 2x_3 = 2 \qquad (2.6.2)$$
$$x_1 + 2x_2 + 2x_3 = 1 \qquad (2.6.3)$$

In the first step of Gauss elimination, we get

$$x_1 + x_2 + x_3 = 1 \qquad (2.6.4)$$
$$0 + 0 + x_3 = 1 \qquad (2.6.5)$$
$$0 + x_2 + x_3 = 0 \qquad (2.6.6)$$

Clearly, there is no problem finding a solution to this problem. Equation (2.6.5) gives $x_3 = 1$, substitution into Eq. (2.6.6) gives $x_2 = -1$, and further substitution into Eq. (2.6.4) gives $x_1 = 1$. However, if we try to use the Gauss elimination algorithm, we run into a problem because $a_{2,2}^{(2)} = 0$.

The diagonal elements are called pivots. To deal with a zero pivot, a simple row interchange (row swap) in the augmented matrix that we use in the forward elimination can suffice. This strategy is called partial pivoting and will be described in detail below. In the example above, swapping Eqs. (2.6.5) and (2.6.6) we get

$$x_1 + x_2 + x_3 = 1 \tag{2.6.7}$$
$$0 + x_2 + x_3 = 0 \tag{2.6.8}$$
$$0 + 0 + x_3 = 1 \tag{2.6.9}$$

The latter system is already in an upper triangular form and allows us to proceed with the solution by backward substitution.

The previous example showed a pathological problem with Gauss elimination. However, due to round-off errors and finite precision in the computer, it is more likely that we will end up with pivots that are small but not zero. Let's change Eq. (2.6.2) slightly so that we now have the system

$$x_1 + x_2 + x_3 = 1 \tag{2.6.10}$$
$$x_1 + 1.0001x_2 + 2x_3 = 2 \tag{2.6.11}$$
$$x_1 + 2x_2 + 2x_3 = 1 \tag{2.6.12}$$

This is a very small change; we just changed $a_{2,2}^{(1)}$ from 1 to 1.0001. After the first step of Gauss elimination, we get

$$x_1 + x_2 + x_3 = 1 \tag{2.6.13}$$
$$0.0001x_2 + x_3 = 1 \tag{2.6.14}$$
$$x_2 + x_3 = 0 \tag{2.6.15}$$

The pivot $a_{2,2}^{(2)}$ is not equal to zero, but it is very small. When we perform the next step of Gauss elimination, we get the upper triangular system

$$x_1 + x_2 + x_3 = 1 \tag{2.6.16}$$
$$0.0001x_2 + x_3 = 1 \tag{2.6.17}$$
$$-9999x_3 = -10\,000 \tag{2.6.18}$$

that we can solve by back substitution. If we keep the maximum precision, the answer is

$$x_3 = 10\,000/9999$$
$$x_2 = -10\,000/9999$$
$$x_1 = 1 \tag{2.6.19}$$

So long as we can express this answer in fractional form, we are fine. The problem is that the computer needs to express the answer in a digital form. If we had a hypothetical computer with 5 digits of accuracy that could handle our original problem, then the result is

$$x_3 = 1.0001$$
$$x_2 = -1.0000$$
$$x_1 = 0.9999 \tag{2.6.20}$$

which is very close to the solution we had calculated before changing the coefficient $a_{2,2}^{(1)}$ from 1 to 1.0001. However, if we only had 4 digits of precision, we would get a completely different answer,

$$x_3 = 1.000$$
$$x_2 = 0.000$$
$$x_1 = 0.000 \qquad (2.6.21)$$

This is a bit of a contrived example, of course, since the 4 digits of precision would be insufficient to hold the value of $a_{2,2}^{(1)}$, but it illustrates our point.

Now let's consider what would happen if we had swapped Eqs. (2.6.14) and (2.6.15) in a pivoting move,

$$x_1 + x_2 + x_3 = 1 \qquad (2.6.22)$$
$$x_2 + x_3 = 0 \qquad (2.6.23)$$
$$0.0001x_2 + x_3 = 1 \qquad (2.6.24)$$

On the next step of forward elimination we would get

$$x_1 + x_2 + x_3 = 1 \qquad (2.6.25)$$
$$x_2 + x_3 = 0 \qquad (2.6.26)$$
$$0.9999x_3 = 1 \qquad (2.6.27)$$

If we only had 4 digits of precision, our answer would be

$$x_3 = 1.000$$
$$x_2 = -1.000$$
$$x_1 = 1.000 \qquad (2.6.28)$$

This is a much better result than we got without swapping the two rows. This discussion illustrates how pivoting can help mitigate the round-off errors due to finite precision arithmetic. Gauss elimination with pivoting therefore provides for greater robustness and greater numerical accuracy, so pivoting should always be employed.

2.6.2   Algorithm

There are several ways this idea can be implemented. The most common scheme is what we just used in the previous example and is called *partial pivoting*. It typically searches each column prior to the elimination steps to find the coefficient with the largest magnitude, and makes it the pivot. So, for the elimination operations to be performed on the kth column, we search in the kth column for the coefficient with the largest absolute value. If

$$|a_{s,k}^{(k)}| > |a_{kk}^{(k)}| \quad \text{for } s = k+1, \dots n \qquad (2.6.29)$$

then rows s and k are switched.

Example 2.4 Solve the following system of equations by Gauss elimination with partial pivoting:

$$
\begin{bmatrix} 2 & 1 & 2 & 0 \\ -2 & 0 & 1 & 4 \\ 4 & 2 & 2 & 1 \\ 4 & 0 & -4 & -4 \end{bmatrix}
\begin{bmatrix} x_1 \\ x_2 \\ x_3 \\ x_4 \end{bmatrix}
=
\begin{bmatrix} 1 \\ 13 \\ 2 \\ -20 \end{bmatrix}
\tag{2.6.30}
$$

Solution

The augmented matrix is

$$
\left[\begin{array}{cccc|c} 2 & 1 & 2 & 0 & 1 \\ -2 & 0 & 1 & 4 & 13 \\ 4 & 2 & 2 & 1 & 2 \\ 4 & 0 & -4 & -4 & -20 \end{array}\right]
\tag{2.6.31}
$$

In the first elimination, we need to swap row 1 and row 3,

$$
\left[\begin{array}{cccc|c} 4 & 2 & 2 & 1 & 2 \\ -2 & 0 & 1 & 4 & 13 \\ 2 & 1 & 2 & 0 & 1 \\ 4 & 0 & -4 & -4 & -20 \end{array}\right]
\tag{2.6.32}
$$

We then eliminate,

$$
\left[\begin{array}{cccc|c} 4 & 2 & 2 & 1 & 2 \\ 0 & 1 & 2 & 4.5 & 14 \\ 0 & 0 & 1 & -0.5 & 0 \\ 0 & -2 & -6 & -5 & -22 \end{array}\right]
\tag{2.6.33}
$$

In the second elimination, we have to swap row 2 and 4,

$$
\left[\begin{array}{cccc|c} 4 & 2 & 2 & 1 & 2 \\ 0 & -2 & -6 & -5 & -22 \\ 0 & 0 & 1 & -0.5 & 0 \\ 0 & 1 & 2 & 4.5 & 14 \end{array}\right]
\tag{2.6.34}
$$

and then eliminate

$$
\left[\begin{array}{cccc|c} 4 & 2 & 2 & 1 & 2 \\ 0 & -2 & -6 & -5 & -22 \\ 0 & 0 & 1 & -0.5 & 0 \\ 0 & 0 & -1 & 2 & 3 \end{array}\right]
\tag{2.6.35}
$$

There is no pivot for the last step, so we just need to eliminate,

$$
\left[\begin{array}{cccc|c} 4 & 2 & 2 & 1 & 2 \\ 0 & -2 & -6 & -5 & -22 \\ 0 & 0 & 1 & -0.5 & 0 \\ 0 & 0 & 0 & 1.5 & 3 \end{array}\right]
\tag{2.6.36}
$$

The resulting upper triangular system can now be solved by back substitution,

$$
\begin{aligned}
x_4 &= \frac{3}{3/2} = 2 \\
x_3 &= \frac{0 + 0.5(2)}{1} = 1 \\
x_2 &= \frac{-22 + 5(2) + 6(1)}{-2} = 3 \\
x_1 &= \frac{2 - 1(2) - 2(1) - 2(3)}{4} = -2
\end{aligned}
\tag{2.6.37}
$$

PROGRAM 2.2   *The implementation of partial pivoting follows directly from the example outlined above. Here, we have modified the naive Gaussian elimination in Program 2.1 to include partial pivoting.*

```
1   function x = linear_gauss_pivot(A,b)
2   %  A = n x n matrix
3   %  b = column vector, n x 1
4   n=length(b);
5   x=zeros(n,1);
6   % Perform the forward elimination
7   for k=1:n-1
8       %see if there is a bigger pivot
9       Amax = A(k,k);
10      swap_row = k;
11      for i = k+1:n
12          if abs(A(i,k)) > abs(Amax)
13              Amax = A(i,k);
14              swap_row = i;
15          end
16      end
17      %exchange rows if true
18      if swap_row ~= k
19          old_pivot(1,:) = A(k,:);
20          old_b = b(k);
21          A(k,:) = A(swap_row,:);
22          A(swap_row,:) = old_pivot;
23          b(k) = b(swap_row);
24          b(swap_row) = old_b;
25      end
26      %eliminate
27      for i=k+1:n
28          m=A(i,k)/A(k,k);
29          for j=k+1:n
30              A(i,j)=A(i,j)-m*A(k,j);
31          end
32          b(i)=b(i)-m*b(k);
33      end
34  end
35  % Perform the back substitution
36  x(n)=b(n)/A(n,n);
```

```
37   for i=n-1:-1:1
38     S=b(i);
39     for j=i+1:n
40         S=S-A(i,j)*x(j);
41     end
42       x(i)=S/A(i,i);
43   end
```

The new content of the program is contained in lines 8–25. In the first block, from lines 9–16, we loop through all of the rows from $k+1$ to n to see if the entry is larger than the value of $A_{k,k}$. If so, we keep that value as the current largest entry Amax *and record the row number as* swap_row. *In the second block from lines 18–25, we switch the rows in* **A** *and the corresponding entry in* **b** *from the partial pivoting if necessary. The remainder of the loops are the same as naive Gauss elimination.*

We have written this program so that it exactly mirrors the approach we used when we did partial pivoting by hand. While this is the easiest way to write the program, it is not very efficient. There are a number of ways that we could speed up this program or shorten the program itself. For example, we could work with the augmented matrix, which would mean we only need to swap rows in the augmented matrix rather than doing it for both **A** *and* **b**. *Moreover, when we do the swap in this program, we are swapping the entire row in the memory. Recall that we are never adjusting any of the entries below the diagonal. As a result, we could swap the section that we need, from* i = k:n *instead of the entire row. An even better solution would be not to swap the rows at all in the memory, but keep track of which rows have been swapped in a separate list and do the math that way. For very large systems, this last method is a huge improvement because the time required to copy the entries back and forth can be substantial.*

This brief discussion of how to improve an algorithm brings up an important point about programming. Often, it is best to make the first pass through a program with the goal of making something simple and easy to debug, rather than making the program as fast as possible. Once you have a working program, like we do here, you can then go about making the improvements listed above. At each step, you can check to make sure that your new program still produces the same answers. At the end, your final program may not be as easy to read as the original one, but you will be assured that it is producing the right results. Wherever possible, we will aim in this book to provide you with the simplest version of the program, and we encourage you to try to modify the programs to make them more efficient.

Partial pivoting is not the only type of algorithm that we can use to avoid inaccuracies caused by small pivots. A useful variant is called *threshold pivoting*; in this scheme the column is searched only to the extent that the magnitude of the pivot is greater than some user-specified value. This scheme can save computational effort and often yields results which are just as accurate as partial pivoting. For especially ill-conditioned systems, a strategy called *full pivoting* can be employed. This approach finds the pivot of largest magnitude among all of the coefficients to be modified during that step by searching through the coefficients of both the remaining rows and columns. Full pivoting is not often employed, however, because the many searches through the submatrices involved in elimination is a time-consuming task and because additional programming complexity arises when matrix columns are interchanged.

2.7   Case Study: Multi-component Flash

2.7.1   Problem Statement

Flash, illustrated in Fig. 2.5, is one of the simplest separation concepts in chemical engineering and the basis behind distillation. In this process, we have a feed of flow rate F containing mole fractions z_i of each of the $i = 1, 2, \ldots, n$ species at some temperature T_f and pressure P_f. In the process, the mixture is taken to a new temperature T or pressure P (or both) such that it reaches a vapor/liquid equilibrium with a liquid flow rate L and vapor flow rate G. The liquid mole fractions x_i are enriched with the low vapor-pressure species, while the vapor mole fractions y_i are enriched with the high vapor-pressure species. At the conceptual level, staged distillation is simply many flash operations combined in series.

In this problem, we will consider an ideal vapor–liquid equilibrium. For each species i, we need to satisfy mass balance

$$x_i L + y_i G = z_i F \tag{2.7.1}$$

and equilibrium

$$x_i P_i^{\text{sat}} = y_i P \tag{2.7.2}$$

In the latter, the vapor pressure is only a function of temperature, and we will model it using the Antoine equation,

$$\log_{10} P_i^{\text{sat}}(\text{mm Hg}) = A_i + \frac{B_i}{T(^\circ\text{C}) + C_i} \tag{2.7.3}$$

where A_i, B_i, and C_i are the Antoine coefficients for species i. The mass balances and equilibrium relations are supplemented by the requirement that the mole fractions sum up to unity

$$\sum_i x_i = 1 \tag{2.7.4}$$

$$\sum_i y_i = 1 \tag{2.7.5}$$

$$\sum_i z_i = 1 \tag{2.7.6}$$

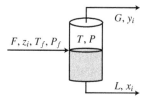

Figure 2.5 Schematic illustration of multi-component flash.

Table 2.1 Antoine coefficients for isomers of xylene.

Component	A	B	C	$T_b(°C)$
o-xylene	6.99891	1474.679	213.69	144.4
m-xylene	7.00908	1462.266	215.11	139.1
p-xylene	6.99052	1453.430	215.31	138.4

For n species, this corresponds to $2n + 3$ equations. A typical flash problem specifies the conditions of the inlets, the pressure and temperature, and then determines the outlet flow rates. There are thus $2n + 2$ equations, since Eq. (2.7.6) is rendered moot by specifying the inlet compositions, and there are $2n$ unknown compositions in the outlet flows along with their two flow rates L and G.

Flash problems, even with ideal equilibrium, are generally nonlinear. To make the problem linear, we can specify the operating temperature and pressure, along with all of the flow rates F, L and G. If we further let $F = 1$, so that L or G now represent the fraction that is liquid or vapor, then our linear equation set becomes

$$x_i P_i^{\text{sat}} - y_i P = 0 \quad \text{for } i = 1, 2, \ldots, n \tag{2.7.7}$$

$$x_i L + y_i G = z_i \quad \text{for } i = 1, 2, \ldots, n - 2 \tag{2.7.8}$$

$$\sum_i x_i = 1 \tag{2.7.9}$$

$$\sum_i y_i = 1 \tag{2.7.10}$$

These $2n$ equations are sufficient to compute the x_i and y_i. We can then compute the remaining two values of z_i by applying Eq. (2.7.8) for the last pair of chemicals.

In this problem, we will consider a three-component system consisting of the three isomers of xylene: ortho-xylene, meta-xylene, and para-xylene. Table 2.1 lists the Antoine coefficients for these species, along with their normal boiling points. We will consider the atmospheric flash ($P = 760$ mm Hg) of the mixture as a function of the feed rate z_i of o-xylene from 10% to 90% and a liquid flow rate of 20% of the feed. The goal of the problem is to determine the temperature range over which we have all three components in both the vapor and liquid phase.

2.7.2 Solution

The first thing we need to be able to do is solve Eqs. (2.7.7)–(2.7.10) for a given value of T and z_o for o-xylene. We first wrote a very short program to convert the data from Table 2.1 into a matrix for easy passage between our programs.

```
function Antoine = write_Antoine
%Antoine coefficients
Antoine(1,:) = [6.99891, 1474.679, 213.69]; %o-xylene
Antoine(2,:) = [7.00908, 1462.266, 215.11]; %m-xylene
Antoine(3,:) = [6.99052, 1453.43,  215.31]; %p-xylene
```

We then wrote a program to generate the system of equations and solve them.

```
1   function [x,y,z] = equilibrium(T,z,Antoine)
2   % T = temperature for the flash
3   % z = mole fraction of n-2 species in the feed
4   % Antoine = Antoine coefficients
5
6   %specifications
7   L = 0.2; %liquid mole fraction
8   P = 760; %pressure mmHg
9   n = length(z)+2;%number of species
10
11  %vapor pressure
12  for i = 1:n
13      A = Antoine(i,1);
14      B = Antoine(i,2);
15      C = Antoine(i,3);
16      Psat(i) = 10^(A - B/(T+C));
17  end
18
19  %unknowns:  1-6  = xi, 7-12 = yi
20  A = zeros(2*n); b = zeros(2*n,1);
21
22  %VLE for all species
23  for i = 1:n
24      %VLE for species i
25      A(i,i) = Psat(i);
26      A(i,i+n) = -P;
27  end
28  %mass balance for n - 2 species
29  for i = 1:n-2
30      A(i+n,i) = L;
31      A(i+n,i+n) = 1-L;
32      b(i+n) = z(i);
33  end
34  %mole fractions
35  for i = 1:n
36      A(2*n-1,i) = 1;
37      A(2*n,i+n) = 1;
38  end
39  b(2*n-1) = 1; b(2*n) = 1;
40
41  %solve problem
42  p = linear_gauss_pivot(A,b);
43
44  %extract data
45  x = p(1:n);
46  y = p(n+1:2*n);
47  for i = n-1:n
48      z(i) = x(i)*L + y(i)*(1-L);
49  end
```

The program is written in a somewhat general form for n species in case you want to modify it to consider other flash problems. Lines 6–9 set up the problem parameters. For a three-component system z is a scalar and line 9 gives n = 3. Once we know the temperature, we can use Eq. (2.7.3) to compute the saturation pressures of each species in the **for** loop of lines 12–17. We are then ready to set up the equations. We choose to write the unknown vector as

$$\mathbf{p} = \begin{bmatrix} x_o \\ x_m \\ x_p \\ y_o \\ y_m \\ y_p \end{bmatrix} \tag{2.7.11}$$

where we use $\mathbf{p}$ to avoid confusion with the liquid mole fractions x. Lines 23–27 set up the equilibrium in Eq. (2.7.7) for each species. Lines 29–33 set up up the mass balance in Eq. (2.7.8) for o-xylene. You may notice that we did this with a **for** loop over $n-2$ species, which only executes once. We did this for generality; if you want to have more than 3 species, you just send in a vector for z and line 9 will adjust the counters automatically. Finally, lines 35–39 write the constraints on the mole fractions in Eq. (2.7.9) and Eq. (2.7.10). At this point, we are ready to solve the problem, which we do by calling Program 2.2 for Gauss elimination with pivoting in line 42. Lines 45–49 extract the data into vectors for x_i, y_i, and z_i, where the two unspecified feed mole fractions for z_m and z_p are computed in line 48 from Eq. (2.7.8).

We now need to write a program that scans through the temperature range and sees if the solution to the linear equations indeed indicates that all three components are in both phases.

```
1  function [Tmin,Tmax] = multicomponent_flash(z,Antoine)
2  % z = mole fraction of o-xylene in feed
3  % Antoine = Antoine coefficients
4  n = length(z)+2;
5  nptsT = 500;
6  Tlow = 138.4;
7  Thigh = 144.4;
8  T = linspace(Tlow,Thigh,nptsT);
9  flash = 0;
10 for i = 1:nptsT
11     [x,y,z_vec] = equilibrium(T(i),z,Antoine);
12     if sum(x > 0) == n && sum(y > 0) == n && sum(z_vec > 0) == n
13         if flash == 0
14             Tmin = T(i);
15             flash = 1;
16         end
17     else
18         if flash == 1
19             Tmax = T(i-1);
20             return
21         end
22     end
23 end
```

For an ideal vapor–liquid equilibrium, we only need to scan the temperature range between the lowest and highest boiling points in Table 2.1; if we are too cold then everything will be in the liquid phase and if we are too hot then everything will be in the vapor phase. Lines 10–23 scan through this temperature range. In line 11, we solve the linear problem and get back the compositions of the liquid, vapor, and the feed. Pay attention to what we did with the feed variable z and the corresponding output of the feed concentrations in line 11. The input to the equilibrium calculation z involves $n - 2$ species, but the output from the equilibrium calculation gives the feed mole fractions for all n species. So we need to define the output z_vec of the equilibrium calculation as a *different* variable than the input z to avoid changing the size of z. If we instead had written in line 11 that [x,y,z] = equilibrium(T(i),z,Antoine); then the variable z would increase in size by two during each iteration of the loop and crash the program. Returning to the overall logic of this program, if every species is in both phases, then all of their mole fractions must be positive, which is checked in line 12. In this line, the MATLAB command sum(x>0) first implements the MATLAB command x > 0, which makes a vector the same length as x with a 1 in each spot in the vector if the statement is true and 0 if the statement is false. In other words, this command is equivalent to writing

```
1  for i = 1:length(x)
2      if x(i) > 0
3          out(i) = 1;
4      else
5          out(i) = 0;
6      end
7  end
```

where out is the output of the command. We then sum up the entries inside this vector — if $x_i > 0$ for each i then the sum will be equal to the length of the vector. This is the case where all of the species are in both phases. If one of the species is entirely in the liquid or the vapor, enforcing Eqs. (2.7.9) and (2.7.10) on the sum of the mole fractions will force one of the x_i or y_i to be negative and, possibly, others to exceed unity. Note that this is not an issue with the math; there is no problem solving the linear *math* problem and getting back negative numbers. The problem is with the equilibrium, meaning that some of the chemical equations themselves are incorrect from the outset.

To see when we cross from unphysical to physical solutions and then back to unphysical solutions for three-component equilibrium, we use the variable flash. Initially, we set this variable to zero and start at a low temperature where we expect the higher boiling species to be entirely in the liquid phase. This corresponds to the **if** statement in line 12 being false. As a result, the **else** block in lines 17–22 will execute. However, since flash = 0, the **if** statement in lines 19–21 is false, so nothing happens and we just go to the next value of the temperature. The first time we get to a temperature that has a physical solution, the **if** statement in line 12 is true and the value of flash = 0, so the **if** statement in lines 13–16 will execute. These lines store the value of this temperature as `Tmin` and set the value of `flash = 1`, indicating that we are inside the physically relevant range of temperatures. When the **for** loop executes the next time, it is quite likely that the **if** condition in line 12 will be true. However, since `flash = 1` now, the **if** block in lines 13–16 will not execute. So we just keep computing new solutions at higher and higher temperatures until the **if** statement in line 12 is false. At this point, since `flash = 1`, the **else** block lines 17–22 will execute. Since we have just identified the first unphysical temperature for three-component equilibrium, the previous temperature was the last physical temperature. Line 19 stores that temperature as `Tmax`. Line 20 resets the variable `flash = 0`. Since we now have the upper and lower bounds on the temperature, line 20 calls **return**, which exits the program.

Note that we are scanning with a very small increment in T, so we need to solve the linear problem hundreds of times. We could have saved time by modifying our equilibrium program so that it only updates the saturation pressures each time and keeps the rest of the matrix **A** the same. For a very large problem, this might be worthwhile because even writing down a large matrix can take quite some time. Our problem is small, so the savings in time by more efficiently writing the matrix **A** are not really worth the additional effort in coding.

Figure 2.6 shows the result of our calculations. Recall from Table 2.1 that o-xylene is the highest boiling isomer of xylene. Thus, when we run at the lower feed rate of o-xylene, there is a window of almost 1 °C where we can have three component equilibrium. As we increase the feed rate of xylene higher and higher, this window narrows because the o-xylene has a higher boiling point and differs somewhat from the

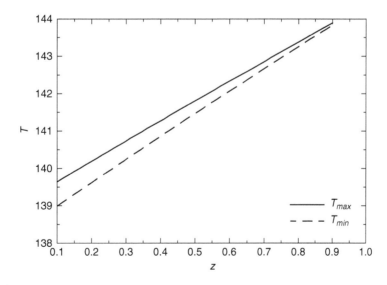

Figure 2.6 Temperature window for three-component equilibrium of xylene isomers for an atmospheric flash of a feed of 10% o-xylene to 20% liquid.

other isomers. Overall, xylene isomer separation by flash (or, indeed, by distillation) is difficult and this is not an optimal separation method.

2.8 Calculation of the Determinant by Gauss Elimination

There is an additional benefit to be gained from Gauss elimination. We can easily calculate the determinant of the original matrix, det $\mathbf{A}$, so that we can be alerted to the case of solution nonexistence or nonuniqueness when det $\mathbf{A} = 0$. How do we do this? The end result of elimination is an upper triangular matrix; computing the determinant of this matrix is exceedingly easy. The determinant of an upper triangular matrix is just the product of its diagonal elements,

$$\det \mathbf{U} = \prod_{i=1}^{n} U_{i,i} \tag{2.8.1}$$

If we transform the original matrix $\mathbf{A}$ into the upper triangular matrix $\mathbf{U}$ using elementary row multiplication and addition operations, then det $\mathbf{A} = $ det $\mathbf{U}$. The tricky part is that the sign of the determinant changes with every row (and column) interchange, therefore we need to track the number of times we perform row (and column) interchanges and adjust the sign of the determinant accordingly. We can finally write

$$\det \mathbf{A} = (-1)^p \det \mathbf{U} = (-1)^p \prod_{i=1}^{n} U_{i,i} \tag{2.8.2}$$

where p is the number of row and column interchanges performed in the process of forming $\mathbf{U}$. Computing the determinant this way scales with n^3. This is much faster than using a co-factor expansion, which would require $n!$ operations. In practice, the determinant is usually computed as a running product of the pivots during the elimination steps. For a large matrix, the determinant can be an extremely large number, so care must be taken to avoid overflow in this running product (e.g., sometimes the logarithm of the determinant is computed instead).

What happens when we apply Gauss elimination with partial pivoting for a case where det $\mathbf{A} = 0$? At the end of the forward elimination step, at least one entry on the diagonal of $\mathbf{U}$ will be zero. We know this must be true from Eq. (2.8.1). For the sake of definiteness, let's assume that the zero entry is $U_{k,k}$ and that this is the only non-zero entry. At the end of Gauss elimination, we will also get a vector

$$\mathbf{b}' = \begin{bmatrix} b_1^{(1)} \\ b_2^{(2)} \\ \vdots \\ b_k^{(k)} \\ \vdots \\ b_n^{(n)} \end{bmatrix} \tag{2.8.3}$$

If $b_k^{(k)} = 0$, then there are infinitely many solutions. If $b_k^{(k)} \neq 0$, then the system has no solution.

PROGRAM 2.3 *We can easily modify Gauss elimination with pivoting in Program 2.2 to compute the determinant.*

```
1  function determinant = linear_gauss_det(A)
2  % A = n x n matrix
3  n=size(A,1);
4  determinant = 1; %initialize det variable
5  % Perform the forward elimination  .
6  for k=1:n-1
7      %see if there is a bigger pivot
8      Amax = A(k,k);
9      swap_row = k;
10     for i = k+1:n
11         if abs(A(i,k)) > abs(Amax)
12             Amax = A(i,k);
13             swap_row = i;
14         end
15     end
16     %exchange rows if true
17     if swap_row ~= k
18         old_pivot(1,:) = A(k,:);
19         A(k,:) = A(swap_row,:);
20         A(swap_row,:) = old_pivot;
21         sign_pivot = -1;
22     else
23         sign_pivot = 1;
24     end
25     determinant = determinant*A(k,k)*sign_pivot; %update det
26     %eliminate
27     for i=k+1:n
28         m=A(i,k)/A(k,k);
29         for j=k+1:n
30             A(i,j)=A(i,j)-m*A(k,j);
31         end
32     end
33 end
34 determinant = determinant*A(n,n); %last entry on diagonal
```

There are a few minor differences to notice between this program and our program for solving $\mathbf{Ax} = \mathbf{b}$*. First, we no longer refer at all to the vector* $\mathbf{b}$*, since we are just computing the determinant, so the number of elimination steps comes from the size of the matrix* $\mathbf{A}$ *in line 3. The value of the determinant is kept as a running product of the pivots, which is done in line 25. To keep track of the number of row swaps, we have slightly modified the* **if** *statements to change the sign if there is a swap and keep the sign the same otherwise. Also do not forget about the last line of the program! In Gauss elimination, we only do elimination to column* $n - 1$*, so the last row is unchanged. Since we are keeping a running total of the product of the pivots, we need to remember to multiply by the diagonal entry on the last line.*

2.9   Banded Matrices

Many engineering problems (for example boundary value ordinary differential equation problems discretized by finite differences) result in coefficient matrices that are sparse, i.e. have a special structure with zeros in many locations. An example is banded matrices, for which non-zero elements are located in a band centered along the principal diagonal. More precisely, an $n \times n$ matrix for which $a_{i,j} = 0$ for $j > i + p$ and $i > j + q$ is called banded, with a bandwidth $p + q + 1$.

Example 2.5   What is the bandwidth of the matrix?

$$
\begin{bmatrix}
1 & 1 & 0 & 0 & 0 \\
1 & 1 & 1 & 0 & 0 \\
0 & 1 & 1 & 1 & 0 \\
0 & 0 & 1 & 1 & 1 \\
0 & 1 & 1 & 1 & 1
\end{bmatrix}
\tag{2.9.1}
$$

Solution
We get $p = 1$ on every row and $q = 3$ from the second column. So the total bandwidth is 5. Note that we only count to the right and down to compute the bandwidth. These are the directions that you need to use to do forward elimination, so the bandwidth gives you a measure of how long it will take to do Gauss elimination.

Usually p and q are equal and much smaller than n. Upper triangular and lower triangular matrices are special cases of banded matrices (try to identify p, q for these), and when $p = q = 1$ the matrix is called tridiagonal.

One can take advantage of the banded structure of a matrix to gain significant computational savings. Consider for example the case where $n \gg 1$, $q = p$ and $p \ll n$. To store every entry of the matrix (including the zeros), we need n^2 words of storage. For the banded matrix, we only need np words. In MATLAB, you can convert a matrix into a sparse matrix by writing A = sparse(A) and the program will store the values of the non-zero entries and their locations, rather than every entry. Memory space can become an issue when you deal with very large systems. In the programs here, we will generate the matrices first and then convert them to sparse form. However, if you want to start with a sparse matrix in the first place, you simply need to declare it as sparse with the right size, e.g. A = sparse(n,n) would make a sparse square matrix of size $n \times n$.

Even for modest systems, the computational time becomes much better if you account for the banded nature of the matrix. Recall that for forward elimination the number of operations scales roughly with n^3, because we had to eliminate n columns, each of which required doing work on n rows with n entries per row. In a banded matrix, we still have to eliminate over n columns. However, we know that we only have to go down q rows and across p columns to take care of all of the non-zero entries. As a result, the number of operations for forward elimination is approximately np^2 if $p = q$. For the back substitution, we also know that we only have p non-zero terms in the sum so the operation count is approximately np. Unlike the full matrix case, where the back substitution operations scale with n^2, much less than the forward elimination operations for $n \gg 1$, in many cases the banded matrix forward elimination step is comparable to the back substitution step because p is usually small compared to n. Nevertheless, the operations involved in Gauss elimination when taking advantage of the banded structure of a matrix scale with np^2.

PROGRAM 2.4 *Let's see how to modify our naive Gauss elimination program (2.1) for a banded system.*

```
1   function x = linear_ngaussel_banded(A,b,p,q)
2   %  A = n x n matrix
3   %  b = column vector, n x 1
4   n=length(b);
5   x=zeros(n,1);
6   % Perform the forward elimination
7   for k=1:n-1
8       for i=k+1:min(k+q,n)
9           m=A(i,k)/A(k,k);
10          for j=k+1:min(k+p,n)
11              A(i,j)=A(i,j)-m*A(k,j);
12          end
13          b(i)=b(i)-m*b(k);
14      end
15  end
16  % Perform the back substitution
17  x(n)=b(n)/A(n,n);
18  for i=n-1:-1:1
19      S=b(i);
20      for j=i+1:min(i+p,n)
21          S=S-A(i,j)*x(j);
22      end
23      x(i)=S/A(i,i);
24  end
```

There are only three changes to the program: line 8, line 10, and line 20. In each case, we count down the rows (for line 8), across the columns (line 10) or use the terms in the sum (line 20) that may be non-zero. We need to use the min(x,y) *function in MATLAB to avoid accidentally counting outside the range of the matrix when we get near the edges. For example, if we have $p = 3$ and we get to the $(n-1)$th row, we might think that the non-zero columns are $j = n-1$, n, $n+1$, and $n+2$. However, since the entries $j > n$ do not exist, we will get an error message. By picking the smallest of either $j = k + p$ or $j = n$, we ensure that we only use entries inside the matrix.*

Taking advantage of the banded structure of a matrix for Gauss elimination can lead to huge savings in number of operations, in the order of $(p/n)^2$. For example, if we have a 1000×1000 system with a p value of 1, which is not uncommon in boundary value problems after finite difference discretization, we would expect a speed up of

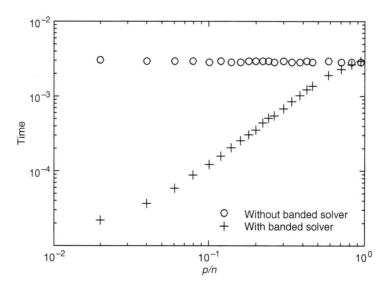

Figure 2.7 Illustration of the scaling of the time for Gauss elimination using a banded solver.

$(10^3)^2 = 10^6$ by using a banded solver! This is a crude calculation, however, since such a small system would probably have a lot of overhead in the computer too in addition to the time to solve the system. We can see this scaling in the log–log plot of the time for the solution versus p/n in Fig. 2.7, but you should be careful for the smaller values of p due to computer overhead.

Taking advantage of a banded matrix requires some thinking ahead. We will talk later in the book about how to set up certain problems in an intelligent way so that you get a banded system in the end, for example when you have coupled boundary value problems.

2.10 Simultaneous Gauss Elimination

Consider the case where we need to solve $\mathbf{Ax} = \mathbf{b}$, for multiple right-hand-side vectors $\mathbf{b_1}, \mathbf{b_2}, \ldots, \mathbf{b_m}$. Let the corresponding solutions be $\mathbf{x_1}, \mathbf{x_2}, \ldots, \mathbf{x_m}$. You may think that it is easiest to treat each system $\mathbf{Ax_k} = \mathbf{b_k}$ separately. If the time for solving a single system scales like $t \sim n^3$, then we would require mn^3 time to solve each of the m systems.

However, the underlying forward elimination operations are identical, so we can save ourselves some time by forming an augmented matrix with all the right-hand-side vectors,

$$\left[\begin{array}{cccc|ccc} a_{1,1} & a_{1,2} & \cdots & a_{1,n} & b_{1,1} & \cdots & b_{1,m} \\ a_{2,1} & a_{2,2} & \cdots & a_{2,n} & b_{2,1} & \cdots & b_{2,m} \\ \vdots & \vdots & \vdots & \vdots & \vdots & \vdots & \vdots \\ a_{n,1} & a_{n,2} & \cdots & a_{n,n} & b_{n,1} & \cdots & b_{n,m} \end{array} \right] \tag{2.10.1}$$

and perform forward elimination simultaneously. Remember, this is the most costly step in Gauss elimination. We now have $n + m$ columns, but we only have n rows. So the time required for the solution is $t \sim n(n + m)^2$. The resulting transformed system will have the form

$$\mathbf{U} \begin{bmatrix} \mathbf{x}_1 & \mathbf{x}_2 & \cdots & \mathbf{x}_m \end{bmatrix} = \begin{bmatrix} \mathbf{b}_1' & \mathbf{b}_2' & \cdots & \mathbf{b}_m' \end{bmatrix} \tag{2.10.2}$$

You can then solve for each $\mathbf{x}_k$ individually by back substitution using the corresponding column vector $\mathbf{b}_k'$.

There is a nice special case. Suppose the column vectors $\mathbf{b}_k$ are the unit vectors

$$\mathbf{e}_k = \begin{bmatrix} 0 \\ \vdots \\ 0 \\ 1 \\ 0 \\ \vdots \\ 0 \end{bmatrix} \tag{2.10.3}$$

where the only non-zero entry is in row k. If we use simultaneous Gauss elimination to solve the n equations

$$\mathbf{A} \begin{bmatrix} \mathbf{x}_1 & \mathbf{x}_2 & \cdots & \mathbf{x}_n \end{bmatrix} = \begin{bmatrix} \mathbf{e}_1 & \mathbf{e}_2 & \cdots & \mathbf{e}_n \end{bmatrix} \tag{2.10.4}$$

the result will be by definition the inverse of the matrix $\mathbf{A}$,

$$\begin{bmatrix} \mathbf{x}_1 & \mathbf{x}_2 & \cdots & \mathbf{x}_m \end{bmatrix} = \mathbf{A}^{-1} \tag{2.10.5}$$

2.11   LU Decomposition

There is a class of methods under this name that are often the ones of choice when solving large systems of equations, possibly with multiple right-hand sides. Suppose we can find a lower triangular matrix $\mathbf{L}$ and an upper triangular matrix $\mathbf{U}$ such that $\mathbf{A}$ can be written as

$$\mathbf{A} = \mathbf{LU} \tag{2.11.1}$$

or

$$\mathbf{A} = \begin{bmatrix} \ell_{11} & & \\ & \ddots & 0 \\ \ell_{ij} & \ddots & \\ & & \ell_{nn} \end{bmatrix} \begin{bmatrix} u_{11} & & \\ & \ddots & u_{ij} \\ 0 & \ddots & \\ & & u_{nn} \end{bmatrix} \tag{2.11.2}$$

Then we can rewrite

$$\mathbf{Ax} = \mathbf{b} \tag{2.11.3}$$

as

$$\mathbf{LUx} = \mathbf{b} \qquad (2.11.4)$$

This may look trivial, but it has important implications for simultaneously solving the system for many different constant vectors $\mathbf{b}$ that is usually better than simultaneous Gauss elimination.

To see how this works, define

$$\mathbf{y} = \mathbf{Ux} \qquad (2.11.5)$$

as an auxiliary vector. Then the original system takes the form

$$\mathbf{Ly} = \mathbf{b} \qquad (2.11.6)$$

Assuming you know $\mathbf{L}$ and $\mathbf{U}$, this gives you a strategy to compute $\mathbf{x}$ for any given $\mathbf{b}$. You first solve for $\mathbf{y}$ from Eq. (2.11.6) by forward substitution (since you have a lower triangular matrix). You then solve for $\mathbf{x}$ from Eq. (2.11.5) by back substitution.

2.11.1   Doolittle's Method

The key to this approach is that you need to know $\mathbf{L}$ and $\mathbf{U}$. For simplicity, let's work with a 3×3 matrix with the understanding that generalization to $n \times n$ matrices is straightforward. We already know how to turn

$$\mathbf{A} = \begin{bmatrix} a_{1,1}^{(1)} & a_{1,2}^{(1)} & a_{1,3}^{(1)} \\ a_{2,1}^{(1)} & a_{2,2}^{(1)} & a_{2,3}^{(1)} \\ a_{3,1}^{(1)} & a_{3,2}^{(1)} & a_{3,3}^{(1)} \end{bmatrix} \qquad (2.11.7)$$

into an upper triangular matrix,

$$\mathbf{U} = \begin{bmatrix} a_{1,1}^{(1)} & a_{1,2}^{(1)} & a_{1,3}^{(1)} \\ 0 & a_{2,2}^{(2)} & a_{2,3}^{(2)} \\ 0 & 0 & a_{3,3}^{(3)} \end{bmatrix} \qquad (2.11.8)$$

by forward elimination. The amazing thing is that the $\mathbf{L}$ matrix which completes the LU decomposition defined by Eq. (2.11.1) with the upper triangular matrix resulting from forward elimination (2.11.8) is

$$\mathbf{L} = \begin{bmatrix} 1 & 0 & 0 \\ m_{2,1} & 1 & 0 \\ m_{3,1} & m_{3,2} & 1 \end{bmatrix} \qquad (2.11.9)$$

where the $m_{i,j}$ are the coefficients we used to convert $\mathbf{A}$ to $\mathbf{U}$ by forward elimination.

Example 2.6   Show Eq. (2.11.9) is true by directly computing

$$\mathbf{LU} = \begin{bmatrix} 1 & 0 & 0 \\ m_{2,1} & 1 & 0 \\ m_{3,1} & m_{3,2} & 1 \end{bmatrix} \begin{bmatrix} a_{1,1}^{(1)} & a_{1,2}^{(1)} & a_{1,3}^{(1)} \\ 0 & a_{2,2}^{(2)} & a_{2,3}^{(2)} \\ 0 & 0 & a_{3,3}^{(3)} \end{bmatrix} \qquad (2.11.10)$$

$$
= \begin{bmatrix}
a_{1,1}^{(1)} & a_{1,2}^{(1)} & a_{1,3}^{(1)} \\
m_{2,1}a_{1,1}^{(1)} & m_{2,1}a_{1,2}^{(1)} + a_{2,2}^{(2)} & m_{2,1}a_{1,3}^{(1)} + a_{2,3}^{(2)} \\
m_{3,1}a_{1,1}^{(1)} & m_{3,1}a_{1,2}^{(1)} + m_{3,2}a_{2,2}^{(2)} & m_{3,1}a_{1,3}^{(1)} + m_{3,2}a_{2,3}^{(2)} + a_{3,3}^{(3)}
\end{bmatrix}
$$

and showing that the entries in Eq. (2.11.10) are identical to those in Eq. (2.11.7).

Solution

This holds trivially for the first row entries. Let's look at the first column. Equating the (2,1) entries gives us

$$
a_{2,1}^{(1)} = m_{2,1}a_{1,1}^{(1)} \tag{2.11.11}
$$

which is also true, since it yields the definition of the coefficient that we used in the first forward elimination step (second row, first column),

$$
m_{2,1} = \frac{a_{2,1}^{(1)}}{a_{1,1}^{(1)}} \tag{2.11.12}
$$

The result is similar for the (3,1) entries,

$$
m_{3,1} = \frac{a_{3,1}^{(1)}}{a_{1,1}^{(1)}} \tag{2.11.13}
$$

Now, let's look at the remaining second row equations. Equating the (2,2) entries gives us

$$
m_{2,1}a_{1,2}^{(1)} + a_{2,2}^{(2)} = a_{2,2}^{(1)} \tag{2.11.14}
$$

If we solve for $a_{2,2}^{(2)}$ we get

$$
a_{2,2}^{(2)} = a_{2,2}^{(1)} - m_{2,1}a_{1,2}^{(1)} \tag{2.11.15}
$$

This is true if

$$
m_{2,1} = \frac{a_{2,1}^{(1)}}{a_{1,1}^{(1)}} \tag{2.11.16}
$$

which is Eq. (2.11.12). The (2,3) entries yield the same type of result,

$$
m_{2,1}a_{1,3}^{(1)} + a_{2,3}^{(2)} = a_{2,3}^{(1)} \tag{2.11.17}
$$

which is true from Eq. (2.11.12).

Let's now work out the two remaining equations in the third row. For the (3,2) entry we have

$$
m_{3,1}a_{1,2}^{(1)} + m_{3,2}a_{2,2}^{(2)} = a_{3,2}^{(1)} \tag{2.11.18}
$$

Rewriting this as

$$
m_{3,2}a_{2,2}^{(2)} = a_{3,2}^{(1)} - m_{3,1}a_{1,2}^{(1)} \tag{2.11.19}
$$

we notice that the term on the right-hand side is just the result after the first step of Gauss elimination,

$$
a_{3,2}^{(1)} - m_{3,1}a_{1,2}^{(1)} = a_{3,2}^{(2)} \tag{2.11.20}
$$

We thus end up with the definition of the coefficient for forward elimination,

$$m_{3,2} = \frac{a_{3,2}^{(2)}}{a_{2,2}^{(2)}} \tag{2.11.21}$$

For the (3,3) entry we have

$$m_{3,1}a_{1,3}^{(1)} + m_{3,2}a_{2,3}^{(2)} + a_{3,3}^{(3)} = a_{3,3}^{(1)} \tag{2.11.22}$$

The logic to show this is true follows from the (3,2) entry. First rewrite the latter result as

$$m_{3,2}a_{2,3}^{(2)} + a_{3,3}^{(3)} = a_{3,3}^{(1)} - m_{3,1}a_{1,3}^{(1)} \tag{2.11.23}$$

and recognize that the right-hand side is just the result after the first step of elimination,

$$m_{3,2}a_{2,3}^{(2)} + a_{3,3}^{(3)} = a_{3,3}^{(2)} \tag{2.11.24}$$

We are then left with the definition of the second step of elimination again,

$$a_{3,3}^{(3)} = a_{3,3}^{(2)} - m_{3,2}a_{2,3}^{(2)} \tag{2.11.25}$$

It turns out that the above LU decomposition result is the same for any $n \times n$ matrix, as long as forward elimination works without having to perform any row rearrangements! The **U** matrix is the one resulting from forward elimination and the **L** matrix is

$$\mathbf{L} = \begin{bmatrix} 1 & 0 & 0 & 0 & \cdots & 0 \\ m_{2,1} & 1 & 0 & 0 & \cdots & 0 \\ m_{3,1} & m_{3,2} & 1 & 0 & \cdots & 0 \\ \vdots & \vdots & \vdots & \vdots & \vdots & \vdots \\ \vdots & \vdots & \vdots & \vdots & \vdots & \vdots \\ m_{n,1} & m_{n,2} & m_{n,3} & \cdots & m_{n,n-1} & 1 \end{bmatrix} \tag{2.11.26}$$

The specific method described here is called Doolittle's method. There are other methods as well which impose additional requirements on the **L** and **U** matrices.

Example 2.7   Solve the following system of equations by LU decomposition:

$$\begin{bmatrix} 2 & 1 & 2 & 0 \\ -2 & 0 & 1 & 4 \\ 4 & 2 & 2 & 1 \\ 4 & 0 & -4 & -4 \end{bmatrix} \begin{bmatrix} x_1 \\ x_2 \\ x_3 \\ x_4 \end{bmatrix} = \begin{bmatrix} 1 \\ 13 \\ 2 \\ -20 \end{bmatrix} \tag{2.11.27}$$

Solution

This is the same problem as Example 2.4. However, since we will do LU decomposition without pivoting, the upper triangular matrix will be different than Eq. (2.6.36). We start with the matrix $\mathbf{A}$

$$\mathbf{A}^{(1)} = \begin{bmatrix} 2 & 1 & 2 & 0 \\ -2 & 0 & 1 & 4 \\ 4 & 2 & 2 & 1 \\ 4 & 0 & -4 & -4 \end{bmatrix} \tag{2.11.28}$$

To do the first step of elimination, we get

$$\mathbf{A}^{(2)} = \begin{bmatrix} 2 & 1 & 2 & 0 \\ 0 & 1 & 3 & 4 \\ 0 & 0 & -2 & 1 \\ 0 & -2 & -8 & -4 \end{bmatrix} \tag{2.11.29}$$

From the coefficients we used to forward eliminate the first column, we now know that

$$\mathbf{L} = \begin{bmatrix} 1 & 0 & 0 & 0 \\ -1 & 1 & 0 & 0 \\ 2 & ? & 1 & 0 \\ 2 & ? & ? & 1 \end{bmatrix} \tag{2.11.30}$$

We now do the next step of elimination, which gives us

$$\mathbf{A}^{(3)} = \begin{bmatrix} 2 & 1 & 2 & 0 \\ 0 & 1 & 3 & 4 \\ 0 & 0 & -2 & 1 \\ 0 & 0 & -2 & 4 \end{bmatrix} \tag{2.11.31}$$

which also gives us two more entries in $\mathbf{L}$,

$$\mathbf{L} = \begin{bmatrix} 1 & 0 & 0 & 0 \\ -1 & 1 & 0 & 0 \\ 2 & 0 & 1 & 0 \\ 2 & -2 & ? & 1 \end{bmatrix} \tag{2.11.32}$$

The last step of elimination gives us the upper triangular matrix,

$$\mathbf{A}^{(4)} = \mathbf{U} = \begin{bmatrix} 2 & 1 & 2 & 0 \\ 0 & 1 & 3 & 4 \\ 0 & 0 & -2 & 1 \\ 0 & 0 & 0 & 3 \end{bmatrix} \tag{2.11.33}$$

along with the last entry in the lower triangular matrix,

$$\mathbf{L} = \begin{bmatrix} 1 & 0 & 0 & 0 \\ -1 & 1 & 0 & 0 \\ 2 & 0 & 1 & 0 \\ 2 & -2 & 1 & 1 \end{bmatrix} \tag{2.11.34}$$

If you are working out a problem by LU decomposition by hand and you are concerned that you might have made a mistake in the algebra, you can always check at the end that $\mathbf{LU} = \mathbf{A}$. In this case, we have

$$
\mathbf{LU} = \begin{bmatrix} 1 & 0 & 0 & 0 \\ -1 & 1 & 0 & 0 \\ 2 & 0 & 1 & 0 \\ 2 & -2 & 1 & 1 \end{bmatrix} \begin{bmatrix} 2 & 1 & 2 & 0 \\ 0 & 1 & 3 & 4 \\ 0 & 0 & -2 & 1 \\ 0 & 0 & 0 & 3 \end{bmatrix} = \begin{bmatrix} 2 & 1 & 2 & 0 \\ -2 & 0 & 1 & 4 \\ 4 & 2 & 2 & 1 \\ 4 & 0 & -4 & -4 \end{bmatrix}
$$

(2.11.35)

so we did the math correctly.

The decomposition is the hard part of the problem and the part that takes most of the time. We can then solve the system for any constant vector $\mathbf{b}$ rather quickly. For this particular example, we have

$$
\begin{bmatrix} 1 & 0 & 0 & 0 \\ -1 & 1 & 0 & 0 \\ 2 & 0 & 1 & 0 \\ 2 & -2 & 1 & 1 \end{bmatrix} \begin{bmatrix} y_1 \\ y_2 \\ y_3 \\ y_4 \end{bmatrix} = \begin{bmatrix} 1 \\ 13 \\ 2 \\ -20 \end{bmatrix}
$$

(2.11.36)

The forward substitution gives us

$$
\begin{aligned}
y_1 &= 1 \\
y_2 &= 13 + 1 = 14 \\
y_3 &= 2 - 2 = 0 \\
y_4 &= -20 - 2 + 2(14) = 6
\end{aligned}
$$

(2.11.37)

We now use back substitution to solve

$$
\begin{bmatrix} 2 & 1 & 2 & 0 \\ 0 & 1 & 3 & 4 \\ 0 & 0 & -2 & 1 \\ 0 & 0 & 0 & 3 \end{bmatrix} \begin{bmatrix} x_1 \\ x_2 \\ x_3 \\ x_4 \end{bmatrix} = \begin{bmatrix} 1 \\ 14 \\ 0 \\ 6 \end{bmatrix}
$$

(2.11.38)

which yields

$$
\begin{aligned}
x_4 &= \frac{6}{3} = 2 \\
x_3 &= \frac{0 - 2}{-2} = 1 \\
x_2 &= 14 - 4(2) - 3(1) = 3 \\
x_1 &= \frac{1 - 2 - 3}{2} = -2
\end{aligned}
$$

(2.11.39)

2.11.2   Implementation

If you think for a bit about the number of operations required for LU decomposition, you will see that it is no faster than simultaneous Gauss elimination. Recall that simultaneous Gauss elimination required $n(n+m)^2$ operations to work on the $n \times (n+m)$ augmented matrix. To decompose $\mathbf{A}$ into $\mathbf{L}$ and $\mathbf{U}$, we require n^3 steps for forward elimination. The substitution for a given vector $\mathbf{b}$ requires n^2 operations. So with m vectors, the total time is $t \sim n^3 + n^2 m = n^2(n+m)$, which is the same approximate result (to within a prefactor) that we had for simultaneous Gauss elimination.

If the time for LU decomposition and simultaneous Gauss elimination are very similar, why is LU decomposition the preferred method? The key advantage in LU decomposition is that we only decompose the matrix once, and then use the results for $\mathbf{L}$ and $\mathbf{U}$ again and again. Thus, once we have saved $\mathbf{L}$ and $\mathbf{U}$ to a file, we can just load them back to the memory when we come across a new value of $\mathbf{b}$. In contrast, we need to know all of the $\mathbf{b}$ vectors in advance to take advantage of simultaneous Gauss elimination. If we solve for a bunch of $\mathbf{b}$ vectors and then realize that we forgot one of them, we need to repeat the forward elimination step again! Since forward elimination is the slowest part, scaling like n^3, this is a major disadvantage for simultaneous Gauss elimination.

PROGRAM 2.5   *It is very easy to modify naive Gauss elimination to instead decompose* $\mathbf{A}$ *into* $\mathbf{L}$ *and* $\mathbf{U}$.

```
1   function A = linear_LU(A)
2   %  A = n x n matrix
3   n=size(A,1);
4   % Perform the forward elimination
5   for k=1:n-1
6       for i=k+1:n
7           m=A(i,k)/A(k,k);
8           A(i,k) = m; %store entry for L
9           for j=k+1:n
10              A(i,j)=A(i,j)-m*A(k,j);
11          end
12      end
13  end
```

The program simply stores the values $m_{i,j}$ *in the lower triangular part of* $\mathbf{A}$*. In this way, you only need to use* n^2 *memory instead of* $2n^2$ *memory to solve the problem. For the* 3×3 *example, at the end of LU decomposition you have on the computer*

$$\begin{bmatrix} a_{1,1}^{(1)} & a_{1,2}^{(1)} & a_{1,3}^{(1)} \\ m_{2,1} & a_{2,2}^{(2)} & a_{2,3}^{(2)} \\ m_{3,1} & m_{3,2} & a_{3,3}^{(3)} \end{bmatrix} \tag{2.11.40}$$

If we wanted to read out the particular matrices $\mathbf{L}$ *and* $\mathbf{U}$ *from* `linear_LU.m`*, we could use a program like the following:*

```
1   function [L,U] = linear_LU_from_A(A)
2   % A = n x n matrix with U as upper part and L as lower part
```

```
3   n = size(A,1);
4   U = zeros(n);
5   L = eye(n); %put ones on the diagonal
6   for i = 1:n
7       for j = i:n
8           U(i,j) = A(i,j);
9       end
10  end
11  for i = 2:n
12      for j = 1:i-1
13          L(i,j) = A(i,j);
14      end
15  end
```

In order to actually solve the problem, we would also need to do the forward and back substitution:

```
1   function x = linear_LU_forward_back(L,U,b)
2   % L = lower triangular matrix from LU decomposition
3   % U = upper triangular matrix from LU decomposition
4   % b = n x 1 vector
5   n = length(b);
6   %Perform the forward substitution
7   y = zeros(n,1);
8   y(1) = b(1);
9   for i = 2:n
10      y(i) = b(i);
11      for j = 1:i-1
12          y(i) = y(i) - L(i,j)*y(j);
13      end
14  end
15  % Perform the back substitution
16  x = zeros(n,1);
17  x(n)=y(n)/U(n,n);
18  for i=n-1:-1:1
19      S=y(i);
20      for j=i+1:n
21          S=S-U(i,j)*x(j);
22      end
23      x(i)=S/U(i,i);
24  end
```

*Note that, during the forward substitution, there is no place where we divide by $L(i,i)$ because we know that $L(i,i) = 1$. This is in contrast with line 23, where we need to divide by the diagonal entry in **U** during back substitution.*

*Similar to some of our previous programs, this is not the most efficient way to code LU decomposition, especially with the step that breaks up the matrix from Gauss elimination into **L** and **U**, but these three functions are easy to follow. You should see if you can figure out a good way to merge them together and remove the redundant operations.*

*You might also wonder why we felt the need to break this program into multiple function files. If you are using LU decomposition to solve many simultaneous problems, you only want to compute L and U once. So you would call `linear_LU` for your matrix **A** and store **L** and **U** in the memory as a single matrix, or you could break it up into two matrices using `linear_LU_from_A`. In either case, you would then call a program similar to `linear_LU_forward_back` to solve the problem for a given value of **b**.*

There is one other very important point to notice in our implementation of LU decomposition, namely that we did the decomposition via Gauss elimination without pivoting. This will work fine so long as the pivots are not small. In general, you should always do Gauss elimination with pivoting to avoid round-off errors. The bookkeeping in LU decomposition then becomes a bit more challenging.

2.12 Case Study: Linear Process Flow Sheet

2.12.1 Problem Statement

Mass and energy balances is the first (and most common) place in chemical engineering courses where you encounter large linear algebra problems. While many real flow sheets for processes result in equations that include nonlinearities due to non-ideal thermodynamics or complicated kinetics and transport processes, it is often useful to consider simplified linear problems first. LU decomposition is a powerful method for analyzing linear flow sheets.

To see why LU decomposition is useful here, let's consider the relatively simple process illustrated in Fig. 2.8. The process consists of two unit operations, a reactor and a separator. The process takes in some flow rates A_1 and B_1 of species A and B. The reactor has a reaction

$$A + B \to C \quad \text{(Reaction 1)} \qquad (2.12.1)$$

that produces an intermediate product C that then needs to be converted to the desired product D by a second reaction

$$A + C \to D \quad \text{(Reaction 2)} \qquad (2.12.2)$$

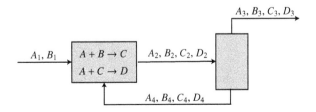

Figure 2.8 Process flow sheet consisting of a reactor and separator with recycle.

The single-pass conversion of the reactor is 90% with a 30% selectivity for Reaction 2. The products of the reaction in stream "2" are fed to a separator, which produces the product stream "3" and a recycle stream "4." The specifications of the separator are a recycle of 65% of D and 85% of C. The flow B_2 is evenly split between the product and recycle streams, while 10% of A_2 is lost to the product stream.

The goal of this case study is to develop the numerical methods required to study the output stream as a function of the inlet flow rates A_1 and B_1.

2.12.2   Solution

We first need to set up the problem using standard concepts from mass balances. We thus define the variables to be the flow rates of each stream along with the extent of reaction ξ_1 for Reaction 1 and ξ_2 for Reaction 2. The steady state mass balances on the reactor are

$$A_4 - A_2 - \xi_1 - \xi_2 = -A_1 \tag{2.12.3}$$
$$B_4 - B_2 - \xi_1 = -B_1 \tag{2.12.4}$$
$$C_4 - C_2 + \xi_1 - \xi_2 = 0 \tag{2.12.5}$$
$$D_4 - D_2 + \xi_2 = 0 \tag{2.12.6}$$

and the mass balances for the separator are

$$A_2 - A_3 - A_4 = 0 \tag{2.12.7}$$
$$B_2 - B_3 - B_4 = 0 \tag{2.12.8}$$
$$C_2 - C_3 - C_4 = 0 \tag{2.12.9}$$
$$D_2 - D_3 - D_4 = 0 \tag{2.12.10}$$

The conversion and selectivity of the reaction give

$$\xi_1 + \xi_2 - 0.9A_4 = 0.9A_1 \tag{2.12.11}$$
$$0.3\xi_1 - 0.7\xi_2 = 0 \tag{2.12.12}$$

and the separator specifications are

$$D_4 - 0.65D_2 = 0 \tag{2.12.13}$$
$$B_3 - B_4 = 0 \tag{2.12.14}$$
$$0.1A_2 - A_3 = 0 \tag{2.12.15}$$
$$0.85C_2 - C_4 = 0 \tag{2.12.16}$$

We wrote the equations so that the unknowns are all on the left-hand side and the inlet flow rates are on the right-hand side. You can see why LU decomposition is useful – we want to solve this system of equations many times for different values of A_1 and B_1, but the left-hand side remains unchanged.

In the formulation of the problem, we do need to be a bit careful about the order of the equations since we only discussed LU decomposition *without* partial pivoting. One possible choice of the equation order that avoids a zero pivot during forward elimination is

$$
\begin{bmatrix}
-1 & 0 & 0 & 0 & 0 & 0 & 0 & 0 & 1 & 0 & 0 & 0 & -1 & -1 \\
0 & -1 & 0 & 0 & 0 & 0 & 0 & 0 & 0 & 1 & 0 & 0 & -1 & 0 \\
0 & 0 & -1 & 0 & 0 & 0 & 0 & 0 & 0 & 0 & 1 & 0 & 1 & -1 \\
0 & 0 & 0 & -1 & 0 & 0 & 0 & 0 & 0 & 0 & 0 & 1 & 0 & 1 \\
1 & 0 & 0 & 0 & -1 & 0 & 0 & 0 & -1 & 0 & 0 & 0 & 0 & 0 \\
0 & 1 & 0 & 0 & 0 & -1 & 0 & 0 & 0 & -1 & 0 & 0 & 0 & 0 \\
0 & 0 & 1 & 0 & 0 & 0 & -1 & 0 & 0 & 0 & -1 & 0 & 0 & 0 \\
0 & 0 & 0 & 1 & 0 & 0 & 0 & -1 & 0 & 0 & 0 & -1 & 0 & 0 \\
0 & 0 & 0 & 0 & 0 & 0 & 0 & 0 & -0.90 & 0 & 0 & 0 & 1 & 1 \\
0 & 0 & 0 & 0 & 0 & 1 & 0 & 0 & 0 & -1 & 0 & 0 & 0 & 0 \\
0 & 0 & 0.85 & 0 & 0 & 0 & 0 & 0 & 0 & 0 & -1 & 0 & 0 & 0 \\
0 & 0 & 0 & -0.65 & 0 & 0 & 0 & 0 & 0 & 0 & 0 & 1 & 0 & 0 \\
0.10 & 0 & 0 & 0 & -1 & 0 & 0 & 0 & 0 & 0 & 0 & 0 & 0 & 0 \\
0 & 0 & 0 & 0 & 0 & 0 & 0 & 0 & 0 & 0 & 0 & 0 & -0.30 & 0.70
\end{bmatrix}
\times
\begin{bmatrix}
A_2 \\ B_2 \\ C_2 \\ D_2 \\ A_3 \\ B_3 \\ C_3 \\ D_3 \\ A_4 \\ B_4 \\ C_4 \\ D_4 \\ \xi_1 \\ \xi_2
\end{bmatrix}
=
\begin{bmatrix}
-10 \\ -20 \\ 0 \\ 0 \\ 0 \\ 0 \\ 0 \\ 0 \\ 9 \\ 0 \\ 0 \\ 0 \\ 0 \\ 0
\end{bmatrix}
\qquad (2.12.17)
$$

You will notice that the specifications for reactor and separator are not in the order given in the problem statement, but that this re-ordering of the equations avoids problems with getting any zeros on the diagonal. Although $a^{(1)}_{13,13} = 0$, this diagonal entry will become non-zero on the first step of forward elimination.

You may also notice that many of the equations are of a very similar form. This suggests a way to write the program to create the matrix **A** in an efficient way.

```
1  function [L,U] = linear_flow_sheet_LU
2  %parameters
3  conv = 0.9;
4  sel = 0.3;
5
6  A = zeros(14);
7  %reactor
```

```
 8  for i = 1:4
 9      A(i,i) = -1; A(i,i+8) = 1;
10  end
11  A(1,13) = -1;      A(2,13) = -1;         A(3,13) = 1;
12  A(1,14) = -1;      A(3,14) = -1;         A(4,14) = 1;
13  %separator
14  for i = 1:4
15      A(i+4,i) = 1; A(i+4,i+4) = -1;    A(i+4,i+8) = -1;
16  end
17  %specifications
18  A(9,9)   = -conv; A(9,13)   = 1;      A(9,14) = 1;
19  A(10,6)  = 1;     A(10,10)  = -1;
20  A(11,3)  = 0.85;  A(11,11)  = -1;
21  A(12,4)  = -0.65; A(12,12)  = 1;
22  A(13,1)  = 0.1;   A(13,5)   = -1;
23  A(14,13) = -sel;  A(14,14)  = (1-sel);
24  %LU decomposition
25  A_LU = linear_LU(A);
26  [L,U] = linear_LU_from_A(A_LU);
```

For the reactor terms in lines 7–12, we first write the in/out terms in a **for** loop. Line 11 adds terms for Reaction 1, and line 12 adds terms for Reaction 2. You may find this way of writing the equations preferable to simply writing each balance separately, since it is easy to debug and see what is going on with the reactions. The separation mass balances are then implemented in a second **for** loop in lines 14–16. The specifications have no simple way to write them, so we just put them in the order of Eq. (2.12.17). The last two lines of the program implement LU decomposition with Program 2.5.

The output of our file is the lower triangular matrix

$$
\mathbf{L} = \begin{bmatrix}
1 & 0 & 0 & 0 & 0 & 0 & 0 & 0 & 0 & 0 & 0 & 0 & 0 & 0 \\
0 & 1 & 0 & 0 & 0 & 0 & 0 & 0 & 0 & 0 & 0 & 0 & 0 & 0 \\
0 & 0 & 1 & 0 & 0 & 0 & 0 & 0 & 0 & 0 & 0 & 0 & 0 & 0 \\
0 & 0 & 0 & 1 & 0 & 0 & 0 & 0 & 0 & 0 & 0 & 0 & 0 & 0 \\
-1 & 0 & 0 & 0 & 1 & 0 & 0 & 0 & 0 & 0 & 0 & 0 & 0 & 0 \\
0 & -1 & 0 & 0 & 0 & 1 & 0 & 0 & 0 & 0 & 0 & 0 & 0 & 0 \\
0 & 0 & -1 & 0 & 0 & 0 & 1 & 0 & 0 & 0 & 0 & 0 & 0 & 0 \\
0 & 0 & 0 & -1 & 0 & 0 & 0 & 1 & 0 & 0 & 0 & 0 & 0 & 0 \\
0 & 0 & 0 & 0 & 0 & 0 & 0 & 0 & 1 & 0 & 0 & 0 & 0 & 0 \\
0 & 0 & 0 & 0 & 0 & -1 & 0 & 0 & 0 & 1 & 0 & 0 & 0 & 0 \\
0 & 0 & -0.85 & 0 & 0 & 0 & 0 & 0 & 0 & 0 & 1 & 0 & 0 & 0 \\
0 & 0 & 0 & 0.65 & 0 & 0 & 0 & 0 & 0 & 0 & 0 & 1 & 0 & 0 \\
-0.10 & 0 & 0 & 0 & 1 & 0 & 0 & 0 & -0.11 & 0 & 0 & 0 & 1 & 0 \\
0 & 0 & 0 & 0 & 0 & 0 & 0 & 0 & 0 & 0 & 0 & 0 & -0.30 & 1
\end{bmatrix}
$$

$$(2.12.18)$$

and the upper triangular matrix

$$\mathbf{U} = \begin{bmatrix} -1 & 0 & 0 & 0 & 0 & 0 & 0 & 0 & 1 & 0 & 0 & 0 & -1 & -1 \\ 0 & -1 & 0 & 0 & 0 & 0 & 0 & 0 & 0 & 1 & 0 & 0 & -1 & 0 \\ 0 & 0 & -1 & 0 & 0 & 0 & 0 & 0 & 0 & 0 & 1 & 0 & 1 & -1 \\ 0 & 0 & 0 & -1 & 0 & 0 & 0 & 0 & 0 & 0 & 0 & 1 & 0 & 1 \\ 0 & 0 & 0 & 0 & -1 & 0 & 0 & 0 & 0 & 0 & 0 & 0 & -1 & -1 \\ 0 & 0 & 0 & 0 & 0 & -1 & 0 & 0 & 0 & 0 & 0 & 0 & -1 & 0 \\ 0 & 0 & 0 & 0 & 0 & 0 & -1 & 0 & 0 & 0 & 0 & 0 & 1 & -1 \\ 0 & 0 & 0 & 0 & 0 & 0 & 0 & -1 & 0 & 0 & 0 & 0 & 0 & 1 \\ 0 & 0 & 0 & 0 & 0 & 0 & 0 & 0 & -0.90 & 0 & 0 & 0 & 1 & 1 \\ 0 & 0 & 0 & 0 & 0 & 0 & 0 & 0 & 0 & -1 & 0 & 0 & -1 & 0 \\ 0 & 0 & 0 & 0 & 0 & 0 & 0 & 0 & 0 & 0 & -0.15 & 0 & 0.85 & -0.85 \\ 0 & 0 & 0 & 0 & 0 & 0 & 0 & 0 & 0 & 0 & 0 & 0.35 & 0 & -0.65 \\ 0 & 0 & 0 & 0 & 0 & 0 & 0 & 0 & 0 & 0 & 0 & 0 & 1.01 & 1.01 \\ 0 & 0 & 0 & 0 & 0 & 0 & 0 & 0 & 0 & 0 & 0 & 0 & 0 & 1 \end{bmatrix}$$

$$(2.12.19)$$

Before we leave this example, it is worth noting the way that we have made our analysis increasingly less concrete but more useful as we proceeded. At the start, we have the description of the process. This is easy to read and helps you visualize what is going on, but it does not really let you compute anything. Writing down all the mass balances and process specifications is a higher level of abstraction, but it is still readable (at least to a chemical engineer!). By the time we get to the form $\mathbf{Ax} = \mathbf{b}$, the abstraction starts to get difficult. You can figure out all of the underlying equations from Eq. (2.12.17), and you could even work backwards and figure out a possible process flow sheet corresponding to those equations, but it is not easy. By the time we produce $\mathbf{L}$ and $\mathbf{U}$, the connection to the original problem has effectively vanished. You would need to be a mass and energy balances savant to look at Eqs. (2.12.18) and (2.12.19) and arrive at Fig. 2.8. What we lose in concreteness we gain in terms of solvability, as the latter equations are extremely useful for studying the flow sheet. We simply need to decide the flow rates A_1 and B_1, and we can readily compute all the other flows by successive forward substitution to solve $\mathbf{Ly} = \mathbf{b}$ and then back substitution to solve $\mathbf{Ux} = \mathbf{y}$. While we probably do not need LU decomposition to solve a small problem like the one posed here, you can imagine that real process flow sheets will have thousands of variables, meaning LU decomposition can lead to considerable savings in time.

2.13 Iterative Methods

So far, we have discussed direct methods to solve linear equations, such as Gauss elimination and LU decomposition. These methods are mostly foolproof. They involve a fixed/finite number of operations and give an exact solution (subject to round-off error) if a solution exists. The problem is that they become computationally expensive as the size of the system grows.

We will now consider a completely different class of methods, called iterative methods. Here, we start with a first approximation of the solution (an initial guess), and try to improve it successively. The procedure terminates when sufficient accuracy is reached. There is no guarantee that these methods will converge to the solution, but when they do they are computationally inexpensive. As a result, they are often preferable for large engineering problems.

Iterative techniques are often thought of as modern ways to solve linear equations, but their history goes back as far as direct solution techniques. Gauss (yes, *that* Gauss of elimination fame) grew frustrated when actually carrying out equation elimination operations by hand due to the great amount of effort required and, especially, when errors made during intermediate steps rendered subsequent operations meaningless. Owing to the sequential nature of the computations, Gauss elimination is especially sensitive to arithmetic errors. (Even on a computer, which does not make arithmetic mistakes, you still need to worry about round-off errors.) Gauss realized that the equations he was solving by hand tended to be diagonally dominant and reasoned that a series of approximate solutions, starting with simply the diagonal elements and correcting for the other components, could be made to successively improve the solution. He was very pleased with the outcome. Not only could he find a solution to a satisfactory level of accuracy very quickly, but errors in the calculation process tended to correct themselves in later iterations. Gauss elatedly wrote to a colleague about how the procedure could be performed when a person is "half asleep or thinking of more important things." No doubt you have done one of these things (or both) at some point during your classes.

Iterative solution methods for linear equations also represent a very different way of thinking about numerical computation. Although the certainty of being able to find a solution and the finite number of steps are gone, let's consider what we may gain. First, the notion of exact versus approximate solutions per the outcomes of Gaussian elimination and iterative techniques, respectively, is made moot if the iterations converge, because iterations can improve the accuracy to an arbitrary level. In fact, an iterative solution technique is able to compute a *more* accurate solution than Gaussian elimination because it is able to overcome round-off errors by subsequent iterations (subject finally to finite precision representation). Even more significant, iterative solvers have the general property of using only matrix-vector multiplications for their operations, so each iteration typically scales as $\sim n^2$ for a full matrix. This is a huge savings, a full factor of n, over Guassian elimination! But, alas, there is an equally huge down side: the iterations may not converge and the method may not be able to find a solution.

2.13.1   Jacobi's Method

We start with perhaps the simplest of these iterative methods for solving linear equation systems, Jacobi's method. Consider again the system of equations we are trying to solve,

$$a_{1,1}x_1 + a_{1,2}x_2 + \cdots + a_{1,n}x_n = b_1$$
$$a_{2,1}x_1 + a_{2,2}x_2 + \cdots + a_{2,n}x_n = b_2$$
$$\vdots \quad \vdots$$
$$a_{n,1}x_1 + a_{n,2}x_2 + \cdots + a_{n,n}x_n = b_n \tag{2.13.1}$$

The idea behind Jacobi's method is to rearrange ("solve") each equation i to isolate x_i,

$$x_1 = \frac{b_1 - a_{1,2}x_2 - \cdots - a_{1,n}x_n}{a_{1,1}}$$

$$x_2 = \frac{b_2 - a_{2,1}x_1 - a_{2,3}x_3 - \cdots - a_{2,n}x_n}{a_{2,2}}$$

$$\vdots \qquad\qquad \vdots$$

$$x_n = \frac{b_n - a_{n,1}x_1 - a_{n,2} - \cdots - a_{n,n-1}x_{n-1}}{a_{n,n}} \tag{2.13.2}$$

These expressions then suggest a natural iterative procedure, whereby previous approximations of the solutions are substituted in the right-hand side to yield the new approximations.

To be more formal, we start with an initial guess

$$\mathbf{x}^{(0)} = \begin{bmatrix} x_1^{(0)} \\ x_2^{(0)} \\ \vdots \\ x_n^{(0)} \end{bmatrix} \tag{2.13.3}$$

If we don't know any better, we can just set all of these entries to zero. We then update this vector using the iterative scheme

$$x_i^{(k+1)} = \frac{1}{a_{i,i}} \left(b_i - \sum_{j=1, j\neq i}^{n} a_{i,j} x_j^{(k)} \right) \tag{2.13.4}$$

provided that $a_{i,i} \neq 0$, until the vector $\mathbf{x}^{(k+1)}$ stops changing significantly. This is usually determined by computing the norm

$$\epsilon_1 = ||\mathbf{x}^{(k+1)} - \mathbf{x}^{(k)}|| \tag{2.13.5}$$

and stopping when ϵ_1 is less than a small number (a tolerance). It is also possible to use the accuracy of the solution as a termination criterion, calculating the residual

$$\epsilon_2 = ||\mathbf{A}\mathbf{x}^{(k+1)} - \mathbf{b}|| \tag{2.13.6}$$

and stopping when ϵ_2 is small enough. While the first approach is most often employed in practice, remember that it is only a valid measure if the iterations are converging. It may happen that the iterations "stall out," in which case the approximate solutions do not change much, even though the residual may still be relatively large. We recommend that Eq. (2.13.6) be checked to determine convergence at the end of the calculation if you do not use it at every step. For both conditions, one may employ whatever vector norm one likes, although most often the L_2 or L_∞ norms are used.

PROGRAM 2.6 *Let's consider how we can directly implement the Jacobi's method given by Eq. (2.13.4).*

```
1   function x = linear_jacobi(A,b,x)
2   %  A = n x n coefficient matrix
3   %  b = n x 1 vector
4   %  x = n x 1 vector of initial guesses
5   n = length(b);
6   tol = 1e-8; %tolerance for convergence
7   count = 0;
8   count_max = 100; %max number of iterations
9   err = norm(A*x-b); %convergence criteria
10  while err > tol
11      xold = x; %store old values
12      i = 1; %first entry
13      S = b(i);
14      for j = 2:n
15          S = S - A(i,j)*xold(j);
16      end
17      x(1) = S/A(i,i);
18      for i = 2:n-1 %interior entries
19          S = b(i);
20          for j = 1:i-1
21              S = S - A(i,j)*xold(j);
22          end
23          for j = i+1:n
24              S = S - A(i,j)*xold(j);
25          end
26          x(i) = S/A(i,i);
27      end
28      i = n; %last entry
29      S = b(i);
30      for j = 1:n-1
31          S = S - A(i,j)*xold(j);
32      end
33      x(n) = S/A(i,i);
34      %update counter and see if too many iterations
35      count = count + 1;
36      if count > count_max
37          fprintf('Did not converge at k = %3d.\n',count)
38          x = 0;
39          break
40      end
41      %compute error and display to screen
42      err = norm(A*x-b);
43      fprintf('k = %3d \t err = %6.4e \n',count,err)
44  end
```

As this is our first iterative program, it is worthwhile to pay attention to some of the most important features. First, we need to define a tolerance for the convergence, which we set as the variable tol. *We then need to compare something to this tolerance, for which we use Eq. (2.13.6). The method requires using a* **while** *loop because we do not know how many iterations we need until we will satisfy the convergence criterion. However, it is also important that we set an upper bound to the number of iterations so that we do not end up stuck in an infinite loop. We do this with the variable* count *and the* **if*** statement in lines 36–40 that checks the number of iterations. We also use line 43 to output the current state of the solution to the screen. This is especially useful in iterative methods so that you can monitor the state of the solution. If you do not display anything to screen, you can end up sitting in front of a blank screen for quite some time until you realize there is a problem!

The programing of Jacobi's method directly from Eq. (2.13.4) requires some care in implementing the sums for the first and last entry. These are handled separately in the program.

We noted above that iterative methods can generally be written in terms of matrix multiplication. To see how to do this, let's convert Program 2.6 into a matrix form.

PROGRAM 2.7   *This program implements Jacobi's method using vector-matrix multiplication:*

```
 1  function x = linear_jacobi_matrix(A,b,x)
 2  %  A = n x n coefficient matrix
 3  %  b = n x 1 vector
 4  %  x = n x 1 vector of initial guesses
 5  n = length(b);
 6  tol = 1e-8; %tolerance for convergence
 7  count = 0;
 8  count_max = 100; %max number of iterations
 9  err = norm(A*x-b); %convergence criteria
10  while err > tol
11      xold = x; %store old values
12      x = b - A*xold;
13      for i = 1:n
14          x(i) = x(i)/A(i,i) + xold(i);
15      end
16      %update counter and see if too many iterations
17      count = count + 1;
18      if count > count_max
19          fprintf('Did not converge at k = %3d.\n',count)
20          x = 0;
21          break
22      end
23      %compute error and display to screen
24      err = norm(A*x-b);
25      fprintf('k = %3d \t err = %6.4e \n',count,err)
26  end
```

You will notice that this program is remarkably more compact than our original program! Line 12 of this new program implements most of Eq. (2.13.4) using a simple matrix-vector multiplication and vector subtraction, except that (i) it also adds $A_{i,i}x_i$ in the sum and (ii) does not divide the result by $A_{i,i}$. We thus loop through the equations in lines 13–15 to correct for these omissions. Also notice that we no longer have to handle the first and last entries separately from the interior entries.

Example 2.8 Solve the system

$$2x_1 + x_2 = 2$$
$$x_1 - 2x_2 = -2 \qquad (2.13.7)$$

by Jacobi's method with an initial guess $(0,0)$.

Solution
This is a very small problem that is easily solved using Cramer's rule in Eq. (2.4.1),

$$x_1 = \frac{2}{5}, \quad x_2 = \frac{6}{5} \qquad (2.13.8)$$

but it makes a nice test problem for looking at Jacobi's method. If we wrote out the equations, the iteration scheme would be

$$x_1^{(k+1)} = \frac{2 - x_2^{(k)}}{2} \qquad (2.13.9)$$

$$x_2^{(k+1)} = \frac{2 + x_1^{(k)}}{2} \qquad (2.13.10)$$

While we could certainly program these equations in MATLAB, it is easier to just use something like Program 2.6 with the matrix

$$\mathbf{A} = \begin{bmatrix} 2 & 1 \\ 1 & -2 \end{bmatrix} \qquad (2.13.11)$$

and vector

$$\mathbf{b} = \begin{bmatrix} 2 \\ -2 \end{bmatrix} \qquad (2.13.12)$$

Figure 2.9 shows the convergence of the solution towards the expected solution.

2.13.2   Gauss–Seidel Method

The Gauss–Seidel method follows from Jacobi's method in a manner described by the aphorism, "If a little bit is good, more is better." The basic idea is to improve the convergence rate of the iterations by using more and presumably better information as soon as it is available. We start with the same approach, namely rearranging the original linear equation as

$$x_i = \frac{1}{a_{i,i}} \left(b_i - \sum_{\substack{j=1 \\ j \neq i}}^{n} a_{i,j} x_j \right) \qquad (2.13.13)$$

for $i = 1, \ldots, n$. However, in the iteration scheme of Gauss–Seidel, new information about the updated solution iterate is used immediately.

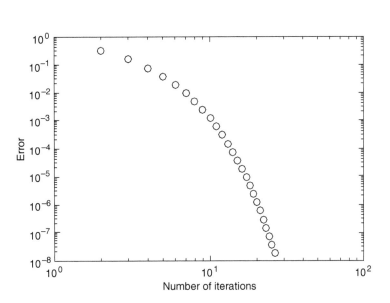

Figure 2.9 Convergence of Jacobi's method for Eq. (2.13.7) with initial guess (0,0).

Consider starting with our initial guess, $\mathbf{x}^{(0)}$. We compute the first component of the next iterate in the same way as Jacobi's method,

$$x_1^{(1)} = \frac{b_1 - a_{1,2}x_2^{(0)} - a_{1,3}x_3^{(0)} - \cdots - a_{1,n}x_n^{(0)}}{a_{1,1}} \qquad (2.13.14)$$

However, we now use the updated value, $x_1^{(1)}$, for computing the second component of the next solution iterate,

$$x_2^{(1)} = \frac{b_2 - a_{2,1}x_1^{(1)} - a_{2,3}x_3^{(0)} - \cdots - a_{2,n}x_n^{(0)}}{a_{2,2}} \qquad (2.13.15)$$

We continue computing each new component of $x_i^{(1)}$ using the most recently computed components, $x_1^{(1)}, x_2^{(1)}, \ldots, x_{i-1}^{(1)}$, and the remainder of the $\mathbf{x}^{(0)}$ values,

$$x_3^{(1)} = \frac{b_3 - a_{3,1}x_1^{(1)} - a_{3,2}x_2^{(1)} - a_{3,4}x_4^{(0)} \cdots - a_{3,n}x_n^{(0)}}{a_{3,3}}$$

$$\vdots \qquad\qquad (2.13.16)$$

$$x_n^{(1)} = \frac{b_n - a_{n,1}x_1^{(1)} - a_{n,2}x_2^{(1)} - \cdots - a_{n,n-1}x_{n-1}^{(1)}}{a_{n,n}}$$

The Gauss–Seidel iteration is expressed in more compact form as,

$$x_i^{(k+1)} = \frac{1}{a_{i,i}} \left(b_i - \sum_{j=1}^{i-1} a_{i,j}x_j^{(k+1)} - \sum_{j=i+1}^{n} a_{i,j}x_j^{(k)} \right) \qquad (2.13.17)$$

for $i = 1, \ldots, n$ and with k denoting iteration number. Iterations continue until we are sufficiently close to the true solution, as discussed in the prior section. The advantage of Gauss–Seidel over Jacobi's method is that the convergence rate is increased with no additional effort.

PROGRAM 2.8   *The programs we used for Jacobi's method are easily changed to implement Gauss–Seidel. For example, the non-matrix version in Program 2.6 becomes:*

```
1   function x = linear_gauss_seidel(A,b,x)
2   %  A = n x n coefficient matrix
3   %  b = n x 1 vector
4   %  x = n x 1 vector of initial guesses
5   n = length(b);
6   tol = 1e-8; %tolerance for convergence
7   count = 0;
8   count_max = 100; %max number of iterations
9   err = norm(A*x-b); %convergence criteria
10  while err > tol
11      i = 1; %first entry
12      S = b(i);
13      for j = 2:n
14          S = S - A(i,j)*x(j);
15      end
16      x(1) = S/A(i,i);
17      for i = 2:n-1 %interior entries
18          S = b(i);
19          for j = 1:i-1
20              S = S - A(i,j)*x(j);
21          end
22          for j = i+1:n
23              S = S - A(i,j)*x(j);
24          end
25          x(i) = S/A(i,i);
26      end
27      i = n; %last entry
28      S = b(i);
29      for j = 1:n-1
30          S = S - A(i,j)*x(j);
31      end
32      x(n) = S/A(i,i);
33      %update counter and see if too many iterations
34      count = count + 1;
35      if count > count_max
36          fprintf('Did not converge at k = %3d.\n',count)
37          x = 0;
38          break
39      end
40      %compute error and display to screen
41      err = norm(A*x-b);
42      fprintf('k = %3d \t err = %6.4e \n',count,err)
43  end
```

All we needed to do to switch to Gauss–Seidel was to remove all references to xold *appearing in Program 2.6, since we want to use the new values of* **x** *as soon as they become available.*

Do you think you can figure out how to convert the matrix form of Jacobi's method in Program 2.6 to implement Gauss–Seidel?

Example 2.9   Repeat the solution of Example 2.8 using Gauss–Seidel.

Solution

If we were to write out the equations explicitly for Gauss–Seidel, we have

$$x_1^{(k+1)} = \frac{2 - x_2^{(k)}}{2} \tag{2.13.18}$$

$$x_2^{(k+1)} = \frac{2 + x_1^{(k+1)}}{2} \tag{2.13.19}$$

The only difference when compared to Jacobi's method is that the old value $x_1^{(k)}$ in Eq. (2.13.10) has been replaced here with the new value, $x_1^{(k+1)}$. Figure 2.10 compares the convergence of Jacobi's method from Example 2.8 to the solution from Gauss–Seidel.

2.13.3   Successive Relaxation Method

The method of successive relaxation, often referred to as SOR in the case of successive over-relaxation, follows from the philosophy of "If more is better, *even* more is best." So, if Gauss–Seidel is working to solve our linear problem, why not take the action occurring in Gauss–Seidel and accelerate it? This is done by introducing an additional degree of freedom, the relaxation factor.

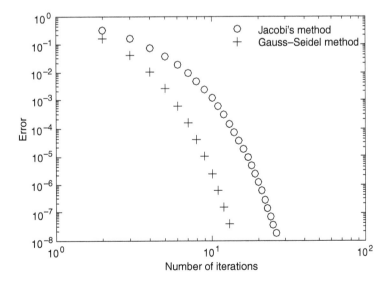

Figure 2.10 Comparison of the convergence of Jacobi's method and Gauss–Seidel for Eq. (2.13.7) with initial guess (0,0).

Let's start from the Gauss–Seidel iteration formula in Eq. (2.13.17). Adding and subtracting $x_i^{(k)}$ from the right-hand side, we obtain an iteration of the form

$$x_i^{(k+1)} = x_i^{(k)} + r_i^{(k)} \qquad (2.13.20)$$

where we have defined the residual

$$r_i = \frac{1}{a_{i,i}} \left(b_i - \sum_{j=1}^{i-1} a_{i,j} x_j^{(k+1)} - \sum_{j=i}^{n} a_{i,j} x_j^{(k)} \right) \qquad (2.13.21)$$

Note that the second sum in Eq. (2.13.21) includes the $j = i$ entry too. If we are at the solution, then $r_i = 0$ for all i.

The idea behind the SOR method is that making the residual zero for one variable may not be the best strategy as a zero residual does not stay zero as we go through the remaining equations. The method suggests an update according to

$$x_i^{(k+1)} = x_i^{(k)} + \omega r_i^{(k)} \qquad (2.13.22)$$

where ω is the so-called relaxation parameter. Different values of ω lead to different results. If we pick $\omega = 1$, then we get back Gauss–Seidel. If you chose $\omega > 1$, this is called over-relaxation, and may accelerate your convergence compared to Gauss–Seidel. If you choose $0 < \omega < 1$, this is called under-relaxation and may dampen oscillations in the solution. It turns out that for $\omega < 0$ or $\omega > 2$, the scheme diverges, whereas the optimal value of ω that maximizes the rate of convergence is problem (matrix) specific and not easy to determine in general. So, often the selection of ω is made by trial and error.

PROGRAM 2.9   *The program below implements the successive relaxation method in the form of Eqs. (2.13.20)–(2.13.22).*

```
1   function x = linear_SR(A,b,x,w)
2   %  A = n x n coefficient matrix
3   %  b = n x 1 vector
4   %  x = n x 1 vector of initial guesses
5   %  w = SOR parameter
6   n = length(b);
7   tol = 1e-8; %tolerance for convergence
8   count = 0;
9   count_max = 100; %max number of iterations
10  err = norm(A*x-b); %convergence criteria
11  while err > tol
12      for i = 1:n
13          S = 0;
14          for j = 1:n
15              S = S + A(i,j)*x(j);
16          end
17          x(i) = x(i) + w*(b(i)-S)/A(i,i);
18      end
19      %update counter and see if too many iterations
20      count = count + 1;
21      if count > count_max
22          fprintf('Did not converge at k = %3d.\n',count)
```

```
23          x = 0;
24          break
25       end
26       %compute error and display to screen
27       err = norm(A*x-b);
28       fprintf('k = %3d \t err = %6.4e \n',count,err)
29    end
```

In this form, successive relaxation is marginally easier to program than Gauss–Seidel in Program 2.8 because we do not need to worry about handling the first and last entries in the sums. Rather, the terms involving x(i) *are handled in line 17.*

Example 2.10   Compute the convergence of successive relaxation for the problem in Eq. (2.13.7) as a function of the relaxation parameter ω.

Solution
Figure 2.11 plots the residual as a function of iteration number.

2.13.4   Convergence

These iterative methods can also be cast into a much more general framework of a *fixed-point* iteration (more on this in the next chapter), where successive iterates are computed using an iteration of the form

$$\mathbf{x}^{(k+1)} = \mathbf{B}\mathbf{x}^{(k)} + \mathbf{c}, \tag{2.13.23}$$

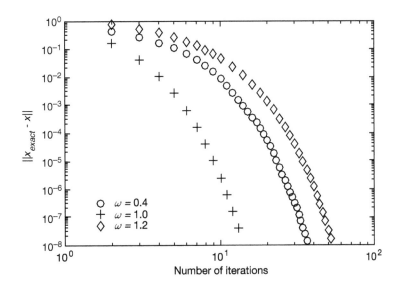

Figure 2.11 Rate of convergence for Eq. (2.13.7) with different relaxation parameters ω for the initial guess (0,0). Note that $\omega = 1$ reduces successive relaxation to Gauss–Seidel, and thus that result corresponds to the plot in Fig. 2.10.

where the exact forms of $\mathbf{B}$ and $\mathbf{c}$ are determined by the specific method. Defining the error in each iteration step as

$$\mathbf{e}_k = \mathbf{x}^{(k)} - \mathbf{x} \qquad (2.13.24)$$

where $\mathbf{x}$ is the actual solution, it follows that

$$\mathbf{e}_1 = \mathbf{B}\mathbf{e}_0 \qquad (2.13.25)$$
$$\mathbf{e}_2 = \mathbf{B}^2\mathbf{e}_0$$
$$\vdots$$
$$\mathbf{e}_k = \mathbf{B}^k\mathbf{e}_0$$

Using the properties of matrix norms, it then follows that

$$||\mathbf{e}_k|| \le ||\mathbf{B}^k||\,||\mathbf{e}_0|| \le ||\mathbf{B}||^k||\mathbf{e}_0|| \qquad (2.13.26)$$

Thus if $||\mathbf{B}|| < 1$, such an iterative scheme is guaranteed to converege, i.e. $\mathbf{e}_k$ will tend to 0 as k tends to ∞. This sufficient condition can be used to generate conditions for convergence for each method. One particularly useful result is that if $\mathbf{A}$ is diagonally dominant, i.e.

$$|a_{i,i}| > \sum_{j=1, j\neq i}^{n} |a_{i,j}| \qquad (2.13.27)$$

then the Jacobi and Gauss–Seidel methods are guaranteed to converge to the solution. This is again a sufficient, but not necessary condition, which means that even if the matrix is not diagonally dominant (which is often the case), the methods may still converge.

Example 2.11   What happens if we switch the order of the equations in Example 2.8?

Solution
The problem now becomes

$$x_1 - 2x_2 = -2 \qquad (2.13.28)$$
$$2x_1 + x_2 = 2 \qquad (2.13.29)$$

so the iterative scheme is

$$x_1^{(k+1)} = 2x_2^{(k)} - 2 \qquad (2.13.30)$$
$$x_2^{(k+1)} = 2 - 2x_1^{(k)} \qquad (2.13.31)$$

This corresponds to switching the rows in Eqs. (2.13.11) and (2.13.12).

Figure 2.12 is quite different from what we saw in Fig. 2.9. Instead of nicely converging to the solution, the solution appears to be running away. What happened? If you look at the matrix in Eq. (2.13.11), it is easy to show that it satisfies Eq. (2.13.27) for diagonal dominance. This is a sufficient condition for convergence. When we switch the rows of the matrix, it is no longer diagonally dominant and convergence cannot be guaranteed.

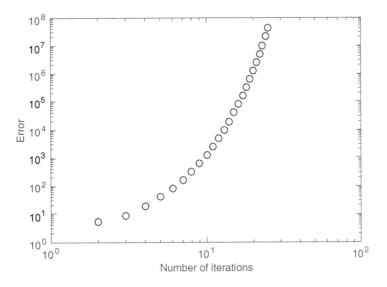

Figure 2.12 Rate of convergence for Eq. (2.13.29) using Jacobi's method for the initial guess (0,0). Compare the result to the convergence of the diagonally dominant form of the equations in Fig. 2.9.

2.14   Application: Polynomial Regression

We conclude our chapter on linear algebra with a very useful application in regression of data, in particular experimental data. Let us first clarify the distinction between regression and interpolation in Fig. 2.13. Regression is the determination of a function that describes a general trend in a data set but does not necessarily pass through any point. This contrasts with interpolation where the function passes through every data point. We will talk about polynomial interpolation more in Chapter 8, since it forms the basis for many numerical integration methods. In the analysis of data, we generally want to use regression to obtain an approximate description of the observations and handle the uncertainty in the data themselves. In contrast, interpolation is useful when we have densely spaced, reasonable accurate data (for example, in a steam table) and require values that are not contained in the table.

In regression, we begin with a set of data y_i obtained at corresponding values x_i. For example, these data could be temperatures T_i obtained at different pressures P_i for a gas in a closed vessel. The goal is then to determine a function $f(x_i)$ that, when evaluated at the different data points x_i, captures the trend in the data. Continuing with our gas data, we might propose that $T_i = aP_i$, as would be the case for an ideal gas.

How do we know if our function is any good? We need some criterion for assessing the deviation between our function $f(x_i)$ and the actual data. To this end, we define the error at point x_i as

$$e_i \equiv y_i - f(x_i) \qquad (2.14.1)$$

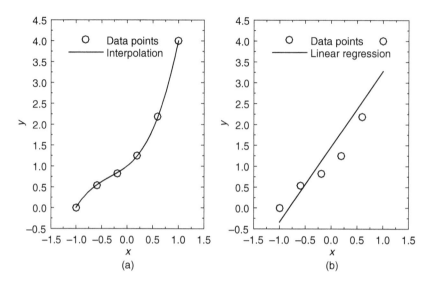

Figure 2.13 Difference between (a) interpolation and (b) regression.

The goal is then to minimize the sum-square error,

$$S_r = \sum_{i=1}^{n} \left[y_i - f(x_i) \right]^2 \tag{2.14.2}$$

Polynomial regression reduces to a linear algebra problem, which can be solved using any of the methods in this chapter. The idea is to use a polynomial function

$$y = a_0 + a_1 x + a_2 x^2 + \cdots a_m x^m \tag{2.14.3}$$

In the latter, n will still be the number of data points $i = 1 : n$ and m will be the order of the polynomial. While the function itself is nonlinear in x, it is linear in the coefficients a_j. So we can still minimize the sum-squared error in Eq. (2.14.2),

$$S_r = \sum_{i=1}^{n} \left(y_i - \sum_{j=0}^{m} a_j x_i^j \right)^2 \tag{2.14.4}$$

The minimum occurs when

$$\frac{\partial S_r}{\partial a_j} = 0 \quad \text{for } j = 0, 1, \dots m \tag{2.14.5}$$

This means that we get a system of m equations to solve for the a_j. If we take this derivative in Eq. (2.14.4), we get

$$\frac{\partial S_r}{\partial a_j} = \sum_{i=1}^{n} 2 \left(y_i - \sum_{j=0}^{m} a_j x_i^j \right) \left(-x_i^j \right) = 0 \tag{2.14.6}$$

We can use Eq. (2.14.6) to derive the equations for any order polynomial. For example, with $m = 3$, we get the system of equations

$$0 = -\sum_{i=1}^{n} y_i + a_0 n + a_1 \sum_{i=1}^{n} x_i + a_2 \sum_{i=1}^{n} x_i^2 + a_3 \sum_{i=1}^{n} x_i^3 \tag{2.14.7}$$

$$0 = -\sum_{i=1}^{n} x_i y_i + a_0 \sum_{i=1}^{n} x_i + a_1 \sum_{i=1}^{n} x_i^2 + a_2 \sum_{i=1}^{n} x_i^3$$

$$+ a_3 \sum_{i=1}^{n} x_i^4 \tag{2.14.8}$$

$$0 = -\sum_{i=1}^{n} x_i^2 y_i + a_0 \sum_{i=1}^{n} x_i^2 + a_1 \sum_{i=1}^{n} x_i^3 + a_2 \sum_{i=1}^{n} x_i^4$$

$$+ a_3 \sum_{i=1}^{n} x_i^5 \tag{2.14.9}$$

$$0 = -\sum_{i=1}^{n} x_i^3 y_i + a_0 \sum_{i=1}^{n} x_i^3 + a_1 \sum_{i=1}^{n} x_i^4 + a_2 \sum_{i=1}^{n} x_i^5$$

$$+ a_3 \sum_{i=1}^{n} x_i^6 \tag{2.14.10}$$

We can write this in the matrix form using shorthand notation,

$$
\begin{bmatrix}
s_{x^0} & s_{x^1} & s_{x^2} & s_{x^3} \\
s_{x^1} & s_{x^2} & s_{x^3} & s_{x^4} \\
s_{x^2} & s_{x^3} & s_{x^4} & s_{x^5} \\
s_{x^3} & s_{x^4} & s_{x^5} & s_{x^6}
\end{bmatrix}
\begin{bmatrix}
a_0 \\
a_1 \\
a_2 \\
a_3
\end{bmatrix}
=
\begin{bmatrix}
s_y \\
s_{xy} \\
s_{x^2 y} \\
s_{x^3 y}
\end{bmatrix}
\tag{2.14.11}
$$

where

$$s_{x^a y^b} = \sum_{i=1}^{n} x_i^a y_i^b \tag{2.14.12}$$

and the notation s_{x^a} implies $s_{x^a y^0}$. Note that $s_{x^0 y^0} = n$, i.e. the number of data points.

The structure of this system of equations for a third-order polynomial should make it easy for you to see how to construct the system for any order polynomial. The entries of the matrix **A** are

$$A_{ij} = s_{x^{i+j-2}} \tag{2.14.13}$$

and the forcing function has the form

$$b_j = s_{x^{i-1} y} \tag{2.14.14}$$

A particularly relevant problem is *linear* regression, where we are fitting with the equation

$$f(x_i) = a_0 + a_1 x_i \qquad (2.14.15)$$

with a_0 and a_1 being constants to be determined. Using Eqs. (2.14.13) and (2.14.14), the problem we need to solve is

$$\begin{bmatrix} S_{x^0} & S_{x^1} \\ S_{x^1} & S_{x^2} \end{bmatrix} \begin{bmatrix} a_0 \\ a_1 \end{bmatrix} = \begin{bmatrix} S_y \\ S_{xy} \end{bmatrix} \qquad (2.14.16)$$

The solution is

$$a_1 = \frac{n S_{xy} - S_x S_y}{n S_{xx} - S_x^2} \qquad (2.14.17)$$

and

$$a_0 = \frac{S_y}{n} - a_1 \frac{S_x}{n} \qquad (2.14.18)$$

The sum of some entries divided by the number of entries is just the average, so we could also write

$$a_o = \bar{y} - a_1 \bar{x} \qquad (2.14.19)$$

The quality of the fit is given by the standard error of the estimate,

$$S_{y/x} = \sqrt{\frac{S_r}{n-2}} \qquad (2.14.20)$$

The factor of $n - 2$ arises from the fact that you need at least three points to assess the error of a line – any line will go through one point, and linear regression gives perfect fit to two points since you have two coefficients. Most often, the quantity reported for the quality of a fit is the correlation coefficient,

$$R^2 = \frac{S_t - S_r}{S_t} \qquad (2.14.21)$$

where

$$S_t = \sum_{i=1}^{n} (y_i - \bar{y})^2 \qquad (2.14.22)$$

The R^2 value tells you how much better your line describes the data when compared with just using the average value of y_i.

PROGRAM 2.10   *To see inside the "black box" of linear regression, the following program computes the regression coefficients and determines the R^2 value.*

```
1   function [a0,a1,R2] = linear_regression(x,y)
2   %  x = n x 1 vector of independent data points
3   %  y = n x 1 vector of dependent data points
4   n = length(x);
5   sx = 0; sy = 0; sxx = 0; sxy = 0;
6   for i = 1:n
7       sx = sx + x(i);
```

```
8       sy = sy + y(i);
9       sxx = sxx + x(i)*x(i);
10      sxy = sxy + x(i)*y(i);
11   end
12   a1 = (n*sxy-sx*sy)/(n*sxx-sx*sx);
13   a0 = sy/n - a1*sx/n;
14   %compute Sr and St
15   Sr = 0; St = 0;
16   yavg = sy/n;
17   for i = 1:n
18       St = St + (y(i) - yavg)^2;
19       Sr = Sr + (y(i) - a0 - a1*x(i))^2;
20   end
21   if Sr == 0
22       R2 = 1; %only two data points
23   else
24       R2 = (St - Sr)/Sr;
25   end
```

We have used a **for** loop to compute terms s_x, s_y, s_{xy}, and s_{xx}, as well as S_r and S_t. In MATLAB, these calculations can be done more efficiently using element-by-element vector multiplication and summation routines that are hard-coded into MATLAB.

In most experimental situations, you are usually interested in not only the regression coefficients, but also how to determine their uncertainty. Do you see how to modify this program to propagate the error for a_0 and a_1?

There are also a number of nonlinear functions that can be converted to a linear form for linear regression.

(1) Power laws are of the form

$$y = ax^b \tag{2.14.23}$$

where a and b are constants. You can convert this into a linear form by taking the natural logarithm of both sides

$$\ln y = \ln ax^b = \ln a + \ln x^b = \ln a + b \ln x \tag{2.14.24}$$

So you can just use linear regression on $y_i \rightarrow \ln y_i$ and $x_i \rightarrow \ln x_i$ to get the coefficients a and b for the power law.

(2) Exponential functions are of the form

$$y = ae^{bx} \tag{2.14.25}$$

where again a and b are constants. This is again converted to a linear form by taking the natural logarithm of both sides,

$$\ln y = \ln ae^{bx} = \ln a + \ln e^{bx} = \ln a + bx \ln e = \ln a + bx \tag{2.14.26}$$

So now you still use linear regression but you are plotting $y_i \rightarrow \ln y_i$.

(3) Saturation kinetics are well fit by a function

$$y = \frac{ax}{b + x} \tag{2.14.27}$$

where a and b are constants. This function arises, for example, in Michaelis–Menten kinetics for enzymes. Interestingly, you can still use linear regression here as well by taking the inverse of the function,

$$\frac{1}{y} = \frac{b+x}{ax} = \frac{b}{a}\left(\frac{1}{x}\right) + \frac{1}{a} \tag{2.14.28}$$

To do linear regression, you are using $y_i \to 1/y_i$ and $x_i \to 1/x_i$.

While Eq. (2.14.28) works, it is not always easy to apply in practice in chemical kinetics. To do good linear regression, we want to have the data evenly spaced in the independent variable, which in this case is $1/x_i$. If the independent variable is time, this means we want lots of short time data and just a bit of long time data. This is hard. In contrast, it is often not very difficult to get logarithmically spaced data for exponential and power law fits in many problems of interest to chemical engineers.

2.15   Implementation in MATLAB

Given that MATLAB stands for "Matrix Laboratory," it should be unsurprising to you that it has a large number of built-in routines for linear algebra, including the solution of linear algebraic equations. The main function for solving equations is `slash`. The syntax is somewhat awkward; if you want to solve the equation $\mathbf{Ax} = \mathbf{b}$, you enter it as `x = A \ b`.

Example 2.12   Use the built-in MATLAB solver to find the solution to

$$\begin{bmatrix} 1 & 2 & 4 \\ 4 & 1 & 2 \\ 2 & -2 & -3 \end{bmatrix}\begin{bmatrix} x_1 \\ x_2 \\ x_3 \end{bmatrix} = \begin{bmatrix} 1 \\ 3 \\ -2 \end{bmatrix} \tag{2.15.1}$$

Solution
A simple MATLAB script to solve this problem is

```
1  function matlab_slash
2  A = [1 2 4; 4 1 2; 2 -2 -3];
3  b = [1; 3; -2];
4  x = A\b
```

This program is so simple that you probably would just enter it at the command line. When you run it, you get back the solution

$$\mathbf{x} = \begin{bmatrix} 0.7143 \\ 6.4286 \\ -3.1429 \end{bmatrix} \tag{2.15.2}$$

The MATLAB slash solver is very handy, and we will use it for the remainder of the book. As we get to very large linear algebraic systems later in the book, you will appreciate the efficiency of the MATLAB solver. The programs that we wrote here were designed to illustrate the numerical methods, but not necessarily to implement them in the most efficient manner.

There is one important thing to keep in mind when you use the MATLAB slash solver. If you want to solve a banded system, you need to have a way to tell MATLAB that the system is banded. In the slash solver, this is implemented by declaring the matrix **A** as a sparse matrix. In the MATLAB syntax, this is implemented by adding the line

```
A = sparse(A)
```

at some point before you solve the problem. Alternatively, you can declare the matrix as sparse at the outset by typing

```
A = sparse(n,n)
```

which makes an $n \times n$ matrix with all zeros. When you declare a matrix as sparse, MATLAB will store the values of each entry in the matrix along with the location of that entry, rather than store all of the entries in the matrix. This is obviously a huge advantage in terms of memory for many problems we will see later in the book. For example, consider a tridiagonal banded matrix with $n = 100$, similar to some of the problems we will encounter in Chapter 6. The full matrix requires 10^4 floating point numbers for storage, but the sparse matrix only requires 300 floating point numbers (for the values) and 600 integers (for the row and column locations). If the sparse matrix is also banded, sending a sparse matrix to MATLAB will force it to use a banded solver. You can easily tell the difference in time if you forget to make **A** a sparse matrix.

In addition to the slash solver, MATLAB implements other routines of the type we've seen here. For example, MATLAB has a built-in routine called `lu` for LU decomposition.

Example 2.13   Write a MATLAB program to do LU decomposition for the matrix in Example 2.12.

Solution
A short program to solve this problem is

```
1  function matlab_LU
2  A = [1 2 4; 4 1 2; 2 -2 -3];
3  [L,U] = lu(A)
```

It is worth paying attention to the output of the MATLAB routine, since it is slightly different than the LU decomposition in Section 2.11. The upper triangular matrix produced by our program looks fine,

$$\mathbf{U} = \begin{bmatrix} 4.0000 & 2.0000 & 1.0000 \\ 0 & 2.5000 & 3.7500 \\ 0 & 0 & 1.0000 \end{bmatrix} \qquad (2.15.3)$$

However, the lower triangular matrix does not look lower triangular,

$$\mathbf{L} = \begin{bmatrix} 0.2500 & 1.0000 & 0 \\ 1.0000 & 0 & 0 \\ 0.5000 & 0.4000 & 1.0000 \end{bmatrix} \qquad (2.15.4)$$

Nevertheless, if you multiply $\mathbf{L}$ times $\mathbf{U}$, you get back $\mathbf{A}$.

The reason that $\mathbf{L}$ is not lower triangular is that the forward elimination of this matrix required pivoting. You can see this right away from the structure of $\mathbf{A}$, since we certainly have to pivot on the first step. During LU decomposition, MATLAB keeps track of a pivot matrix, and then uses that matrix to re-order the rows in $\mathbf{L}$ so that the final result is indeed $\mathbf{LU} = \mathbf{A}$.

MATLAB also has built-in routines for iterative solvers. A particularly powerful solver used in many research applications is GMRES. To solve the system in Example 2.12, we could write a program like the following:

```
1  function matlab_gmres
2  A = [1 2 4; 4 1 2; 2 -2 -3];
3  b = [1; 3; -2];
4  x = gmres(A,b);
```

The result of this program is the expected one in Eq. (2.15.2). You will also see an output line `gmres converged at iteration 3 to a solution with ...` `relative residual 0.` that describes how the program converged. There are many options for the GMRES solver, described in the MATLAB documentation, that you can use to optimize the solver for difficult problems. Clearly, the small problem in Example 2.12 does not require any optimization.

We selected the functions `slash`, `lu`, and `gmres` as examples of MATLAB codes that solve linear algebraic equations. This is merely scratching the surface of the built-in algorithms that are available in MATLAB. The concepts that we discussed in this chapter provide a starting point for understanding not only the MATLAB implementation of the methods described here, but also more advanced methods.

2.16   Further Reading

There are many excellent books on linear algebra and matrix methods. One of the earlier ones with a historical significance to chemical engineers (and an obvious connection to us!) is

- N. R. Amundson, *Mathematical Methods in Chemical Engineering – Matrices and their Application*, Prentice Hall, 1966 whereas a more recent and more advanced text is

- H. T. Davis and K. T. Thomson, *Linear Algebra and Linear Operators in Engineering*, Academic Press, 2000.

A very readable and insightful linear algebra text from the mathematics literature is

- G. Strang, *Linear Algebra and its Applications*, Harcourt, Brace, Jovanovich, 1988

whereas an excellent, more advanced text is

- G. H. Golub and C. F. Van Loan, *Matrix Computations*, Johns Hopkins University Press, 1996.

The subject finds extensive coverage in general purpose numerical methods books such as

- G. Dahlquist and A. Bjork, *Numerical Methods*, Prentice Hall, 1974
- J. D. Hoffman and S. Frankel, *Numerical Methods for Engineers and Scientists*, McGraw Hill, 2001
- C. Pozrikidis, *Numerical Computation in Science and Engineering*, Oxford University Press, 2008.

A compact, yet excellent discussion on more theoretical topics such as ill-conditioning, convergence of iterative methods, etc., can be found in

- G. H. Golub and J. M. Ortega, *Scientific Computing and Differential Equations*, Academic Press, 1992.

More advanced texts covering iterative methods include

- L. A. Hageman and M. Young, *Applied Iterative Methods*, Academic Press, 1981
- Y. Saad, *Iterative Methods for Sparse Linear Systems*, SIAM, 2003
- H. A. van der Vorst, *Iterative Krylov Methods for Large Linear Systems,* Cambridge University Press, 2003.

A particularly insightful review paper on the ubiquitous presence of matrices in modern scientific computing in chemical engineering is

- I. G. Kevrekidis, "Matrices are Forever: On Applied Mathematics and Computing in Chemical Engineering ," *Chemical Engineering Science* **50**, 4005, 1995.

The case study problems here are found in many introductory chemical engineering text books such as

- R. Murphy, *Introduction to Chemical Engineering Processes: Principles, Applications, Synthesis*, McGraw-Hill, 2007.

Physical property data, such as the Antoine coefficients appearing here, can be found in such introductory textbooks. Another excellent source of physical property data is

- D. R. Lide, *CRC Handbook of Chemistry and Physics*, CRC Press, 1994.

Finally, our historical discussion about Gauss and relaxation methods is discussed in the book

- H. A. van der Vorst, *Iterative Krylov Methods for Large Linear Systems*, Cambridge University Press, 2003.

Problems

2.1 Write the determinant of the following 3×3 matrix $\mathbf{A}$ as a sum of the determinants of 2×2 matrices using co-factor expansion on the third row ($i = 3$):

$$\mathbf{A} = \begin{bmatrix} 6 & 3 & 5 \\ 2 & 1 & 5 \\ 2 & 3 & 2 \end{bmatrix}$$

2.2 Express the determinant of the following matrix as a sum of 2×2 determinants using co-factor expansion on the first row. You do not need to evaluate the 2×2 determinants.

$$\begin{bmatrix} -2 & 1 & 3 \\ 4 & -1 & 3 \\ -1 & 1 & 1 \end{bmatrix}$$

2.3 What is the determinant of the following matrix?

$$\begin{bmatrix} 4 & -3 & 4 & 2 \\ 0 & 1 & 2 & 1 \\ 0 & 0 & -0.25 & 3 \\ 0 & 0 & 0 & 10 \end{bmatrix}$$

2.4 Write the determinant of the following 3×3 matrix $\mathbf{A}$ as a sum of the determinants of 2×2 matrices using co-factor expansion on the second row ($i = 2$):

$$\mathbf{A} = \begin{bmatrix} 6 & 3 & 5 \\ 2 & 1 & 5 \\ 2 & 3 & 2 \end{bmatrix}$$

2.5 Use co-factor expansion to find

$$\det \begin{bmatrix} 3 & 5 & 6 \\ 2 & 4 & 5 \\ 1 & 2 & 4 \end{bmatrix}$$

2.6 Compute the determinant of the following matrix by co-factor expansion:

$$\mathbf{A} = \begin{bmatrix} 6 & 1 & -2 & 3 \\ 2 & -2 & 4 & -1 \\ 1 & -1 & 1 & -1 \\ 0 & -2 & 2 & 1 \end{bmatrix}$$

2.7   Compute the determinant of the following matrix by co-factor expansion:

$$\mathbf{A} = \begin{bmatrix} 2 & -3 & 2 & 1 \\ 1 & 8 & 3 & 2 \\ -5 & -4 & 2 & -1 \\ 11 & 2 & 11 & 2 \end{bmatrix}$$

2.8   Use co-factor expansion to reduce the determinant of

$$\mathbf{A} = \begin{bmatrix} 3 & 2 & 1 & 0 & 1 \\ 5 & -3 & 2 & 1 & 4 \\ -1 & 0 & 1 & 0 & 3 \\ 8 & -7 & 2 & 3 & 4 \\ -4 & 3 & 4 & 1 & 2 \end{bmatrix}$$

to the sum of the fewest number of 4×4 matrices.

2.9   Does the following system have a unique solution, no solution, or infinitely many solutions?

$$\begin{bmatrix} 4 & 2 & -1 & 1 \\ 0 & 0 & 2 & 1 \\ 0 & 0 & 2 & 4 \\ 0 & 0 & 0 & 8 \end{bmatrix} \begin{bmatrix} x_1 \\ x_2 \\ x_3 \\ x_4 \end{bmatrix} = \begin{bmatrix} 0 \\ 1 \\ -1 \\ 4 \end{bmatrix}$$

2.10   Use co-factor expansion to prove that the determinant of an $n \times n$ upper triangular matrix $\mathbf{U}$ is the product of the entries along its diagonal,

$$\det \mathbf{U} = \prod_{i=1}^{n} U_{ij}$$

2.11   Use Cramer's rule to find the solution to

$$\begin{bmatrix} 2 & 1 & 3 \\ 1 & 2 & 1 \\ 2 & 3 & 4 \end{bmatrix} \begin{bmatrix} x \\ y \\ z \end{bmatrix} = \begin{bmatrix} 13 \\ 8 \\ 20 \end{bmatrix}$$

2.12   Use Cramer's rule to compute the solution to the system of equations

$$2x + y - z = 1$$
$$-x + 2z = 4$$
$$4x - 2y + 3z = 0$$

2.13   Use Cramer's rule to solve the system of equations

$$3x + 2y + z = 5$$
$$x - y + 2z = 5$$
$$x - 3z = -1$$

2.14   Solve the system of equations

$$
\begin{array}{rcrcrcl}
3x_1 & - & 2x_2 & + & 3x_3 & = & 8 \\
4x_1 & + & 2x_2 & - & 2x_3 & = & 2 \\
x_1 & + & x_2 & + & 2x_3 & = & 9
\end{array}
$$

using

(a) Cramer's rule
(b) Gauss elimination (show clearly all steps in forward elimination and back-substitution).

2.15   Cramer's rule grows like $n!$ with the size n of the matrix. In this problem, you will construct the key steps of the proof of this scaling. The limiting step for Cramer's rule is to compute the determinants of the matrices. So you want to show that the effort to compute the determinant of a matrix by cofactor expansion is factorial in the number of rows (columns) n of the matrix. To arrive at this answer, you should make the following explicit calculations.

(a) Determine the number of operations (addition, subtraction, multiplication, division) required to compute the determinant of a 2×2 matrix. Call this number q for future reference.
(b) Repeat for a 3×3 matrix using co-factor expansion. You can ignore the effort required to set the sign of the coefficients. Report the answer in terms of q.
(c) Repeat for a 4×4 matrix using co-factor expansion. You can ignore the effort required to set the sign of the coefficients. Report the answer in terms of the size of the matrix $n = 4$ and q.
(d) Repeat for a 5×5 matrix using co-factor expansion. Report the answer in terms of the size of the matrix $n = 5$ and q.
(e) Use the formula from the last step to estimate the number of operations required for a large $n \times n$ matrix using co-factor expansion. You only need to report the term that dominates as $n \to \infty$.

2.16   Use Gauss elimination to find the solution to

$$
\begin{bmatrix}
4 & 2 & 3 & 1 \\
3 & 2 & 1 & 0 \\
0 & 2 & 3 & 5 \\
4 & -2 & 3 & 1
\end{bmatrix}
\begin{bmatrix}
w \\ x \\ y \\ z
\end{bmatrix}
=
\begin{bmatrix}
0 \\ -3 \\ 4 \\ 0
\end{bmatrix}
$$

2.17   Determine the value of the matrix element $a_{4,3}^{(2)}$ during the naive Gauss elimination of the matrix $\mathbf{A}$

$$
\mathbf{A} =
\begin{bmatrix}
2 & 6 & 3 & 5 & 4 & 2 \\
2 & 1 & 5 & 3 & 2 & 8 \\
3 & 2 & 3 & 2 & 1 & 1 \\
3 & 2 & 4 & 2 & 1 & 8 \\
7 & 5 & 3 & 5 & 8 & 2 \\
3 & 2 & 8 & 3 & 2 & 1
\end{bmatrix}
$$

2.18   Use Gauss elimination to solve the system of equations

$$
\begin{array}{rrrrr}
2x_1 & - & 3x_2 & + & x_3 & = & 7 \\
x_1 & - & x_2 & - & 2x_3 & = & -2 \\
3x_1 & + & x_2 & - & x_3 & = & 0
\end{array}
$$

2.19   Use simultaneous Gauss elimination to solve the system of equations $Ax = c$ with

$$
A = \begin{bmatrix} 5 & 2 & 1 \\ 4 & 1 & -1 \\ -2 & 3 & -3 \end{bmatrix}
$$

for $c_1 = [3 \ -3 \ 5]^T$ and $c_2 = [2 \ -2 \ 8]^T$

2.20   Use simultaneous Gauss elimination to solve the system of equations $Ax = c$ with

$$
A = \begin{bmatrix} 5 & 6 & -3 \\ 1 & 3 & 1 \\ 4 & 2 & -6 \end{bmatrix}
$$

for $c_1 = [1 \ 0 \ 0]^T$, $c_2 = [0 \ 1 \ 0]^T$, and $c_3 = [0 \ 0 \ 1]^T$. Based on your results, what is A^{-1}?

2.21   At the current step in Gauss elimination, show the result after partial pivoting for the augmented matrix

$$
\begin{bmatrix}
1 & 2 & 7 & 2 & 4 \\
0 & 2 & 3 & 1 & 8 \\
0 & 3 & 4 & 1 & 5 \\
0 & 4 & 1 & 8 & 4
\end{bmatrix}
$$

2.22   Which two rows (if any) should be swapped in the current step of Gauss elimination with the following matrix:

$$
\begin{bmatrix}
8 & 2 & 3 & -2 & 5 & 5 & 1 & 2 \\
0 & 5 & -5 & 2 & 4 & -4 & 1 & 1 \\
0 & 0 & 0 & 2 & 4 & 2 & 1 & 2 \\
0 & 0 & 0 & -1 & 4 & -1 & 2 & 3 \\
0 & 0 & 0 & -1 & 2 & -1 & 4 & 2 \\
0 & 0 & 2 & 1 & 2 & 1 & 2 & 5 \\
0 & 0 & 0 & 1 & 1 & 2 & 4 & -1 \\
0 & 0 & 0 & -3 & 1 & 2 & 1 & 4
\end{bmatrix}
$$

2.23   Use Gauss elimination with partial pivoting to find the solution to

$$
\begin{bmatrix}
4 & 2 & 2 & 1 \\
2 & 1 & 3 & 0 \\
0 & 2 & 2 & 3 \\
-1 & 3 & 2 & 3
\end{bmatrix}
\begin{bmatrix}
w \\ x \\ y \\ z
\end{bmatrix}
=
\begin{bmatrix}
-2 \\ -9 \\ 2 \\ -1
\end{bmatrix}
$$

2.24 Use Gauss elimination with pivoting to solve

$$\begin{bmatrix} -1 & 3 & -2 & -3 & 2 \\ 10 & 5 & 2 & 3 & -5 \\ -3 & 2 & 9 & 0 & 4 \\ 3 & 1 & -3 & 2 & 7 \\ 1 & 1 & 1 & 1 & 1 \end{bmatrix} \begin{bmatrix} u \\ w \\ x \\ y \\ z \end{bmatrix} = \begin{bmatrix} -5 \\ 11 \\ 1 \\ 14 \\ 4 \end{bmatrix}$$

2.25 Solve the system

$$\begin{bmatrix} 3 & 2 & 4 & -5 \\ 3 & 4 & 1 & 2 \\ -1 & 0 & 3 & -2 \\ -2 & 4 & 3 & 1 \end{bmatrix} \begin{bmatrix} x_1 \\ x_2 \\ x_3 \\ x_4 \end{bmatrix} = \begin{bmatrix} 0 \\ 8 \\ 2 \\ 14 \end{bmatrix}$$

using Gauss elimination with pivoting. Determine the solution $\mathbf{x}$ and the determinant of the original matrix, $\mathbf{A}$.

2.26 Consider some matrix $\mathbf{A}$. After using Gauss elimination with three partial pivoting steps, you obtain the upper triangular matrix

$$\mathbf{U} = \begin{bmatrix} 1 & 6 & 3 & 5 & 4 & 2 \\ 0 & 1 & 3 & 5 & 6 & 8 \\ 0 & 0 & 2 & 4 & 2 & 1 \\ 0 & 0 & 0 & 1 & 3 & 8 \\ 0 & 0 & 0 & 0 & 2 & 9 \\ 0 & 0 & 0 & 0 & 0 & 1 \end{bmatrix}$$

What is the determinant of the original matrix $\mathbf{A}$?

2.27 Use Gauss elimination with pivoting to find the determinant of

$$\mathbf{A} = \begin{bmatrix} 2 & 3 & 4 \\ 1 & 1 & 1 \\ 4 & 6 & 4 \end{bmatrix}$$

2.28 The code below takes the matrix $\mathbf{A}$ and creates the upper triangular matrix $\mathbf{U}$.

```
1  for k=1:n-1
2      for i=k+1:n
3          m=A(i,k)/A(k,k);
4          for j=k+1:n
5              A(i,j)=A(i,j)-m*A(k,j);
6          end
7      end
8  end
```

We have modified the code so that it does banded Gauss elimination $p = q$ when the function is given the value of p as an input argument.

```
1   for k=1:B
2      for i=C:D
3         m=A(i,k)/A(k,k);
4         for j=E:F
5            A(i,j)=A(i,j)-m*A(k,j);
6         end
7      end
8   end
9   for k=G:n-1
10     for i=H:L
11        m=A(i,k)/A(k,k);
12        for j=M:N
13           A(i,j)=A(i,j)-m*A(k,j);
14        end
15     end
16  end
```

Determine the entires for B through H and L through N.

2.29 What is the bandwidth of the following matrix?

$$\begin{bmatrix} 0 & 4 & 2 & 2 & 1 & 0 & 0 \\ 0 & 0 & 2 & 1 & 3 & 0 & 0 \\ 0 & 0 & 2 & 2 & 3 & 1 & 0 \\ 0 & 0 & 0 & 1 & 3 & 2 & 0 \\ 0 & 0 & 0 & 2 & 3 & 4 & 0 \\ 0 & 0 & 0 & 0 & 5 & 3 & 2 \\ 0 & 0 & 0 & 1 & 4 & 2 & 3 \end{bmatrix}$$

2.30 What is the bandwidth of the following matrix?

$$\begin{bmatrix} 8 & 2 & 3 & -2 & 5 & 5 & 1 & 2 \\ 0 & 5 & -5 & 2 & 4 & -4 & 1 & 1 \\ 0 & 0 & 0 & 2 & 4 & 2 & 1 & 2 \\ 0 & 0 & 0 & -1 & 4 & -1 & 2 & 3 \\ 0 & 0 & 0 & -1 & 2 & -1 & 4 & 2 \\ 0 & 0 & 2 & 1 & 2 & 1 & 2 & 5 \\ 0 & 0 & 0 & 1 & 1 & 2 & 4 & -1 \\ 0 & 0 & 0 & -3 & 1 & 2 & 1 & 4 \end{bmatrix}$$

2.31 Compute the bandwidth of the following matrix:

$$\mathbf{A} = \begin{bmatrix} 1 & 6 & 3 & 5 & 0 & 0 \\ 2 & 3 & 0 & 1 & 0 & 0 \\ 0 & 0 & 2 & 4 & 2 & 1 \\ 1 & 0 & 0 & 0 & 1 & 0 \\ 0 & 0 & 0 & 0 & 4 & 2 \\ 0 & 0 & 0 & 0 & 2 & 1 \end{bmatrix}$$

2.32   In LU decomposition, determine the missing value of **L** for

$$A = \begin{bmatrix} 2 & 6 & 2 \\ 4 & 13 & 8 \\ 2 & 9 & 17 \end{bmatrix} \quad L = \begin{bmatrix} 1 & 0 & 0 \\ 2 & 1 & 0 \\ 1 & L_{3,2} & 1 \end{bmatrix} \quad U = \begin{bmatrix} 2 & 6 & 2 \\ 0 & 1 & 4 \\ 0 & 0 & 3 \end{bmatrix} \quad x = \begin{bmatrix} x_1 \\ x_2 \\ x_3 \end{bmatrix}$$

2.33   Perform LU decomposition on the matrix

$$\begin{bmatrix} 1 & -1 & 2 \\ -2 & 4 & -5 \\ 1 & 1 & 2 \end{bmatrix}$$

2.34   Compute the LU decomposition of

$$A = \begin{bmatrix} 2 & 4 \\ 6 & 13 \end{bmatrix}$$

2.35   Using the LU decomposition in Problem 2.32, determine the value of x_2 for $Ax = b$ if the forcing function is

$$b = \begin{bmatrix} 5 \\ 10 \\ 0 \end{bmatrix}$$

2.36   Use LU decomposition to solve the system of equations

$$2x_1 - 3x_2 + x_3 = 7$$
$$1x_1 - x_2 - 2x_3 = -2$$
$$3x_1 + x_2 - x_3 = 0$$

2.37   Use LU decomposition to solve the system of equations

$$\begin{bmatrix} 1 & -1 & 2 & 1 \\ 3 & 1 & 4 & -1 \\ 1 & 0 & 2 & 0 \\ 1 & 0 & 0 & 1 \end{bmatrix} \begin{bmatrix} x_1 & y_1 & z_1 \\ x_2 & y_2 & z_2 \\ x_3 & y_3 & z_3 \\ x_4 & y_4 & z_4 \end{bmatrix} = \begin{bmatrix} 11 & -3 & -12 \\ 17 & 5 & -10 \\ 9 & 0 & -7 \\ 6 & 2 & -6 \end{bmatrix}$$

2.38   Solve the set of linear algebraic equations

$$\begin{bmatrix} 1 & 1 & 1 \\ 2 & 1 & 2 \\ 3 & 2 & 2 \end{bmatrix} \begin{bmatrix} x_1 \\ x_2 \\ x_3 \end{bmatrix} = \begin{bmatrix} 0 & 9 & 0 \\ 0 & 15 & -3 \\ -1 & 20 & -1 \end{bmatrix}$$

using LU decomposition.

2.39 You are solving a problem using LU decomposition. At the end of the solution, you have come up with the upper triangular matrix

$$\mathbf{U} = \begin{bmatrix} 1 & 3 & 2 \\ 0 & -2 & 1 \\ 0 & 0 & 2 \end{bmatrix}$$

and the lower triangular matrix

$$\mathbf{L} = \begin{bmatrix} 1 & 0 & 0 \\ 2 & 1 & 0 \\ -2 & -1 & 1 \end{bmatrix}$$

The solution to the problem is

$$\mathbf{x} = \begin{bmatrix} 1 \\ -1 \\ 2 \end{bmatrix}$$

Determine the original system of equations $\mathbf{Ax} = \mathbf{b}$ and the vector $\mathbf{y}$ from LU decomposition.

2.40 Use LU decomposition with Doolittle's method to solve the set of equations given by

$$\begin{bmatrix} 1 & -1 & 1 & 2 \\ 1 & 2 & -2 & 3 \\ 2 & 1 & 3 & 1 \\ 1 & 5 & -5 & 9 \end{bmatrix} \begin{bmatrix} x_{11} & x_{12} & x_{13} \\ x_{21} & x_{22} & x_{23} \\ x_{31} & x_{32} & x_{33} \\ x_{41} & x_{42} & x_{43} \end{bmatrix} = \begin{bmatrix} 2 & 3 & 9 \\ 6 & 15 & 1 \\ 12 & -2 & 10 \\ 15 & 42 & -2 \end{bmatrix}$$

2.41 Use LU decomposition (Doolitle's method) to solve the system of equations

$$\begin{array}{rcrcrcl} 4x_1 & + & 3x_2 & - & x_3 & = & 1 \\ -2x_1 & - & 4x_2 & + & 5x_3 & = & 2 \\ 1x_1 & + & 2x_2 & + & 6x_3 & = & 3 \end{array}$$

2.42 Use LU decomposition to compute the inverse of

$$\begin{bmatrix} 2 & 4 & 1 \\ -2 & 1 & 4 \\ 4 & 3 & 1 \end{bmatrix}$$

2.43 Determine the solution to

$$\begin{bmatrix} 2 & 2 & 3 \\ 4 & 5 & 7 \\ 2 & 4 & 6 \end{bmatrix} \begin{bmatrix} x_1 \\ x_2 \\ x_3 \end{bmatrix} = \begin{bmatrix} 5 \\ 11 \\ 8 \end{bmatrix}$$

using

(a) Gauss elimination without pivoting
(b) LU decomposition. Show the values of $\mathbf{U}$, $\mathbf{L}$, and $\mathbf{y}$.

2.44   Solve the following problem

$$\begin{bmatrix} 2 & 3 \\ 1 & 2 \end{bmatrix} \begin{bmatrix} x \\ y \end{bmatrix} = \begin{bmatrix} 3 \\ 2 \end{bmatrix}$$

(a) using Cramer's rule
(b) using Gauss elimination with partial pivoting
(c) using LU decomposition.

2.45   Solve the following system of equations

$$\begin{bmatrix} 1 & 2 & 3 \\ 4 & 5 & 6 \\ 1 & 3 & 2 \end{bmatrix} \begin{bmatrix} x \\ y \\ z \end{bmatrix} = \begin{bmatrix} 6 \\ 15 \\ 6 \end{bmatrix}$$

using

(a) naive Gauss elimination
(b) Gauss elimination with pivoting
(c) LU decomposition.

2.46   Compute the L_1 norm for the matrix

$$A = \begin{bmatrix} 2 & 6 & -2 \\ -4 & -13 & 8 \\ 2 & 9 & 10 \end{bmatrix}$$

2.47   What is the condition number on the system below using the Euclidean norm?

$$3x + 2y + z = 5$$
$$x - y + 2z = 5$$
$$x - 3z = -1$$

2.48   Consider the matrix

$$A = \begin{bmatrix} 2 & x \\ -1 & 4 \end{bmatrix}$$

Determine the values of x such that the Euclidean norm of the matrix is larger than the 2-norm.

2.49   Compute the L_1, L_2, and L_∞ norms for the vector x = [10 3 –4 –1 5].

2.50   For the matrix

$$A = \begin{bmatrix} 6 & 1 & -2 & 3 \\ 2 & -2 & 4 & -1 \\ 1 & -1 & 1 & -1 \\ 0 & -2 & 2 & 1 \end{bmatrix}$$

which definition of the norm gives the largest value and which definition gives the smallest value? You only need to consider the L_1, L_e, and L_∞ norms.

2.51 For the matrix

$$\mathbf{A} = \begin{bmatrix} 2 & -3 & 2 & 1 \\ 1 & 8 & 3 & 2 \\ -5 & -4 & 2 & -1 \\ 11 & 2 & 11 & 2 \end{bmatrix}$$

which definition of the norm gives the largest value and which definition gives the smallest value? You only need to consider the L_1, L_e, and L_∞ norms.

2.52 Compute the L_1, L_e, and L_∞ norms for the matrix

$$\begin{bmatrix} 1 & 2 & 3 & 2 \\ -5 & 4 & 6 & -3 \\ 1 & -4 & 5 & 2 \\ 10 & 2 & 1 & 6 \end{bmatrix}$$

2.53 Use the 1-norm to determine the condition number of

$$\mathbf{A} = \begin{bmatrix} 1 & 1 & 1 \\ 2 & 1 & 2 \\ 4 & 0 & 2 \end{bmatrix}$$

2.54 Compute the condition number for

$$\begin{bmatrix} 1 & 1 & 2 \\ 1 & 3 & 1 \\ 0 & 1 & 2 \end{bmatrix}$$

using the Euclidean norm.

2.55 Compute the condition number for the matrix

$$\mathbf{A} = \begin{bmatrix} 1 & 2 \\ 0 & 1 \end{bmatrix}$$

using the Euclidean norm.

2.56 Compute the 1-norm, Euclidean norm, spectral norm, and infinity norm of

$$\mathbf{A} = \begin{bmatrix} 2 & 3 \\ 1 & 2 \end{bmatrix}$$

2.57 Consider the 2×2 system of equations

$$2x_1 + x_2 = 4$$
$$x_1 + x_2 = 3$$

What is the value of the 1-norm of $||\mathbf{A}\mathbf{x}^{(1)} - \mathbf{b}||_1$ after one iteration of Jacobi's method with the initial guess $x_1^{(0)} = 1$, $x_2^{(0)} = 0$?

2.58 Perform the first two iterations of Jacobi's method for the system of equations

$$
\begin{bmatrix}
6 & 1 & 0 & -3 \\
-2 & 8 & 2 & 3 \\
1 & -6 & 10 & -2 \\
2 & 1 & 3 & 9
\end{bmatrix}
\begin{bmatrix}
x_1 \\ x_2 \\ x_3 \\ x_4
\end{bmatrix}
=
\begin{bmatrix}
-9 \\ 9 \\ 17 \\ 13
\end{bmatrix}
$$

using an initial guess $\mathbf{x} = \mathbf{0}$.

2.59 Perform the first two iterations of Gauss–Seidel for the system of equations

$$
\begin{bmatrix}
6 & 1 & 0 & -3 \\
-2 & 8 & 2 & 3 \\
1 & 6 & 10 & -2 \\
2 & 1 & 3 & 9
\end{bmatrix}
\begin{bmatrix}
x_1 \\ x_2 \\ x_3 \\ x_4
\end{bmatrix}
=
\begin{bmatrix}
-9 \\ 9 \\ 17 \\ 13
\end{bmatrix}
$$

using an initial guess $\mathbf{x} = \mathbf{0}$.

2.60 Determine $x_2^{(1)}$ for the system of equations

$$
\begin{bmatrix}
3 & 2 & 4 \\
1 & 3 & 0 \\
2 & 4 & 1
\end{bmatrix}
\begin{bmatrix}
x_1 \\ x_2 \\ x_3
\end{bmatrix}
=
\begin{bmatrix}
6 \\ 3 \\ 4
\end{bmatrix}
$$

using Gauss–Seidel with an initial guess of $x_1 = 0$, $x_2 = 0$, $x_3 = 0$.

2.61 Solve

$$
\begin{bmatrix}
3 & 2 & 4 \\
1 & 3 & 0 \\
2 & 4 & 1
\end{bmatrix}
\begin{bmatrix}
x_1 \\ x_2 \\ x_3
\end{bmatrix}
=
\begin{bmatrix}
6 \\ 3 \\ 4
\end{bmatrix}
$$

using successive relaxation with a parameter $w = 3/2$ and an initial guess of $x_1 = 0$, $x_2 = 0$, $x_3 = 0$.

2.62 Do one iteration of successive relaxation with $w = 3/2$ using the system in Problem 2.58 and an initial guess $\mathbf{x} = [-0.9\ 0.1\ 2.2\ 1.1]$.

2.63 Is the matrix from Problem 2.46 diagonally dominant?

2.64 Show that the matrix $\mathbf{A}$ in Problem 2.58 is diagonally dominant.

2.65 Perform linear regression for the following data set.

x	1.1	1.6	2.0	2.3	2.8
y	2.4	3.8	5.2	6.0	7.9

2.66 For numerical calculations, the computer is limited by two major constraints: the amount of memory available and the speed of the processor. Let's examine how these constraints affect our ability to perform Gauss elimination on an old Mac. The Mac SE/30s was equipped with 2 MB (megabytes) of Random Access Memory (RAM).

A fairly large part of it is occupied by the operating system and the QuickBASIC application. Let's assume that 1 MB is available for our use.

(a) What is the largest system of equations $\mathbf{Ax} = \mathbf{b}$ that we could solve in double precision on the Mac according to our memory limitation? Assume that we only need to fit the matrix $\mathbf{A}$ (n^2 numbers) and the vector $\mathbf{b}$ (n numbers) into memory.
   1 MB = 1 megabyte = 1024 KB
   1 KB = 1 kilobyte = 1024 bytes
   1 byte = 8 bits
   1 bit holds one binary piece of information (0 or 1)
   1 double precision variable requires 64 bits

(b) Repeat this calculation for the Cray-2 supercomputer, which has 512 megawords (MW) of memory. Assume that 500 MW is available for our use.
   1 MW = 1 megaword = 1024 × 1024 words
   1 word = 64 bits
   (Notice that, on the Cray, the default variable size is double precision.)

(c) What is the largest system of equations $\mathbf{Ax} = \mathbf{b}$ that we could reasonably solve on a Mac SE/30 if computer speed is our only limitation?
   Assume that we will be using Gauss elimination and that "reasonable" means in less than one hour.
   A rough estimate of the speed of a compiled program on a Mac SE/30 is 0.01 MFLOPS (for double precision).
   1 MFLOPS = 1 million floating-point operations per second.
   Gauss elimination takes approximately $n^3/3$ operations.

(d) Repeat this calculation for the Cray-2 supercomputer, if its nominal speed is assumed to be 100 MFLOPS.

(e) Based on your answers to (a) through (d), what is the limiting factor for Gauss elimination on a Mac? On a Cray-2?

2.67 If I want to solve a 200 × 200 matrix with a bandwidth of 5 and $p = q$, what is the ratio of the time required for full Gauss elimination relative to using a banded Gauss elimination solver?

2.68 Let's assume we have a computer that can solve a 1000 × 1000 system in 2 seconds using Gauss elimination. If this is in fact a banded system with $p = q = 2$ and I can solve the same problem in five iterations using Jacobi's method, estimate how long will it take. Which method would you prefer to use?

2.69 Let's assume we have a computer that can solve a 100 × 100 system in 1 second using Gauss elimination. Estimate the time required to solve the following problems on the same computer. You must indicate the scaling as well as the time.

(a) 1000 × 1000 system using Gauss elimination
(b) 1000 × 1000 system with a band structure $p = q = 2$ using banded Gauss elimination
(c) The original 100 × 100 system using an iterative method with 10 iterations.

2.70   What are the criteria for the number of iterations such that Jacobi's method is faster than Gauss elimination for solving a linear problem that is not banded?

2.71   Determine the value of y such that $[3,y]$ is an eigenvector of the matrix

$$\mathbf{A} = \begin{bmatrix} 1 & 3 \\ 4 & 2 \end{bmatrix}$$

for the eigenvalue $\lambda = 5$.

2.72   What are the magnitudes of the eigenvalues of

$$\mathbf{A} = \begin{bmatrix} 1 & -1 \\ 2 & 1 \end{bmatrix}$$

2.73   Consider the matrix

$$\begin{bmatrix} 1 & -3 & \pi & 2 & \ln 3 \\ 0 & 2 & -50 & \ln 2 & \sqrt{3} \\ 0 & 0 & 3 & -1 & 2 \\ 0 & 0 & 0 & 1/2 & 0 \\ 0 & 0 & 0 & 0 & 1/3 \end{bmatrix}$$

(a)   What is the determinant of the matrix $\mathbf{A}$?
(b)   What is the bandwidth of the matrix $\mathbf{A}$?
(c)   What is the 1-norm of the matrix $\mathbf{A}$?
(d)   How many eigenvalues do you expect to have for $\mathbf{A}$? (If you know the eigenvalues, you can include them too but this is not required.)
(e)   Is matrix $\mathbf{A}$ diagonally dominant?

2.74   Answer the following questions about this 12×12 matrix:

$$\mathbf{A} = \begin{bmatrix} 1 & 2 & 3 & 1 & 1 & 0 & 0 & 0 & 0 & 0 & 0 & 2 \\ 0 & 1 & 3 & 2 & 1 & 0 & 0 & 0 & 0 & 0 & 0 & 0 \\ 2 & 4 & 9 & 2 & 2 & 0 & 0 & 0 & 0 & 0 & 0 & 0 \\ 0 & 0 & 10 & 3 & 2 & 1 & 4 & 0 & 0 & 0 & 0 & 0 \\ 0 & 0 & 0 & 0 & 1 & 2 & 0 & 0 & 0 & 0 & 0 & 0 \\ 0 & 0 & 0 & 3 & 2 & 1 & 0 & 0 & 2 & 0 & 0 & 0 \\ 0 & 0 & 0 & 0 & 5 & 2 & 3 & 4 & 0 & 0 & 0 & 0 \\ 0 & 0 & 0 & 0 & 4 & 3 & 0 & 3 & 0 & 0 & 0 & 0 \\ 0 & 0 & 0 & 0 & 0 & 0 & 0 & 3 & 0 & 0 & 0 & 0 \\ 0 & 0 & 0 & 0 & 0 & 0 & 4 & 8 & 6 & 1 & 0 & 0 \\ 0 & 0 & 0 & 0 & 0 & 0 & 0 & 0 & 1 & 2 & 1 & 2 \\ 0 & 0 & 0 & 0 & 0 & 0 & 0 & 0 & 2 & 2 & 3 & 1 \end{bmatrix}$$

(a)   What is the bandwidth?
(b)   Assuming this system has a solution for a given vector $\mathbf{b}$, are you guaranteed to have convergence using successive relaxation? Explain your answer.
(c)   If you perform Gauss elimination with partial pivoting, what is the entry $a_{3,2}^{(2)}$?

2.75   What numerical method is implemented by the code listed below (be specific).

```
1  function out = problem2_75
2  n= size(A,1);
3  for k=1:n-1
4      for i=k+1:n
5          m=A(i,k)/A(k,k);
6          for j=k+1:n
7              A(i,j)=A(i,j)-m*A(k,j);
8          end
9      end
10 end
11 out = A;
```

2.76   Consider the following program:

```
1  function problem2_76
2  A = [2,3,1;-2,3,2;1,2,5];
3  b = [1;4;2];
4  x = [0;0;0];
5  err = norm(A*x -b);
6  k = 0;
7  while err > 1e-6;
8      k = k+1;
9      x(1) = (b(1) - A(1,2)*x(2) - A(1,3)*x(3))/A(1,1);
10     x(2) = (b(2) - A(2,1)*x(1) - A(2,3)*x(3))/A(2,2);
11     x(3) = (b(3) - A(3,1)*x(1) - A(3,2)*x(2))/A(3,3);
12     err = norm(A*x-b);
13     if k > 100
14         err = 0;
15         x = 0;
16     end
17 end
18 if err == 0
19     fprintf('Did not converge.\n')
20 else
21     fprintf('Value after %3d steps:\n',k)
22     fprintf('x1 = %8.6f \t x2 = %8.6f \t x3 = %8.6f ...
            \n.',x(1),x(2),x(3))
23 end
```

What system of equations is being solved by this method? What method is used for the solution? Will this program converge to a solution? To save you time, we have already determined that the condition number is 2.6473. You should be as specific as possible in your answer using your knowledge of both linear algebra and this type of numerical method.

2.77   Answer the following questions about this MATLAB program. You can assume that the norm computed by MATLAB is the 1-norm. (We are using the 1-norm to simplify the problem, not because MATLAB actually uses the 1-norm.)

```
1   function problem2_77
2   x1 = 0;
3   x2 = 0;
4   R = residual(x1,x2);
5   error = norm(R);
6   count = 0;
7   fprintf('============STARTING CALCULATION=============\n')
8   fprintf('k \t x1 \t x2 \t norm(R)\n')
9   fprintf('%d \t %9.6f \t %9.6f \t %6.4e \n',count,x1,x2,error)
10  while error > 10^(-4)
11      count = count + 1;
12      x1_old = x1;
13      x1 = 2*(1-x2)/3;
14      x2 = (x1_old+1);
15      R = residual(x1,x2);
16      error = norm(R);
17      fprintf('%d \t %9.6f \t %9.6f \t %6.4e \n',count,x1,x2,error)
18      if count > 20
19          fprintf('Did not converge.\n')
20          break;
21      end
22  end
23
24  function out = residual(x1,x2)
25  out(1,1) = 3*x1+2*x2-2;
26  out(2,1) = x1-x2+1;
```

(a) What *mathematical* problem is solved by this program?

(b) What *numerical method* is implemented by this program?

(c) What are the criteria for stopping the while loop?

(d) When you run this program in MATLAB (assuming that norm is the 1-norm), what is the third line printed to the screen?

(e) When you run this program in MATLAB (assuming that norm is the 1-norm), what is the fourth line printed to the screen?

2.78   Answer the following questions about this program:

```
1   function problem2_78
2   clc
3   A = [2 1 3; 4 2 1; 1 1 1];
4   n = 3;
5   q = 1;
6   for k=1:n-1
7       Amax = A(k,k);
8       swap_row = k;
9       for i = k+1:n
10          if abs(A(i,k)) > abs(Amax)
11              Amax = A(i,k);
12              swap_row = i;
13          end
14      end
15      if swap_row ~= k
16          old_pivot(1,:) = A(k,:);
```

```
17        A(k,:) = A(swap_row,:);
18        A(swap_row,:) = old_pivot;
19        s = -1;
20    else
21        s = 1;
22    end
23    q = q*A(k,k)*s;
24    for i=k+1:n
25        m=A(i,k)/A(k,k);
26        for j=k+1:n
27            A(i,j)=A(i,j)-m*A(k,j);
28        end
29    end
30 end
31 q = q*A(n,n);
32 fprintf('q = %8.6f\n',q)
```

(a) What *mathematical problem* is solved by this program?

(b) What *numerical method* is used to solve the problem?

(c) What does the variable n represent?

(d) What does the variable q represent?

(e) What does the variable s represent?

(f) What mathematical operation is implemented by the **for** loop in lines 24–29?

(g) What is the output of line 32? You must provide the numerical value.

2.79   Answer the following questions about this program:

```
1  function problem2_79
2  clc
3  n = 3;
4  x = zeros(n,1);
5  A = [3 10 -5; -4 1 2; 1 1 -4];
6  b = [-2; 1; 0];
7  R = norm(A*x-b);
8  k = 0;
9  fprintf('k \t err \n')
10 fprintf('%2d \t %6.4e \n',k,R)
11 while R > 1e-7
12     for i = 1:n
13         s = 0;
14         for j = 1:n
15             s = s + A(i,j)*x(j);
16         end
17         x(i) = x(i) + 0.8*(b(i)-s)/A(i,i);
18     end
19     R = norm(A*x-b);
20     k = k + 1;
21     fprintf('%2d \t %6.4e \n',k,R)
22     if k > 50
23         fprintf('Did not converge.\n')
24         R = -1;
25     end
26 end
27 if R ~= -1
```

```
28      fprintf('The converged result is:\n')
29      for i = 1:n
30          fprintf('%8.6f\n',x(i))
31      end
32  end
```

(a) What *mathematical* problem does this program solve?
(b) What *numerical method* does this program use?
(c) What is the initial condition used to start the numerical method?
(d) What is the criteria for convergence?
(e) What is the reason you might choose the value 0.8 in line 17?
(f) What is the significance of changing line 17 to: x(i)= x(i)+ (b(i)-s)/A(i,i);
(g) What happens if you change line 17 to: x(i)= x(i)+ 2.5*(b(i)-s)/A(i,i);
(h) If we forget to include line 13, the program does not work. Explain the problem with this bug.
(i) How would you modify the mathematical problem from part (a) (with the same initial guess) to ensure that this numerical method will converge? Explain the reason behind your answer. Your answer can be in words, and you do not need to rewrite the program.

Computer Problems

2.80   Consider the process flow diagram shown in Fig. 2.14. We can write this out as the system of equations

$$m_1 = m_2 + m_3 + m_4 + m_5$$
$$m_2 = m_9 + m_{10} + m_{11}$$
$$m_5 = m_8 + m_7 + m_6$$

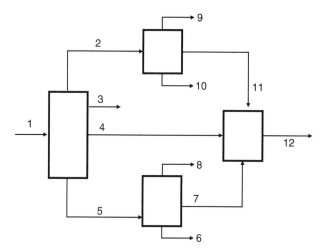

Figure 2.14 Process schematic for Problem 2.80.

$$m_{12} = m_4 + m_7 + m_{11}$$
$$m_1 = 100$$
$$m_5 = 5m_8$$
$$0.84m_{12} = m_4 + m_7$$
$$0.7m_1 = m_2 + m_3$$
$$0.55m_1 = m_9 + m_{12}$$
$$0.2m_9 = m_{10}$$
$$0.85m_2 = m_9 + m_{11}$$
$$3.2m_6 = m_7 + m_8$$

The first four equations are the mass balances; the next eight equations are the process specifications.

(a) Write a MATLAB program that generates the matrix **A** and vector **b** required to solve the problem for this set of equations.

(b) Write a MATLAB program that solves this system using naive Gauss elimination. What happens? How do you fix it?

2.81 A convenient way to visualize the evolution of a two-dimensional system is to make a phase plane plot. We will do more of this type of calculation in Chapter 5. For now, we would like to see how the solution to the system of equations

$$6x + y = 13$$
$$4x - 8y = 0$$

changes when you use Jacobi's method with an initial guess $x^{(0)} = 5$ and $y^{(0)} = 0$. Write a MATLAB program that computes the values of $x^{(k)}$ and $y^{(k)}$ using Jacobi's method until the error in the solution, defined as $||\mathbf{x}^{(k)} - \mathbf{x}^{(k-1)}||$ is less than 10^{-4}. Your program should store all of the values of $x^{(k)}$ and $y^{(k)}$. In a "normal" plot, you might want to look at x vs. k or y vs. k. In the phase plane, you want to plot the points $(x^{(k)}, y^{(k)})$ on an x, y plane and connect them by lines. The corresponding trajectory will show you how the solution method evolves to the final answer. Your program should automatically produce two plots. In the first plot, show all of the points at each value of k on an x, y plane. In the second plot, zoom in on the data close to the solution to see the details.

2.82 We saw that the time t required to solve a banded matrix of size $n \times n$ with a bandwidth $p = q$ scales like $t \sim np^2$. In this problem, we will explore the ability of MATLAB's internal solver to handle large, banded matrices. Write a MATLAB program that determines the time required to solve a 500×500 system of the form $\mathbf{Ax} = \mathbf{b}$ using the slash command for values of $p = q$ from 1:499. To check that you have the right matrix, you should compute the number of non-zero elements in your matrix using the `find` and `size` functions in MATLAB. To get the time required for a given value of p in your program, you use the command `tic` to start to clock before you solve the matrix system and `toc` to stop the clock after you solve the matrix system. If you write `t = toc`, the value of the time will be stored in the variable `t`. The non-zero elements of the matrix $\mathbf{A}$ and the 500×1 forcing vector $\mathbf{b}$ should be randomly chosen for each value of p using the `rand` function in MATLAB.

Make a plot of the number of non-zero elements in your matrix versus the value of p and make a log–log plot of the time required for the matrix solution versus the value of p. State how you would expect the slope of the log–log plot to look for Gauss elimination and whether the time required for the MATLAB solution method scales faster, the same, or slower than Gauss elimination with a banded solver.

2.83   In this problem, we want to figure out how the rate of convergence to the successive over-relaxation (SOR) solution of a linear system of equations depends on the relaxation parameter for $w = 0.1, 0.2, \ldots, 1.9$. You can download the 100×100 matrix $\mathbf{A}$ and the 100×1 forcing function $\mathbf{b}$ as a zip file and put them in the memory by using the `load` command in MATLAB. Write a MATLAB program that determines the number of iterations to reach and error smaller than 10^{-4}, where we will define the error at the end of step k as $||\mathbf{A}\mathbf{x}^{(k)} - \mathbf{b}||$. Use the initial condition $x_i^{(0)} = 0$ for all i. Your program should automatically make a plot of the number of iterations for convergence versus the value of w. Include a brief explanation of what you are seeing.

2.84   In this problem, you are going to investigate the work required to solve a matrix system using Gauss–Seidel. The matrix system is defined by the entries

$$a_{i,j} = \frac{i+j}{ij}, \quad (i \neq j),$$
$$a_{i,j} = 8 + 0.2i, \quad (i = j),$$

and the forcing vector entries are

$$b_i = in.$$

Your program should first check to see if the matrix is diagonally dominant and output a value of 1 (if dominant) or 0 (if not dominant). Your program should then implement the Gauss–Seidel scheme with an initial guess $\mathbf{x}^{(0)} = \mathbf{0}$. The criterion for convergence for iteration k is $||\mathbf{A}\mathbf{x}^{(k)} - \mathbf{b}|| < 10^{-8}$. You can use the `norm` command in MATLAB for this calculation. Add a conditional statement into your iterative scheme that will terminate the loop if you reach 100 iterations. The output of the loop should be the number of iterations and the value of the norm.

Your program should perform this calculation for matrix sizes $n = 5, 10, \ldots, 150$. It should make a plot of the number of iterations k versus the size of the matrix n. On the plot, also include a line $k = n$ to indicate the scaling for the work required by Gauss elimination. Set the x-scale to be [0,150] and the y-scale to be [0,100]. If all of your values of $k < 100$, then you know that the scheme converged. Label the axes, provide a title, and save the graph as a jpg. You should also compile your output into a matrix that you can save to file using `dlmwrite`. The matrix should have 30 rows and the following four columns: (i) the size of the matrix, (ii) the result of the diagonal dominance calculation (a zero or a one), (iii) the norm at the end of your Gauss–Seidel calculation, and (iv) the number of iterations at the end of your Gauss–Seidel calculation.

2.85 We want to explore the accuracy of MATLAB's solution to the following set of equations

$$\begin{bmatrix} 3 & 1 & 2 \\ 6 & 3 & 3 \\ 3 & 1 & 2+\epsilon \end{bmatrix} \begin{bmatrix} x_1 \\ x_2 \\ x_3 \end{bmatrix} = \begin{bmatrix} 7 \\ 10 \\ 8 \end{bmatrix}$$

for values of ϵ from 10^{-1} to 10^{-15}. For this problem, you are allowed to use the slash command to solve the system of equations and the `cond` function to compute the condition number.

Write a MATLAB function that computes the condition number and the relative error in the value of x_3 as a function of ϵ. The error in x_3 should be computed by the formula

$$\text{error} = \frac{|x_3^{\text{compute}} - x_3^{\text{exact}}|}{x_3^{\text{exact}}}$$

where x_3^{compute} is the value obtained in MATLAB and x_3^{exact} is the exact solution for x_3 that you obtained by solving this system by hand using Gauss elimination. Your program should make (i) a log–log plot of the condition number versus epsilon and (ii) a log–log plot of the relative error versus the condition number.

3   Nonlinear Algebraic Equations

3.1   Introduction

In this chapter we will cover the solution of nonlinear algebraic equations, starting from scalar ones and proceeding to systems of equations. You have encountered nonlinear equations in many of your previous classes. For example, consider the Antoine equation that we used in Section 2.7 for the multi-component flash problem,

$$\log_{10} P^{\text{sat}} = A - \frac{B}{T + C} \tag{3.1.1}$$

This is an equation describing the saturation pressure, P^{sat}, of some chemical species as a function of temperature, T, where A, B, and C are parameters set by the chemistry. This is the standard way that the equation appears in thermodynamics textbooks, where the left-hand side has something that you want to compute (e.g., the saturation pressure) as a function of some parameter (e.g., the temperature). For the numerical methods that we will explore here, it is important that we write these equations in the general form

$$f(x) = 0 \tag{3.1.2}$$

where $f(x)$ is the nonlinear function of some variable x. This is easy; for Eq. (3.1.1) we could just write

$$f(P^{\text{sat}}) = \log_{10} P^{\text{sat}} - A + \frac{B}{T + C} \tag{3.1.3}$$

The situation is the same for a system of equations, where we now have a vector of unknowns $\mathbf{x}$ and we write

$$\mathbf{R}(\mathbf{x}) = \mathbf{0} \tag{3.1.4}$$

where $\mathbf{R}$ denotes a vector function that we will call the residual.

Nonlinearity is the rule rather than the exception in nature and, of course, in engineering problems. Sources of nonlinearity include reaction rate laws, thermodynamic relations, transport correlations, etc. Linearity typically comes from approximations, either physical or mathematical. For example, a vapor–liquid equilibrium is often modeled by the system of equations

$$x_1 \gamma_1(x_1, T) P_1^{\text{sat}}(T) = y_1 P \tag{3.1.5}$$

$$(1 - x_1) \gamma_2(x_1, T) P_2^{\text{sat}}(T) = (1 - y_1) P \tag{3.1.6}$$

where x_i is the mole fraction of species i in the liquid phase, $\gamma_i(x_1, T)$ is the activity coefficient for species i in the liquid phase, which is generally a nonlinear function of x_i and temperature, P_i^{sat} is the saturation pressure of species i at the temperature T, and y_i is the mole fraction of species i in the vapor phase. (We have assumed an ideal vapor phase for simplicity, since these equations are already hard enough to solve!) If we want to create a temperature–composition phase diagram (T–x–y), Eqs. (3.1.5) and (3.1.6) lead to a nonlinear system of equations of the form of Eq. (3.1.4) if we rewrite them as

$$\mathbf{R}(\mathbf{x}) = \begin{bmatrix} x_1\gamma_1(x_1, T)P_1^{sat}(T) - y_1 P \\ (1 - x_1)\gamma_2(x_1, T)P_2^{sat}(T) - (1 - y_1)P \end{bmatrix} = \begin{bmatrix} 0 \\ 0 \end{bmatrix} \tag{3.1.7}$$

We will work out the numerical solution for a non-ideal equilibrium problem in our case study at the end of the chapter. Before we can start working on chemical engineering problems, we first need to spend some time learning how to solve nonlinear equations in general.

3.1.1   Superposition

How do you know if an equation is linear or nonlinear? When asked this question, many students provide some variation on Justice Potter Stewart's famous quotation about pornography in the case *Jacobellis v. Ohio*: "I know it when I see it." We can certainly be more precise about mathematics!

Linear systems obey the principle of superposition, whereas nonlinear ones do not. In the context of algebraic functions this means that if $f(x)$ is a nonlinear function of a scalar variable, then

$$f(\alpha x_1 + \beta x_2) \neq \alpha f(x_1) + \beta f(x_2) \tag{3.1.8}$$

for α, β scalars. Note that this result only applies for homogeneous equations, where there are no constant terms in $f(x)$.

Example 3.1   Show that

$$f(x) = x \tag{3.1.9}$$

is a linear equation.

Solution
The superposition principle requires that

$$f(\alpha x_1 + \beta x_2) = \alpha x_1 + \beta x_2 \tag{3.1.10}$$

be equal to

$$\alpha f(x_1) + \beta f(x_2) = \alpha x_1 + \beta x_2 \tag{3.1.11}$$

This is true for any α and β. Note that the superposition principle is only used for homogeneous equations. If we had instead written an inhomogeneous problem

$$f(x) = x + 1 \tag{3.1.12}$$

we would find that superposition is only true if $\alpha + \beta = 1$.

Example 3.2   Show that

$$f(x) = x^2 \tag{3.1.13}$$

is a nonlinear function.

Solution
The superposition principle requires that

$$f(\alpha x_1 + \beta x_2) = (\alpha x_1 + \beta x_2)^2 = \alpha^2 x_1^2 + 2\alpha\beta x_1 x_2 + \beta^2 x_2^2 \tag{3.1.14}$$

be equal to

$$\alpha f(x_1) + \beta f(x_2) = \alpha x_1^2 + \beta x_2^2 \tag{3.1.15}$$

This is not true for any α and β, so $f(x) = x^2$ is a nonlinear function.

3.1.2   Differences Between Linear and Nonlinear Equations

In Section 2.2, we showed that systems of linear equations have only three possible outcomes: (i) no solution, (ii) a unique solution, or (iii) infinitely many solutions. If the determinant of the matrix is non-zero, we are *guaranteed* the get a unique solution to the problem. As a result, we put our effort into figuring out how to make sure that we got a solution (e.g., by using pivoting) and that we arrived at the solution in the fastest possible manner (e.g., by using a banded solver).

In making the shift from linear to nonlinear equations, several complications arise. Perhaps the most important change is that there is no general solvability criterion for nonlinear algebraic equations similar to the one expressed in terms of the determinant of matrix for linear equations. This may just sound like mathematical "mumbo jumbo" to you; many engineers find applied math to be interesting and useful but dismiss the more theoretical aspects. In this case, the theory has huge implications for actually solving engineering problems – if you do not even know whether a solution exists, how can you know if you have computed it?

For example, let's try to play the same graphical construction game that we played in Fig. 2.2, just focusing on a single nonlinear equation. Consider first the quadratic equation

$$f(x) = x^2 - 1 = 0 \tag{3.1.16}$$

This equation has two real roots, $x = \pm 1$, as illustrated in Fig. 3.1. What if the constant term changed sign? Now we have

$$f(x) = x^2 + 1 = 0 \tag{3.1.17}$$

This equation has no real roots; the solutions are the imaginary numbers $x = \pm \imath$. You might be tempted to dismiss this as a contrived mathematical example, but the appearance and disappearance of solutions is an important topic in nonlinear systems. We will

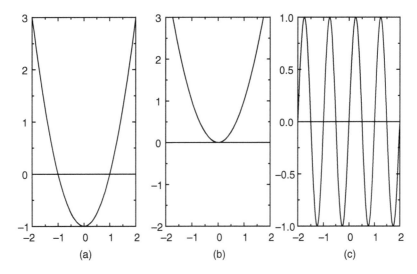

Figure 3.1 Illustration of the real roots for Eqs. (3.1.16), (3.1.19), and (3.1.20).

explore this concept, known as a bifurcation, in more detail in Chapter 5. In the present context, imagine that the system is really of the form

$$f(x) = x^2 + p \tag{3.1.18}$$

where p is some parameter that depends on, say, the temperature while x is the output that you are trying to measure. In this mathematical model, the output could disappear and then reappear as the temperature fluctuates. There is a special case when this problem crosses over from two real roots to no real roots, as p turns from negative to positive. The $p = 0$ case gives

$$f(x) = x^2 \tag{3.1.19}$$

which has a repeated root at $x = 0$. This is known as the bifurcation point for Eq. (3.1.18).

In addition to having a finite number of roots (3.1.16), no real roots (3.1.17), and repeated real roots (3.1.19), a single nonlinear equation can have an infinite number of roots. For example, the nonlinear equation

$$f(x) = \sin 2\pi x \tag{3.1.20}$$

has an infinite number of roots for any integer

$$x = -\infty, \ldots, -2, -1, 0, 1, 2, \ldots \infty \tag{3.1.21}$$

An interesting part about this problem is that there are infinitely many roots, but you can count them. Mathematicians sometimes like to distinguish between a function with a countable, infinite number of roots, such as $f(x) = \sin 2\pi x$, and a function with an uncountable, infinite number of roots, such as $f(x) = 0$. We will not worry about such subtle differences here.

Compare all of the different possibilities for a single nonlinear equation to the underwhelming result for the single linear equation

$$f(x) = ax + b \tag{3.1.22}$$

which has but a single root, $x = -b/a$.

Needless to say, the situation for systems of nonlinear equations only gets more complicated. While a system of linear equations only has three possible outcomes, even trying to find a single root of a system of nonlinear equations can be a Herculean task. The goal of this chapter is to show you how to find such solutions – if you can.

3.2   Solution Approaches

For linear algebraic equations, there is an analytical expression for the solution of the problem by Cramer's rule (2.4.1). While we would never use Cramer's rule to solve anything but the smallest system of equations, at least such a result exists. The situation is much worse for nonlinear equations. There are a few cases where there is an analytical expression for the problem. For example, you know that a quadratic equation

$$f(x) = ax^2 + bx + c \tag{3.2.1}$$

has the solution

$$x = \frac{-b \pm \sqrt{b^2 - 4ac}}{2a} \tag{3.2.2}$$

There are also cases where the "solution" to the problem is a tabulated function. For example, consider the function

$$f(x) = \ln x + 2 \tag{3.2.3}$$

If you were asked to solve this equation, you would probably write down

$$x = e^{-2} \tag{3.2.4}$$

and be satisfied because if you punch this result into your calculator, you get back

$$x = 0.135\,335\,283 \tag{3.2.5}$$

This is hardly a solution to the problem analogous to the quadratic formula in Eq. (3.2.2). Rather, the function $f(x) = e^x$ is tabulated (or approximated) by your calculator. In going from Eq. (3.2.3) to Eq. (3.2.4), we have just exchanged one transcendental function for another.

In this chapter, we will show you how to find the roots of nonlinear equations. As you will see next, the processes to find these roots are intrinsically iterative in nature.

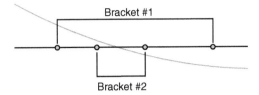

Figure 3.2 Illustration of using a bracketing method for Eq. (3.1.16). As we make each successive bracket, the region for the solution becomes narrower and narrower.

3.2.1   Bracketing Methods

Bracketing is the most intuitive approach for finding the roots of a nonlinear equation. In these methods (often illustrated graphically, as in Fig. 3.2) the idea is to find an interval in the vicinity of a root where the function changes sign, and narrow down this interval, essentially trapping the solution. In other words, you use the fact that if $f(a) < 0$ and $f(b) > 0$ for some scalars a, b, then a root x^* lies between $x = a$ and $x = b$.

PROGRAM 3.1   *It is relatively simple to write a program to implement bracketing for an initial guess that we think will contain a root. We pick the value of a and b according to the criteria we defined for bracketing, with a being the lower bound on x and b being the upper bound on x. The best way to program bracketing is to use a recursive algorithm, where the function calls itself.*

```
1   function out = nonlinear_bracketing_recursion(a,b,dx,tol)
2   %  a = left boundary
3   %  b = right boundary
4   %  dx = change in boundary
5   %  tol = criteria for convergence
6   %  Requires function f(x). Outputs q so f(q) is close to 0.
7   dx_left = dx;
8   dx_right = -dx;
9   if abs(getf((a+b)/2)) < tol
10      out=(a+b)/2;
11      return
12  else
13      a = a + change_dx(a,dx_left);
14      b = b + change_dx(b,dx_right);
15      out=nonlinear_bracketing_recursion(a,b,dx,tol);
16  end
17
18  function out = change_dx(q,dx)
19  %  q = value of x
20  %  dx = change in boundary
21  if getf(q)*getf(q+dx) > 0
22      out=dx;
23      return
24  else
25      out=change_dx(q,dx/2);
26  end
```

In this program, we provide a tolerance variable `tol` *that tells us how close we want to be to the root such that the average of* $f(a)$ *and* $f(b)$ *is smaller than the tolerance. We also provide an initial step size* `dx` *to use to change the values of a and b. Our program requires that a and b actually be the brackets of* $f(x)$*; there is no error checking. The first thing we do is define new variables in lines 7 and 8 for the step sizes on the left and right, since we will let them be different as we hone in on the root. The* `if-else` *block in lines 9–16 sets up the bracketing scheme. If the bracketing is good enough, line 10 computes the answer and line 11 sends back the result. If we are not close enough, lines 13 and 14 pick the new values of a and b using the subfunction* `change_dx`*, which we will describe in a moment. Line 15 implements the recursive part of the algorithm, where the program calls itself. If we did not get a good enough answer, we call the bracketing program again with the new bounds of a and b. This logic could also be implemented with a* **while** *loop, but it is much more efficiently written as recursion.*

The subfunction for changing `dx_left` *and* `dx_right` *first computes the product of the old value of a (or b) and the new value; if this quantity is positive then they have the same sign and thus we are properly moving the bracket towards the root. If they have opposite signs, then our value of* `dx` *was too large and we passed through the root. To deal with this problem, line 25 again uses recursion to halve the value of* `dx` *and try to move the bracket again.*

Example 3.3   Use bracketing for Eq. (3.1.16) using the initial guesses $a = -1.51$ and $b = 0.01$ with the initial step size of $\Delta x = 0.1$.

Solution
As should be clear from Fig. 3.2, giving a better bound for the root ($a = -1.51$) leads to a faster convergence of the bracketing solution. Also note that there is a pathological problem with this program. Try to run it where you guess $a = -1.5$ and $b = 0$. With $\Delta x = 0.1$, it should converge directly to the root but it (usually) only comes close. Can you figure out what is going on?

We will find bracketing to be a useful approach to generate initial guesses for the more robust methods listed below. A good initial guess for the root is very helpful for finding roots of a single equation, and usually essential for finding the roots of systems of nonlinear equations. However, bracketing is not a universally useful method. For example, think about what would happen if we use bracketing to find the roots of $f(x) = x^2$, which we discussed in Eq. (3.1.19). It immediately becomes clear that we cannot use bracketing for a root at $x = 0$; the solution never changes sign, so we cannot even identify an initial guess to provide to Program 3.1. Bracketing also has trouble with functions such as the one in Fig. 3.3. If we tried to use our program, we would be able to find a root that is somewhere, but our algorithm is designed to find a single root in an interval, not a range of roots.

Figure 3.3 Example of a piecewise continuous function where there are an infinite number of roots between the two circles.

Bracketing is practical for a single nonlinear equation – we just plot the solution and look to see where it crosses zero to generate the initial guesses for a and b. What happens when we go to higher dimensions? We would need to extend this idea to n different initial brackets $x_j \in [a_j, b_j]$ for $j = 1, 2, \ldots, n$ and then try to zoom in on each of the bracketed regions. There are two major problems with this method. First, we need to systematically search an n-dimensional space, which is going to be slow. Second, and perhaps even more important, we need to generate the initial brackets for each of the dimensions. In principle, this requires making an n-dimensional plot for every dimension near its root, and then adjusting each one of those plots as we try to improve our estimate in each dimension. If this sounds confusing and difficult as you are reading it, imagine trying to modify Program 3.1 to implement a multi-dimensional bracketing method on a computer.

3.2.2 Fixed-Point (Open) Methods

Fixed-point methods, sometimes called open methods, are the most popular and useful class of methods for solving nonlinear algebraic equations, as they remove all of the complications inherent in the bracketing methods. Let's illustrate the idea using a single equation, with the aim to generalize the concept to systems of nonlinear algebraic equations in Section 3.7. In a fixed point method, the idea is to rewrite $f(x) = 0$ as

$$x = g(x) \tag{3.2.6}$$

We then use this as an update scheme. If we start from some initial guess $x^{(0)}$, at each step we compute $x^{(k+1)}$ from the old value $f(x^{(k)})$ by

$$x^{(1)} = g\left(x^{(0)}\right)$$
$$x^{(2)} = g\left(x^{(1)}\right)$$
$$\vdots \quad \vdots$$
$$x^{(k+1)} = g\left(x^{(k)}\right) \tag{3.2.7}$$

There are two issues to address here. The obvious question is how to determine the function $g(x)$, which we will address shortly. For the moment, let's answer the second question, namely how to figure out when we should stop our application of Eq. (3.2.7) to generate new values of $x^{(k+1)}$. If $x^{(k+1)} \rightarrow x^{(k)}$, then we say that the method is converging. To be more precise, if

$$\lim_{k \to \infty} x^{(k)} \rightarrow x^* \tag{3.2.8}$$

then

$$x^* = g\left(x^*\right) \tag{3.2.9}$$

and

$$f\left(x^*\right) = 0 \tag{3.2.10}$$

In this case x^* is a root. It is also called a *fixed point* because, if you insert x^* into $g(x)$, you get back x^* by Eq. (3.2.9).

Recall that we saw the same philosophy in the iterative solution of systems of linear algebraic equations in Section 2.13. As we did for those methods, we need to specify some termination criteria in any numerical program since we will never exactly satisfy Eq. (3.2.10). One possible criterion is that the value of x has not changed much between successive iterations, i.e. that

$$\left| \frac{x^{(k+1)} - x^{(k)}}{x^{(k+1)}} \right| < \epsilon_1 \tag{3.2.11}$$

where ϵ_1 is a small number. The risk with this criterion is that we may get a very small change away from the actual root over a given iteration if convergence is very slow. Another option is to use the residual (the value of the function itself) to verify that we are sufficiently close to the root,

$$\left| f\left(x^{(k)}\right) \right| < \epsilon_2 \tag{3.2.12}$$

where ϵ_2 is another small number. This criterion is usually easy to check since, for the methods we will discuss, you have to evaluate the function anyway in a fixed point method.

Example 3.4   Construct a fixed-point method to find the root of the function

$$f(x) = e^{-x} - x \tag{3.2.13}$$

using an initial guess $x^{(0)} = 1$.

Solution

A natural way to write this in the form of Eq. (3.2.6) is

$$x^{(k+1)} = e^{-x^{(k)}} \tag{3.2.14}$$

If we use an initial guess of $x^{(0)} = 1$, then we see in Fig. 3.4 that the solution eventually converges to $x^* = 0.566$, which is indeed a root.

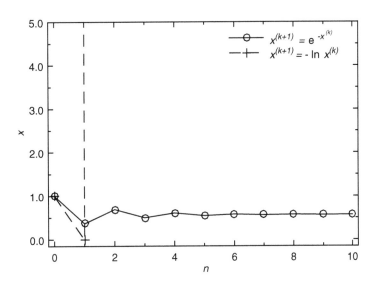

Figure 3.4 Evolution of the fixed-point method in Eq. (3.2.14) and Eq. (3.2.15) for $x^{(0)} = 1$. The vertical dashed line indicates where the iteration using $x^{(k+1)} = -\ln x^{(k)}$ exceeds the bounds of the y-axis.

While Eq. (3.2.14) seems like an obvious choice to set up the fixed-point method, it is not the only choice. Indeed, the function $g(x)$ is not unique, which gives us some opportunities to be creative with our solution method. For example, we could have written Eq. (3.2.13) as

$$x^{(k+1)} = -\ln x^{(k)} \tag{3.2.15}$$

This particular choice of the function has some interesting properties. Obviously, our first problem is that we need to be careful with our initial guess $x^{(0)}$ so that we get a real, finite value for $x^{(1)}$. Even more importantly, we see from Fig. 3.4 that the fixed-point method in Eq. (3.2.15) does not even converge!

3.3   Fixed-Point Methods for Single Nonlinear Algebraic Equations

While we have outlined an intuitive approach to develop fixed point methods, we naturally would prefer to have a more systematic method. Moreover, we definitely want to know if we are going to run into problems with the solution similar to what we saw in Eq. (3.2.15). In general, whether the iteration will converge or not, how fast, and whether there will be monotonic or oscillatory convergence will depend on the function $g(x)$ that we use and the initial guess.

In this section, we will explore Picard's method and Newton's method, which are two ways to systematically pick the function $g(x)$. Once we have a thorough understanding of how these methods work for a single equation, we will move onto the much more challenging problem of solving a system of nonlinear equations.

3.3.1   Picard's Method

Picard's method, which is also known as the method of successive substitution, involves simply adding x to both sides of $f(x) = 0$. The iterative scheme is

$$x^{(k+1)} = f(x^{(k)}) + x^{(k)} \tag{3.3.1}$$

Picard's method is exactly what we did in Example 3.4. It is very easy to implement, but it may be very slow to converge (more on this shortly).

PROGRAM 3.2   *Picard's method is very simple to implement as a program:*

```
1  function [x,flag] = nonlinear_picard(x,tol)
2  %  x = initial guess for x
3  %  tol = criteria for convergence
4  %  requires function f = getf(x)
5  %  flag = 0 if failed, 1 if converged
6  f = getf(x);
7  k = 0; %counter
8  kmax = 20; %max number of iterations allowed
9  while abs(f) > tol
10      x = x + f; %Picard's method
11      k = k + 1; %update counter
12      f = getf(x); %f for while loop and counter
13      if k > kmax
14          fprintf('Did not converge.\n')
15          break
16      end
17  end
18  if k > kmax || x == Inf || x == -Inf
19      flag = 0;
20  else
21      flag = 1;
22  end
```

This program requires two inputs: (i) the initial guess for the root, $x^{(0)}$; and (ii) the distance you want to be from the root at the end, which is the tolerance variable tol. *Notice that we use the stop criterion in Eq. (3.2.12) since we need to calculate $f(x)$ anyway for Picard's method. The program will return two variables in a vector. The first entry is the value of $x^{(k+1)}$ at the end of the program. The second entry is a variable called* flag *that tells you whether or not the program found a root. We have hard-coded in the maximum number of iterations as* kmax = 20, *which is a very large number if we give a reasonable guess. If Picard's method keeps going past* kmax *iterations, it will return a flag value of 0. We also return a flag value of 0 if x exceeds the maximum value of a floating point number on the computer, which can happen before we reach* kmax *iterations. If we use this function in another program, we need to remember to check the flag value to see if the result that came back is garbage. In many cases, Picard's method will diverge and the value of x will eventually exceed the maximum value allowed by the computer.*

The program also assumes that there is a function getf(x) *also in the working directory. (Alternatively, you can cut-and-paste this function into the same m-file.) For example, if we wanted to use Picard's method to solve Eq. (3.2.13), we would need to also have the function*

```
1  function f = getf(x)
2  f = exp(-x) - x;
```

Indeed, this is exactly what we did to generate the data in Fig. 3.4 for Eq. (3.2.14).

You might be wondering why we felt the need to write a second function `getf(x)` *for this problem, since we could have just replaced line 10 of* `nonlinear_picard` *with* `x = exp(-x)`. *Our approach here is an example of modular programming, where we have created a module for performing Picard's method for any function $f(x) = 0$. If we want to find the roots of a different function, we just need to change the function* `getf(x)`. *The concept of modular programming is a crucial one to numerical methods, and you will see it throughout this book. Indeed, when we find the need to solve nonlinear equations later in the book, we will write programs that call the functions in this chapter as modules.*

Example 3.5   Use Picard's method to find the roots of

$$f(x) = x^2 - 3x + 2 = 0 \qquad (3.3.2)$$

for the initial guesses -0.1, 0.6, 1.99, and 2.01.

Solution
The roots are $x^* = 1$ and $x^* = 2$, which we can obtain quicky from factorization. In Picard's method we have

$$x^{(k+1)} = x^{(k)} + \left[(x^{(k)})^2 - 3x^{(k)} + 2 \right] = (x^{(k)})^2 - 2x^{(k)} + 2 \qquad (3.3.3)$$

To see what happens, we could just pull out our calculators and successively apply Eq. (3.3.3) many times. However, it would be much nicer to use Program 3.2 with the additional function

```
1  function f = getf(x)
2  f = x^2-3*x+2;
```

For this problem, we also modified Program 3.2 to stop after ten iterations because there are some problems with the solution. You can see the results of the calculation in Fig. 3.5. For $x^{(0)} = -0.1$, you might have expected to converge on $x^* = 1$, but the solution diverges! Likewise, for $x^{(0)} = 2.01$, you might have expected to converge on $x^* = 2$, but the solution again diverges. For $x^{(0)} = 0.6$, the solution nicely converges to $x^* = 1$. Perhaps most interestingly, if you give this problem a guess very close to $x^* = 2$ but slightly less, such as the case $x^{(0)} = 1.99$, the solution converges to the other root, $x^* = 1$. We will revisit this problem in Section 3.4 to try to understand this strange behavior better.

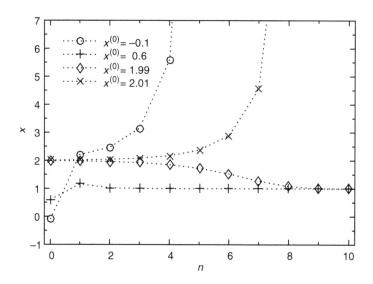

Figure 3.5 Result of up to ten iterations of Picard's method for Eq. (3.3.2) for the initial guesses $x^{(0)} = -0.1, 0.6, 1.99$, and 2.01.

3.3.2   Newton's Method

Picard's method is simple, but it has some serious problems with speed and convergence, which we explore in Section 3.4. Newton's method is by far the most common method for solving a single nonlinear equation. Newton's method is a fixed point method of the form of Eq. (3.2.6), but the function $g(x)$ is not obvious *a priori*. To derive the form of $g(x)$, let's first expand $f(x)$ in a Taylor series around $x^{(k)}$,

$$f(x) = f(x^{(k)}) + f'(x^{(k)})(x - x^{(k)}) + \frac{1}{2}f''(x^{(k)})(x - x^{(k)})^2 + \cdots \tag{3.3.4}$$

If x is a root, then $f(x) = 0$. Let's instead imagine now that x is close to the root, such that $x - x^{(k)}$ is a small number and $f(x) \approx 0$. In this case, the higher order terms become small since they depend on higher powers of a small number. Neglecting these higher-order terms and keeping just the linear ones, we have

$$0 \approx f(x^{(k)}) + f'(x^{(k)})(x - x^{(k)}) \tag{3.3.5}$$

We can then solve for x

$$x = x^{(k)} - \frac{f(x^{(k)})}{f'(x^{(k)})} \tag{3.3.6}$$

and use this solution to build the iterative scheme,

$$x^{(k+1)} = x^{(k)} - \frac{f(x^{(k)})}{f'(x^{(k)})} \tag{3.3.7}$$

This iterative scheme is called Newton's method, and the term $-f(x^{(k)})/f'(x^{(k)})$ acts as a correction to the current iterate. If we compare Eq. (3.3.7) to Eq. (3.2.6), the right-hand side of Eq. (3.3.7) can be thought of as the $g(x)$ function of the fixed-point iteration.

PROGRAM 3.3   *We can very easily revise our program for Picard's method to instead implement Newton's method.*

```
1   function [x,flag] = nonlinear_newton(x,tol)
2   %  x = initial guess for x
3   %  tol = criteria for convergence
4   %  requires function f = getf(x) and df = getdf(x)
5   %  flag = 0 if failed, 1 if converged
6   f = getf(x);
7   k = 0; %counter
8   kmax = 20; %max number of iterations allowed
9   while abs(f) > tol
10      df = getdf(x);
11      x = x - f/df; %Newton'ǍŽs method
12      k = k + 1; %update counter
13      f = getf(x); %f for while loop and counter
14      if k > kmax
15          fprintf('Did not converge.\n')
16          break
17      end
18  end
19  if k > kmax || x == Inf || x == -Inf
20      flag = 0;
21  else
22      flag = 1;
23  end
```

There are only two differences between this program and Program 3.2. In line 10 of this program, we call a subfunction getdf(x) *that computes the derivative for the current value of x. Line 11 then implements Newton's method (compare to line 10 of Program 3.2).*

Similar to Program 3.2, we have constructed our Newton's method in a modular fashion, where it calls the subsections getf *and* getdf *for the function and its derivative. These files need to be in the same working directory, or appended to the end of the Newton's method function.*

Example 3.6   Use Newton's method to find the root of Eq. (3.2.13) using an initial guess $x^{(0)} = 1$.

Solution

This is the Newton's method version of Example 3.4. If we go back to Eq. (3.2.13), Newton's method is

$$x^{(k+1)} = x^{(k)} - \frac{e^{-x^{(k)}} - x^{(k)}}{-e^{x^{(k)}} - 1} \tag{3.3.8}$$

For the numerical calculation, we used Program 3.3 with the additional functions

```
1   function f = getf(x)
2   f = exp(-x)- x;
```

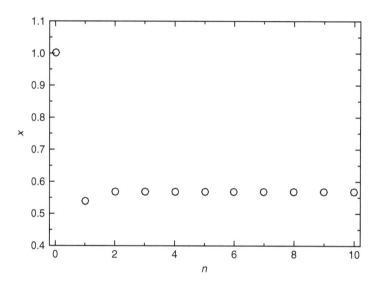

Figure 3.6 Convergence of Newton's method for Eq. (3.2.13) using an initial guess $x^{(0)} = 1$. Compare to Picard's method in Fig. 3.4.

and

```
1  function df = getdf(x)
2  df = -exp(-x) - 1;
```

Figure 3.6 shows the evolution of the solution using Newton's method. It is useful to compare the result to Picard's method in Fig. 3.4, as well as the other fixed point method that we derived in Eq. (3.2.15). Newton's method is clearly superior – not only does it converge, but it does so in much fewer iterations than Picard's method.

If we compare Newton's method to Picard's method in Eq. (3.3.1), it should be clear to you that Newton's method is harder to implement because you need to compute f'. This can be done analytically or numerically (using a finite difference approximation, which we will discuss in Chapter 6). The upside, however, is that this method tends to converge much more rapidly, as we saw in Example 3.6. The reduction in the number of iterations tends to be more important than the additional work required to compute $f'(x)$.

On balance, Newton's method is almost always preferred over Picard's. Let's see why this is so in the next example.

Example 3.7 Repeat Example 3.5 using Newton's method.

Solution

If we want to write out Newton's method exactly for this problem, we have

$$x^{(k+1)} = x^{(k)} - \frac{(x^{(k)})^2 - 3x^{(k)} + 2}{2x^{(k)} - 3} \tag{3.3.9}$$

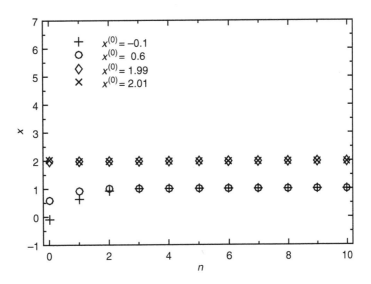

Figure 3.7 Result of Newton's method for Eq. (3.3.2) for the initial guesses $x^{(0)} = -0.1, 0.6, 1.99,$ and 2.01. Compare these results to Fig. 3.5 using Picard's method.

However, it is again better to just use the generic program we wrote for Newton's method (Program 3.3) with the function

```
1  function f = getf(x)
2  f = x^2-3*x+2;
```

for $f(x)$ and the function

```
1  function df = getdf(x)
2  df = 2*x-3;
```

for its derivative.

The results of this calculation are presented in Fig. 3.7. As we can see, all of the initial guesses now converge nicely to one of the roots. Moreover, since the function $f(x)$ is very smooth, we converge to the nearest root.

Newton's method has a nice geometric interpretation, which we illustrate in Fig. 3.8. Starting from an initial guess $x^{(0)}$, $x^{(1)}$ is obtained as the "zero" of the linearized function at $x^{(0)}$. Graphically, we are drawing a tangent to the curve $f(x)$ at $x = x^{(0)}$ and then following that line until it crosses the x-axis. In general, we correct each estimate by moving in the direction of the tangent of the function $f(x)$ evaluated at the current estimate.

Let's see how this geometric interpretation affects the results that we have studied so far in an example.

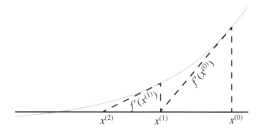

Figure 3.8 Geometric interpretation of Newton's method.

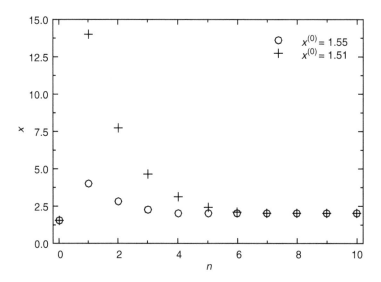

Figure 3.9 Result of Newton's method for Eq. (3.3.2) for the initial guesses $x^{(0)} = 1.55$ and 1.51.

Example 3.8   Investigate the behavior of Newton's method for the previous example, Eq. (3.3.2), for initial guesses near $x = 1.5$.

Solution

This is the dangerous region for Newton's method, since the derivative of the function vanishes at $x = 1.5$. As a result, we can get a value of $x^{(1)}$ that is very large due to the second term in Eq. (3.3.9). However, for such a simple function, Newton's method still behaves quite nicely. As we see in Fig. 3.9, the error initially gets quite large. However, once the solution has moved away from the singularity at $f' = 0$, we still get nice convergence back to a root. Indeed, even the very bad guess $x^{(0)} = 1.51$ will converge in eight iterations.

3.4   Convergence of General Fixed-Point Iteration

In the particular example that we studied in Examples 3.4 and 3.6, it was clear that the convergence of Newton's method was much faster than Picard's method. Moreover, we saw that it is possible to come up with a fixed point method such as Eq. (3.2.15), start doing calculations, and never find the root!

We will now make these assertions more concrete for the two methods we discussed. In general there are two questions that we want to ask:

(1) Will we find a solution?
(2) If so, how fast will we find it?

Let's address these two questions in turn.

3.4.1 Will We Find a Solution?

We will first examine conditions for convergence of the general fixed-point iteration scheme. Consider a function $f(x) = 0$ which we have rearranged to derive the iteration formula

$$x^{(k+1)} = g(x^{(k)}) \tag{3.4.1}$$

Let x^* denote the exact solution such that

$$f(x^*) = 0 \tag{3.4.2}$$

Since x^* is a fixed point, we also have that

$$x^* = g(x^*) \tag{3.4.3}$$

Let us subtract the fixed point (3.4.3) from the iterative scheme (3.4.1),

$$x^{(k+1)} - x^* = g(x^{(k)}) - g(x^*) \tag{3.4.4}$$

This can be written equivalently as

$$x^{(k+1)} - x^* = \frac{g(x^{(k)}) - g(x^*)}{x^{(k)} - x^*}(x^{(k)} - x^*) \tag{3.4.5}$$

Recall now the mean value theorem of calculus illustrated in Fig. 3.10: if $g(x)$ and $g'(x)$ are continuous over $[a, b]$, then there exists at least one value $\xi \in [a, b]$ such that

$$g'(\xi) = \frac{g(b) - g(a)}{b - a} \tag{3.4.6}$$

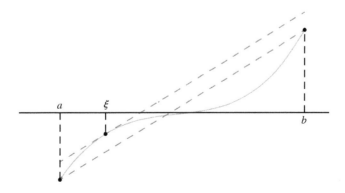

Figure 3.10 Illustration of the mean value theorem of calculus.

For some value on the interval, we can find a tangent to the curve that is the same as the "rise-over-run" for the interval, which are the two dashed lines in Fig. 3.10. Note that the mean value theorem does not provide the value of ξ, just an assurance that there exists some value ξ that satisfies Eq. (3.4.6).

For our problem in Eq. (3.4.5), we recognize that the correspondence with the mean value theorem is $a \to x^{(k)}$ and $b \to x^*$. As a result, we can use the mean value theorem of calculus to say that there exists at least one value $\xi_k \in (x^{(k)}, x^*)$ such that

$$x^{(k+1)} - x^* = g'(\xi_k)(x^{(k)} - x^*) \qquad (3.4.7)$$

If we now define the error in the solution as the distance from the root,

$$e_j = x^{(j)} - x^* \qquad (3.4.8)$$

we recognize that $x^{(k+1)} - x^*$ is the error for $j = k + 1$ and $x^{(k)} - x^*$ is the error for $j = k$. We then get

$$e_{k+1} = g'(\xi_k)e_k, \quad \xi_k \in (x^{(k)}, x^*) \qquad (3.4.9)$$

and taking absolute values, we arrive at the key result

$$|e_{k+1}| = |g'(\xi_k)||e_k|, \quad \xi_k \in (x^{(k)}, x^*) \qquad (3.4.10)$$

We can use these relations to determine several features about the nature of the convergence of the method. Most importantly, if $|g'(x)| < 1$ for all x in an interval about the solution x^*, then if $x^{(0)}$ is chosen within this interval, the iteration will converge. This can be seen easily from Eq. (3.4.10). For example, let's assume that $|g'(x)| = 0.1$ over the whole interval (and not worry about how such a constant value arises). Then $e_{k+1} = 0.1e_k$, $e_{k+2} = 0.01e_k$, $e_{k+3} = 0.001e_k$, and so forth. The error will decrease in each iteration and ultimately

$$\lim_{k \to \infty} e_k = 0 \qquad (3.4.11)$$

This result is often referred to as *contraction mapping*. Note that it is just a sufficient condition and it is also a local result. This means that only initial conditions in the interval where the $|g'(x)| < 1$ are guaranteed to result in convergence.

Contraction mapping for nonlinear equations, which is a local result, is markedly different from the condition for convergence that we stated in Section 2.13.4 for linear equations, which was valid independent of the initial guess. This local feature is typical of nonlinear systems and highlights the importance of choosing good initial guesses.

Equation (3.4.9) can also help us infer additional features of the nature of the convergence. When we are close to the root, $|g'(x)|$ is small. If we move a bit further away from the root for a condition $0 < g'(x) < 1$, the error decreases monotonically. In contrast, if we are similarly far from the root but $-1 < g'(x) < 0$, the errors alternate in sign, i.e. we have oscillatory convergence.

One final comment regarding the condition for convergence that we derived. It is elegant and powerful, but since we do not know the solution x^* *a priori*, it is also an uncheckable solution! In practice, we could check that $|f(x^{(k+1)})| < |f(x^{(k)})|$ as the iterations proceed.

Application to Picard's Method

The previous analysis can be adopted directly to analyze the convergence of Picard's method. Recall that from Eq. (3.3.1), the function $g(x)$ is

$$g(x) = x + f(x) \qquad (3.4.12)$$

The derivative of g is then

$$g'(x) = 1 + f'(x) \qquad (3.4.13)$$

and the convergence condition becomes $f'(x) < 0$ and $f'(x) > -2$.

Example 3.9 Use contraction mapping to explain the behavior of the solution to Example 3.5.

Solution
The fixed-point method is

$$g(x) = x^2 - 2x + 2 \qquad (3.4.14)$$

The derivative is then

$$g' = 2x - 2 \qquad (3.4.15)$$

If we want the absolute value to be less than 1,

$$|2x - 2| < 1 \qquad (3.4.16)$$

and we get the result

$$\frac{1}{2} < x < \frac{3}{2} \qquad (3.4.17)$$

Interestingly, the root $x^* = 2$ itself is outside this interval that guarantees convergence. (The other root $x^* = 1$ is fine.) It is also interesting that the negative initial guess fails to find the root $x^* = 1$, or that the initial guess $x^{(0)} = 1.99$ ends up at $x^* = 1$ instead of just going off to infinity. The condition for convergence based on contraction mapping cannot be used to explain these cases definitively, as it is a sufficient condition only and a local result. However, we can get insights on what is happening by observing the numerical solution, which is what we are doing in Fig. 3.5. For this problem, it turns out that solutions that go past $x = 2$ diverge, but those that stay below $x = 2$ will converge to $x^* = 1$. The problem with the negative guess is that the first step of Picard's method sends the solution past $x = 2$, whereupon the solution diverges.

Application to Newton's Method

When we write Newton's method (3.3.7) in the form of a fixed-point method, we have

$$g(x) = x - \frac{f(x)}{f'(x)} \qquad (3.4.18)$$

The derivative of $g(x)$ is therefore

$$g' = 1 - \frac{(f')^2 - ff''}{(f')^2} = \frac{ff''}{(f')^2} \tag{3.4.19}$$

Assuming that $f'(x^*) \neq 0$ and that in an interval about x^*, $f'(x) \neq 0$ and $|g'(x)| < 1$, the method will converge if the initial guess is in this interval. This is another local convergence result – away from the root anything is possible!

The convergence criterion for Newton's method in Eq. (3.4.19) has an intriguing twist. Since $f(x^*) = 0$ at the root x^*, then necessarily $|g'(x^*)| = 0$. (We also need to be careful that $f \to 0$ more quickly than $f' \to 0$ if the derivative vanishes at the root.) So by virtue of the continuity of the function g', there will necessarily be a neighborhood of x^* such that $|g'(x)| < 1$. In other words, the iteration will always converge if the initial guess (or an iterate) is sufficiently close to the root. This is an important intrinsic property of Newton's method that makes it very appealing.

3.4.2   Rate of Convergence

We complete our discussion of convergence by deriving important results about the rate of convergence for our two methods. Once we know if the method will converge, it is then worth considering how fast you get to the solution.

Recall the equation that we derived relating the errors between successive iterations (3.4.9),

$$e_{k+1} = g'(\xi_k)e_k, \quad \xi_k \in [x^{(k)}, x^*] \tag{3.4.20}$$

As $x^{(k)} \to x^*$, we also have that $g'(\xi_k) \to g'(x^*)$. In other words the ξ_ks are squeezed in smaller and smaller intervals around the solution x^*, and $g'(\xi_k)$ is almost constant. This suggests that close to a root, the error at the $(k + 1)$th iteration is proportional to the error at the kth iteration. This is called linear convergence, and is considered slow and a disadvantage of Picard's method. In the case of Newton's method, the condition $|g'(x^*)| = 0$ implies by continuity that $|g'(x)|$ is very small close to x^*, which in turn suggests a fast rate of convergence.

Let's be more quantitative about the convergence of Newton's method. Recall the Taylor series expansion that we used in our derivation,

$$f(x) = f(x^{(k)}) + f'(x^{(k)})(x - x^{(k)}) + \frac{1}{2}f''(x^{(k)})(x - x^{(k)})^2 + \cdots \tag{3.4.21}$$

For $x = x^*$, we get $f(x^*) = 0$ and

$$0 = f(x^{(k)}) + f'(x^{(k)})(x^* - x^{(k)}) + \frac{1}{2}f''(x^{(k)})(x^* - x^{(k)})^2 + \cdots \tag{3.4.22}$$

To derive Newton's method we kept only the linear term to obtain $x^{(k+1)}$,

$$0 = f(x^{(k)}) + f'(x^{(k)})(x^{(k+1)} - x^{(k)}) \tag{3.4.23}$$

If we subtract Newton's method (3.4.23) from the Taylor series evaluated at the root (3.4.22), we get

$$0 = f'(x^{(k)})(x^* - x^{(k+1)}) + \frac{1}{2}f''(x^{(k)})(x^* - x^{(k)})^2 + \cdots \qquad (3.4.24)$$

Neglecting the higher-order terms and rearranging gives

$$x^* - x^{(k+1)} = -\frac{f''(x^{(k)})}{2f'(x^{(k)})}(x^* - x^{(k)})^2 \qquad (3.4.25)$$

With the definition for the error (3.4.8), we can write this as

$$e_{k+1} = \frac{f''(x^{(k)})}{2f'(x^{(k)})}e_k^2 \qquad (3.4.26)$$

As we converge, $x^{(k)} \to x^*$, and the error at step $k+1$ becomes proportional to the square of the error at step k. This is called quadratic convergence and is a huge advantage of Newton's method, as the convergence accelerates rapidly near the solution. This feature is often used as a test of correct implementation of the method.

Example 3.10   Demonstrate that Newton's method exhibits quadratic convergence for the problem in Example 3.7.

Solution
To demonstrate quadratic convergence, we first need to know the value of the roots and then compute the distance from the roots at each iteration. For this simple quadratic equation, we already know the roots. (Indeed, there is no reason to use a numerical method to solve this problem other than to investigate the properties of the numerical solution.) So once we know which root is obtained for a given initial guess, we can easily compute the error at each iteration.

Figure 3.11 shows the error for the different initial guesses as a function of the iteration number. Note that, to look for a quadratic scaling, we should use a log–log plot and test the slope. (If you don't remember this point, take a look back at linear regression of power laws in Section 2.14.) When we are far from the root, the local result for convergence does not hold. Indeed, we already saw for the initial guesses near $x = 1.5$ that the error goes up at the start of the solution before eventually decaying. However, once the solution gets close to the root, the decay is quadratic in all cases.

The last example may make you wonder how to demonstrate quadratic convergence in general. This is one of many "chicken-and-the-egg" type problems in numerical methods – how can I figure out the distance from the root during the solution if I don't know the value of the root in the first place? You have two options, both of which are just approximations. The first option is to compute the root numerically to some level of accuracy, store the value in the memory, and then use that value to determine the convergence. This method works, but you have to be careful since you are using a numerical

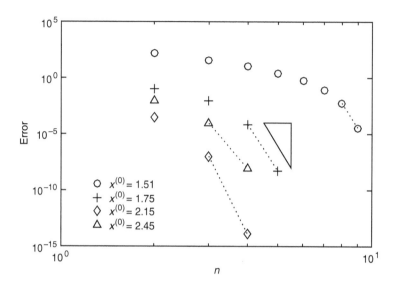

Figure 3.11 Demonstration of quadratic convergence for Newton's method using the solutions from Example 3.7. The triangle indicates a slope of 2 on the log–log plot.

result to test a numerical result. Obviously, if you use the same number of iterations to test for quadratic convergence that you used to compute the root in the first place, you will end up with zero error at the end. This will be a problem when you try to make the log–log plot, since the logarithm of the error will diverge! The second option is to use the value of $f(x^{(k+1)})$ as a "proxy" for the error in the solution. While the quadratic convergence of Newton's method only applies to $x^{(k+1)} - x^*$, you will normally see that the value of $f(x^{(k+1)})$ tends to exhibit quadratic convergence as well.

3.5   Case Study: Eigenvalue Calculation in Unsteady Diffusion

3.5.1   Problem Statement

In Section 1.2.6, we discussed the problem of unsteady diffusion with constant boundary conditions. Let's consider a slightly different problem here where we allow for a chemical reaction at the right surface. This is going to complicate the problem in two important ways. First, from a physical standpoint, we cannot arbitrarily set the reaction rate and still achieve a steady state since the flux from the reaction must be balanced by the flux on the left side. Second, from a mathematical standpoint, we are not going to be able to compute the eigenvalues of the diffusion equation analytically. This is qualitatively different from the problem in Section 1.2.6.

Let us begin by stating the equations we need to solve. The transport equation inside the boundary is still the unsteady diffusion equation,

$$\frac{\partial c}{\partial t} = D \frac{\partial^2 c}{\partial x^2} \qquad (3.5.1)$$

where D is the solute's diffusion coefficient. Let's fix the concentration at the left boundary to zero,

$$c(0, t) = 0 \tag{3.5.2}$$

At the other boundary, $x = L$, we set the diffusive flux to be equal to the rate of production,

$$D \left. \frac{\partial c}{\partial x} \right|_{x=L} = kc \tag{3.5.3}$$

In the latter, k is the rate of reaction.

The goal in this problem is to determine the time scale over which the concentration profile evolves from its initial profile $c(x, 0)$ to the non-trivial steady state. To do so, it will prove convenient to rewrite the equation in a dimensionless form with the dimensionless variables

$$\kappa = \frac{c}{c_0}, \quad \xi = \frac{x}{L}, \quad \tau = \frac{Dt}{L^2} \tag{3.5.4}$$

where c_0 is some reference concentration. Equation (3.5.1) becomes

$$\frac{\partial \kappa}{\partial \tau} = \frac{\partial^2 \kappa}{\partial \xi^2} \tag{3.5.5}$$

with the left boundary condition (3.5.2) now

$$\kappa(0, \tau) = 0 \tag{3.5.6}$$

and the right boundary condition (3.5.3) becoming

$$\left. \frac{\partial \kappa}{\partial \xi} \right|_{\xi=1} = \phi^2 \kappa \tag{3.5.7}$$

where

$$\phi = \sqrt{\frac{kL^2}{D}} \tag{3.5.8}$$

is the Thiele modulus. In dimensionless form, we can restate the problem as determining the time scale for the transients for $\phi = 1$, which is the only case where a non-trivial steady state exists.

3.5.2 Solution

This problem can be solved by separation of variables. If we follow the approach in Section 1.2.6, we have the standard result

$$f(\xi) = c_1 \cos(\lambda \xi) + c_2 \sin(\lambda \xi) \tag{3.5.9}$$

The left boundary condition (3.5.6) requires that $c_1 = 0$, so the solution is a sine series. The right boundary condition gives

$$c_2 \lambda \cos(\lambda) = c_2 \sin(\lambda) \tag{3.5.10}$$

which leads to the eigenvalue equation

$$\sin\lambda - \lambda\cos\lambda = 0 \tag{3.5.11}$$

One solution is $\lambda = 0$, which just corresponds to the steady state, $\kappa_s(x)$. The solution to the problem is then

$$\kappa = \kappa_s(x) + \sum_{i=1}^{\infty} c_n \sin(\lambda_n\xi)e^{-\lambda_n^2\tau} \tag{3.5.12}$$

where the constants c_n are determined from the initial condition.

Since the exponentials decay faster and faster as λ_n increases, the problem boils down to computing the smallest positive eigenvalue λ_1 from Eq. (3.5.11). We can find λ_1 by using Program 3.3 for Newton's method with the function

```
1  function f = getf(x)
2  f = sin(x) - x*cos(x);
```

and the derivative function

```
1  function df = getdf(x)
2  df = x*sin(x);
```

We know that one of the roots is at $\lambda = 0$, and we also know that there are infinitely many roots. So we need to be sure to pick a good initial guess that is near the first positive root. Using a good guess $\lambda_1^{(0)} = 4$, we find that the converged solution is $\lambda_1^* = 4.4934$.

If we want to think about this result in dimensional terms, what we are saying is that the initial condition tends to disappear on a time scale

$$t^* \approx \frac{L^2}{\lambda_1^2 D} = 0.0495\frac{L^2}{D} \tag{3.5.13}$$

For $t \gg t^*$, we expect that the system is close to the steady state. While it is common to think about the time scale for diffusion problems to be L^2/D, the addition of the boundary reaction plays an important role here – the prefactor for this scaling is very small!

3.6   Case Study: Extension of a Tethered Polymer by a Force

3.6.1   Problem Statement

Polymer chains form random coils in solution to maximize their configurational entropy. As a result, we need to apply an external force to a polymer to stretch it out. This idea is illustrated in Fig. 3.12. In the simplest model, we can assume that the chain follows ideal random walk (Gaussian) statistics. As a result, the force F required to extend the chain to a distance X is

$$F = k_s X \tag{3.6.1}$$

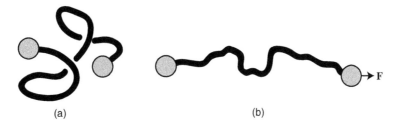

(a)   (b)

Figure 3.12 Illustration of the force–extension of a polymer chain.

You may recognize this as a Hookean spring with a spring constant k_s. In the case of polymers, this spring constant is related to the entropy lost by the chain extension,

$$k_s = \frac{3k_BT}{Lb} \tag{3.6.2}$$

where k_B is Boltzmann's constant (and hence the connection to entropy), T is the temperature, L is the total length of the chain, and b is the Kuhn length that represents one step of the random walk. The segment length depends on the chemistry, but the rest of this model is rather generic.

You may see a problem with the Gaussian spring model. Let's first rewrite the problem in a dimensionless form,

$$\frac{Fb}{k_BT} = \frac{3X}{L} \tag{3.6.3}$$

The maximum possible extension of the chain is $X/L = 1$. As a result, applying a force greater than

$$F^* = \frac{3k_BT}{b} \tag{3.6.4}$$

leads to the chain being extended beyond its contour length! This is physically impossible. In many cases of interest, the forces acting on the polymers are small, and the Gaussian model works just fine. However, for polymers in flow (which is important for material processing) this is an issue.

There are a number of approaches to deal with the so-called "finite elasticity" problem for polymers. Let's focus on DNA, which has a Kuhn length $b = 100$ nm in a typical biological fluid. From the Gaussian model, this means that DNA would reach its maximum extension for a force

$$F^* = \frac{3(1.38 \times 10^{-23}\,\text{J/mol K})(298\ \text{K})}{10^{-7}\ \text{m}} = 0.12\ \text{pN} \tag{3.6.5}$$

which actually is a very small force in biological situations, such as DNA transcription. John Marko and Eric Siggia developed a famous formula for the actual force–extension behavior of DNA, based on a wormlike chain model, which has the form

$$\frac{Fb}{k_BT} = \frac{1}{2}\left(1 - \frac{X}{L}\right)^{-2} - \frac{1}{2} + \frac{2X}{L} \tag{3.6.6}$$

and is a remarkably good fit for the force–extension behavior of DNA up to around 60 pN of force. (Above 60 pN, DNA undergoes a structural transition.) In this problem, we would like to write a program that allows us to compute the extension of λ-DNA, a virus DNA, with $L = 16.8\ \mu$m, for a given value of the force.

3.6.2   Solution

The first thing we need to do is convert the force to dimensionless form. To do this, we can write a simple program that takes in the force in pN and provides the dimensionless force in Eq. (3.6.6).

```
1  function F = convertF(F)
2  %  converts F from pN to Fb/kT
3  k = 1.38e-23; %Boltzmann constant in J/mol K
4  T = 298; %temperature in K
5  b = 100e-9; %segment length in m
6  FN = F*1e-12; %force in N
7  F = FN*b/k/T;
```

The problem is then to find the root of the force–extension curve written in the form

$$f(X/L) = \frac{Fb}{k_BT} - \frac{1}{2}\left(1 - \frac{X}{L}\right)^{-2} + \frac{1}{2} - \frac{2X}{L} \tag{3.6.7}$$

This means we can write a file for $f(x)$ like this one:

```
1  function f = getf(x,F)
2  f = F - 0.5*(1-x)^(-2) + 0.5 - 2*x;
```

We also need to compute the derivative of Eq. (3.6.7) with respect to the unknown X/L,

$$f'(X/L) = -\left(1 - \frac{X}{L}\right)^{-3} - 2 \tag{3.6.8}$$

which is easily coded into MATLAB.

```
1  function df = getdf(x,F)
2  df = -(1-x)^(-3) - 2;
```

Since the parameter Fb/k_BT is passed to both getf and getdf, we need to make a small modification to the Newton's method in Program 3.3 to send the parameter:

```
1  function [x,flag] = nonlinear_newton_DNA(x,tol,F)
2  %  x = initial guess for x
3  %  tol = criteria for convergence
4  %  requires function f = getf(x) and df = getdf(x)
5  %  flag = 0 if failed, 1 if converged
6  f = getf(x,F);
7  k = 0; %counter
8  kmax = 20; %max number of iterations allowed
```

```
 9  while abs(f) > tol
10      df = getdf(x,F);
11      x = x - f/df; %Newton's method
12      k = k + 1; %update counter
13      f = getf(x,F); %f for while loop and counter
14      if k > kmax
15          fprintf('Did not converge.\n')
16          break
17      end
18  end
19  if k > kmax || x == Inf || x == -Inf
20      flag = 0;
21  else
22      flag = 1;
23  end
```

Note that this new Newton's method function takes in the parameter as an input, called F, and sends it to getf and getdf whenever these functions are invoked.

That's all we need to do the calculation, provided we give a decent initial guess. An easy way to generate a good guess is to plot the function and do bracketing by eye. For example, let's consider a force of 30 pN, which is way above the upper bound F^* for the Gaussian model in Eq. (3.6.5). Figure 3.13 shows Eq. (3.6.7) for this force.

From this figure, a guess of $X/L = 0.9$ seems reasonable. However, you need to be very careful! This function has a very shallow slope for most values of X/L, and then a very, very steep slope for large values of X/L to prevent overstretching the polymer. Figure 3.14 shows you how this can go horribly wrong. If you pick a value that is too small, then you will shoot off to infinity and have trouble converging. (Newton's method does not know that you cannot exceed $X/L = 1$.) Likewise, if you pick a value very close

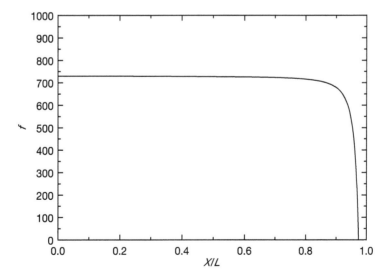

Figure 3.13 Force extension equation (3.6.7) for DNA with $F = 30$ pN.

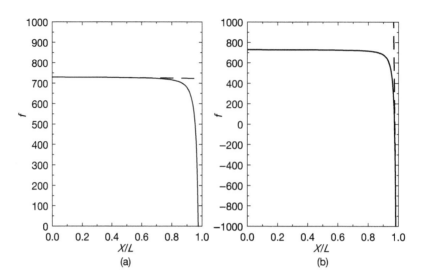

Figure 3.14 Examples of making the first step in Newton's method for (a) a small value of X/L and (b) a value of $X/L \approx 1$ for Fig. 3.13.

to $X/L = 1$, then the strong negative slope will send you back to the small values of X/L, and your solution will fail again.

To get a very good guess, we should zoom in on the force–extension curve to at least be in the right range to within a few decimal places. We found that $X/L = 0.97$ is a good guess, converging to $X/L = 0.9738$. If we were looking at the extension of λ-DNA at this force, it would be extended to 16.44 μm.

3.7   Systems of Nonlinear Algebraic Equations

If solving a single nonlinear equation is hard, you can only imagine how much more challenging it will be to solve systems of nonlinear equations. While the actual numerical solutions can be difficult, the mathematics required to generalize from single equations to systems of equations are straightforward.

We will now focus on systems of nonlinear algebraic equations of the general form

$$
\begin{aligned}
R_1(x_1, x_2, \ldots, x_n) &= 0 \\
R_2(x_1, x_2, \ldots, x_n) &= 0 \\
&\vdots \\
R_n(x_1, x_2, \ldots, x_n) &= 0
\end{aligned}
\tag{3.7.1}
$$

Just as we saw in the single equation case, we need to be careful to write the equations such that they are equal to zero in order to apply the fixed point methods. In what follows, it will be convenient for us to write the system of nonlinear equations in the compact notation

$$
\mathbf{R}(\mathbf{x}) = \mathbf{0}
\tag{3.7.2}
$$

with $\mathbf{x}$ being the vector of unknowns

$$\mathbf{x} = \begin{bmatrix} x_1 \\ x_2 \\ \vdots \\ x_n \end{bmatrix} \qquad (3.7.3)$$

and

$$\mathbf{R} = \begin{bmatrix} R_1(x_1, x_2, \ldots, x_n) \\ R_2(x_1, x_2, \ldots, x_n) \\ \vdots \\ R_n(x_1, x_2, \ldots, x_n) \end{bmatrix} \qquad (3.7.4)$$

a vector function which we will also refer to as the *residual*.

3.7.1   Picard's Method

Picard's iteration for this problem is a straightforward extension of the single-equation one in Eq. (3.3.1); instead of adding the scalar x to both sides of the equation $f(x) = 0$, we add the vector $\mathbf{x}$ to both sides of Eq. (3.7.2) to get the iterative scheme.

$$\mathbf{x}^{(k+1)} = \mathbf{x}^{(k)} + \mathbf{R}(\mathbf{x}^{(k)}) \qquad (3.7.5)$$

PROGRAM 3.4   *We can easily modify Program 3.2 from a single equation to a system of equations.*

```
1   function [x,flag] = nonlinear_picard_system(x,tol)
2   %   x = initial guess for x
3   %   tol = criteria for convergence
4   %   requires function R = getR(x)
5   %   flag = 0 if failed, 1 if converged
6   R = getR(x);
7   k = 0; %counter
8   kmax = 100; %max number of iterations allowed
9   while norm(R) > tol
10          x = x + R; %Picard'ÁŽs method
11          k = k + 1; %update counter
12          R = getR(x); %f for while loop and counter
13          if k > kmax
14                  fprintf('Did not converge.\n')
15                  break
16          end
17   end
18   if k > kmax || max(x) == Inf || min(x) == -Inf
19          flag = 0;
20   else
21          flag = 1;
22   end
```

If you compare this program to Program 3.2, you will notice that the key difference is that we have changed the function `f = getf(x)` *to* `R = getR(x)`. *This is solely* *for aesthetic reasons – we prefer to use $f(x)$ to denote a single equation and $\mathbf{R}(\mathbf{x})$ to denote a system of equations. Since we made this notational change, we also needed to change all of the references to f in Program 3.2 to R. Furthermore, to determine convergence, we also need to switch from evaluating the absolute value of a function, $|f(x)|$, to computing the residual of the norm, $||\mathbf{R}||$. Recall from Section 2.3.2 that the norm provides a measure of the size of a vector, which is the quantity we want to drive to zero during the iterative method.*

Picard's method is simple, but rarely used for the same reasons we found it unattractive for the single equation case (poor convergence properties). Let's see why with a simple example.

Example 3.11   Use Picard's method to find a root of the system of equations

$$f_1(x_1, x_2) = e^{-x_1} - x_2 \tag{3.7.6}$$

$$f_2(x_1, x_2) = x_1 + x_2^2 - 3x_2 \tag{3.7.7}$$

using the initial guesses $x_1^{(0)} = 0$ and $x_2^{(0)} = 0$.

Solution
For Picard's method, we have

$$x_1^{(k+1)} = x_1^{(k)} + e^{-x_1^{(k)}} - x_2^{(k)} \tag{3.7.8}$$

$$x_2^{(k+1)} = x_2^{(k)} + x_1^{(k)} + (x_2^{(k)})^2 - 3x_2^{(k)} \tag{3.7.9}$$

Let's do one iteration by hand. With the initial guess, we get

$$x_1^{(1)} = 0 + e^0 - 0 = 1 \tag{3.7.10}$$

$$x_2^{(1)} = 0 + 0 + (0)^2 - 3(0) = 0 \tag{3.7.11}$$

The 2-norm of the residual after the first iteration is

$$||\mathbf{R}||_2 = \sqrt{(e^{-1})^2 + 1^2} = 1.065 \tag{3.7.12}$$

To go further, it is useful to use MATLAB. We used Program 3.4 with the residual function

```
1  function R = getR(x)
2  R = zeros(2,1);
3  R(1) = exp(-x(1))- x(2);
4  R(2) = x(1) + x(2)^2 - 3*x(2);
```

As we can see in Fig. 3.15, the solution eventually reaches a root at $x_1 = 0.982753$ and $x_2 = 0.374279$. However, it takes quite some time – 32 iterations to reach a residual with $||\mathbf{R}|| < 10^{-9}$. Moreover, the solution starts getting worse at the first few iterations before it moves towards a root.

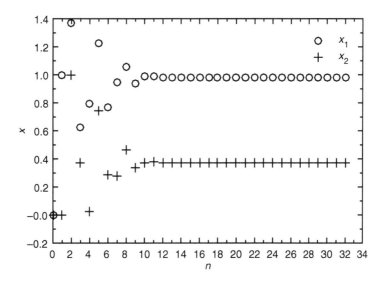

Figure 3.15 Convergence of Picard's method for Eqs. (3.7.6) and (3.7.7).

Example 3.12 Now try to solve the equations from Example 3.11 using the initial guesses $x_1^{(0)} = 0$ and $x_2^{(0)} = 0$, but write the equations in the opposite order,

$$f_1(x_1, x_2) = x_1 + x_2^2 - 3x_2 \qquad (3.7.13)$$

$$f_2(x_1, x_2) = e^{-x_1} - x_2 = 0 \qquad (3.7.14)$$

Solution

To make this change, all we need to do is rewrite the function `getR`,

```
1  function R = getR(x)
2  R = zeros(2,1);
3  R(2) = exp(-x(1))- x(2);
4  R(1) = x(1) + x(2)^2 - 3*x(2);
```

The change is hard to see – we just changed the entry `R(1)` on line 3 to `R(2)`, and vice versa on line 4.

As we can see in Fig. 3.16, the iterative method fails to converge even though we gave the same initial guess. This sensitivity to the order of the equations, along with the slow convergence even when it works, makes Picard's method effectively useless for solving systems of equations.

3.7.2   Newton–Raphson

The method of choice for solving nonlinear systems of equations is called Newton–Raphson, and it relies on the same idea as Newton's method but now generalized to

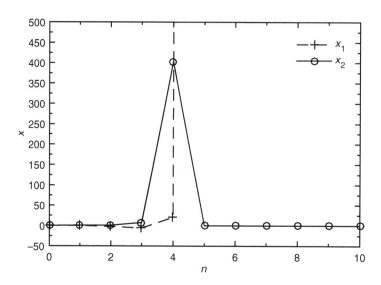

Figure 3.16 Lack of convergence of Picard's method for Eqs. (3.7.6) and (3.7.7) when the equation order is reversed. The vertical dashed line indicates the point where x_1 exceeds the bounds of the y-axis.

n-dimensions. To derive Newton–Raphson, we start again by expanding each function in $\mathbf{R}$ in a (multi-variable) Taylor series about an iterate,

$$
\mathbf{x}^{(k)} = \begin{bmatrix} x_1^{(k)} \\ x_2^{(k)} \\ \vdots \\ x_n^{(k)} \end{bmatrix}
\tag{3.7.15}
$$

Keeping only the linear terms, this gives

$$
\begin{aligned}
R_1(\mathbf{x}) &= R_1(\mathbf{x}^{(k)}) + \left.\frac{\partial R_1}{\partial x_1}\right|_{\mathbf{x}^{(k)}}(x_1 - x_1^{(k)}) + \cdots + \left.\frac{\partial R_1}{\partial x_n}\right|_{\mathbf{x}^{(k)}}(x_n - x_n^{(k)}) \\
&\vdots \qquad\qquad\qquad\qquad \vdots \\
R_n(\mathbf{x}) &= R_n(\mathbf{x}^{(k)}) + \left.\frac{\partial R_n}{\partial x_1}\right|_{\mathbf{x}^{(k)}}(x_1 - x_1^{(k)}) + \cdots + \left.\frac{\partial R_n}{\partial x_n}\right|_{\mathbf{x}^{(k)}}(x_n - x_n^{(k)})
\end{aligned}
\tag{3.7.16}
$$

It will be easier for us to work with this equation in the matrix form

$$
\mathbf{R}(\mathbf{x}) = \mathbf{R}(\mathbf{x}^{(k)}) + \mathbf{J}(\mathbf{x}^{(k)})(\mathbf{x} - \mathbf{x}^{(k)})
\tag{3.7.17}
$$

where we have introduced the Jacobian matrix,

$$
\mathbf{J} = \begin{bmatrix} \dfrac{\partial R_1}{\partial x_1} & \cdots & \dfrac{\partial R_1}{\partial x_n} \\ \vdots & \vdots & \vdots \\ \dfrac{\partial R_n}{\partial x_1} & \cdots & \dfrac{\partial R_n}{\partial x_n} \end{bmatrix}
\tag{3.7.18}
$$

which is evaluated at the value $\mathbf{x}^{(k)}$. In component form, the elements of the Jacobian are the partial derivatives

$$J_{ij} = \frac{\partial R_i}{\partial x_j} \tag{3.7.19}$$

At the root $\mathbf{x}^*$ (and there can be multiple roots!), we know that

$$\mathbf{R}(\mathbf{x}^*) = \mathbf{0} \tag{3.7.20}$$

Similar to Newton's method, this gives us the iterative scheme

$$\mathbf{J}(\mathbf{x}^{(k)})(\mathbf{x}^{(k+1)} - \mathbf{x}^{(k)}) = -\mathbf{R}(\mathbf{x}^{(k)}) \tag{3.7.21}$$

It is common to use a shorthand notation

$$\boldsymbol{\delta}^{(k+1)} = \mathbf{x}^{(k+1)} - \mathbf{x}^{(k)} \tag{3.7.22}$$

for the difference between the previous values of $\mathbf{x}$ and the new values of $\mathbf{x}$. The iterative scheme then becomes

$$\mathbf{J}(\mathbf{x}^{(k)})\boldsymbol{\delta}^{(k+1)} = -\mathbf{R}(\mathbf{x}^{(k)}) \tag{3.7.23}$$

Note that we have reduced solving the system of nonlinear equations into an iterative method where, at each step, we need to solve a system of linear equations. Since $\mathbf{x}^{(k)}$ changes at each time step, we need to solve the linear system for different right-hand-side constant vectors. The methods we discussed in the previous chapter should come in handy!

The implementation of this method goes through the following steps.

(1) Create a vector with the residuals. The order of the equations is important – it sets the order of the equations in the Jacobian.
(2) Compute the elements in the Jacobian. It is always best to make this calculation analytically if possible. Symbolic manipulation can help in case of complicated functions. We can also do it numerically (using a finite difference approximation), but this introduces a numerical error which affects the rate of convergence.
(3) Pick an initial guess $\mathbf{x}^{(0)}$.
(4) Iterate until either $||\mathbf{R}||$ or $||\boldsymbol{\delta}||$ is small:
    (1) compute $\mathbf{R}(\mathbf{x}^{(k)})$
    (2) compute $\mathbf{J}(\mathbf{x}^{(k)})$
    (3) compute $\boldsymbol{\delta}^{(k+1)} = -\mathbf{J}^{-1}\mathbf{R}$
    (4) update the values: $\mathbf{x}^{(k+1)} = \mathbf{x}^{(k)} + \boldsymbol{\delta}^{(k+1)}$
    (5) compute $||\mathbf{R}||$ and/or $||\boldsymbol{\delta}||$
    (6) check if $||\mathbf{R}||$ and/or $||\boldsymbol{\delta}||$ are small.

A useful (and practical) way to think about Newton–Raphson is as an optimization algorithm for minimizing the error between the current value of the solution, $\mathbf{x}^{(k)}$, and the optimal solution, $\mathbf{x}^*$. For example, consider how the solution to a system of ten equations in Fig. 3.17 evolves as a function of the iteration number k. In this example, we have selected as our initial guess for the ten equations $x_i^{(0)} = 1$ for each unknown x_i. This is a good guess for the smallest values of i, but a poor one for the larger values of i. After

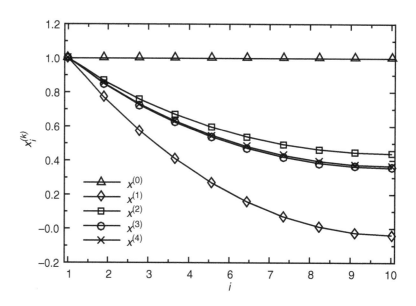

Figure 3.17 Evolution of the solution to a system of nonlinear equations towards the root using Newton–Raphson. Each of the curves plots the current state of each component $x_i^{(k)}$ of the vector $\mathbf{x}^{(k)}$ for the ten unknowns for this problem.

the first iteration, we see that the solution has changed quite substantially, with different values of $x_i^{(1)}$ now having different values. However, it appears that Newton–Raphson overshoots the root for the solution on the first iteration, since the next iteration keeps the same general shape but moves the solution upwards. This iteration turns out to be an overcorrection, as the next iteration moves the solution downward and the subsequent one moves it upward again. Note, however, that the change in the solution gets smaller with each iteration. Indeed, the "x" and "o" symbols are almost overlapping. Given the very fast quadratic convergence to the root, there is no need to make another iteration – $\mathbf{x}^{(5)}$ is essentially indistinguishable from $\mathbf{x}^{(4)}$.

You may be wondering why the plot of $x_i^{(k)}$ versus i makes such a smooth plot in Fig. 3.17, as there is no reason for this to be the case in general when we have a system of nonlinear equations. The reason for this nice form of the solution is that the plot in Fig. 3.17 was generated as the solution to the nonlinear boundary value problem

$$\frac{d^2x}{dy^2} = e^{x^2} \tag{3.7.24}$$

subject to $x(0) = 1$ and $dx/dy = 0$ at $y = 1$ using the finite differences method. We will discuss this method in Chapter 6, where Newton–Raphson will play a crucial role in the numerical solution of boundary value problems. Indeed, as we proceed through the book, we will encounter Newton–Raphson many times. Its ubiquity at some point in the solution of many nonlinear problems makes it arguably the most important method in this book. However, since Newton–Raphson requires repeatedly solving a linear system of equations, you could just as well argue that the techniques in Chapter 2 are even more important.

PROGRAM 3.5   *Similar to how we went about writing Program 3.3 for Newton's method, our program for Newton–Raphson is a modification of Program 3.4 for Picard's method:*

```
1  function [x,flag] = nonlinear_newton_raphson(x,tol)
2  %  x = initial guess for x
3  %  tol = criteria for convergence
4  %  requires function R = getR(x)
5  %  requires function J = getJ(x)
6  %  flag = 0 if failed, 1 if converged
7  R = getR(x);
8  k = 0; %counter
9  kmax = 100; %max number of iterations allowed
10  while norm(R) > tol
11      J = getJ(x); %compute Jacobian
12      del = -J\R; %Newton Raphson
13      x = x + del; %update x
14      k = k + 1; %update counter
15      R = getR(x); %f for while loop and counter
16      if k > kmax
17          fprintf('Did not converge.\n')
18          break
19      end
20  end
21  if k > kmax || max(x) == Inf || min(x) == -Inf
22      flag = 0;
23  else
24      flag = 1;
25  end
```

The key difference is on line 12, where we implement Eq. (3.7.23) for Newton's method. For the linear problem, we use MATLAB's "slash" operator. This is not a division operator! Rather, the slash operator calls a relatively sophisticated linear solver that implements a method that is appropriate for the system at hand. After we solve for δ, line 13 updates the values of $\mathbf{x}^{(k+1)}$ for the next iteration. Naturally, since Newton–Raphson requires the Jacobian, we introduced a new subfunction getJ(x) *that computes the Jacobian for the current value of* **x**. *This function is called on line 11 to be ready to do one iteration of Newton–Raphson.*

Let's see how Newton's method compares to Picard's method for the problems in Examples 3.11 and 3.12.

Example 3.13   Solve the system of Eqs. (3.7.6) and (3.7.7) using Newton's method with an initial guess $x_1^{(0)} = 0$ and $x_2^{(0)} = 0$.

Solution

The residual vector for this problem is

$$\mathbf{R} = \left[\begin{array}{c} e^{-x_1} - x_2 \\ x_1 + x_2^2 - 3x_2 \end{array} \right] \tag{3.7.25}$$

If we take the logical choice for the vector of unknowns to be

$$\mathbf{x} = \begin{bmatrix} x_1 \\ x_2 \end{bmatrix} \qquad (3.7.26)$$

then the Jacobian is

$$\mathbf{J} = \begin{bmatrix} -e^{-x_1} & -1 \\ 1 & 2x_2 - 3 \end{bmatrix} \qquad (3.7.27)$$

Let's do the first iteration by hand to see how the algorithm works. For our initial guess

$$\mathbf{x}^{(0)} = \begin{bmatrix} 0 \\ 0 \end{bmatrix} \qquad (3.7.28)$$

we need to solve

$$\begin{bmatrix} -1 & -1 \\ 1 & -3 \end{bmatrix} \begin{bmatrix} \delta_1 \\ \delta_2 \end{bmatrix} = - \begin{bmatrix} 1 \\ 0 \end{bmatrix} \qquad (3.7.29)$$

Let's solve this system with Gauss elimination using the augmented matrix

$$\begin{bmatrix} -1 & -1 & | & -1 \\ 1 & -3 & | & 0 \end{bmatrix} \qquad (3.7.30)$$

With the one step of elimination we get

$$\begin{bmatrix} -1 & -1 & | & -1 \\ 0 & -4 & | & -1 \end{bmatrix} \qquad (3.7.31)$$

So back substitution gives us

$$\delta_2 = \frac{1}{4} \qquad (3.7.32)$$

and

$$\delta_1 = \frac{-1 + \delta_2}{-1} = \frac{3}{4} \qquad (3.7.33)$$

The new values of x_1 and x_2 are just δ_1 and δ_2 since our initial guesses were zero. After one step, the norm of the residual is now

$$||\mathbf{R}||_2 = \sqrt{\left(e^{-3/4} - \frac{1}{4}\right)^2 + \left(\frac{3}{4} + \left(\frac{1}{4}\right)^2 - \frac{3}{4}\right)^2} = 0.23 \qquad (3.7.34)$$

If we want to go further, we should clearly start using a computer. We used Program 3.5 with the residual function

```
1  function R = getR(x)
2  R = zeros(2,1);
3  R(1) = exp(-x(1)) - x(2);
4  R(2) = x(1) + x(2)^2 -3*x(2);
```

and the Jacobian function

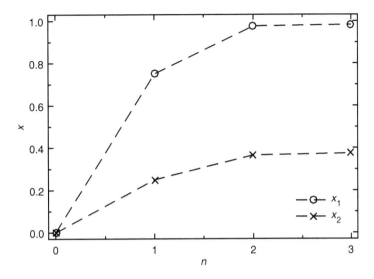

Figure 3.18 Trajectory of $x_1^{(k)}$ and $x_2^{(k)}$ as a function of iteration number for the solution of Eqs. (3.7.6) and (3.7.7).

```
1   function J = getJ(x)
2   J = zeros(2);
3   J(1,1) = -exp(-x(1));
4   J(1,2) = -1;
5   J(2,1) = 1;
6   J(2,2) = 2*x(2) - 3;
```

Figure 3.18 shows that both variables rapidly converge to the solution in only four iterations. Compare this to more than 30 iterations for Picard's method in Example 3.11, and it is obvious that Newton–Raphson is worth the extra effort to compute the Jacobian.

Example 3.14   Solve the same problem as Example 3.13 with the equations in the opposite order.

Solution
The residual vector is now

$$\mathbf{R} = \begin{bmatrix} x_1 + x_2^2 - 3x_2 \\ e^{-x_1} - x_2 \end{bmatrix} \tag{3.7.35}$$

and the Jacobian becomes

$$\mathbf{J} = \begin{bmatrix} 1 & 2x_2 - 3 \\ -e^{-x_1} & -1 \end{bmatrix} \tag{3.7.36}$$

Note that, since we swapped the rows in the residual vector, we just need to swap the same rows in the Jacobian.

For the numerical solution, we again use Program 3.5 with the residual function

```
1  function R = getR(x)
2  R = zeros(2,1);
3  R(2) = exp(-x(1)) - x(2);
4  R(1) = x(1) + x(2)^2 -3*x(2);
```

and the Jacobian function

```
1  function J = getJ(x)
2  J = zeros(2,2);
3  J(2,1) = -exp(-x(1));
4  J(2,2) = -1;
5  J(1,1) = 1;
6  J(1,2) = 2*x(2) - 3;
```

If you compare with the previous example, you will see that we have just switched the indexing of the rows in **R** and **J**.

The solution again proceeds to the root in only four iterations. Seeing that Picard's method failed when we switched the equations in Example 3.12, Newton's method is certainly preferred!

Newton–Raphson has the same convergence properties as Newton's method: it will converge if you are near the root and it does so quadratically. Indeed, looking for quadratic convergence is a good way to test your code. It is easy to make a mistake in the values for the Jacobian, for example, that might lead to a converged solution that is incorrect or a rate of convergence that is not quadratic.

Example 3.15   Show that the solution in Example 3.13 converges quadratically.

Solution
We used the two-step method for demonstrating quadratic convergence by first finding the root and then measuring the distance from the root. As we can see in Fig. 3.19, the convergence is indeed quadratic. Notice that we need to go to a very high tolerance to see the quadratic convergence. Newton's method and Newton–Raphson converge so quickly that, for moderate values of the tolerance, they converge in a single step once they get near the root and thus do not even allow us to see the quadratic convergence.

Choosing a good initial guess is another challenge, as with increasing dimension n insight on where physically meaningful solutions may lie becomes harder. Let's do an example to see how we might end up in trouble with our choice of initial guess.

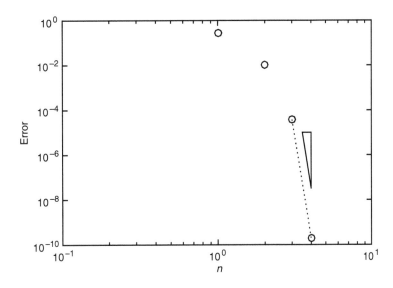

Figure 3.19 Demonstration of quadratic convergence for the solution of Eqs. (3.7.6) and (3.7.7) by Newton–Raphson. The triangle indicates a slope of 2 on the log–log plot.

Example 3.16   Find the root at $x = 0$, $y = 0$, and $z = 1$ for

$$x^2 + y^2 = 0 \tag{3.7.37}$$

$$e^{xy} = z \tag{3.7.38}$$

$$z^3 = 1 \tag{3.7.39}$$

using Newton–Raphson.

Solution

Remember that we first need to write the problem in the standard residual form,

$$\mathbf{R} = \begin{bmatrix} x^2 + y^2 \\ e^{xy} - z \\ z^3 - 1 \end{bmatrix} = \begin{bmatrix} 0 \\ 0 \\ 0 \end{bmatrix} \tag{3.7.40}$$

If we pick to write our unknowns in the obvious order

$$\mathbf{x} = \begin{bmatrix} x \\ y \\ z \end{bmatrix} \tag{3.7.41}$$

then this means that the Jacobian has the form

$$\mathbf{J} = \begin{bmatrix} 2x & 2y & 0 \\ ye^{xy} & xe^{xy} & -1 \\ 0 & 0 & 3z^2 \end{bmatrix} \tag{3.7.42}$$

Let's see what happens if we make an initial guess $\mathbf{x}^{(0)} = \mathbf{0}$, which is usually a reasonable choice in the absence of other insight into the solution. The value of the residual is

$$\mathbf{R} = \begin{bmatrix} 0 \\ 1 \\ -1 \end{bmatrix} \qquad (3.7.43)$$

and the Jacobian is

$$\mathbf{J} = \begin{bmatrix} 0 & 0 & 0 \\ 0 & 0 & -1 \\ 0 & 0 & 0 \end{bmatrix} \qquad (3.7.44)$$

We have a problem! To compute $\mathbf{x}^{(k+1)}$, we need to solve the linear system in Eq. (3.7.23). However, $\det \mathbf{J} = 0$, so there is no solution.

Let's make a different guess

$$\mathbf{x}^{(0)} = \begin{bmatrix} 2 \\ 0 \\ 1 \end{bmatrix} \qquad (3.7.45)$$

Now the residual is

$$\mathbf{R} = \begin{bmatrix} 4 \\ 0 \\ 0 \end{bmatrix} \qquad (3.7.46)$$

and the Jacobian is

$$\mathbf{J} = \begin{bmatrix} 4 & 0 & 0 \\ 0 & 1 & -1 \\ 0 & 0 & 3 \end{bmatrix} \qquad (3.7.47)$$

The Jacobian is upper triangular so we can easily compute its determinant, $\det \mathbf{J} = 12 \neq 0$, so there is going to be a solution on the first iteration. The problem that we need to solve is

$$\begin{bmatrix} 4 & 0 & 0 \\ 0 & 1 & -1 \\ 0 & 0 & 3 \end{bmatrix} \begin{bmatrix} \delta_1 \\ \delta_2 \\ \delta_3 \end{bmatrix} = - \begin{bmatrix} 4 \\ 0 \\ 0 \end{bmatrix} \qquad (3.7.48)$$

The solution is obtained by Gauss elimination (or any other method) as

$$\begin{bmatrix} \delta_1 \\ \delta_2 \\ \delta_3 \end{bmatrix} = \begin{bmatrix} -1 \\ 0 \\ 0 \end{bmatrix} \qquad (3.7.49)$$

We then update the values of $\mathbf{x}$,

$$\begin{bmatrix} x_1^{(k+1)} \\ x_2^{(k+1)} \\ x_3^{(k+1)} \end{bmatrix} = \begin{bmatrix} 2 \\ 0 \\ 1 \end{bmatrix} + \begin{bmatrix} -1 \\ 0 \\ 0 \end{bmatrix} = \begin{bmatrix} 1 \\ 0 \\ 1 \end{bmatrix} \qquad (3.7.50)$$

We are going in the right direction!

Table 3.1 Antoine coefficients for ethylbenzene and toluene.

Chemical	A	B	C
Ethylbenzene	6.95719	1424.255	213.21
Toluene	6.95464	1344.8	219.48

We will discuss in Section 3.9 the concept of continuation, which may help in choosing a good initial guess if we know a root of a slightly modified problem (e.g., a linear approximation of our problem, or one for a slightly different value of a parameter). For now, let's take a look at a useful application of Newton–Raphson in thermodynamics.

3.8 Case Study: Nonlinear Flash

3.8.1 Problem Statement

In Section 2.7, we saw a case study on a flash separation where the formulation of the problem resulted in a linear set of equations. Here we will consider a nonlinear flash problem. Let's consider the flash of ethylbenzene and toluene. If we assume that both the liquid and vapor phases are ideal, then the equilibrium condition requires that

$$x_i P_i^{\text{sat}} = p_i \tag{3.8.1}$$

where x_i is the mole fraction of species i in the liquid phase, P_i^{sat} is its saturation pressure, and p_i is its partial pressure in the vapor phase. We will approximate the saturation pressures using the Antoine equation

$$\log_{10} P^{\text{sat}} = A - \frac{B}{T + C} \tag{3.8.2}$$

where the saturation pressure is in mm Hg and the temperature is in celsius. The Antoine coefficients are in Table 3.1. We thus have two equilibrium equations (3.8.1) with four possible unknowns: the partial pressures of each species in the vapor phase, the liquid-phase mole fraction of species 1 (since $x_2 = 1 - x_1$), and the temperature.

By far the hardest problem to solve for this ideal thermodynamic system is to compute the temperature and liquid-phase composition of the flash for a given partial pressure of each species. Let's figure out how to solve that problem using Newton–Raphson.

3.8.2 Solution

The equilibrium equations we need to solve are

$$x_{\text{EB}} P_{\text{EB}}^{\text{sat}} - p_{\text{EB}} = 0 \tag{3.8.3}$$

$$(1 - x_{\text{EB}}) P_{\text{T}}^{\text{sat}} - p_{\text{T}} = 0 \tag{3.8.4}$$

where EB subscripts stand for ethylbenzene and T subscripts stand for toluene. This residual can be written in MATLAB as

```
1  function R = getR(z)
2  %data from problem statement
3  AntoineEB = [6.95719, 1424.255, 213.21];
4  AntoineT = [6.95464, 1344.8, 219.48];
5  pEB = 250;
6  pT = 343;
7  %unpack z for convenience
8  x = z(1);
9  T = z(2);
10 %write the residual
11 R = zeros(2,1);
12 R(1) = x*getPsat(AntoineEB,T) - pEB;
13 R(2) = (1-x)*getPsat(AntoineT,T) - pT;
```

Note that we have hard-coded the problem parameters directly into getR, including the saturation pressure of 250 mmHg for ethylbenzene and 343 mm Hg for toluene. This avoids the need to send around the parameters to the function or declare them as global variables. However, this does reduce the flexibility of our program since we need to go into this function and change all of the data if we want to use the program to solve a different problem (or even a different set of partial pressures). We will see how to construct a more flexible program with parameter passing in a later case study in Section 3.10.

Notice also that getR requires another function to compute the saturation pressure. To avoid writing functions for each species, we just wrote a single function to do the calculation:

```
1  function Psat = getPsat(antoine,T)
2  %handy function to get Psat
3  A = antoine(1);
4  B = antoine(2);
5  C = antoine(3);
6  RHS = A - B/(T+C); %RHS of Antoine equation
7  Psat = 10^RHS;
```

Once we have the residual, we need to compute the Jacobian too. If we write the unknown vector as

$$\mathbf{x} = \begin{bmatrix} x_{EB} \\ T \end{bmatrix} \tag{3.8.5}$$

then the Jacobian is

$$\mathbf{J} = \begin{bmatrix} P_{EB}^{sat} & x_{EB}\dfrac{dP_{EB}^{sat}}{dT} \\ -P_{T}^{sat} & (1 - x_{EB})\dfrac{dP_{T}^{sat}}{dT} \end{bmatrix} \tag{3.8.6}$$

This is easily coded into another function:

```
1  function J = getJ(z)
2  %data from problem statement
3  AntoineEB = [6.95719, 1424.255, 213.21];
4  AntoineT = [6.95464, 1344.8, 219.48];
```

```
5   pEB = 250;
6   pT = 343;
7   %unpack z for convenience
8   x = z(1);
9   T = z(2);
10  %write the Jacobian
11  J = zeros(2,2);
12  J(1,1) = getPsat(AntoineEB,T);
13  J(1,2) = x*getdPsat(AntoineEB,T);
14  J(2,1) = -getPsat(AntoineT,T);
15  J(2,2) = (1-x)*getdPsat(AntoineT,T);
```

You will notice that getJ is constructed very similar to getR, with the data hard-coded into the function. We also make use of getPsat to compute the saturation pressures that appear in the Jacobian. We also wrote a second function getdPsat that computes the derivative of the saturation pressure. If you forget how to do these types of derivatives, recall that the trick is to rewrite the function as

$$P = \exp\left(\ln\left[10^{A - \frac{B}{T + C}}\right]\right) \tag{3.8.7}$$

which you can then convert to

$$P = \exp\left(\left[A - \frac{B}{T + C}\right]\ln 10\right) \tag{3.8.8}$$

The derivative is then

$$\frac{dP}{dT} = P\left[\frac{B}{(T + C)^2}\right]\ln 10 \tag{3.8.9}$$

The MATLAB code to compute these derivatives is then:

```
1   function dPsat = getdPsat(antoine,T)
2   %handy function to get dPsat/dT
3   A = antoine(1);
4   B = antoine(2);
5   C = antoine(3);
6   Psat = getPsat(antoine,T);
7   chain = B*log(10)/(T+C)^2; %part from the chain rule
8   dPsat = Psat*chain;
```

This function brings up a very important coding issue – the function log(x) is the natural logarithm of x. If you want to compute $\log_{10}$ in MATLAB, the function is log10(x).

Once we have the above files, we can just use the Newton–Raphson program (3.5) to find that $x_{EB} = 0.6016$ and the temperature is $T = 115.07\,°C$.

3.8.3   A Variation on the Problem

The problem we just solved is the hardest one for an ideal vapor–liquid equilibrium. Let's take a moment to think about how one would go about solving the problem if we were given the liquid-phase mole fractions and the total pressure of the system. The equilibrium equations are then

$$x_{EB}P_{EB}^{sat} - y_{EB}P = 0 \tag{3.8.10}$$

$$(1 - x_{EB})P_{T}^{sat} - (1 - y_{EB})P = 0 \tag{3.8.11}$$

where we have introduced the vapor-phase mole fractions y_i. This problem can still be solved by Newton–Raphson using the same strategy as before. The unknowns are $y_{EB} = 1 - y_T$ and T. The Jacobian would then become

$$J = \begin{bmatrix} -P & x_{EB}\dfrac{dP_{EB}^{sat}}{dT} \\ P & -x_{EB}\dfrac{dP_{T}^{sat}}{dT} \end{bmatrix} \tag{3.8.12}$$

Maybe you noticed that the pressure is known, so the first column in the Jacobian is just a constant. This suggests that we might be able to reduce the problem to a single nonlinear equation. Indeed, if we add Eqs. (3.8.10) and (3.8.11), we get a single equation for temperature,

$$f(T) = x_{EB}P_{EB}^{sat} + (1 - x_{EB})P_{T}^{sat} - P \tag{3.8.13}$$

To use Newton's method, we would also need the derivative

$$f'(T) = x_{EB}\left(\dfrac{dP_{EB}^{sat}}{dT} - \dfrac{dP_{T}^{sat}}{dT}\right) \tag{3.8.14}$$

The question is now which method is better? On one hand, Newton's method is nice because it is a single equation. Indeed, we could just plot Eq. (3.8.13) and get a decent guess for the temperature from bracketing. On the other hand, we already wrote a program to solve the harder problem by Newton–Raphson. So it would not be much work to revise our program for a slightly different Jacobian and residual. So the answer to our question is that there is no answer! It is a matter of what you think is easiest, what work you have already done, and the difficulty in finding initial guesses for the system of equations in Newton–Raphson versus the single equation for Newton's method. Let's now turn our attention to this last issue in the context of continuation methods.

3.9   Continuation Methods

Here, we are interested in efficient ways to calculate solutions for problems where the vector function $\mathbf{R}(\mathbf{x})$ depends on one or more parameters p that can take on different values. We will denote this system as

$$\mathbf{R}(\mathbf{x}, p) = 0 \tag{3.9.1}$$

For example, consider the following nonlinear system that depends on some parameter p,

$$x^2 + py^2 = 1 \tag{3.9.2}$$

$$xy + y^3 = 2 \tag{3.9.3}$$

In this case the residual vector is

$$\mathbf{R} = \begin{bmatrix} x^2 + py^2 - 1 \\ xy + y^3 - 2 \end{bmatrix} \tag{3.9.4}$$

If we define the unknown vector in the obvious way,

$$\mathbf{x} = \begin{bmatrix} x \\ y \end{bmatrix} \tag{3.9.5}$$

then the Jacobian is

$$\mathbf{J} = \begin{bmatrix} 2x & 2py \\ y & x + 3y^2 \end{bmatrix} \tag{3.9.6}$$

The solution to this problem requires iteratively solving Eq. (3.7.23), which is the linear algebraic system

$$\begin{bmatrix} 2x^{(k)} & 2py^{(k)} \\ y^{(k)} & x^{(k)} + 3(y^{(k)})^2 \end{bmatrix} \begin{bmatrix} x^{(k+1)} - x^{(k)} \\ y^{(k+1)} - y^{(k)} \end{bmatrix} = - \begin{bmatrix} (x^{(k)})^2 + p(y^{(k)})^2 - 1 \\ x^{(k)}y^{(k)} + (y^{(k)})^3 - 2 \end{bmatrix} \tag{3.9.7}$$

We already know how to solve this for a given value of p. The basic idea behind a continuation method is that if we know the converged solution for some value of p, we can use it as a guide for the initial guess at a different (but close) value of p, namely $p + \Delta p$.

3.9.1   Zero-Order Continuation

The most common continuation method is the zero-order method, which is the one you might have guessed if you thought about it for a moment. If the solution is $\mathbf{x}^*$ at some value of p, we use that value as the initial guess for the solution at $p + \Delta p$,

$$\mathbf{x}^{(0)}(p + \Delta p) = \mathbf{x}^*(p) \tag{3.9.8}$$

This method is easy to implement, but it requires a very small value of Δp to be effective.

3.9.2   First-Order Continuation

This is a more sophisticated method, which relies on using a linear approximation of our vector function as the basis for updating the initial guess. Consider again the system

$$\mathbf{R}(\mathbf{x}, p) = 0 \tag{3.9.9}$$

with $x(p)$ denoting the solution for a specific value of the parameter p, and $p' = p + \Delta p$ the new value of the parameter for which we are seeking an initial guess. A Taylor series expansion of $\mathbf{R}$ around the point $(\mathbf{x}(p), p)$ keeping only the linear terms yields

$$\mathbf{R}(\mathbf{x}(p'), p') = \mathbf{R}(\mathbf{x}(p), p) + \frac{\partial \mathbf{R}}{\partial x}\bigg|_{(\mathbf{x}(p), p)} [\mathbf{x}(p') - \mathbf{x}(p)]$$

$$+ \frac{\partial \mathbf{R}}{\partial p'}\bigg|_{(\mathbf{x}(p), p)} (p' - p) \tag{3.9.10}$$

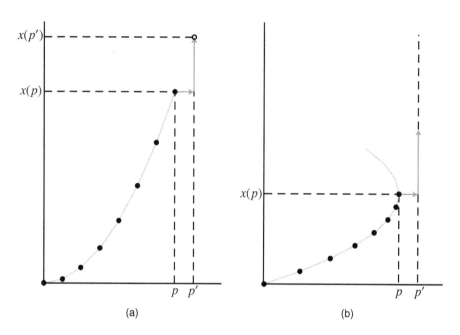

Figure 3.20 Illustration of (a) first-order continuation and (b) the failure of continuation methods at a turning point.

By setting this truncated expansion to zero and taking into account that the first term in the right-hand side is also zero, we can calculate the initial guess $\mathbf{x}^{(0)}(p')$ from

$$\left.\frac{\partial \mathbf{R}}{\partial x}\right|_{(\mathrm{x}(p),p)} [\mathbf{x}^{(0)}(p') - \mathbf{x}(p)] = - \left.\frac{\partial \mathbf{R}}{\partial p'}\right|_{(\mathrm{x}(p),p)} (p' - p) \qquad (3.9.11)$$

The term $\partial \mathbf{R}/\partial x|_{(\mathrm{x}(p),p)}$ is known, whereas the right-hand side can be approximated using a finite difference as simply $-\big[\mathbf{R}(p') - \mathbf{R}(p)\big]$. In the case of a scalar function of a single variable x the above reduces to

$$x^{(0)}(p') = x(p) + \left(\frac{\partial x}{\partial p}\right)_{x(p)} (p' - p) \qquad (3.9.12)$$

and has a clear interpretation: the new initial guess is obtained as a linear extrapolation of the previous solution. This idea is illustrated in Fig. 3.20a.

Both the zero and first-order continuation methods fail if there is a turning point in the solutions, such as Fig. 3.20b, requiring more sophisticated approaches. We will see an example of a turning point when we study the non-isothermal CSTR in Section 5.7.

3.10   Case Study: Liquid–Liquid Phase Diagram from the van Laar Model

3.10.1   Problem Statement

Let's use our knowledge of Newton–Raphson and continuation methods to construct a liquid–liquid equilibrium phase diagram for a hypothetical pair of hydrocarbons that exhibit a purely enthalpic interaction in the liquid mixture. We will assume that the excess free energy is described by the van Laar model,

$$\frac{G^{\text{EX}}}{RT} = \frac{\alpha\beta x_1 x_2}{\alpha x_1 + \beta x_2},$$

(3.10.1)

where R is the ideal gas constant, T is the absolute temperature, x_1 is the mole fraction of species 1, $x_2 = 1 - x_1$ is the mole fraction of species 2, and α and β are constants. This model for the excess Gibbs free energy leads to the activity coefficients

$$\ln \gamma_1 = \frac{\alpha}{\left[1 + \dfrac{\alpha x_1}{\beta x_2}\right]^2}$$

(3.10.2)

$$\ln \gamma_2 = \frac{\beta}{\left[1 + \dfrac{\beta x_2}{\alpha x_1}\right]^2}$$

(3.10.3)

Since the interactions are purely enthalpic, we can rewrite the constants as

$$\alpha = \frac{a}{T}, \quad \beta = \frac{b}{T}$$

(3.10.4)

where $a > 0$ and $b > 0$ are constants with units of degrees Kelvin. At equilibrium the equality of the chemical potential of each species in phase I and phase II leads to the pair of equations

$$x_1^{\text{I}}\gamma_1(T, x_1^{\text{I}}, x_2^{\text{I}}) = x_1^{\text{II}}\gamma_1(T, x_1^{\text{II}}, x_2^{\text{II}})$$

(3.10.5)

$$x_2^{\text{I}}\gamma_2(T, x_1^{\text{I}}, x_2^{\text{I}}) = x_2^{\text{II}}\gamma_2(T, x_1^{\text{II}}, x_2^{\text{II}})$$

(3.10.6)

with the ancillary relationships $x_1^{\text{I}} + x_2^{\text{I}} = 1$ and $x_1^{\text{II}} + x_2^{\text{II}} = 1$.

Since the mixture has positive values of α and β, its phase diagram has an upper-consolute temperature. Let's assume that we have measured the upper consolute temperature as $110\,^\circ\text{C}$ and we found that it occurs at a composition $x_1^{\text{I}} = x_1^{\text{II}} = 0.4$. Using this information, we want to construct a phase diagram of the liquid–liquid mixture over the range $T = 25\text{–}110\,^\circ\text{C}$.

3.10.2   Solution

We begin by computing the coefficients a and b. Recall from classical thermodynamics that the total Gibbs free energy of the mixture is

$$\underline{G} = x_1 \underline{G}_1(T, P) + x_2 \underline{G}_2(T, P) + RT\,(x_1 \ln x_1 + x_2 \ln x_2) + \underline{G}^{\text{EX}}$$

(3.10.7)

and the upper-consolute temperature corresponds to the maximum value of the temperature at which

$$\frac{\partial^2 \underline{G}}{\partial x_1^2} = 0$$

(3.10.8)

Any temperature for which the latter equation is true is a critical temperature for the fluid stability. For the van Laar model, the differentiation yields the critical temperature

$$T_c = \frac{2(ab)^2 x_1 x_2}{(ax_1 + bx_2)^3} \tag{3.10.9}$$

If we want the particular value of T_c corresponding to the upper-consolute temperature, we need to differentiate Eq. (3.10.9) with respect to x_1 and set the derivative equal to zero. This gives us a second equation,

$$\frac{(a-b)x_1^2 - 2ax_1 + b}{(ax_1 + bx_2)^4} = 0 \tag{3.10.10}$$

The latter equation provides a relationship between the parameters a and b,

$$b = a\delta \tag{3.10.11}$$

where

$$\delta = \frac{x_1(1+x_2)}{x_2(1+x_1)} \tag{3.10.12}$$

Substituting Eqs. (3.10.11) and (3.10.12) into Eq. (3.10.9) and solving for a yields

$$a = \frac{T_c (x_1 + x_2\delta)^3}{2x_1 x_2 \delta^2} \tag{3.10.13}$$

Now that we know how to convert the upper consolute temperature and composition into the van Laar parameters, let's begin the programming part of the problem by writing a function that returns these parameters for a particular consolute temperature and composition:

```
1  function [a,b] = getAB(Tc,x1)
2  %  compute the van Laar coefficients
3  %  from the upper consolute condition
4  Tc = Tc + 273.15; %convert to Kelvin
5  x2 = 1-x1;
6  abfactor = x1*(1+x2)/x2/(1+x1);
7  a = Tc*(x1+x2*abfactor)^3/(2*x1*x2*abfactor^2);
8  b = a*abfactor;
```

Looking forward, we are going to write the main program to work in relative temperature, so we have to be sure to convert to absolute temperature for the calculations of the van Laar parameters. This is the reason for the conversion factor in line 4 of the latter program.

We are now ready to construct our numerical solution for the phase diagram. It will prove convenient for us to define $x = x_1^{\mathrm{I}}$ and $y = x_1^{\mathrm{II}}$. The residual equations are then

$$R_1 = x\gamma_1^{\mathrm{I}} - y\gamma_1^{\mathrm{II}} \tag{3.10.14}$$
$$R_2 = (1-x)\gamma_2^{\mathrm{I}} - (1-y)\gamma_2^{\mathrm{II}} \tag{3.10.15}$$

where γ_i^{I} corresponds to the activity of component i with $x_1 = x$ and γ_i^{II} corresponds to the activity of component i with $x_1 = y$. This is the easiest form to use for the program. To see how this works, let's first write short functions that compute the activity coefficients in each phase:

```
1  function out = vanLaar1(a,b,x)
2  %computes the activity coefficient for species 1
3  denom = 1 + a*x/(b*(1-x));
4  out = exp(a/denom/denom);
5  end
```

```
1  function out = vanLaar2(a,b,x)
2  %computes the activity coefficient for species 2
3  denom = 1 + b*(1-x)/(a*x);
4  out = exp(b/denom/denom);
5  end
```

Note that these functions do not require specifying which phase, just the chemical identity. As a result, we can write a reasonably compact function for the residual:

```
1  function R = getR(z,a,b)
2  R = zeros(2,1);
3  x = z(1); y = z(2); %unpack the inputs
4  %compute the residuals for the equilibrium calculations
5  R(1) = x*vanLaar1(a,b,x) - y*vanLaar1(a,b,y);
6  R(2) = (1-x)*vanLaar2(a,b,x) - (1-y)*vanLaar2(a,b,y);
7  end
```

If you compare this function to previous ones for Newton–Raphson, you will notice that we are passing the values of α and β to the function, since we will be evaluating the residual at different temperatures. Also, pay careful attention to how we are able to re-use the functions for the activity coefficients in each phase.

For the equation order we used and variable order x, y, the elements of the Jacobian are

$$\mathbf{J} = \begin{bmatrix} \gamma_1^I[1 + xt_1(x)] & -\gamma_1^{II}[1 + yt_1(y)] \\ -\gamma_2^I[1 - (1-x)t_2(x)] & \gamma_2^{II}[1 - (1-y)t_2(y)] \end{bmatrix} \qquad (3.10.16)$$

where

$$t_1(k) = \left(\frac{-2\alpha^2}{\beta(1-k)^2}\right)\left(1 + \frac{\alpha k}{\beta(1-k)}\right)^{-3} \qquad (3.10.17)$$

$$t_2(k) = \left(\frac{2\beta^2}{\alpha k^2}\right)\left(1 + \frac{\alpha k}{\beta(1-k)}\right)^{-3} \qquad (3.10.18)$$

The MATLAB function to compute the Jacobian thus looks like:

```
1  function J = getJ(z,a,b)
2  J = zeros(2,2);
3  x = z(1); y = z(2); %unpack the inputs
4  J(1,1) = vanLaar1(a,b,x)*(1+x*Jterm1(a,b,x));
5  J(1,2) = -vanLaar1(a,b,y)*(1+y*Jterm1(a,b,y));
```

```
6  J(2,1) = vanLaar2(a,b,x)*(-1 + (1-x)*Jterm2(a,b,x));
7  J(2,2) = vanLaar2(a,b,y)*(1-(1-y)*Jterm2(a,b,y));
8  end
```

There are some similarities between this function and the one we wrote for the residual. First, we are sending α and β to the function as local variables. Second, we are again re-using the van Laar activity coefficient functions `vanLaar1` and `vanLaar2`. We also need to write functions to evaluate the terms t_1 and t_2 for each phase. Let's follow the strategy that we took for the van Laar activity coefficients and make flexible functions of the form

```
1  function out = Jterm1(a,b,x)
2  %computes the extra terms in J for species 1
3  denom = 1 + a*x/(b*(1-x));
4  term1 = -2*a/denom^3;
5  term2 = a/b/(1-x)^2;
6  out = term1*term2;
7  end
```

and

```
1  function out = Jterm2(a,b,x)
2  %computes the extra terms in J for species 2
3  denom = 1 + b*(1-x)/(a*x);
4  term1 = -2*b/denom^3;
5  term2 = -b/a/x^2;
6  out = term1*term2;
7  end
```

Since we are passing α and β around as local variables between the residual and Jacobian, we also need to make some minor adjustments to Newton–Raphson in Program 3.5:

```
1  function [x,flag] = newton_raphson_LLE(x,tol,a,b)
2  %  x = initial guess for x
3  %  tol = criteria for convergence
4  %  requires function R = getR(x)
5  %  requires function J = getJ(x)
6  %  flag = 0 if failed, 1 if converged
7  %  a,b are parameters supplied for the LLE program
8  R = getR(x,a,b);
9  k = 0; %counter
10 kmax = 100; %max number of iterations allowed
11 while norm(R) > tol
12     J = getJ(x,a,b); %compute Jacobian
13     del = -J\R; %Newton Raphson
14     x = x + del; %update x
15     k = k + 1; %update counter
16     R = getR(x,a,b); %f for while loop and counter
17     if k > kmax
18         fprintf('Did not converge.\n')
19         break
20     end
21 end
```

```
22  if k > kmax || max(x) == Inf || min(x) == -Inf
23      flag = 0;
24  else
25      flag = 1;
26  end
```

In this revised program, we are now sending α and β to the solver, and the solver is sending these parameters to `getR` and `getJ`. You may find all of this parameter passing to be a bit annoying, in which case you could declare α and β as global variables. We are going to try to avoid declaring global variables in our programs, since they require some care in implementation. Local variables are simple – they exist inside the function and disappear later, so they cannot "infect" your other functions.

We are now in good shape to write the actual program to compute the phase diagram via zero-order continuation starting from the upper consolute condition.

```
1   function [xI,xII,Temps] = nonlinear_LLE_main(Tc,Tmin,xc)
2   %   Tc = upper consolute temperature in centigrade
3   %   Tmin = lowest temperature for the plot
4   %   xc = upper consolute composition
5
6   %get van Laar coefficients
7   [a,b] = getAB(Tc,xc);
8
9   %make space to store the solution
10  npts = (Tc-Tmin);
11  xI = zeros(npts+1,1); %left curve, species 1 in phase 1
12  xII = zeros(npts+1,1); %right curve, species 1 in phase 2
13  Temps = Tc:-1:Tmin; %temperatures in centigrade
14
15  %put the consolute point into the output vectors
16  xI(1,1) = xc;
17  xII(1,1) = xc;
18
19  %Start a bit away from the consolute point
20  z = [xc-0.1; xc+ 0.1];
21  T = Tc-1; %starting point
22  tol = 1e-8; %for Newton-Raphson
23  for i = 1:npts
24      alpha = a/(T+273.15); %%alpha parameter
25      beta = b/(T+273.15); %beta parameter
26      %zero-order continuation method
27      [z,flag] = newton_raphson_LLE(z,tol,alpha,beta);
28      %extract the results
29      if flag == 1
30          xI(i+1,1) = z(1);
31          xII(i+1,1) = z(2);
32      else
33          fprintf('Newton-Raphson failed.\n')
34      end
35      T = T - 1; %step down by 1 degree C for next iteration
36  end
```

This program takes in as input the upper-consulate conditions, as well as a lower bound in the temperature. It returns three vectors: (i) the mole fraction $x_1^{(I)}$ of species 1 in

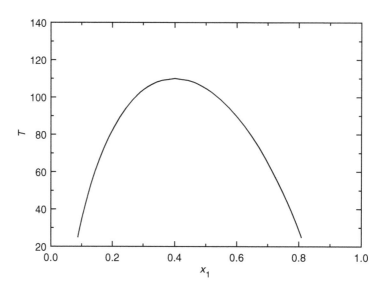

Figure 3.21 Liquid–liquid phase diagram for a van Laar model with an upper-consolute temperature of $110\,°C$ and composition $x_1 = 0.4$.

phase I; (ii) the mole fraction $x_1^{(II)}$ of species 1 in phase II; and (iii) the corresponding temperatures. These data are sufficient to create a phase diagram like the one shown in Fig. 3.21.

Let's look a bit more closely at how the main program works. We first determine the parameters a and b from the upper consolute temperature. We then have a block of lines that make space in the memory to store the results of the calculation. In most programming languages, you are required to declare all variables at the outset and say how much memory you need. MATLAB is flexible in this respect, but your programs will usually run faster if you make space in the memory for vectors and matrices first. The main part of the program implements a zero-order continuation method. For the first step, we take a temperature 1 degree lower than the consolute temperature, and we pick a guess at a slightly lower mole fraction and a slightly higher mole fraction for the composition variables. We have to be a bit careful on the first step, since we want to be sure to pick *different* initial concentrations so that the solution easily converges onto the two branches. We also only want to pick a temperature *slightly* below the upper consolute temperature so that we know that the concentrations in the different phases will not change much. After getting the first solution, we then use zero-order continuation to compute the rest of the phase diagram. At the end of the **for** loop, we decrement the temperature for the next iteration. Then, when it comes time to solve the problem on line 27, the old value of the composition from the last iteration appears as the input to the function `newton_raphson_LLE`.

3.11   Implementation in MATLAB

MATLAB has two built-in solvers, `fzero` for finding the roots of a single equation and `fsolve` for finding the roots of a system of equations.

Example 3.17   Solve Eq. (3.5.11) using `fzero`.

Solution

A MATLAB script to solve this problem is

```
1  function matlab_fzero
2  x = fzero(@(x) getf(x),4)
3
4  function f = getf(x)
5  f = sin(x) - x*cos(x);
```

The syntax for `fzero` is cumbersome, since it requires the use of anonymous functions in MATLAB. The idea is that there is some function `getf(x)` that is the function we want to solve. The additional text before the function call, `@(x)` tells MATLAB that the unknown in this function is the variable x. The second entry in the call to `fzero` is the initial guess. In addition to providing a single value as an initial guess, you can also provide a range for the initial guess:

```
1  function matlab_fzero_range
2  x = fzero(@(x) getf(x),[4,5])
3
4  function f = getf(x)
5  f = sin(x) - x*cos(x);
```

The only restriction on the range is that the function must change sign between the starting point and the ending point. In other words, the range that you give MATLAB must be a bracket for the solution.

In some of the problems we solved in this chapter, and in many problems in future chapters, we also want to pass parameters along with the unknown values. This is accomplished in `fzero` through the anonymous function. For example, consider this small modification to Example 3.17:

```
1  function matlab_fzero_parameter
2  x = fzero(@(x) getf(x,1),4)
3
4  function f = getf(x,c)
5  f = sin(x) - c*x*cos(x);
```

The problem we are solving is now

$$f(x) = \sin x - cx \cos x \qquad (3.11.1)$$

where c is some parameter. When we call `fzero`, the anonymous function handle tells MATLAB that x is the variable in the nonlinear equation. We then just need to send the value of the parameter c. The particular choice $c = 1$ is exactly the problem in Example 3.17.

The syntax for solving systems of equations with `fsolve` is very similar to `fzero`, including the use of anonymous functions and the ability to pass parameters.

Example 3.18   Solve the system of equations

$$f_1(x_1, x_2) = x_1^2 + x_2 \tag{3.11.2}$$

$$f_2(x_1, x_2) = x_1 x_2 + 8 \tag{3.11.3}$$

using `fsolve`.

Solution

This problem is easily solved by hand to give $x_1 = 2$ and $x_2 = -4$. If we want to solve it using `fsolve`, we can use the following program:

```
1  function matlab_fsolve
2  x = fsolve(@(x) getR(x),[1.5,1])
3
4  function R = getR(x)
5  R(1,1) = x(1)^2 + x(2);
6  R(2,1) = x(1)*x(2)+8;
```

Note that the function call is very similar to `fzero`, except that now we need to send a vector of initial guesses, $[x_1^{(0)} = 1.5, x_2^{(0)} = 1]$, rather than a single value.

You will notice in both of these examples that it is not required to provide the derivative of the function (for Newton's method) or the Jacobian (for Newton–Raphson). These quantities can be computed numerically using approximations for the derivative that we will discuss in Chapter 6.

3.12   Further Reading

Nonlinear algebraic equations are covered extensively in all general purpose numerical methods books such as

- G. Dahlquist and A. Bjork, *Numerical Methods*, Prentice Hall, 1974
- G. H. Golub and J. M. Ortega, *Scientific Computing and Differential Equations*, Academic Press, 1992
- J. D. Hoffman and S. Frankel, *Numerical Methods for Engineers and Scientists*, McGraw Hill, 2001
- C. Pozrikidis, *Numerical Computation in Science and Engineering*, Oxford University Press, 2008.

The topic of continuation was covered at a very elementary level in this book. It has attracted a lot of attention from both the mathematics and engineering communities. Some classic references on more advanced methods are

- H. B. Keller, "Numerical Solution of Bifurcation and Nonlinear Eigenvalue Problems," in *Applications of Bifurcation Theory*, P. Rabinowitz, ed., Academic Press, 1977
- E. L. Allgower and K. Georg, *Introduction to Numerical Continuation Methods*, SIAM, 2003.

The problem of DNA force–extension behavior is based on the paper

- C. Bustamante, J. F. Marko, E. D. Siggia and S. Smith, "Entropic Elasticity of λ-Phage DNA," *Science* **265**, 1599–1600, 1994.

The vapor–liquid phase equilibrium problem in the case studies is discussed in most thermodynamic textbooks, such as

- S. I. Sandler, *Chemical, Biochemical, and Engineering Thermodynamics*, Wiley, 2006.

Problems

3.1 Consider the Newton's method calculation of the roots of the nonlinear equation $f(x) = x^3$. If the initial guess is $x^{(0)} = 2$, what is $x^{(1)}$?

3.2 For the nonlinear functions that follow, provide the Newton iteration formula that you would use to find the roots of $f(x) = 0$:

$$\begin{aligned} f(x) &= 32x^5 - 64x + 31 \\ f(x) &= e^{-3x} - x \\ f(x) &= \cos x - x \end{aligned}$$

3.3 For the nonlinear functions that follow, provide the Newton iteration formula that you would use to find the roots of $f(x) = 0$:

(a) $f(x) = x^3 - 3x - 2$
(b) $f(x) = xe^{-x}$
(c) $f(x) = (x - 1)\ln(x)$

3.4 Consider using Newton's method to find the real root of

$$f(x) = \ln x + 2$$

with an initial guess $x^{(0)} = 2$. What is $x^{(1)}$? Do you see any potential problem here?

3.5 For what values of $x^{(0)}$ do you expect to have a problem with the Newton's method solution to compute the roots of $f(x) = \tan x$?

3.6 Determine the number of steps required to find the root of $f(x) = x^2$ by Newton's method as a function of the initial guess x_0 if the goal is to reach a tolerance $|f(x^*)| \le \epsilon$.

3.7 Consider the roots of the equation

$$\sin x - x \cos x = 0$$

(a) Write Newton's method for computing $x^{(k+1)}$ from the previous value $x^{(k)}$ for this problem.

(b) Simplify your result for large $x^{(k)}$.

(c) Assume you have computed root x_n^* of this equation for some large value of x using Newton's method. Use the result from part (b) to explain why $x_{n+1}^{(0)} = x_n^* + \pi$ is a good guess for the next root when using Newton's method. You may (or may not) find the following trigonometric identities useful.

$$\sin(a + b) = \sin(a)\cos(b) + \cos(a)\sin(b)$$
$$\sin(a - b) = \sin(a)\cos(b) - \cos(a)\sin(b)$$
$$\cos(a + b) = \cos(a)\cos(b) - \sin(a)\sin(b)$$
$$\cos(a - b) = \cos(a)\cos(b) + \sin(a)\sin(b)$$

3.8 What *mathematical problem* is solved by the following MATLAB code?

```
1  function x = problem3_8
2  itmax = 100; tol = 0;, x0 = 1;
3  for k=1:itmax
4      x=x-f(x)/fprime(x);
5      r=f(x);
6      if abs(r)<=tol
7          break
8      end
9  end
10 function y=f(x)
11 y=exp(-x)-x;
12 function yp=fprime(x)
13 yp=-exp(-x)-1;
```

3.9 Consider the code below that implements Newton's method and answer the following questions.

```
1  function problem3_9
2  fprintf('x \t error \n')
3  error = abs(x*log(x)-x);
4  fprintf('%6.9f \t %6.6e \n',x,error)
5  while error > 10^(-8)
6      x = x - (x*log(x)-x)/log(x);
7      error = abs(x*log(x)-x);
8      fprintf('%6.9f \t %6.6e \n',x,error)
9  end
```

(a) What is the largest real root of the function studied in this problem?

(b) What is the criterion for convergence in terms of the value $x^{(k)}$ on the kth iteration?

(c) When I run the program with an initial guess $x^{(0)} = 1.1$, I get the following:

x	error
1.100000000	9.951588e−01
11.541264556	1.668785e+01
4.718561059	2.602305e+00

3.041282127   3.414727e−01
2.734279473   1.604463e−02
2.718328446   4.661767e−05
2.718281829   3.997207e−10

Explain the behavior of the error (i) near the initial guess and (ii) near the root.

3.10 Answer the following question about this MATLAB function:

```
1  function out = problem3_10
2  x = 0;
3  err = 100;
4  while err > 10^(-6)
5      x = x - (x^3-2*x^2-3*x+1)/(3*x^2-4*x-3);
6      err = x^3-2*x^2-3*x+1;
7  end
8  out = x;
```

I get the output x = 0.3333 after one iteration, which is not the correct answer. What is required to fix this code?

3.11 If I am using Newton's method and I am near the root with a value $|f(x^{(5)})| = 10^{-2}$, what would I expect to have for $|f(x^{(7)})|$? You only need an approximate answer.

3.12 Show that Newton's method for $f(x) = x^3$ does not exhibit quadratic convergence.

3.13 What is the Jacobian needed for the Newton–Raphson solution of the system of equations

$$R_1 = x^3 - y^2 = 0$$
$$R_2 = xy - 3 = 0$$

with the unknown vector listed as

$$\mathbf{x} = \begin{bmatrix} x \\ y \end{bmatrix}$$

3.14 Consider the following nonlinear vector function:

$$\mathbf{F(x)} = \begin{bmatrix} f_1(\mathbf{x}) \\ f_2(\mathbf{x}) \\ f_3(\mathbf{x}) \end{bmatrix} = \begin{bmatrix} x_1 x_2 x_3 - x_1^2 + x_2^2 - 1.34 \\ x_1 x_2 - x_3^2 - 0.09 \\ e^{x_1} - e^{x_2} + x_3 - 0.41 \end{bmatrix}$$

Show the form of all the equations needed to perform the Newton–Raphson iteration scheme for solving $\mathbf{F(x)} = \mathbf{0}$.

3.15 Consider the nonlinear system of equations

$$f_1(x, y) = xy$$
$$f_2(x, y) = x(1 - y)$$

(a) What are the roots for this problem?

(b) What is the Jacobian if we write our unknown vector as [x,y]?

(c) When I try to solve this with Newton–Raphson, I run into a problem. What happens?

3.16   Consider a system of $i = 1, 2, \ldots, n$ residuals of the form

$$R_i = \left(\sum_{j=1}^{n} a_{ij} x_j \right) - b_i$$

where a_{ij} and b_i are constants. Show that applying Newton–Raphson to this system of equations will converge to a solution in one iteration, independent of the initial guess $x_j^{(0)}$.

3.17   This problem deals with the Newton–Raphson solution to the system of equations

$$\mathbf{R} = \begin{bmatrix} bx^2 y + 4 \\ \ln(x - y) \end{bmatrix}$$

with the unknown vector

$$\mathbf{x} = \begin{bmatrix} x \\ y \end{bmatrix}$$

If you give this problem the initial guess

$$\mathbf{x}^{(0)} = \begin{bmatrix} 2 \\ 1 \end{bmatrix}$$

and get the vector

$$\mathbf{x}^{(1)} = \begin{bmatrix} 3 \\ 2 \end{bmatrix}$$

after one iteration, what is the value of b?

3.18   Consider the system of nonlinear equations

$$\mathbf{R} = \begin{bmatrix} x_1^2 x_2 + x_1 x_3 + x_3 - 2 \\ x_1 + x_2 x_3 + x_3^5 - 2 \\ x_1 x_2 x_3 + x_1 + x_3 - 2 \end{bmatrix}$$

A root of this system is

$$\mathbf{x}^* = \begin{bmatrix} x_1^* \\ x_2^* \\ x_3^* \end{bmatrix} = \begin{bmatrix} 1 \\ 0 \\ 1 \end{bmatrix}$$

Show that the convergence of Newton–Raphson is independent of the value of ϵ for an initial guess

$$\mathbf{x}^{(0)} = \mathbf{x}^* + \begin{bmatrix} \epsilon \\ 0 \\ 0 \end{bmatrix}$$

for any real number ϵ.

3.19   Recall the derivation of the Newton iteration formula discussed in the lecture; it was based on a Taylor series expansion of the function $f(x)$ around an estimate of the root $x^{(k-1)}$, truncated after the *linear* term. You are asked to derive a more accurate iteration scheme as follows: Start from the Taylor series expansion of $f(x)$ around $x^{(k-1)}$, and truncate it after the *quadratic* term; then derive a general iteration formula for $x^{(k)}$, and explain how you would use it.

Computer Problems

3.20   Consider the simple equation $f(x) = x^3 - x^2$. Do the following analysis for this equation:

(a) Determine the three roots by hand.
(b) Perform the first five iterations by hand for Picard's method for an initial guess of 0.1.
(c) Write a MATLAB program to compute the solution by Picard's method. How many iterations are required to get to within 0.01 of the nearest root for an initial guess of 0.1?
(d) How many iterations of Newton's method are required to reach the same level of accuracy? You should do this by hand and by a MATLAB program.

3.21   We want to explore the ability of Newton's method to find roots to the equation $\sin(\pi x) = 0$. You should already know the roots of this equation. Write a MATLAB program that determines the roots of the equation for initial guesses $x = 0, 0.01, 0.02,$ $\ldots$, 1.99, 2.00 by using Newton's method. Your program should generate a plot of the root produced by Newton's method as a function of the initial guess. You should explain your plot in the context of what you should know about convergence of Newton's method.

3.22   Write a MATLAB program that uses Newton–Raphson to compute the solution to the following system of equations

$$\frac{x^3 - x^2}{z} + \sin(\pi y) = 0$$
$$xy^3 - \cos(\pi z) - z = 0$$
$$\frac{x}{z} + \cos[\pi(x - z)] - 2e^y = 0$$

for an initial guess of $x = y = z = 0.2$. How many iterations are required such that $||\delta|| < 0.001$?

3.23   Consider the system of equations

$$x(x - y) = 0$$
$$x^2 - 2y = -2$$

Find all of the real roots of this system of equations by hand. Write a MATLAB program that demonstrates quadratic convergence towards each of the real roots of the system.

Your program should display the $||\mathbf{x} - \mathbf{x}^*||$ to screen to demonstrate quadratic equation for a suitably small tolerance.

3.24   For a packed bed, the Ergun equation provides a relationship between the pressure drop per unit length and the properties of the bed,

$$\frac{\Delta P}{L} = 150 \left(\frac{\eta v_0}{D_p^2} \right) \frac{(1 - \epsilon)^2}{\epsilon^3} + \frac{7}{4} \left(\frac{\rho v_0^2}{D_p} \right) \frac{1 - \epsilon}{\epsilon^3}$$

where η is the fluid viscosity, v_0 is the superficial velocity, D_p is the particle diameter, ρ is the fluid density, and ϵ is the void fraction of the bed. Consider a 1.5 m long packed bed containing 5 cm diameter particles. Fluid of density 2 g/cm^3 and viscosity 1 cP flows through this bed at a superficial velocity of 0.1 m/s. The pressure drop in the bed is 416 Pa.

(a) Write a program that uses Newton's method to determine the void fraction of the bed and outputs it to screen along with the number of iterations and the value of $|f(x)|$. Provide the equations needed for Newton's method along with evidence from your program that the solution has quadratic convergence. (You should pick a bad initial guess and a very small tolerance to see the quadratic convergence.)

(b) Now let's use the program from the first part to see how the solution speed is affected by the initial guess. Modify your program from part (a) to compute the number of iterations required for convergence with a tolerance $|f(\epsilon)| < 10^{-14}$ for initial guesses $\epsilon^{(0)} = 0.01, 0.02, \ldots, 0.99$. You program should automatically plot the number of iterations versus the initial guess for ϵ. Does this plot make sense?

3.25   In this problem, you are going to look at how the numerical accuracy of the derivative affects the rate of convergence in Newton's method. Consider the calculation of the root of

$$f(x) = e^{x^2} - 2x^3 + \sin(3\pi x)$$

with an initial guess $x^{(0)} = 1$. In the most accurate case, you would use an analytical calculation of $f'(x)$. However, you can also approximate the derivative using a centered-finite difference formula,

$$f'(x) \approx \frac{f(x + \Delta x) - f(x - \Delta x)}{2\Delta x}$$

where Δx is the discretization. We will discuss finite differences in much more detail when we cover boundary value problems in Chapter 6. You can think of the analytical case as the limit $\Delta x \to 0$.

Write a program that implements Newton's method for values of $\Delta x = 0, 0.01, 0.02, \ldots, 0.50$ using the initial guess $x^{(0)} = 1$. For $\Delta x = 0$, you should use the analytical calculation of $f'(x)$ in Newton's method; for the other cases, you should use the finite-difference approximation. Set your convergence criteria to $|f(x^{(k)})| < 10^{-6}$ for iteration k. The error in the derivative can become very severe as you increase Δx, so also add a check in your Newton's method to stop the iterations if $k \geq 250$. You should record the number of iterations and the root x^* for each value of Δx. If the scheme does not converge, record a value of zero for the number of iterations and the root. Your program should automatically perform this calculation for $\Delta x = 0, 0.01, 0.02, \ldots, 0.50$. It should make a plot of the number of iterations k versus Δx, using $k = 0$ if the solution does not converge. In addition to making a plot, compile your output into a matrix that you can save to file using `dlmwrite`. The matrix should have 51 rows and the following three columns: (i) Δx, (ii) the value of the root, and (iii) the number of iterations for convergence.

3.26 A vapor–liquid mixture of ethylbenzene and toluene has a partial pressure of 250 mm Hg of ethylbenzene and 343 mm Hg of toluene. Write a MATLAB program that will compute the composition of the liquid phase and the temperature of this mixture if we assume ideal gas and liquid behavior. You can approximate the saturation pressures using the Antoine equation

$$\log_{10} P^{\text{sat}} = A - \frac{B}{T + C}$$

where the saturation pressure is in mm Hg and the temperature is in Celsius. The Antoine coefficients are in the table below.

Chemical	A	B	C
Ethylbenzene	6.957 19	1424.255	213.21
Toluene	6.954 64	1344.8	219.48

Your program should output the temperature and liquid-phase mole fractions of each species. What are the equations used for the residual and the Jacobian?

3.27 This problem involves the analysis of a counter-current heat exchanger. The design equation for a counter-current heat exchanger is

$$Q = UA\Delta T_{lm}$$

where Q is the heat transferred between streams, A is the area of the exchanger, and the log-mean temperature is defined as

$$\Delta T_{lm} = \frac{(T_2' - T_2) - (T_1' - T_1)}{\ln(T_2' - T_2) - \ln(T_1' - T_1)}$$

where T_1 is the inlet temperature of the inner stream, T_2 is the outlet temperature of the inner stream, T_2' is the inlet temperature of the outer stream, and T_1' is the outlet temperature of the inner stream. The first law of thermodynamics requires that

$$Q = \dot{m}C_p(T_2 - T_1)$$

where $\dot{m}$ is the flow rate of the inner stream, C_p is the heat capacity of the inner stream, and we have chosen to set $T_2 > T_1$ so that Q is positive. A similar energy balance applies to the outer stream.

(a) We will begin by reviewing the analysis of a counter-current heat exchanger. The inner fluid has a flow rate of 3 kg/s and a heat capacity of 2.3 kJ/kg °C and the outer fluid has a flow rate of 5 kg/s and a heat capacity of 4 kJ/kg °C. The heat exchanger cools the inner fluid from 100 °C to 50 °C and the outer fluid enters at 15 °C. The overall heat transfer coefficient is 1 kW/m^2 °C. Determine the outlet temperature of the cooling fluid and the area of the heat exchanger. Use of MATLAB is *not* required.

(b) We now want to figure out the outlet temperature of the inner fluid if we change the flow rate of the cooling fluid. Write a MATLAB program that uses Newton's method to compute this quantity. Test your program using the same parameters as part (a) with an initial guess of 1000 °C (which makes no sense physically) and demonstrate quadratic convergence. Include all relevant calculations in your written solution (e.g., the nonlinear equation and its derivative) and include a printout of the data demonstrating quadratic convergence.

(c) Use your Newton's method algorithm from part (b) to compute the outlet temperature of the inner fluid as a function of the cooling fluid flow rate for logarithmically spaced flow rates from $10^{-1/2}$ to 10^3 kg/s. Your program should automatically generate a semilog-x plot of temperature versus cooling flow rate.

3.28 Consider the partial differential equation

$$\frac{\partial c}{\partial t} = \frac{\partial^2 c}{\partial x^2}$$

subject to the constant concentration boundary condition $c(0, t) = 0$ and the reaction–diffusion boundary condition

$$c - \frac{\partial c}{\partial x} = 0 \quad \text{at } x = 1$$

and the initial condition $c(x, 0) = 1$. The solution to this problem is

$$c(x, t) = \frac{3}{2}x + \sum_{n=1}^{\infty} c_n \sin \lambda_n x e^{-\lambda_n^2 t}$$

where λ_n are the positive roots of

$$\sin \lambda_n - \lambda_n \cos \lambda_n = 0$$

the Fourier coefficients for $n = 1, 2, \ldots$ are

$$c_n = \frac{4(1 - \cos(\lambda_n))}{2\lambda_n - \sin(2\lambda_n)}$$

The goal here is to figure out how to make a plot of the concentration profile.

(a) Write a MATLAB program that plots the eigenvalue equation from $\lambda \in [-10, 10]$. This will help you with finding the roots since you can see roughly where the first few positive roots are located. Estimate for λ_1 by looking at the graph.

(b) Write a MATLAB program that finds λ_1 using Newton's method using the guess you obtained from part (a). The program should output the value of λ_1 to the screen. What is the function $f(\lambda)$ and the derivative $f'(\lambda)$ needed for Newton's method?

(c) Now let's see if we can find many of the eigenvalues using your program from part (b). In the first part of your program, you should automatically find the first ten positive eigenvalues and print them to screen. To be sure that you got the right values, it may be useful for you to use the program from part (a) to plot the eigenvalue equation and put your results on top of the function as circles. (This is not required, but it is very helpful to make sure that my program works.) In the second part of your program, you should automatically find the first 150 positive eigenvalues and determine how $\lambda_{n+1} - \lambda_n$ depends on the eigenvalue number n. Plot this result and explain the asymptotic value for large n.

(d) Write a MATLAB program that automatically generates a plot of the function $c(x, t)$ for the times $t = 0.0005, 0.01, 0.05, 0.07$, and 0.2.

3.29 In this problem, you will use your knowledge of solving systems of nonlinear equations and continuation methods to determine both the composition and the temperature distribution in an ideal staged rectification column. Recall from mass and energy balances that the operating line for a rectifying column is given by

$$y_{n+1} = \frac{R_D}{1 + R_D} x_n + \frac{x_D}{1 + R_D}$$

where R_D is the reflux ratio, x_D is the concentration of the distillate, y_{n+1} is the concentration of the gas entering stage n, and x_n is the concentration of the liquid exiting stage n. The concentrations refer to the lighter boiling component. We will be considering a total condensor, whereupon $x_0 = y_1 = x_D$. In an ideal column, the liquid composition x_n and gas composition y_n exiting stage n are assumed to be in equilibrium. We will further assume that this equilibrium is given by Raoult's law

$$x_n P_1^{sat} = y_n P$$
$$(1 - x_n) P_2^{sat} = (1 - y_n) P$$

where species 1 is the lighter boiling component, P_i^{sat} is the saturation pressure of species $i = 1, 2$, and P is the pressure of the column. The saturation pressures (in mm Hg) are correlated using the Antoine equation,

$$\log_{10} P_i^{sat} = A_i - \frac{B_i}{T_n + C_i}$$

where T_n is the temperature of stage n in centigrade A_i, B_i, and C_i are the Antoine coefficients for species i.

In mass and energy balances, you learned how to make a graphical solution to this problem using the McCabe–Thiele method on an x–y phase diagram for the lighter boiling component. In this problem, you will figure out how to make this solution numerically without creating the phase diagram in advance. As an added bonus, the numerical solution will also furnish the temperature of each equilibrium stage. To help guide your solution to the problem, there are several subtasks to complete prior to making the full calculation.

For this problem, we will consider the rectification of benzene and toluene at atmospheric pressure. The Antoine coefficients are given in the table below.

	A	B	C
Benzene	6.905 65	1211.033	220.790
Toluene	6.954 64	1344.800	219.480

We will consider a rectifying column that produces a distllate of 99.95% pure benzene using a reflux ratio $R_D = 1.95$. The column contains 15 stages.

(a) Recast the problem into a pair of coupled nonlinear equations that you can use to compute x_n and T_n for given values of y_n and P.

(b) Determine the Jacobian for this 2×2 system.

(c) Write a MATLAB function called `computeJ` that computes the elements of the Jacobian. Confirm that this function produces the correct output for the values $T_n = 90\,°C$ and $x_n = 0.7$ if you input $y_n = 0.7$. Report the Jacobian matrix for these values in your written submission. Save the function as an m-file in your working directory for use in later programs.

(d) Write a MATLAB function called `computeR` that computes the elements of the residual. Confirm that this function produces the correct output for the values $T_n = 90\,°C$ and $x_n = 0.7$ if you input $y_n = 0.7$. Report the residual vector for these values. Save the function as an m-file in your working directory for use in later programs.

(e) Write a MATLAB function called `NewtonRaphson` that performs the Newton–Raphson method to solve your linear system. This function should call `computeJ` and `computeR`. Use the norm of δ for convergence and stop the iterations when this is less than 10^{-4}. You are allowed to use a built-in MATLAB function to do the matrix inversion. Test your solution when $y_n = 0.7$ using an initial guess of $T_n = 90\,°C$ and $x_n = 0.7$. Report the values of T_n and x_n produced by your program. Also confirm, by hand, that these values are the correct solution to the nonlinear equations. Save the function as an m-file in your working directory for use in later programs.

(f) Write a MATLAB function called `rectification` that executes a zero-order continuation method to determine the values of x_n, y_n, and T_n for each stage of the column. The function should call `NewtonRaphson` for the equilibrium calculation. The output of the function should be an $N \times 3$ matrix with the values of x_n in the first column, y_n in the second column, and T_n in the third column. Your function should also plot each variable versus the stage number n.

3.30 Consider non-ideal liquid phases where the activity coefficient is given by the one-parameter model

$$\ln \gamma_i = A(1 - x_i)^2$$

The saturation pressure of species 1 is 800 mm Hg, and the saturation pressure of species 2 is 1000 mm Hg.

(a) Write a MATLAB program that uses Newton's method to compute the azeotrope pressure as a function of the composition of the liquid phase, x_1, for values of $A \in [1.6, 2.0]$. (These are below the liquid–liquid phase separation value of A.) What is the nonlinear algebraic equation you need to solve and the derivative needed for Newton's method? Your program should automatically plot the azeotrope pressure versus the composition of the liquid phase.

(b) Develop a system of equations that you need to solve to compute the P–x–y diagram for $A = 2$. In your written solution, include the unknown vector, the residual equations, and the Jacobian for this system.

(c) Use the result from part (b) to write a program that automatically produces the P–x–y and x–y diagrams for this system (where the compositions refer to species 1).

3.31 This problem involves computing the liquid–vapor P–x–y diagram from the two parameter Margulies model. In this model, the activity coefficients are given by

$$\ln \gamma_1 = \alpha_1 x_2^2 + \beta_1 x_2^3$$

and

$$\ln \gamma_2 = \alpha_2 x_1^2 + \beta_2 x_1^3$$

The parameters α and β are already in dimensionless form in units of RT and are restricted to have the values

$$\alpha_i = A + 3(-1)^{i+1}B$$

and

$$\beta_i = 4(-1)^i B$$

For this problem, we will use $A = 1$ and $B = 0.3$. This is sufficiently non-ideal to exhibit an azeotrope but not so bad that you will also get liquid–liquid phase separation. Recall from thermodynamics that the equilibrium is given by

$$x_1 \gamma_1 P_1^{\text{sat}} = y_1 P$$

and

$$x_2 \gamma_2 P_2^{\text{sat}} = y_2 P$$

where we have assumed an ideal vapor phase. For this problem, let's assume the saturation pressures are $P_1^{\text{sat}} = 1100$ mm Hg and $P_2^{\text{sat}} = 800$ mm Hg. The ultimate goal is to plot the P–x–y diagram, but we are going to lead you through the problem in steps to make it easier.

(a) The first step is to determine the location of the azeotrope because this is a turning point in the continuation method – we will need to construct solutions on the left and right of the azeotrope. Use you knowledge of thermodynamics to come up with a single equation that allows you to compute the azeotrope composition. Write a MATLAB program that computes both the value of x_{azeo} and the corresponding pressure P_{azeo}. Report these values along with the function f and derivative f' needed for Newton's method.

(b) Write a MATLAB program that implements Newton–Raphson to compute the equilibrium compositions. It is easiest to do this with four unknowns

$$\mathbf{x} = \begin{bmatrix} y_1 \\ y_2 \\ x_1 \\ y_1 \end{bmatrix}$$

to avoid having to take very complicated derivatives. What is the residual vector and Jacobian needed for Newton–Raphson? Show that your answer from part (a) gives a very small value of $\|\mathbf{R}\|$ and compute the condition number of the Jacobian with the answer from part (a) using MATLAB's function `cond`. What does this mean for implementing Newton–Raphson? You do not need to execute the `while` loop here if you can answer the previous question without an "experiment" – you can make it not execute by giving it a very loose tolerance value.

(c) Now write a MATLAB program that uses the results so far to make the P–x–y diagram. You already have a big hint; the azeotrope is a turning point so you will need to make the diagram in two pieces, one before the turning point and one after. Your program should produce the P–x–y diagram.

3.32 The overall goal of this problem is to compute the P–V and P–T equilibrium diagrams for a single component fluid described by the van der Waals equation of state. Let us recall the key things we need to know from thermodynamics, following the notation from Sandler's textbook. The van der Waals equation of state is

$$P = \frac{RT}{V - b} - \frac{a}{V^2}$$

where P is the pressure, R is the ideal gas constant, T is the temperature, V is the molar volume, and a and b are constants. Since we are looking at phenomena that will include supercritical values, it proves convenient to use MPa for the pressure and L/mol for the volume. For this problem, we will consider a fluid with $a = 2.5$ MPa L^2/mol^2 and $b = 0.2$ L/mol. The critical properties of the van der Waals fluid are readily related to these parameters,

$$V_c = 3b, \quad P_c = \frac{a}{27b^2}, \quad T_c = \frac{8a}{27Rb}$$

The van der Waals equation of state also has a very nice expression for the departure functions,

$$H^{dep} = RT(Z - 1) - \frac{a}{V}$$

and

$$S^{dep} = R \ln Z + R \ln \frac{V - b}{V}$$

where $Z = PV/RT$ is the compressibility.

(a) We will begin with an easy problem where we have a supercritical fluid with $T = 470$ K and $P = 3$ MPa. Have your program use Newton's method to determine V

at these conditions. To help you out, you should first make a plot of the isotherm to get a guess for the answer for $\underline{V}$. Print out the plot of the isotherm and include the value of $\underline{V}$ on your plot. You should also include a brief explanation of your solution method, such as the algorithm for Newton's method with the relevant equations.

(b) While it is fairly easy to calculate volumes for supercritical fluids, it is harder to calculate the two-phase region. Let us first consider a case where $T = 425$ K and $P = 2$ MPa. Have your program make a plot of the isotherm to start so that you can see what it looks like and include it in your written solution. To make a nice plot, use $0.21 \leq \underline{V} \leq 3$ and $0 \leq P \leq 6$ for the boundaries of the plot. We now want to see how the value of the volume in Newton's method, $\underline{V}^{(k)}$, depends on the number of iterations k for a given initial guess, $\underline{V}^{(0)}$. First let's consider $\underline{V}^{(0)} = 0.23$ L/mol, which is an initial guess very close to b. Have your program execute Newton's method for this initial guess and produce the plot of $\underline{V}^{(k)}$ vs. k. Which root did you find? Demonstrate that the convergence is quadratic using $|f(x)| < 10^{-8}$ as the condition.

(c) Now let's use $T = 425$ K and $P = 2$ MPa and consider a really bad guess, $\underline{V}^{(0)} = 0.6$. What happened to the evolution of your solution? You should be able to explain this result with a combination of your knowledge of numerical methods and thermodynamics. As we approach the critical point, what happens to the roots? How does this affect your numerical solution?

(d) Now let's compute an isotherm. We know that in the two phase region, there should be a tie-line connecting the vapor and liquid molar volumes at the saturation pressure, $P^{sat}(T)$, for a given temperature. We will do the isotherm for $T = 395$ K. An algorithm to find the saturation pressure and the corresponding molar volumes using a departure function approach is:

(1) make an initial guess for P^{sat}
(2) compute the molar volumes of the vapor and liquid at this pressure
(3) Compute $\Delta G^{dep} = G_{liquid}^{dep} - G_{vapor}^{dep}$.
(4) If $|\Delta G^{dep}| < \epsilon$, where ϵ is a small number, then the saturation pressure is correct. If not, update P^{sat} and repeat steps (2)–(4).

Reformulate this into a Newton's method algorithm for computing the saturation pressure. Include the details of your calculation in the written assignment, and add the algorithm to your program. Now let's make the real isotherm. To produce a good initial guess for P^{sat}, you should make a plot of the isotherm from the van der Waals equation of state and find a pressure that looks like it will satisfy the equal area rule (a graphical way to find the saturation pressure). What is your rationale for the initial guess? Your program should then make the plot of the isotherm from the van der Waals equation of state and the real isotherm, which has the tie-line between the two equilibrium conditions. What are the values of the liquid and vapor molar volumes?

(e) You are now in a position to make the P–V and P–T diagrams, since they "only" require computing isotherms for a number of different temperatures. The challenge here is making good guesses, which should be clear from Problem 3.2. The numerical approach is to use a continuation method, where you use the values of P^{sat}, $\underline{V}_{liquid}$, and $\underline{V}_{vapor}$ at some value of T to compute the new values of the pressure and volumes at a temperature $T = T + \Delta T$, where ΔT is a small change in temperature. The definition of "small" depends on how quickly things are changing in the numerical solution, so you will need to do some exploration to find good values of ΔT. Your program should generate a the P–V and P–T plots for $T = 365$ K to $T = T_c$. Label the regions where you have liquid, vapor, or liquid–vapor equilibrium.

3.33   Let's look at another way to compute phase diagrams from an equation of state without using departure functions but still requiring a numerical solution. The van der Waals equation of state can be written in a convenient, dimensionless form as

$$P = \frac{8T}{3V - 1} - \frac{3}{V^2}$$

where the reduced pressure is denoted by $P \equiv \hat{P}/P_c$, the reduced temperature is denoted by $T \equiv \hat{T}/T_c$, and the reduced volume as $V \equiv \hat{V}/V_c$, where hats denote real quantities and the subscript cs denote the values at the critical point. These dimensionless quantities will be used from here on forward for ease and clarity of calculation.

   Construct a phase diagram for a van der Waals material for values of $0 < V \leq 10$ and $0 < P \leq 1.4$. Your plot should show the phase boundary and isotherms at reduced temperatures given below.

(a) Start by constructing the phase boundary. There are multiple ways to achieve this, but please use the following. The following conditions must be met along the phase boundary:

$$P_1 = P_2$$
$$T_1 = T_2$$
$$\mu_1 = \mu_2$$

where μ is the chemical potential and 1 and 2 denote the liquid and gas phases respectively.
Consider the thermodynamic relation,

$$d\mu = VdP - SdT$$

where S is entropy. Along an isotherm (constant T)

$$\left(\frac{\partial \mu}{\partial V}\right)_T = V \left(\frac{\partial P}{\partial V}\right)_T$$

Integrating this equation, substituting the vander Waals equation of state, and evaluating gives

$$\mu = C(T) - \frac{8T}{3} \ln(3V - 1) + \frac{8TV}{3V - 1} - \frac{6}{V}$$

where $C(T)$ is some function of T.

The specific steps for implementation are:

(1)   choose a temperature T^*
(2)   solve for V_1 and V_2 using the equations along with the specific temperature
(3)   use T^* along with V_1 or V_2 to solve for P^*
(4)   plot the points (V_1, P^*) and (V_2, P^*), on the phase boundary
(5)   increment T^* to a new value and employ a zeroth-order continuation to solve for the new values of V_1, V_2, and P^*.

(b)   On this phase diagram, plot isotherms corresponding to $T = 0.7$, 0.8, 0.9, 1, 1.5, 2, 2.5, and 3. Remember that isotherms on the phase diagram should be perfectly straight and horizontal through the two-phase region.

4 Initial Value Problems

4.1 Introduction

The next three chapters focus on the numerical solution of Ordinary Differential Equations (ODEs). These problems comprise the vast majority of the mathematics that arise in undergraduate chemical engineering courses in transport phenomena, separations, reactions, and process control. As a result, you may be tempted to conclude that you already know how to solve ODEs – after all, how could you complete any of these courses without knowing the requisite mathematics? It is probably more accurate to say that you know how to solve *linear* ODEs. In many cases, the underlying nonlinear processes, such as the inertia in a fluid flow or the temperature dependence of a chemical reaction rate, are approximated as linear phenomena (or constant or even nonexistent phenomena!) to arrive at a problem that you can readily solve using the pencil-and-paper techniques from an elementary differential equations class. In this part of the book, we will show you how to numerically solve problems from kinetics, transport phenomena, and elsewhere where you do not need to make these approximations. In this way, we will show you how to take the physics and chemistry that you learned in these core chemical engineering classes and apply them to more realistic problems.

4.1.1 Classification of ODE Problems

Before we get into the details, it is useful for us to provide an unambiguous definition of ordinary differential equation problems and how to classify them. While this may seem like some sort of mathematical taxonomy with little use in the "real world," it is critical to their numerical solution, since certain numerical methods only work for certain types of problems. This represents a significant departure from the approach you probably used in your differential equations class, where you first solve the differential equation (without worrying about its classification) and then apply the conditions to determine the unknown coefficients.

The general form of a single ODE is given by

$$F\left(x, y, \frac{dy}{dx}, \frac{d^2y}{dx^2}, \ldots, \frac{d^ny}{dx^n}\right) = 0 \qquad (4.1.1)$$

where x is the independent variable and y is the dependent variable. The function F is (in general) a nonlinear function of x, y, and the derivatives of y with respect to x, and n is the order of equation, corresponding to the highest-order derivative. There may also be multiple, coupled ODEs (with multiple dependent variables), that we need to solve simultaneously. Equation (4.1.1) is a non-autonomous form of an ODE, where the function F depends on the independent variable x. In many cases, the non-autonomous nature of F does not cause us any trouble. However, we will encounter a problem when we consider the solution of systems of equations in Section 4.6. Don't worry; we will show you then how to solve the problem.

The solution of Eq. (4.1.1) is a function $y(x)$, and in order to be able to specify it uniquely, we need auxiliary conditions on the dependent variable y. This leads to two types of problems.

- **Initial value problems (IVP)**: Here, all of the auxiliary conditions for the dependent variables are given at the same value of the independent variable x. These conditions are usually called initial conditions. This is different from the term "initial guesses" that we used before. Initial conditions are not approximations; they reflect true values of y at a specific value of x. Often, you will see these conditions given at $x = 0$, which is a rather literal interpretation of an "initial" value. However, we can put the conditions at any single value $x = x_0$. If you like to have the data at the origin, you can always make a change of variable $\tilde{x} = x - x_0$ such that the initial conditions are at the value $\tilde{x} = 0$.

- **Boundary value problems (BVP)**: Here, the auxiliary conditions are given at different values of the independent variable. They are called boundary conditions. Similar to the initial value problems, where you probably associate the initial conditions with the specification of a physical system at the start of a process, you most likely think of boundary value problems as having conditions specified at the boundaries of some physical system, such as fixing the velocity to be zero on the walls of a tube. Although this will indeed be the case for the vast majority of problems that we will encounter here, a boundary value problem has no more need to specify conditions on a physical boundary than an initial value problem needs to specify conditions at $x = 0$. The concept of a boundary value problem is more general and can be thought of as "not an initial value problem."

We will use this classification system of IVPs and BVPs to guide our exploration of the numerical solution of ODEs. In the present chapter, we will focus on IVPs. We begin with single equations, which will allow us to introduce the concepts in a fairly straightforward way. This is not to say that single equations are unimportant, and we will demonstrate the utility of numerically solving a single equation in the context of a continuous stirred tank reactor. We will then move on to solving systems of first-order ODEs. This discussion will be very brief, as we simply need to address the autonomous equation problem that we mentioned at the outset. We will also show you how to use these techniques to solve a single higher-order IVP, which can be reduced to a system of coupled first-order IVPs. We will use this knowledge to analyze a non-isothermal plug-flow reactor, which is a nice example of a coupled system of IVPs that does not have time as the independent variable. When we are talking about time-dependent problems with multiple dependent variables, we are referring to dynamical systems. These problems

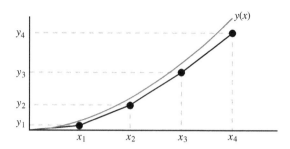

Figure 4.1 Illustration of the basic idea behind solving an ordinary differential equation. The continuous function $y(x)$ is approximated by a discrete set of values y_i at the points x_i. Owing to the approximations used for the numerical solution, there is usually some difference between the exact result and the numerical one.

are so interesting that we have devoted all of Chapter 5 to discussing their analysis in conjunction with a detailed case study on the non-isothermal continuous stirred tank reactor. We will then take up the solution of BVPs in Chapter 6. The solution methods for BVPs are quite different than those for IVPs, so it is important to remember how to classify different types of ODE problems.

4.1.2   General Solution Approach

While IVPs and BVPs will take advantage of different solution methods that benefit from having the conditions all at the same point in x (IVPs) or different points in x (BVPs), the basic ideas behind their numerical solutions are the same. The key idea is that, since we cannot find the solution functions for these equations analytically, we seek approximations of these functions. For the single ODE in Eq. (4.1.1), we seek approximate values of $y(x)$, y_i, at discrete values of the independent variable, x_i, as illustrated in Fig. 4.1. The goal is of course to have y_i be close to $y(x_i)$.

Here lies the connection with the numerical methods that we discussed so far for algebraic equations: the usual strategy to get these approximate solutions is to approximate the derivatives with finite differences, and hence reduce the differential equations to algebraic ones! There are several twists and turns on how to do this best, and these will be discussed in the next sections. But you should keep in mind that after these approximations are employed, we will often have to resort to the methods in Chapter 2 and Chapter 3 to solve the resulting algebraic equations.

4.2   First-Order ODE IVPs

A first-order version of Eq. (4.1.1) corresponds to the case $n = 1$, whereupon the general form of an ODE reduces to

$$F\left(x, y, \frac{dy}{dx}\right) = 0 \qquad (4.2.1)$$

By definition, first-order ODEs can only be initial value problems since we only need one condition – there is no way to put this one condition at two different places. In general, we can specify the initial condition to be

$$y(x = x_0) = y_0 \tag{4.2.2}$$

You can probably imagine that it is not always possible to solve Eq. (4.2.1) for the first derivative; if you could always solve a nonlinear algebraic equation then there would be no need for Chapter 3. However, for most physical problems, Eq. (4.2.1) takes the form

$$F\left(x, y, \frac{dy}{dx}\right) = \frac{dy}{dx} - f(x, y) \tag{4.2.3}$$

which we can write in the familiar form

$$\frac{dy}{dx} = f(x, y) \tag{4.2.4}$$

that we will consider from now on. Furthermore, very often the independent variable x is time, t. So we could have also written Eq. (4.2.4) as

$$\frac{dy}{dt} = f(t, y) \tag{4.2.5}$$

We will tend to use x as the independent variable when we discuss these methods in general, and switch to t when we are referring in particular to dynamic systems, where time is the independent variable.

Regardless of whether the independent variable is x or t, the solution strategy for the above systems involves marching forward from the initial condition to generate approximations $y_1, y_2, \ldots$, of the function $y(x)$, for specific values of the independent variable, $x_1, x_2, \ldots$. For simplicity, we will use evenly spaced steps between the values of x_i and discuss the limitations of this strategy when we consider stiff systems in Section 4.8.2.

4.2.1 Explicit (Forward) Euler

This is the simplest numerical approximation method for IVPs, in which we approximate the derivative with a forward finite difference. Assume that we know y_i, which is an approximate value of $y(x)$ for $x = x_i$. We then seek to find the value of y_{i+1}, which is a new approximation of $y(x)$ for $x = x_{i+1}$. At the point (x_i, y_i), the derivative can be approximated as

$$\left.\frac{dy}{dx}\right|_{x_i} \approx \frac{y_{i+1} - y_i}{x_{i+1} - x_i} \tag{4.2.6}$$

If we use this approximation in the ODE in Eq. (4.2.4), we get

$$\frac{y_{i+1} - y_i}{x_{i+1} - x_i} = f(x_i, y_i) \tag{4.2.7}$$

This algebraic equation can now be solved explicitly for y_{i+1} to obtain

$$y_{i+1} = y_i + (x_{i+1} - x_i)f(x_i, y_i) \tag{4.2.8}$$

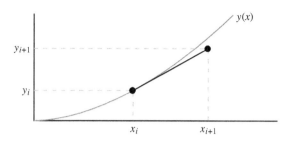

Figure 4.2 Graphical illustration of explicit (forward) Euler. The next approximation y_{i+1} at x_{i+1} is obtained by extrapolating from the old values (x_i, y_i) using the slope of the function $f(x, y)$ at that point.

The increment between two successive values of x, $(x_{i+1} - x_i)$ is called the integration step size and is denoted by h. So we could also write the forward Euler method iteration as

$$y_{i+1} = y_i + hf(x_i, y_i) \qquad (4.2.9)$$

The idea behind explicit Euler is most easily understood in the graphical form of Fig. 4.2. The method of Eq. (4.2.9) consists of using a linear extrapolation over h, using the slope at the beginning of the interval. Continuing similarly, we generate a series of approximate values for each value of y_i. Let's see how this can be programmed in MATLAB.

PROGRAM 4.1   *The implementation of explicit Euler is straightforward. Given the initial condition at x_0, we want to make n steps of size h to obtain values of y_i from $x = x_0$ to $x = x_0 = nh$.*

```matlab
function [xout,yout] = ivp_explicit_euler(x0,y0,h,n_out,i_out)
%   x0 = initial value for x
%   y0 = initial value for y
%   h = step size
%   i_out = frequency to output x and y to [xout, yout]
%   n_out = number of output points
%   requires the function getf(x,y)
xout = zeros(n_out+1,1); xout(1) = x0;
yout = zeros(n_out+1,1); yout(1) = y0;
x = x0; y = y0;
for j = 2:n_out+1
    for i = 1:i_out
        y = y + h*getf(x,y);
        x = x + h;
    end
    xout(j) = x;
    yout(j) = y;
end
```

This program takes as its inputs the initial condition, (x_0, y_0), the step size h, and two arguments for writing the output to two vectors xout *and* yout*: (i) the steps between* *output writing, i_out, and (ii) the number of writing steps, n_out. In general, we will want to use a very small time step to make the solution accurate (which we will discuss next), but often the end result is a plot of the outputs y_i for each x_i. If we have a million data points, we probably only need a few hundred to make a satisfactory plot. The variable i_out keeps the amount of data at a manageable level.*

*The initial lines of the program (8 and 9) set up the output vectors to be the correct size, thereby avoiding the slowdown of the program by continuously increasing the size of the vectors as we add a new point. Line 10 initializes the values of the variables, whereupon we enter a pair of nested **for** loops that execute explicit Euler for the desired number of time steps. Notice the way these loops are constructed. The outer loop sets the number of times we want to write the output to the vectors, while the inner loop does the number of steps between each time we write. The result is that we integrate for a total of n_out * i_out steps, writing the output every i_out steps. Note that the loop runs from j = 2:n_out+1 because we store the initial condition in the first entry of the output vectors.*

Clearly, the smaller the step size h, the better the approximation; yet, more calculations are needed to march over the same distance. As a result, we need to make a good balance between the wall clock time required to solve the problem and the accuracy of the solution. Moreover, we need to be careful about the accumulation of round-off errors if we use a very small step size.

To help us decide how to pick the step size, let's make the analysis of the accuracy of the method and its dependence on h more quantitative. To this end, note that the function $y(x)$ can be expressed in a Taylor series about the point x_i as

$$y(x) = y(x_i) + \frac{dy}{dx}\bigg|_{x_i} (x - x_i) + \frac{1}{2} \frac{d^2 y}{dx^2}\bigg|_{x_i} (x - x_i)^2 + \cdots \qquad (4.2.10)$$

or

$$y(x) = y(x_i) + \frac{dy}{dx}\bigg|_{x_i} (x - x_i) + R_1 \qquad (4.2.11)$$

where all the terms beyond the linear ones have been grouped into the remainder R_1. Using the original ODE,

$$\frac{dy}{dx} = f(x, y) \qquad (4.2.12)$$

the Taylor series expansion becomes

$$y(x) = y(x_i) + f(x_i, y(x_i))(x - x_i) + R_1 \qquad (4.2.13)$$

When we evaluate the Taylor series at $x = x_{i+1}$, the above equation becomes

$$y(x_{i+1}) = y(x_i) + f(x_i, y(x_i))h + R_1 \qquad (4.2.14)$$

Note that quantities like $y(x_i)$ are the values of $y(x)$ at the points x_i; during numerical integration we are using the approximate values y_i, so we could rewrite this result as

$$y_{i+1} = y_i + hf(x_i, y_i) + R_1 \qquad (4.2.15)$$

The first part of the latter expression is exactly the algorithm for explicit Euler in Eq. (4.2.9), i.e. Euler's method truncates the Taylor series expansion at the R_1 term. For small h, the magnitude of R_1 is $\mathcal{O}(h^2)$; the quadratic terms dominate. This results in a local truncation error of $\mathcal{O}(h^2)$ in each step of the method.

What is more important from a practical standpoint is the global truncation error, $y(x_N) - y_N$, where x_N is the value of x where the integration terminates. Since the total number of integration steps is given by

$$N = \frac{x_N - x_0}{h} \tag{4.2.16}$$

the total number of steps is proportional to h^{-1}. For very small h, the total error accumulating from each step scales with $h^{-1}\mathcal{O}(h^2)$, i.e. Euler's method has a global truncation error of $\mathcal{O}(h)$. This is just a sketch of the proof of this very important result on the accuracy of the method. The formal proof is beyond the scope of this text.

In addition to truncation error, there is also round-off error. So the overall error in the solution will be a combination of both effects. If you take big step sizes, you will have big truncation errors but small round-off errors. Conversely, if you take really small steps, the truncation error is small but you have to make many additions so the sum of the round-off errors can be large. There is usually a sweet spot in the middle that minimizes the total error.

Example 4.1   Use explicit Euler to compute the concentration profile for a second-order reaction in batch reactor, with an initial reactant concentration of $c_0 = 3$ mole/liter and a reaction rate constant $k = 2$ L/mol s. The differential equation describing this process is

$$\frac{dc}{dt} = -kc^2 \tag{4.2.17}$$

Solution
Our independent variable is the time, t, and the dependent variable is the concentration, c. Equation (4.2.17) is already written in the standard form of Eq. (4.2.4), where we identify that $f(t,c) = -kc^2$ is only a function of concentration. The explicit Euler scheme (4.2.9) is then

$$c_{i+1} = c_i - 2hc_i^2 \tag{4.2.18}$$

with the initial condition $c_0 = 3$.

This problem can be solved by separation of variables, with the concentration profile

$$c = \frac{c_0}{1 + kc_0 t} \tag{4.2.19}$$

which, with our initial concentration and reaction rate, becomes

$$c = \frac{3}{1 + 6t} \tag{4.2.20}$$

We can easily understand the local truncation error of explicit Euler by taking one step using Eq. (4.2.18) and comparing the result to the exact result in Eq. (4.2.20). For example, if we have a relatively small step $h = 0.001$, the numerical result is

$$c_1 = 3 - 2(0.001)(3)^2 = 2.982\,000\,0 \qquad (4.2.21)$$

while the exact result gives

$$c(t = 0.001) = \frac{3}{1 + 6(0.001)} = 2.982\,100\,7 \qquad (4.2.22)$$

For such a small step, the numerical solution is equal to the exact solution out to the number of significant digits of the step size. This makes sense since the local truncation error is of order $h^2 = 1 \times 10^{-6}$.

Now what happens if we take a large step size, such as $h = 0.1$. The numerical result is then

$$c_1 = 3 - 2(0.1)(3)^2 = 1.2 \qquad (4.2.23)$$

and the exact result is

$$c(t = 0.1) = \frac{3}{1 + 6(0.1)} = 1.875 \qquad (4.2.24)$$

This result also makes sense; the local truncation error is of order $h^2 = 0.01$.

Note that you can run into nonsensical results if you go to even larger step sizes. For example, if you use $h = 1$, the numerical solution gives $c_1 = -15$, which is not even physically realistic.

If we want to assess the global truncation error, we should use Program 4.1 to perform the integration. For this problem, the function `getf` should be:

```
1  function out = getf(t,c)
2  out = -2*c^2;
```

There are two things to notice about this function relative to Program 4.1. First, we have switched from the notation (x, y) in the main function for explicit Euler to (t, c) in our function file for the right-hand side of the ODE. This does not cause any problems for the program, since the variables within a function are defined locally – the main function sends the current values of x and y, which are received by the subfunction as variables called t and c. The important point is that the independent and dependent variables appear in the same order in the main function and the subfunction. Second, the original differential equation (4.2.17) is autonomous, where the forcing function only depends on the concentration. In order to get our file `getf` to work seamlessly with Program 4.1, we send both the values of the time and concentration to the subfunction. You should notice, however, that line 2 of the subfunction uses only the value of the concentration.

Figure 4.2 shows the results from the numerical integration. It is clear from the main panel that the accuracy of the integration decreases with the step size h. Moreover, we can see the accumulation of these errors as a function of time, as the numerical solution gradually diverges from the exact result for larger values of h. The inset shows the

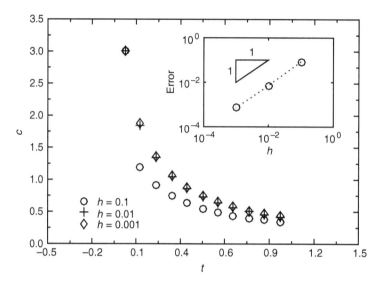

Figure 4.3 Concentration profiles for a batch reactor with second-order kinetics (Eq. (4.2.17)) as a function of time for different values of the step size h from explicit Euler. The inset shows the error between the numerical results and the exact solution at the end of the integration, illustrating the $\mathcal{O}(h)$ global truncation error.

difference between the exact solution and the numerical integration at the longest time. By plotting these data on a log–log plot, we can see that the error indeed has a slope that is linear in the step size.

4.2.2   Implicit (Backward) Euler

As you were reading our description of the explicit Euler method, you may have thought that our choice to evaluate the derivative at the current value of (x_i, y_i) seemed arbitrary. Couldn't we instead evaluate the derivative at the new values (x_{i+1}, y_{i+1})? This is indeed possible and known as implicit Euler or backward Euler. In this method, we use a backward finite difference approximation, whereby

$$\left.\frac{dy}{dx}\right|_{x_{i+1}} \approx \frac{y_{i+1} - y_i}{x_{i+1} - x_i} \tag{4.2.25}$$

Inserting this approximation into the original ODE in Eq. (4.2.4) gives

$$\frac{y_{i+1} - y_i}{x_{i+1} - x_i} = f(x_{i+1}, y_{i+1}) \tag{4.2.26}$$

An equivalent way to arrive at the same result is to express the function $y(x)$ in a Taylor series about the point x_{i+1} as

$$y(x) = y(x_{i+1}) + \left.\frac{dy}{dx}\right|_{x_{i+1}} (x - x_{i+1}) + R_1 \tag{4.2.27}$$

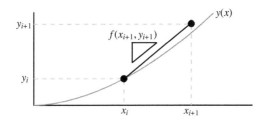

Figure 4.4 Schematic illustration of the implicit Euler method.

or

$$y(x) = y(x_{i+1}) + f(x_{i+1}, y(x_{i+1}))(x - x_{i+1}) + R_1 \qquad (4.2.28)$$

Evaluating at $x = x_i$, truncating R_1, and using the approximate values of the function $y(x)$ we get again

$$\frac{y_{i+1} - y_i}{x_{i+1} - x_i} = f(x_{i+1}, y_{i+1}) \qquad (4.2.29)$$

This second derivation shows that both implicit and explicit Euler have the same truncation error, since they both neglect the $\mathcal{O}(h^2)$ terms in R_1.

Switching the location at which we evaluate the derivative introduces a new complication that we illustrate graphically in Fig. 4.4. In the illustration of the explicit Euler scheme in Fig. 4.2, we were able to evaluate the slope at (x_i, y_i) and thus march forward in time. This feature is the "explicit" part of explicit Euler. If we express the implicit Euler scheme as

$$y_{i+1} = y_i + hf(x_{i+1}, y_{i+1}) \qquad (4.2.30)$$

we see the problem – in general, y_{i+1} cannot be calculated analytically (explicitly). Rather, it needs to be computed from the solution of the algebraic equation

$$R(y_{i+1}) = y_{i+1} - y_i - hf(x_{i+1}, y_{i+1}) = 0 \qquad (4.2.31)$$

which in general will be nonlinear. The method of choice for its solution is, as we know, Newton's method, and it has to be applied in each step of the integration! Clearly, the implicit Euler method is more difficult to implement than the explicit Euler method, and we already saw that the local truncation error is the same for both methods. So, at a first glance, there appears to be no particular motivation to use implicit Euler. We will settle this issue later in Section 4.3.

PROGRAM 4.2 *We can quickly modify Program 4.1 to perform implicit Euler by changing the way we make a step.*

```
1  function [xout,yout] = ivp_implicit_euler(x0,y0,h,n_out,i_out)
2  %   x0 = initial value for x
3  %   y0 = initial value for y
4  %   h = step size
5  %   i_out = frequency to output x and y to [xout, yout]
6  %   n_out = number of output points
```

```
7    %  requires the function getf(y,yold,x,h)
8    xout = zeros(n_out+1,1); xout(1) = x0;
9    yout = zeros(n_out+1,1); yout(1) = y0;
10   x = x0; y = y0; tol = 1e-8;
11   for j = 2:n_out+1
12       for i = 1:i_out
13           yold = y;
14           f = getf(y,yold,x,h);
15           while abs(f) > tol
16               y = y - f/getdf(y,x,h);
17               f = getf(y,yold,x,h);
18           end
19           x = x + h;
20       end
21       xout(j) = x;
22       yout(j) = y;
23   end
```

In this modified version, each iteration of the outer time stepping loop has an inner loop that does Newton's method. Notice a few items in this modified program. First, on line 10, we added the tolerance variable for Newton's method. Second, on line 13, we save the value of y_{i+1} from the previous loop as the value of y_i for the current iteration. This is an important step, since we are solving for the new y_{i+1} but the old value y_i appears in Eq. (4.2.31). Third, the subfunction getf *requires sending the current value of the unknown y_{i+1}, along with the old value of y_i, x_i and the time step h. However, since the old value y_i only appears by itself in the implicit Euler scheme in Eq. (4.2.31), it disappears when we differentiate with respect to y_{i+1}. As a result, the subfunction* getdf *for the derivative only requires the values of y_{i+1}, x, and h. Fourth, notice that the way we are starting up Newton's method at each time step is a zero-order continuation method. We are using the old value of the dependent variable y as the initial guess. Normally, this continuation method works very well with convergence in a few iterations. The reason is that we already know that we need to take a very small step size h to avoid large truncation errors. If the step size is small, then the value of y_{i+1} is not going to be very different than the value of y_i. As a result, zero-order continuation produces a good initial guess if the solution is going to be accurate in the first place.*

Example 4.2   Use implicit Euler to integrate

$$\frac{dy}{dx} = x^2 y^{1/2} \qquad (4.2.32)$$

subject to the initial condition $y(0) = 1$.

Solution

The implicit Euler formula would be

$$y_{i+1} = y_i + h x_{i+1}^2 y_{i+1}^{1/2} \qquad (4.2.33)$$

In this equation, we already know the value of y_i from the previous step, and we also know $x_{i+1} = x_i + h$. We can thus write the implicit Euler scheme as the root of

$$R(y_{i+1}) = y_{i+1} - y_i - h(x_i + h)^2 y_{i+1}^{1/2} = 0 \tag{4.2.34}$$

The algebra in this problem is relatively simple, so it is a worthwhile exercise for us to write out the iterative scheme in its entirety. To solve for the unknown variable y_{i+1} using Newton's method, we start with an initial guess $y_{i+1}^{(0)}$ and iterate using the formula

$$y_{i+1}^{(k+1)} = y_{i+1}^{(k)} - \frac{y_{i+1}^{(k)} - y_i - h x_{i+1}^2 \left(y_{i+1}^{(k)}\right)^{1/2}}{1 - \frac{1}{2} h x_{i+1}^2 \left(y_{i+1}^{(k)}\right)^{-1/2}} \tag{4.2.35}$$

This problem also has a solution by separation of variables, which we can use to evaluate the accuracy of our numerical integration. We thus write

$$\frac{dy}{y^{1/2}} = x^2 dx \tag{4.2.36}$$

Integrating both sides gives

$$2y^{1/2} = \frac{x^3}{3} + c \tag{4.2.37}$$

and the initial condition leads to $c = 2$. We thus have

$$y = \left(\frac{x^3}{6} + 1\right)^2 \tag{4.2.38}$$

To solve this problem using implicit Euler, we used Program 4.2 along with the function

```
1  function f = getf(y,yold,x,h)
2  f = y - yold - h*(x+h)^2*y^(0.5);
```

and

```
1  function df = getdf(y,x,h)
2  df = 1 - 0.5*h*(x+h)^2*y^(-0.5);
```

Figure 4.5 shows the resulting integration for different values of h. As was the case in explicit Euler, the implicit Euler solution improves as the step size decreases. Likewise, we are able to clearly show that the global truncation error at the end of the integration scales like the step size h, exactly the same as we had for explicit Euler.

Example 4.3 Write a program to solve

$$\frac{dy}{dt} = y^{1/2}, \quad y(0) = 2 \tag{4.2.39}$$

using implicit Euler.

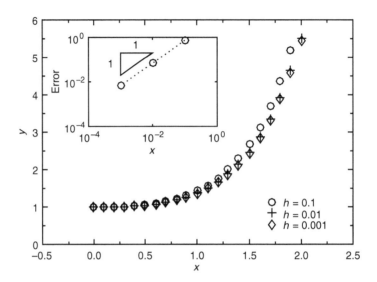

Figure 4.5 Implicit Euler solution to Eq. (4.2.32) with initial condition $y(0) = 1$ for different step sizes h. The inset shows that the global truncation error is $\mathcal{O}(h)$.

Solution

At each time step we need to solve the equation

$$y_{i+1} = y_i + hy_{i+1}^{1/2} \tag{4.2.40}$$

We thus need an inner loop where we use Newton's method to solve

$$R = y_{i+1} - y_i - hy_{i+1}^{1/2} \tag{4.2.41}$$

We also need the derivative for Newton's method,

$$\frac{\partial R}{\partial y_{i+1}} = 1 - \frac{h}{2}y_{i+1}^{-1/2} \tag{4.2.42}$$

Thus, the inner loop of the program involves iterating over k for

$$y_{i+1}^{(k+1)} = y_{i+1}^{(k)} - \frac{y_{i+1}^{(k)} - y_i - h(y_{i+1}^{(k)})^{1/2}}{1 - \frac{h}{2}(y_{i+1}^{(k)})^{-1/2}} \tag{4.2.43}$$

The numerical solution to this problem can still use Program 4.2 even though the forcing function is autonomous. We use the subfunctions

```
1  function f = getf(y,yold,x,h)
2  f = y - yold - h*y^(0.5);
```

and

```
1  function df = getdf(y,x,h)
2  df = 1 - 0.5*h*y^(-0.5);
```

While both functions take as an input the independent variable x, to provide compatibility with Program 4.2 for implicit Euler, we simply ignore them in the computation of f or f'. Obviously we could modify Program 4.2 so that the function calls to `getf` and `getdf` only require the dependent variable y. However, it is nice to be able to reuse our existing programs without doing any extra work.

Similar to Example 4.2, this problem also has a solution by separation of variables. We rewrite

$$\frac{dy}{y^{1/2}} = dt \tag{4.2.44}$$

and integrate to get

$$2y^{1/2} = t + c \tag{4.2.45}$$

The initial condition gives

$$2\sqrt{2} = c \tag{4.2.46}$$

So the result is

$$y^{1/2} = \frac{t}{2} + \sqrt{2} \tag{4.2.47}$$

or, if you prefer,

$$y = \left(\frac{t}{2} + \sqrt{2}\right)^2 \tag{4.2.48}$$

Figure 4.6 shows the results from the numerical integration and a comparison with the exact solution. Note that the overall trend in the result is very similar to Fig. 4.5, but the errors are not exactly the same. This is what we mean when we say that the global truncation errors are of order h – they are not exactly equal to h, just of that magnitude.

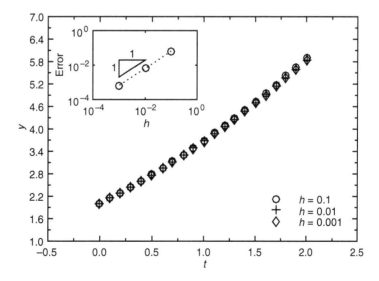

Figure 4.6 Implicit Euler solution to Eq. (4.2.39) for different step sizes h. The inset shows that the global truncation error is $\mathcal{O}(h)$.

Both of the methods that we discussed so far were derived by keeping just the first-order terms in the Taylor series expansion of the function towards deriving a finite difference approximation of its derivative. The implication is that the finite difference approximation that we used is within $\mathcal{O}(h)$ of the actual derivative for small h. This property carries over to the global accuracy of the methods which also scales with $\mathcal{O}(h)$, as we saw in all of the examples so far. We will next discuss higher-order methods that have better accuracy properties.

4.2.3 Predictor–Corrector

This method combines the explicit and implicit Euler methods, by using information on the slope both at x_i and x_{i+1}. One naive way to do this would be just to add the explicit Euler and implicit Euler formulas, which gives

$$y_{i+1} = y_i + \frac{h}{2}[f(x_i, y_i) + f(x_{i+1}, y_{i+1})]$$
(4.2.49)

In essence here, we are using an average of the slopes in the right-hand side of Eq. (4.2.4). This approach is sometimes called the trapezoidal rule for integration of IVPs, as it is analogous to the trapezoidal rule for computing an integral that we will see in Chapter 8. The problem with this approach is that it results in an implicit method, and it is thus rarely used.

A better idea is to use the explicit Euler step to "predict" the value of $f(x_{i+1}, y_{i+1})$, which can then be used to "correct" the value of y_{i+1} predicted by explicit Euler. The algorithm thus consists of a predictor step,

$$y_{i+1}^P = y_i + hf(x_i, y_i)$$
(4.2.50)

which is identical to explicit Euler, and an *explicit* corrector step,

$$y_{i+1}^C = y_i + hf(x_{i+1}, y_{i+1}^P)$$
(4.2.51)

The new value y_{i+1} is then taken as the average of the predictor and corrector steps,

$$y_{i+1} = \frac{y_{i+1}^P + y_{i+1}^C}{2}$$
(4.2.52)

Since we already know the value of y_{i+1}^P from the predictor step, the corrector value y_{i+1}^C can be calculated explicitly, and the method is classified as explicit. This method has a local truncation error of $\mathcal{O}(h^3)$ and a global truncation error of $\mathcal{O}(h^2)$, as we will derive shortly.

PROGRAM 4.3 *It is relatively easy to modify Program 4.1 for explicit Euler to perform the predictor–corrector method.*

```
1  function [xout,yout] = ivp_predictor_corrector(x0,y0,h,n_out,i_out)
2  %   x0 = initial value for x
3  %   y0 = initial value for y
4  %   h = step size
5  %   i_out = frequency to output x and y to [xout, yout]
```

```
6   %   n_out = number of output points
7   %   requires the function getf(x,y)
8   xout = zeros(n_out+1,1); xout(1) = x0;
9   yout = zeros(n_out+1,1); yout(1) = y0;
10  x = x0; y = y0;
11  for j = 2:n_out+1
12      for i = 1:i_out
13          yp = y + h*getf(x,y);
14          yc = y + h*getf(x+h,yp);
15          y = 0.5*(yp + yc);
16          x = x + h;
17      end
18      xout(j) = x;
19      yout(j) = y;
20  end
```

In this revised program, we have replaced line 13 from Program 4.1 for explicit Euler with new lines 13–15 that implement predictor–corrector. Note that line 13 is exactly the same as we had in explicit Euler, except that we decided to change notation for the output of the line to yp to emphasize that this is the predictor value of y.

Example 4.4 Use predictor–corrector to integrate

$$\frac{dy}{dx} = x^2 y^{1/2} \qquad (4.2.53)$$

subject to the initial condition $y(0) = 1$.

Solution
This is the same problem that we saw in Example 4.2 using implicit Euler, which gave a global truncation error of $\mathcal{O}(h)$. We solved this problem for predictor–corrector using Program 4.3 and the function file

```
1  function out = getf(x,y)
2  out = x^2*y^(0.5);
```

The results for different step sizes are shown in Fig. 4.7. As expected, these results are more accurate than we saw in Fig. 4.5 with a global truncation error that scales like h^2.

4.2.4 Runge–Kutta Methods

The predictor–corrector method and Euler's method are special cases of a broader class of higher-order explicit methods known as Runge–Kutta methods. Their more general form is

$$y_{i+1} = y_i + h\phi(x_i, y_i, h) \qquad (4.2.54)$$

The function ϕ is defined as

$$\phi = a_1 k_1 + \cdots + a_n k_n \qquad (4.2.55)$$

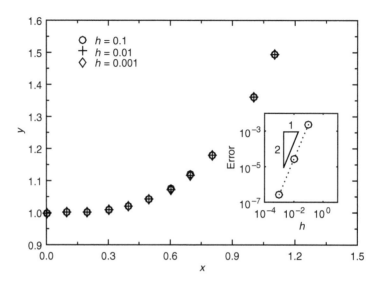

Figure 4.7 Predictor–corrector solution to Eq. (4.2.53) for different step sizes h. The inset shows that the global truncation error is $\mathcal{O}(h^2)$.

where a_is are weights and k_is are evaluations of $f(x, y)$ at different points along the interval $[x_i, x_{i+1}]$. The latter can be interpreted as "slopes" or "directions" that are weighted suitably, and are specified through a set of recursive relations of the form

$$k_1 = f(x_i, y_i) \tag{4.2.56}$$

$$k_2 = f(x_i + p_1 h, y_i + q_{11} k_1 h) \tag{4.2.57}$$

$$\vdots \quad \vdots \tag{4.2.58}$$

$$k_n = f\left(x_i + p_{n-1} h, y_i + q_{n-1,1} k_1 h + \cdots + q_{n-1,n-1} k_{n-1} h\right) \tag{4.2.59}$$

The constants p_i, q_{ij} are also weights, which along with a_i, are determined so as to enforce a desired accuracy. The number of such terms, n, refers to the order of the method. It also is the number of terms used in the Taylor series expansion used in the derivation of the method, and the global truncation error is $\mathcal{O}(h^n)$.

We can thus "dial in" the desired accuracy by increasing the value of n, albeit at a cost of increasing the number of function evaluations that we use to make a single time step. Let's see how this works for the first few levels of accuracy.

First-Order Runge–Kutta

A Runge–Kutta method with $n = 1$ where $a_1 = h$ results in

$$\phi = hf(x_i, y_i) \tag{4.2.60}$$

If we substitute this form of ϕ into Eq. (4.2.54), we get back the explicit Euler formula. Thus, explicit Euler is the first-order Runge–Kutta method.

Second-Order Runge–Kutta

Let's begin by expanding $y(x)$ in a Taylor series about the point x_i,

$$y(x) = y(x_i) + \left.\frac{dy}{dx}\right|_{x_i}(x - x_i) + \frac{1}{2}\left.\frac{d^2y}{dx^2}\right|_{x_i}(x - x_i)^2 + \cdots \qquad (4.2.61)$$

but now, in contrast to what we did for the Euler methods, let us also retain the second-order terms and group the remaining higher-order terms in R_2,

$$y(x) = y(x_i) + \left.\frac{dy}{dx}\right|_{x_i}(x - x_i) + \frac{1}{2}\left.\frac{d^2y}{dx^2}\right|_{x_i}(x - x_i)^2 + R_2 \qquad (4.2.62)$$

Evaluating y at $x = x_{i+1}$, the above equation becomes

$$y(x_{i+1}) = y(x_i) + \left.\frac{dy}{dx}\right|_{x_i}h + \frac{1}{2}\left.\frac{d^2y}{dx^2}\right|_{x_i}h^2 + R_2 \qquad (4.2.63)$$

For small h, we thus know that R_2 is $\mathcal{O}(h^3)$ since the first term that we neglected was the cubic term in the Taylor series. We also know from the differential equation that

$$\frac{dy}{dx} = f(x, y) \qquad (4.2.64)$$

This can be used to substitute for both the first and second derivatives above, where for the second derivative we make use of the chain rule,

$$\frac{d^2y}{dx^2} = \frac{\partial f}{\partial x} + \frac{\partial f}{\partial y}\frac{dy}{dx} \qquad (4.2.65)$$

or in simpler notation,

$$\frac{d^2y}{dx^2} = f_x + f_y f \qquad (4.2.66)$$

The resulting equation is then

$$y(x_{i+1}) = y(x_i) + f(x_i, y(x_i))h + \frac{h^2}{2}\left(f_x + f_y f\right)_{(x_i, y(x_i))} + R_2 \qquad (4.2.67)$$

Truncating at the R_2 term and using the approximate values of $y(x)$, we get

$$y_{i+1} = y_i + f(x_i, y_i)h + \frac{h^2}{2}\left(f_x + f_y f\right)_{(x_i, y_i)} \qquad (4.2.68)$$

where, by neglecting R_2, we have introduced a truncation error of $\mathcal{O}(h^3)$.

The Runge–Kutta method for $n = 2$ has the form

$$y_{i+1} = y_i + (a_1 k_1 + a_2 k_2)h \qquad (4.2.69)$$

where

$$k_1 = f(x_i, y_i) \qquad (4.2.70)$$
$$k_2 = f\left(x_i + p_1 h, y_i + q_{11} k_1 h\right) \qquad (4.2.71)$$

We need to select the weights a_1, a_2, p_1, q_{11} so that this form matches Eq. (4.2.68). First, we expand k_2 around (x_i, y_i) to obtain

$$k_2 = f(x_i, y_i) + p_1 h f_x(x_i, y_i) + q_{11} h k_1 f_y(x_i, y_i) + \mathcal{O}(h^2) \qquad (4.2.72)$$

or equivalently,

$$k_2 = f(x_i, y_i) + p_1 h f_x(x_i, y_i) + q_{11} h f(x_i, y_i) f_y(x_i, y_i) + \mathcal{O}(h^2) \qquad (4.2.73)$$

Since k_2 is multiplied by h in Eq. (4.2.69), we only want to keep terms up to $\mathcal{O}(h)$ in the latter equation to ensure that in both expressions we neglect terms of $\mathcal{O}(h^3)$ and higher. If we substitute Eqs. (4.2.70) and (4.2.73) into Eq. (4.2.69) and only retain terms up to $\mathcal{O}(h^2)$, the resulting Runge–Kutta formula becomes

$$y_{i+1} = y_i + a_1 f(x_i, y_i) h + a_2 [f(x_i, y_i) + p_1 h f_x(x_i, y_i) + q_{11} h (ff_y)(x_i, y_i)] h \qquad (4.2.74)$$

Matching up the coefficients of equal powers of h, we get

$$a_1 + a_2 = 1 \qquad (4.2.75)$$

$$a_2 p_1 = \frac{1}{2} \qquad (4.2.76)$$

$$a_2 q_{11} = \frac{1}{2} \qquad (4.2.77)$$

This system of equations is underspecified, so there are infinitely many solutions and thus infinitely many second-order Runge–Kutta methods.

The predictor–corrector method in Section 4.2.3 corresponds to the choice $a_1 = 1/2$, which implies that $a_2 = 1/2$ and $p_1 = q_{11} = 1$. In this case, we get

$$y_{i+1} = y_i + \frac{h}{2} f(x_i, y_i) h + \frac{h}{2} f(x_i + h, y_i + h f(x_i, y_i)) \qquad (4.2.78)$$

which reduces to Eq. (4.2.52).

Another example of a second-order Runge–Kutta method is the choice $a_1 = 0$, which is called the mid-point method. With the latter, we then have $a_2 = 1$ and $p_1 = q_{11} = 1/2$. Using these results, we get

$$y_{i+1} = y_i + h f \left(x_i + \frac{h}{2}, y_i + \frac{h}{2} f(x_i, y_i) \right) \qquad (4.2.79)$$

Fourth-Order Runge–Kutta

By far the most popular Runge–Kutta method is the particular fourth-order method

$$y_{i+1} = y_i + \frac{h}{6}(k_1 + 2k_2 + 2k_3 + k_4) \qquad (4.2.80)$$

with the function evaluations at

$$k_1 = f(x_i, y_i) \qquad (4.2.81)$$

$$k_2 = f\left(x_i + \frac{h}{2}, y_i + \frac{h}{2} k_1 \right) \qquad (4.2.82)$$

$$k_3 = f\left(x_i + \frac{h}{2}, y_i + \frac{h}{2}k_2\right) \tag{4.2.83}$$

$$k_4 = f\left(x_i + h, y_i + hk_3\right) \tag{4.2.84}$$

This particular method is very efficient and easy to program, and represents a good compromise between the number of functional evaluations and the global accuracy, which is $\mathcal{O}(h^4)$. Given its ubiquitousness, this particular fourth-order Runge–Kutta method is normally called "RK4."

The derivation of RK4 follows the same approach as the derivation we used for the second-order methods, except that we now need to expand up to the fourth-order terms in the Taylor series and match them to the Runge–Kutta scheme. The algebra is tedious but the concept is straightforward.

Higher-order methods or implicit variants of Runge–Kutta methods are also possible, but we will not discuss them here.

PROGRAM 4.4   *As was the case with predictor–corrector, it is relatively easy to modify the explicit Euler program 4.1 to instead execute RK4.*

```
function [xout,yout] = ivp_RK4(x0,y0,h,n_out,i_out)
%   x0 = initial value for x
%   y0 = initial value for y
%   h = step size
%   i_out = frequency to output x and y to [xout, yout]
%   n_out = number of output points
%   requires the function getf(x,y)
xout = zeros(n_out+1,1); xout(1) = x0;
yout = zeros(n_out+1,1); yout(1) = y0;
x = x0; y = y0;
for j = 2:n_out+1
    for i = 1:i_out
        k1 = getf(x,y);
        k2 = getf(x + 0.5*h,y + 0.5*h*k1);
        k3 = getf(x + 0.5*h,y + 0.5*h*k2);
        k4 = getf(x + h, y + h*k3);
        y = y + (h/6)*(k1 + 2*k2 + 2*k3 + k4);
        x = x + h;
    end
    xout(j) = x;
    yout(j) = y;
end
```

In this program, we replaced line 13 of the original program (which performed one time step of explicit Euler) with lines 13–17 here. In the first four lines, we sequentially compute the values of k_i, using the value of k_{i-1} to simplify the coding. Line 17 then provides the appropriate summation to get the next time step of RK4.

Example 4.5   Write out the equations required to integrate

$$\frac{dy}{dt} = y^2 + t, \quad y(0) = 1 \tag{4.2.85}$$

using RK4.

Solution

The RK4 algorithm is

$$y_{i+1} = y_i + \frac{h}{6}(k_1 + 2k_2 + 2k_3 + k_4) \tag{4.2.86}$$

For this problem, $f(t, y) = y^2 + t$. So for the first evaluation we get

$$k_1 = f(t_i, y_i) = y_i^2 + t_i \tag{4.2.87}$$

For the second evaluation, we have

$$\begin{aligned} k_2 &= f\left(t_i + \frac{h}{2}, y_i + \frac{h}{2}k_1\right) \\ &= f\left(t_i + \frac{h}{2}, y_i + \frac{h}{2}\left[y_i^2 + t_i\right]\right) \\ &= \left[y_i + \frac{h}{2}\left(y_i^2 + t_i\right)\right]^2 + \left(t_i + \frac{h}{2}\right) \end{aligned} \tag{4.2.88}$$

For the third evaluation, we have

$$\begin{aligned} k_3 &= f\left(t_i + \frac{h}{2}, y_i + \frac{h}{2}k_2\right) \\ &= f\left(t_i + \frac{h}{2}, y_i + \frac{h}{2}\left\{\left[y_i + \frac{h}{2}\left(y_i^2 + t_i\right)\right]^2 + \left(t_i + \frac{h}{2}\right)\right\}\right) \\ &= \left[y_i + \frac{h}{2}\left\{\left[y_i + \frac{h}{2}\left(y_i^2 + t_i\right)\right]^2 + \left(t_i + \frac{h}{2}\right)\right\}\right]^2 \\ &\quad + \left(t_i + \frac{h}{2}\right) \end{aligned} \tag{4.2.89}$$

For the fourth evaluation, we have

$$\begin{aligned} k_4 &= f\left(t_i + h, y_i + hk_3\right) \\ &= f\left(t_i + h, y_i + h\left\{\left[y_i + \frac{h}{2}\left\{\left[y_i + \frac{h}{2}\left(y_i^2 + t_i\right)\right]^2 + \left(t_i + \frac{h}{2}\right)\right\}\right]^2 + \left(t_i + \frac{h}{2}\right)\right\}\right) \\ &= \left(y_i + h\left\{\left[y_i + \frac{h}{2}\left\{\left[y_i + \frac{h}{2}\left(y_i^2 + t_i\right)\right]^2 + \left(t_i + \frac{h}{2}\right)\right\}\right]^2 + \left(t_i + \frac{h}{2}\right)\right\}\right)^2 \\ &\quad + t_i + h \end{aligned} \tag{4.2.90}$$

Clearly, even for a relatively simple form of $f(t, y)$, you should not try to write out the equations for RK4 explicitly! Rather, you should use the approach illustrated in Program 4.4.

Example 4.6 Use RK4 to integrate

$$\frac{dy}{dx} = x^2 y^{1/2} \qquad (4.2.91)$$

subject to the initial condition $y(0) = 1$.

Solution
This is the same problem that we saw in Example 4.2 and using implicit Euler, which gave a global truncation error of $\mathcal{O}(h)$, and Example 4.4 using predictor–corrector, which gave a global truncation error of $\mathcal{O}(h^2)$. We solved this problem here using Program 4.4 and the function file

```
1  function out = getf(x,y)
2  out = x^2*y^(0.5);
```

The results for different step sizes are shown in Fig. 4.8. As expected, these results are more accurate than we saw in Fig. 4.5 or Fig. 4.7, with a global truncation error that scales like h^4. Indeed, we can hardly distinguish between the different points in the plot in Fig. 4.8, even at the largest step size.

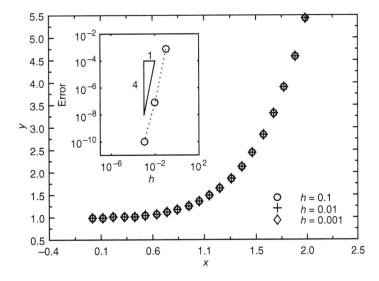

Figure 4.8 RK4 solution to Eq. (4.2.91) for different step sizes h. The inset shows that the global truncation error is $\mathcal{O}(h^4)$.

4.3   Numerical Stability

We will now discuss another important feature of these numerical integration methods, the one of numerical stability. Consider the following simple example,

$$\frac{dy}{dx} = -y, \quad y(0) = 1 \tag{4.3.1}$$

The solution to this separable equation is

$$y = e^{-x} \tag{4.3.2}$$

As $x \to \infty$, $y \to 0$ so the solution is bounded.

Now let's solve this equation using explicit Euler with a step size of $h = 4$,

$$y_{i+1} = y_i - hy_i = y_i(1 - h) = -3y_i \tag{4.3.3}$$

If we evaluate the first few terms, we get the results in Fig. 4.9. The solution approximation becomes unbounded (diverges), in contrast to the actual solution obtained analytically. Apparently, the problem was caused by the numerical method!

A numerical method is numerically stable if it produces a bounded solution approximation when the exact solution is bounded. Otherwise, it is numerically unstable. Numerical stability and accuracy are critical attributes of numerical methods, and we will investigate the former a bit more in the next subsections.

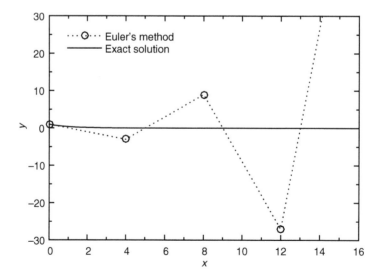

Figure 4.9 The first four steps of the explicit Euler integration of Eq. (4.3.1) with a step size $h = 4$.

4.3.1   Linear IVPs

We begin by considering the linear model problem

$$\frac{dy}{dx} = -\lambda y, \quad y(0) = y_0, \quad \lambda > 0 \tag{4.3.4}$$

The solution to this problem is

$$y = y_0 e^{-\lambda x} \tag{4.3.5}$$

and is bounded and convergent to zero. Under what conditions are the numerical methods we have discussed so far numerically stable, i.e. also produce bounded/convergent approximations? Let's go through them one by one.

Explicit Euler

For this problem, we have

$$y_{i+1} = y_i - h\lambda y_i = y_i(1 - h\lambda) \tag{4.3.6}$$

The approximations that we get are therefore

$$y_1 = y_0(1 - h\lambda) \tag{4.3.7}$$

$$y_2 = y_1(1 - h\lambda) = y_0(1 - h\lambda)^2 \tag{4.3.8}$$

$$\vdots \quad \vdots$$

$$y_n = y_0(1 - h\lambda)^n \tag{4.3.9}$$

For y_n to be bounded as $n \to \infty$, we must have

$$|1 - h\lambda| \leq 1 \tag{4.3.10}$$

Since both h and λ are positive, this means that

$$h \leq \frac{2}{\lambda} \tag{4.3.11}$$

This result shows that the explicit Euler method for this simple linear ODE requires a sufficiently small step in order to be numerically stable. For the previous simple example where $\lambda = 1$, our choice of $h = 4$ violated this stability limit. We should have picked $h < 2$. If λ is large (which would indicate a fast rate of decay in the actual solution), then we would need a very small time step to generate convergent approximations (at the expense of course of many function evaluations in the implementation of the method). Without considering this fact, we may generate an unbounded solution and mistakenly conclude that it represents the actual behavior of our system.

Implicit Euler

Now let's consider what happens with the implicit Euler method,

$$y_{i+1} = y_i - h\lambda y_{i+1} \tag{4.3.12}$$

We can easily solve this equation analytically for y_{i+1},

$$y_{i+1} = \left(\frac{1}{1+h\lambda}\right) y_i \tag{4.3.13}$$

Now, when we compute the solution approximations we get

$$y_1 = y_0(1 + h\lambda)^{-1} \tag{4.3.14}$$

$$y_2 = y_1(1 + h\lambda)^{-1} = y_0(1 + h\lambda)^{-2} \tag{4.3.15}$$

$$\vdots \quad \vdots$$

$$y_n = y_0(1 + h\lambda)^{-n} \tag{4.3.16}$$

In order for them to stay bounded, we must have

$$\left|\frac{1}{1+h\lambda}\right| \le 1 \tag{4.3.17}$$

Since $h > 0$ and $\lambda > 0$, this is always satisfied!

Implicit Euler is thus unconditionally stable. This is the big advantage of implicit methods: you can choose any step size h and not worry about the solution blowing up when it should stay bounded. The price you pay of course is computational complexity. Remember, though, that even if the method is stable for any step size h, the step size still matters as it affects the accuracy.

Predictor–corrector

The stability of the predictor–corrector method is somewhat harder to assess but the same approach can be followed. The predictor–corrector formula for Eq. (4.3.4) is

$$y_{i+1}^P = y_i - h\lambda y_i = (1 - h\lambda)y_i \tag{4.3.18}$$

$$y_{i+1}^C = y_i - h\lambda y_{i+1}^P = y_i(1 - h\lambda + h^2\lambda^2) \tag{4.3.19}$$

$$y_{i+1} = \frac{y_{i+1}^P + y_{i+1}^C}{2} = y_i \left(1 - h\lambda + \frac{1}{2}h^2\lambda^2\right) \tag{4.3.20}$$

For boundedness of the approximations we need

$$\left|1 - h\lambda + \frac{1}{2}h^2\lambda^2\right| \le 1 \tag{4.3.21}$$

Note that the argument in the absolute value above is always positive and the minimum of the function is 1/2 at $h\lambda = 1$. As a result, we need

$$1 - h\lambda + \frac{1}{2}h^2\lambda^2 \le 1 \tag{4.3.22}$$

which we can rewrite as

$$h\lambda \left(\frac{1}{2}h\lambda - 1\right) \le 0 \tag{4.3.23}$$

Since $h\lambda > 0$, the condition becomes

$$h \le \frac{2}{\lambda} \tag{4.3.24}$$

Remarkably, the stability condition for predictor–corrector in Eq. (4.3.24) is identical to the one for the explicit Euler method in Eq. (4.3.11). While predictor–corrector is no more stable than explicit Euler, remember that the predictor–corrector method has the advantage of higher global accuracy.

Fourth-Order Runge–Kutta

The stability of RK4 involves even more tortured algebra but we can work it out with some patience. Like in the previous derivations, we need to rewrite the original formula

$$y_{i+1} = y_i + \frac{h}{6}(k_1 + 2k_2 + 2k_3 + k_4) \tag{4.3.25}$$

in a form

$$y_{i+1} = cy_i \tag{4.3.26}$$

To make our life simpler, let us denote

$$\delta = h\lambda \tag{4.3.27}$$

Then we have

$$k_1 = -\lambda y_i \tag{4.3.28}$$

$$k_2 = -\lambda \left(y_i + \frac{h}{2}k_1 \right) = -\lambda y_i \left(1 - \frac{\delta}{2} \right) \tag{4.3.29}$$

$$k_3 = -\lambda \left(y_i + \frac{h}{2}k_2 \right) = -\lambda y_i \left(1 - \frac{\delta}{2} + \frac{\delta^2}{4} \right) \tag{4.3.30}$$

$$k_4 = -\lambda \left(y_i + hk_3 \right) = -\lambda y_i \left(1 - \delta + \frac{\delta^2}{2} - \frac{\delta^3}{4} \right) \tag{4.3.31}$$

Putting everything together and factoring out the y_i term gives us

$$y_{i+1} = y_i \left[1 - \frac{\delta}{6} \left(6 - 3\delta + \delta^2 - \frac{\delta^3}{4} \right) \right] \tag{4.3.32}$$

The function in brackets is positive everywhere. So for boundedness of the solutions we need to have

$$1 - \frac{\delta}{6} \left(6 - 3\delta + \delta^2 - \frac{\delta^3}{4} \right) \leq 1 \tag{4.3.33}$$

The equation

$$6 - 3\delta + \delta^2 - \frac{\delta^3}{4} = 0 \tag{4.3.34}$$

has a single root, which can be found using Newton's method to be

$$\delta = 2.785 \tag{4.3.35}$$

The stability condition then becomes

$$h \leq \frac{2.785}{\lambda} \tag{4.3.36}$$

Interestingly, RK4 allows a slightly larger step size than the previous explicit methods, and is more accurate.

4.3.2   Nonlinear IVPs

The analysis in the previous subsection was for a model linear ODE that has an analytical solution. Although this may seem like a trivial example, it turns out to have substantial value when we try to characterize numerical stability when solving the nonlinear ODE

$$\frac{dy}{dx} = f(x, y) \tag{4.3.37}$$

To see why, consider again a Taylor series expansion of $f(x, y)$ around a solution point (x_s, y_s),

$$f(x, y) \approx f(x_s, y_s) + \left.\frac{\partial f}{\partial x}\right|_{(x_s, y_s)} (x - x_s) + \left.\frac{\partial f}{\partial y}\right|_{(x_s, y_s)} (y - y_s) + \cdots \tag{4.3.38}$$

For (x, y) close to (x_s, y_s) we can neglect the higher-order terms and write

$$\begin{aligned}\frac{dy}{dx} &= \left(f(x_s, y_s) - \left.\frac{\partial f}{\partial x}\right|_{(x_s, y_s)} x_s - \left.\frac{\partial f}{\partial y}\right|_{(x_s, y_s)} y_s \right) \\ &\quad + \left.\frac{\partial f}{\partial x}\right|_{(x_s, y_s)} x + \left.\frac{\partial f}{\partial y}\right|_{(x_s, y_s)} y\end{aligned} \tag{4.3.39}$$

which has the form

$$\frac{dy}{dx} = \left.\frac{\partial f}{\partial y}\right|_{(x_s, y_s)} y + F(x) \tag{4.3.40}$$

where the term $F(x)$ has no influence on stability. Defining

$$\lambda = -\left.\frac{\partial f}{\partial y}\right|_{y=y_s} \tag{4.3.41}$$

reduces Eq. (4.3.40) exactly to the linear form in Eq. (4.3.4) that we studied before. As a result, the same stability limits will hold, although only locally at the point (x_s, y_s).

As the solution marches along x, λ will change (in a way that we do not know *a priori*!) and so will the actual stability limits if we use an explicit method. This makes the numerical stability of nonlinear ODE initial value problems considerably more difficult to guarantee using explicit methods, calling for either very small steps or the use of implicit methods.

4.4   Case Study: CSTR with a Fluctuating Inlet

4.4.1   Problem Statement

The continuous stirred tank reactor, illustrated in Fig. 4.10, is one of the classic chemical reactor models that we discussed in Section 1.2.4. In this model, we have an inlet flow of some concentration of the reactant, c_{in}, at a volumetric flow rate q. The reactor has a

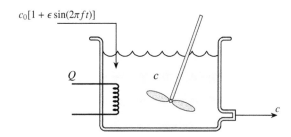

Figure 4.10 Schematic illustration of a continuous stirred tank reactor with a fluctuating inlet concentration given by Eq. (4.4.2).

volume V, so in order to maintain this constant volume we also have an outlet volumetric flow rate q. The reactor is assumed to be well mixed, so the outlet flow is at the same concentration c as the inside of the reactor. We also supply heat at a rate Q to maintain isothermal operation. The reactants inside the reactor are depleted at a rate r per unit volume. If we assume a second-order reaction, we have $r = kc^2$, where k is the rate constant of the reaction with units of L/mol s.

The mass balance for the reactant is given by

$$V\frac{dc}{dt} = q(c_{in} - c) - kc^2 V \tag{4.4.1}$$

In this problem, let's assume that the reactor initially has no reactant, $c(0) = 0$ and see how it approaches the steady state concentration.

When this model is typically used in chemical engineering, we also assume that the inlet concentration is constant at some value $c_{in} = c_0$. In practice, the inlet concentration will fluctuate due to other processing parameters. While we would like to understand the stochastic problem (which can certainly be done by modifying the work below), let's make a simplifying model and assume that the inlet concentration has a periodic oscillation of the form

$$c_{in} = c_0 \left[1 + \epsilon \sin(2\pi ft)\right] \tag{4.4.2}$$

where f is the frequency of the oscillations and $\epsilon < 1$ controls the magnitude of the oscillations.

The goal of this problem is to analyze how the output concentration depends on the relevant dimensionless numbers for this process, in particular the magnitude of the inlet fluctuations and their frequency.

4.4.2   Solution

We first need to determine the relevant dimensionless numbers. We already have one of them, ϵ, which must be dimensionless from the form of Eq. (4.4.2). The typical scale for the concentration is c_0; if there was no reaction then this would be the steady state concentration of the reactor. As is normally the case in the analysis of a CSTR, let us use the residence time V/q as the time scale. When the residence time is short, the reactants flow through the reactor very quickly and the reaction is suppressed. The extent of the reaction with respect to the residence time is controlled by the Dämkohler number, $Da = kc_0V/q$. With these scales, we can rewrite Eq. (4.4.1) as

$$\frac{d\theta}{d\tau} = 1 + \epsilon \sin(2\pi\omega\tau) - \theta - Da\,\theta^2, \quad \theta(0) = 0 \tag{4.4.3}$$

where $\theta = c/c_0$ is the dimensionless concentration, $\tau = tq/V$ is the dimensionless time, and $\omega = fV/q$ is the dimensionless frequency of the inlet oscillations relative to the residence time.

To solve this problem, we modified the RK4 program to also pass the parameters ϵ, ω, and Da:

```
1  function [xout,yout] = ivp_RK4(x0,y0,h,n_out,i_out,omega,Da,epsilon)
2  %  x0 = initial value for x
3  %  y0 = initial value for y
4  %  h = step size
5  %  i_out = frequency to output x and y to [xout, yout]
6  %  n_out = number of output points
7  %  requires the function getf(x,y)
8  xout = zeros(n_out+1,1); xout(1) = x0;
9  yout = zeros(n_out+1,1); yout(1) = y0;
10 x = x0; y = y0;
11 for j = 2:n_out+1
12     for i = 1:i_out
13         k1 = getf(x,y,omega,Da,epsilon);
14         k2 = getf(x + 0.5*h,y + 0.5*h*k1,omega,Da,epsilon);
15         k3 = getf(x + 0.5*h,y + 0.5*h*k2,omega,Da,epsilon);
16         k4 = getf(x + h, y + h*k3,omega,Da,epsilon);
17         y = y + (h/6)*(k1 + 2*k2 + 2*k3 + k4);
18         x = x + h;
19     end
20     xout(j) = x;
21     yout(j) = y;
22 end
```

The function that we need to use is just the right-hand side of Eq. (4.4.3):

```
1  function f = getf(t,c,omega,Da,epsilon)
2  f = 1 + epsilon*sin(2*pi*omega*t) - c - Da*c^2;
```

In writing this function, we took the liberty of using local variables t and c in place of x and y.

Let's first see how the reactor output varies with the frequency of the fluctuations. As we can see in Fig. 4.11, slowly oscillating solutions barely perturb the approach to steady state for a constant inlet concentration. As we increase the oscillation frequency, we get oscillations about the steady state behavior after the initial transient. Interestingly, the transient behavior is affected by the oscillations. We also note that, although the oscillations are of the same magnitude, their frequency affects the magnitude of the concentration fluctuations about the steady state.

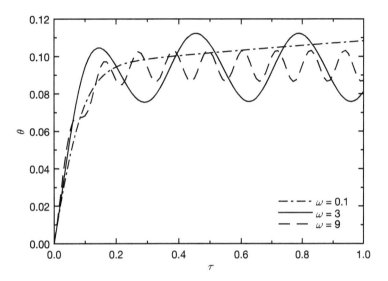

Figure 4.11 Dimensionless concentration as a function of time for different frequencies ω of the forcing function for $\epsilon = 0.5$ and Da $= 100$.

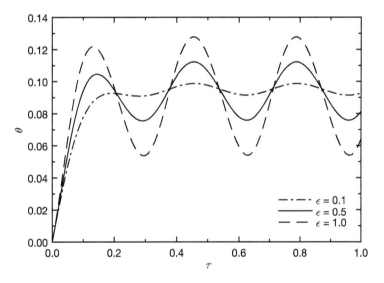

Figure 4.12 Dimensionless concentration as a function of time for different values of the strength of the oscillation, ϵ, for $\omega = 3$ and Da $= 100$.

The effect of the oscillation strength is somewhat different. As we can see in Fig. 4.12, for small values of ϵ, the system rises smoothly to the steady state that we would expect from the classic CSTR model with a fixed input. In contrast, for larger values of ϵ, the system concentration oscillates on the way towards its plateau value and reaches a state in which the concentration steadily oscillates about its mean concentration.

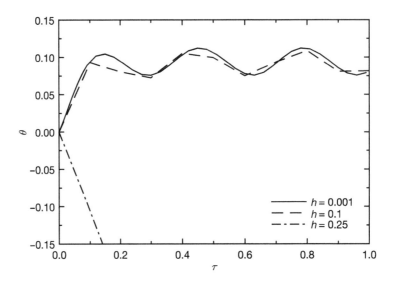

Figure 4.13 Illustration of the numerical stability of the solution for the oscillating CSTR using RK4 with different step sizes, h, for $\epsilon = 0.5$, $\omega = 3$, and Da $= 100$.

The overshoot before the steady oscillations in $c(t)$ depends on ϵ, with stronger inlet concentration fluctuations leading to a higher overshoot.

We will also use this case study to examine the numerical stability of RK4 for this nonlinear problem. In all of the solutions shown above, we picked a sufficiently small time step so that the solution stays stable. (From our knowledge of chemical engineering, we know that the physical system is bounded. The worst case scenario is that the reaction is super fast and the concentration goes to zero. The concentration cannot become negative, and a reaction that consumes the species cannot give a concentration that diverges to infinity.) In Fig. 4.13, we have taken one particular case from our previous parametric study and varied the values of h. For $h = 0.5$, the solution still tracks the more accurate solution at $h = 0.1$ with the additional truncation error that we would expect from RK4. However, when we further increase the time step to $h = 1.0$, the solution becomes unstable. This step size is close to the stability limit, as we are losing control over the integration but the solution has not yet exploded.

4.5   Case Study: Liquid Level Dynamics and Control

4.5.1   Problem Statement

In the previous case study, we considered the dynamics of a reactor when the inlet concentration is fluctuating. In many chemical processes, we want to be able to control the outputs of a process in the presence of changes in the inlets. In this case study, we will see how our knowledge about integrating initial value problems can be applied to a problem in process control.

Rather than continue with the relatively complicated problem of controlling a chemical reactor, we consider a very simple process, a cylindrical storage tank where liquid enters at a volumetric flow rate q and exits at a rate q_o. The liquid level in the tank is denoted by h, A is the cross-sectional area of the tank and ρ is the density of the liquid. There is no reaction occurring in the tank. The unsteady state total mass balance is given by

$$\frac{d(Ah\rho)}{dt} = q\rho - q_o\rho \qquad (4.5.1)$$

or, assuming that ρ and A do not change,

$$A\frac{dh}{dt} = q - q_o \qquad (4.5.2)$$

At steady state, the inlet and outlet flows will be equal, $q = q_o = q_s$, and the liquid level will be constant at its steady state value h_s. We are interested in studying what happens to the level when the inlet flow rate changes due to an upstream disturbance. This will clearly lead to a deviation of the liquid level from its steady state value. We assume that the outlet stream flows freely, driven by gravity, and we know from fluid mechanics that the flow rate will depend on the square root of the liquid level in the tank, i.e. $q_o = k\sqrt{h}$ where k captures the resistance to the flow. Our model thus becomes

$$A\frac{dh}{dt} = q - k\sqrt{h} \qquad (4.5.3)$$

or

$$\frac{dh}{dt} = \frac{q}{A} - \frac{k}{A}\sqrt{h} \qquad (4.5.4)$$

with the corresponding steady state height

$$h_s = \left(\frac{q}{k}\right)^2 \qquad (4.5.5)$$

In this problem, we will consider two separate problems often encountered in dynamic analysis of process behavior.

The first issue is the linearity of the process. In Eq. (4.5.4), the square root dependence of the outlet flow rate on the height makes the initial value problem nonlinear. If the perturbation to q is small, then it is reasonable to simply put Eq. (4.5.4) into a linearized form. To do so, we expand the nonlinear term in a Taylor series around h_s and keep only the linear term, to obtain

$$\sqrt{h} = \sqrt{h_s} + \frac{1}{2h_s^{1/2}}(h - h_s) \qquad (4.5.6)$$

The resulting linear ODE

$$\frac{dh}{dt} = \frac{q}{A} - \frac{k}{A}\left[\sqrt{h_s} + \frac{1}{2h_s^{1/2}}(h - h_s)\right] \qquad (4.5.7)$$

would be a good approximation of the system behavior for small deviations from the steady state. Thus, the first question we want to address is when this linearization approximation is valid.

The second issue is how we can control the system in response to a sudden change in the inlet flow rate q. In other words, what if we would like to operate our tank at (or near) the original steady state level *despite* any changes in the inlet flow rate? In order to achieve this we would need a feedback control system. For example, we could measure the liquid level in the tank and adjust automatically the outlet flow rate through a control valve, to try to correct for any disturbances in the inlet flow rate that would change the level. The way that the valve will change the flow rate will be determined by the controller. You will encounter many such problems in the Process Control course where you will study different types of controllers that can be used to regulate process behavior. The simplest control action in our case would be to change the outlet flow rate proportionally to the deviation of the height from its steady state value, i.e. program our controller to implement the following action

$$q_o = q_{os} + K_c(h - h_s) \tag{4.5.8}$$

where q_{os} is the flow rate of the old steady state (i.e., the old value of q_s before the perturbation). This is called (rather unimaginatively) a proportional controller and K_c is called the controller gain. This should work as long as K_c is positive, since it will tend to correct deviations of the height from its steady state value by changing the outlet flow rate in the right direction (e.g. increase it when the height exceeds its steady state value, and decrease it when the height decreases). The second question we want to address is the effectiveness of this control strategy. For that we would need to solve the differential equation describing the behavior of the tank with the controller in place, i.e.

$$\frac{dh}{dt} = \frac{q}{A} - \frac{q_{os} + K_c(h - h_s)}{A} \tag{4.5.9}$$

To make the problem more concrete, let's consider the parameters $A = 1$ ft^2, $k = 20$ ft^2/min, and $q_s = 60$ ft^3/min. From Eq. (4.5.5), the steady state height is then $h_s = 9$ ft. For our first question, let's consider the difference between the linear model in Eq. (4.5.7) and the nonlinear model in Eq. (4.5.3) corresponding to a 10% drop in the inlet flow rate and an 85% drop in the inlet flow rate. In other words, we want to solve these equations subject to the initial condition

$$h(0) = h_s = 9 \text{ ft} \tag{4.5.10}$$

with $q = 54$ ft^3/min and $q = 9$ ft^3/min, respectively. For our second question, let's consider the 85% drop in the inlet flow rate with controller constants $K_c = 10, 20,$ and 30 ft^2/min.

4.5.2   Solution

From a numerical methods standpoint, the problem is relatively simple – we just want to solve the initial value problems in Eqs. (4.5.7), (4.5.3), and (4.5.9) subject to the initial condition in Eq. (4.5.10) and the suitable value of q. Let's use RK4 with an appropriate modification for the problem:

```
1   function [xout,yout] = ivp_RK4_control(x0,y0,h,n_out,i_out,p,model)
2   %   x0 = initial value for x
3   %   y0 = initial value for y
4   %   h = step size
5   %   i_out = frequency to output x and y to [xout, yout]
6   %   n_out = number of output points
7   %   p = vector of problem parameters
8   %   model = string to select correct function
9   %   requires the function getf(x,y)
10  xout = zeros(n_out+1,1); xout(1) = x0;
11  yout = zeros(n_out+1,1); yout(1) = y0;
12  x = x0; y = y0;
13  for j = 2:n_out+1
14      for i = 1:i_out
15          k1 = getf(x,y,p,model);
16          k2 = getf(x + 0.5*h,y + 0.5*h*k1,p,model);
17          k3 = getf(x + 0.5*h,y + 0.5*h*k2,p,model);
18          k4 = getf(x + h, y + h*k3,p,model);
19          y = y + (h/6)*(k1 + 2*k2 + 2*k3 + k4);
20          x = x + h;
21      end
22      xout(j) = x;
23      yout(j) = y;
24  end
```

If you compare this program with Program 4.4, you will see only a few changes. First, our control program takes in two additional inputs. The vector p contains all of the parameters for the problem. The string model tells us which model we are trying to solve. We will allow three different strings: 'linear' for Eq. (4.5.7), 'nonlinear' for Eq. (4.5.3) and 'controller' for Eq. (4.5.9). We then adjusted the calls for the RK4 quantities k_i in lines 16–19 to pass p and model to the function evaluation.

We coupled our modified RK4 program to a flexible program to compute the forcing function for the initial value problem:

```
1   function f = getf(x,h,p,model)
2   A = p(1);
3   k = p(2);
4   Q = p(3);
5   hs = p(4);
6   Qs = p(5);
7   K = p(6);
8   if strcmp(model,'linear') == 1
9       f = Q/A - k/A*(sqrt(hs) + 0.5*(h-hs)/sqrt(hs));
10  elseif strcmp(model,'nonlinear') == 1
11      f = Q/A - k/A*sqrt(h);
12  elseif strcmp(model,'controller') == 1
13      f = Q/A - (Qs + K*(h-hs))/A;
14  else
15      f = 0;
16  end
```

Lines 2–7 of this program unpack the vector p into the different problem parameters. While this is not necessary to get the program to execute, it simplifies the readability

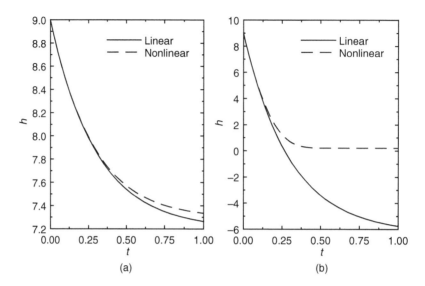

Figure 4.14 Comparison of the linear model in Eq. (4.5.7) and the nonlinear model in Eq. (4.5.3) for a sudden decrease of 10% (a) and 85% (b) in inlet flow rate.

of the code. Lines 8–16 then execute an **if-else** block to select the correct value of $f(h)$ based on the string in model. The MATLAB function strcmp(s1,s2) compares string s1 to s2 and returns a logical output if they are the same (1) or different (0). If the RK4 program sends an unknown model, then we just send back $f = 0$.

Note that it is relatively easy to add additional control strategies to our program by simply adding more cases to getf.m. If we have lots of different strategies, it makes sense to switch from an **if-else** structure to a **switch-case** structure. We will see the **switch** operator in Chapter 8 when we write programs to use different methods of numerical integration.

Let's begin by analyzing the validity of the linear model in Eq. (4.5.7). The linear model is quite useful, since it can be solved exactly. (Nevertheless, since this book is about numerical methods, let's solve it numerically.) Figure 4.14 shows the predicted height of the liquid in the tank for the case of (a) a decrease to the inlet flow rate by 10% and (b) a decrease by 85%. Note that in the case of a small change in the inlet flow rate, both the linear and the nonlinear model give very similar results, showing that the liquid level will eventually settle to a new smaller steady state value. Of course, you could have determined the difference in the steady state directly from the equations without knowing anything about their dynamics. If we denote f as the drop in the flow rate, then the nonlinear model has a steady state

$$h_{s,nl} = \left[\frac{(1-f)q}{k} \right]^2 \tag{4.5.11}$$

while the linear model has the steady state

$$h_{s,l} = \frac{q}{k}\left(\frac{q}{k} - 2f \right) \tag{4.5.12}$$

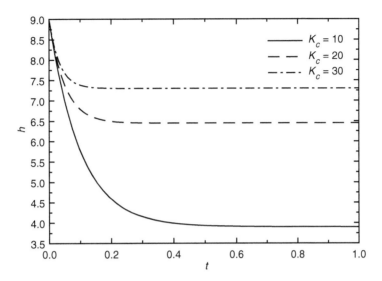

Figure 4.15 Dynamics of the liquid height using the proportional controller in Eq. (4.5.9) for a sudden decrease of 85% in inlet flow rate for $K_C = 10$, 20, and 30 ft^2/min.

By integrating the equations, you can see the time at which the two models start to diverge towards their different steady states. When f is small, the difference between Eq. (4.5.11) and Eq. (4.5.12) is small. When the inlet flow rate change is large, however, Eq. (4.5.12) for the linear model predicts a negative height – essentially that the tank will dry out, which is wrong given the solution of the nonlinear model. This illustrates clearly the inadequacy of the approximate linear model in predicting the actual process behavior when the deviations from the steady around which we linearized the nonlinear function are large. From the dynamics in Fig. 4.14b, we see that the two models differ at a much shorter time for $f = 0.85$ than $f = 0.1$.

Now let's see how the proportional controller can help maintain the height of fluid in the tank. Figure 4.15 shows the change in height for the case of an 85% decrease in the inlet flow rate, for three different values of the proportional gain. As we can see, the higher the gain, the smaller the deviation between the final steady state value and the desired one, i.e. the more effective the control action. If we wanted to eliminate this deviation completely, we would need to resort to more sophisticated control action.

It is also important to recognize that the controller in Eq. (4.5.9) leads to a linear dependence of the outlet flow rate on the current height in the tank. As a result, it has the same problems as we saw in Fig. 4.14 if we do not set the proportionality constant large enough. Indeed, if we analyze the steady state solution to Eq. (4.5.9), we see that we will predict a new steady state liquid level that is negative if the gain satisfies

$$K_c < \frac{fq}{h_s} \qquad (4.5.13)$$

where, again, f is the fraction that the flow rate has dropped. For our particular problem $f = 0.85$ and we need to choose $K_c > 17/3$ ft^2/min to ensure that a positive liquid level is maintained for this particular perturbation in the inlet flow rate.

4.6   Systems of ODE IVPs

We now want to discuss the solution of coupled systems of initial value problems. These types of situations arise frequently in chemical engineering. A very common example is a non-isothermal reaction, where the species conservation equation is coupled to the energy conservation through the temperature-dependent kinetics and the heat of reaction. Fortunately, if you know how to solve a single IVP, solving a system of IVPs is only marginally more difficult.

In general, we want to know how to solve a system of equations of the form of Eq. (4.2.1) where there are now n independent variables y_j,

$$\frac{dy_1}{dx} = f_1(y_1, y_2, \ldots, y_n) \tag{4.6.1}$$

$$\frac{dy_2}{dx} = f_2(y_1, y_2, \ldots, y_n) \tag{4.6.2}$$

$$\vdots \quad \vdots$$

$$\frac{dy_n}{dx} = f_n(y_1, y_2, \ldots, y_n) \tag{4.6.3}$$

The functions $f_j(y_1, y_2, \ldots, y_n)$ couple these equations together. In this general form, we have assumed that each equation is completely coupled to every other equation, but this does not necessarily need to be the case. In addition to the differential equations, we also need to specify the initial conditions for each of the independent variables,

$$
\begin{aligned}
y_1(x = x_0) &= y_{10} \\
y_2(x = x_0) &= y_{20} \\
\vdots \quad &\quad \vdots \\
y_n(x = x_0) &= y_{n0}
\end{aligned}
\tag{4.6.4}
$$

Importantly, since this is a system of IVPs, the initial condition for each of the y_i needs to be specified at exactly the same value of the dependent variable, $x = x_0$.

It is often easier for us to work with the system of IVPs in a more compact vector form

$$\frac{d\mathbf{y}}{dx} = \mathbf{f}(\mathbf{y}) \tag{4.6.5}$$

with the initial condition

$$\mathbf{y}(0) = \mathbf{y_0} \tag{4.6.6}$$

In the latter, $\mathbf{y}$ and $\mathbf{y_0}$ are $n \times 1$ vectors of the dependent variables and their initial conditions, and $\mathbf{f}$ is an n-dimensional vector function.

Note that we choose to focus on autonomous systems, rather than the more natural generalization of the scalar case

$$\frac{d\mathbf{y}}{dx} = \mathbf{f}(x, \mathbf{y}) \tag{4.6.7}$$

The reason is that a non-autonomous system of n equations like Eq. (4.6.7) can be converted directly into an autonomous system of $n + 1$ equations by introducing a new variable,

$$y_{n+1} = x \tag{4.6.8}$$

The differential equation for y_{n+1} is obtained trivially as

$$\frac{dy_{n+1}}{dx} = 1 \tag{4.6.9}$$

with the initial condition

$$y_{n+1}(0) = x_0 \tag{4.6.10}$$

Thus, there is no need to consider systems of non-autonomous systems.

4.6.1   Explicit Methods

The methods that we have described for single equations can be readily adapted for systems of equations, and the accuracy properties of the methods are unchanged. In particular, the adaptation for explicit methods is really straightforward. All that changes is that you have to step through a system of equations instead of a single equation. As a result, their implementation follows directly from the programs that we saw for a single IVP. The only change is that instead of passing a single variable y, we are passing a vector $\mathbf{y}$ to the subfunction `getf`, which will now return a vector of forcing functions $\mathbf{f}$. Likewise, since we only deal with autonomous systems, we do not need to send the independent variable x to `getf`. However, if we want to simply use the programs that are included for a single IVP, we can follow the same strategy that we used in Example 4.3 when we solved an autonomous problem by implicit Euler; the subfuction `getf` can receive the independent variable x but then simply ignore it in the computation of the output. There are also some minor changes we have to make to the output to handle the vector $\mathbf{y}$, which we will explain in Program 4.5.

Explicit Euler
The explicit Euler method in this case has the form

$$\mathbf{y}_{i+1} = \mathbf{y}_i + h\mathbf{f}(\mathbf{y}_i) \tag{4.6.11}$$

Example 4.7   Write the equations required to solve

$$\frac{dy_1}{dt} = y_2 e^{-t} \tag{4.6.12}$$

$$\frac{dy_2}{dt} = -y_1 \tag{4.6.13}$$

by explicit Euler.

Solution

First, we need to convert this problem to an autonomous system by introducing a new variable $y_3 = t$. This gives

$$\frac{dy_1}{dt} = y_2 e^{-y_3} \tag{4.6.14}$$

$$\frac{dy_2}{dt} = -y_1 \tag{4.6.15}$$

$$\frac{dy_3}{dt} = 1 \tag{4.6.16}$$

If we apply the explicit Euler formula in Eq. (4.6.11), the update scheme becomes

$$\begin{bmatrix} y_{1,i+1} \\ y_{2,i+1} \\ y_{3,i+1} \end{bmatrix} = \begin{bmatrix} y_{1,i} \\ y_{2,i} \\ y_{3,i} \end{bmatrix} + h \begin{bmatrix} y_{2,i} e^{-y_{3,i}} \\ -y_{1,i} \\ 1 \end{bmatrix} \tag{4.6.17}$$

Example 4.8 Set up a forward Euler scheme to compute the concentration for a batch reactor with

$$A + B \underset{k_{-1}}{\overset{k_1}{\rightleftharpoons}} C \tag{4.6.18}$$

$$B + C \underset{k_{-2}}{\overset{k_2}{\rightleftharpoons}} D \tag{4.6.19}$$

Solution

If we operate on the assumption that these are elementary chemical reactions, then the system of equations we need to solve is

$$\frac{dc_A}{dt} = -k_1 c_A c_B + k_{-1} c_C \tag{4.6.20}$$

$$\frac{dc_B}{dt} = -k_1 c_A c_B + k_{-1} c_C - k_2 c_B c_C + k_{-2} c_D \tag{4.6.21}$$

$$\frac{dc_C}{dt} = k_1 c_A c_B - k_{-1} c_C - k_2 c_B c_C + k_{-2} c_D \tag{4.6.22}$$

$$\frac{dc_D}{dt} = k_2 c_B c_C - k_{-2} c_D \tag{4.6.23}$$

The corresponding explicit Euler equations are

$$c_{A,i+1} = c_{A,i} + h(-k_1 c_{A,i} c_{B,i} + k_{-1} c_{C,i}) \tag{4.6.24}$$

$$\begin{aligned} c_{B,i+1} = c_{B,i} + h(-k_1 c_{A,i} c_{B,i} + k_{-1} c_{C,i} \\ -k_2 c_{B,i} c_{C,i} + k_{-2} c_{D,i}) \end{aligned} \tag{4.6.25}$$

$$\begin{aligned} c_{C,i+1} = c_{C,i} + h(k_1 c_{A,i} c_{B,i} - k_{-1} c_{C,i} \\ -k_2 c_{B,i} c_{C,i} + k_{-2} c_{D,i}) \end{aligned} \tag{4.6.26}$$

$$c_{D,i+1} = c_{D,i} + h(k_2 c_{B,i} c_{C,i} - k_{-2} c_{D,i}) \tag{4.6.27}$$

Predictor–corrector

The predictor–corrector method is now

$$\mathbf{y}^P_{i+1} = \mathbf{y}_i + h\mathbf{f}(\mathbf{y}_i) \qquad (4.6.28)$$

$$\mathbf{y}^C_{i+1} = \mathbf{y}_i + h\mathbf{f}(\mathbf{y}^P_{i+1}) \qquad (4.6.29)$$

$$\mathbf{y}_{i+1} = \frac{\mathbf{y}^P_{i+1} + \mathbf{y}^C_{i+1}}{2} \qquad (4.6.30)$$

Fourth-order Runge–Kutta

The RK4 scheme becomes

$$\mathbf{y}_{i+1} = \mathbf{y}_i + \frac{h}{6}(\mathbf{k}_1 + 2\mathbf{k}_2 + 2\mathbf{k}_3 + \mathbf{k}_4) \qquad (4.6.31)$$

where

$$\mathbf{k}_1 = \mathbf{f}_1(\mathbf{y}_i) \qquad (4.6.32)$$

$$\mathbf{k}_2 = \mathbf{f}_2\left(\mathbf{y}_i + \frac{h}{2}\mathbf{k}_1\right) \qquad (4.6.33)$$

$$\mathbf{k}_3 = \mathbf{f}_3\left(\mathbf{y}_i + \frac{h}{2}\mathbf{k}_2\right) \qquad (4.6.34)$$

$$\mathbf{k}_4 = \mathbf{f}_4(\mathbf{y}_i + h\mathbf{k}_3) \qquad (4.6.35)$$

Since we are only solving autonomous systems of ODEs, there is no x dependence in the functions $\mathbf{k}$.

4.6.2   Implicit Euler

For implicit methods the concepts for solving systems of ODEs are similar, but the bookkeeping is a bit more complicated. The form of the implicit Euler method becomes now

$$\mathbf{y}_{i+1} = \mathbf{y}_i + h\mathbf{f}(\mathbf{y}_{i+1}) \qquad (4.6.36)$$

which is a system of nonlinear algebraic equations. When we had a single ODE the next step involved solving this single nonlinear algebraic equation using Newton's method. In this case, we need to solve the above system of algebraic equations (at each step) using Newton–Raphson. We therefore form the residual vector

$$\mathbf{R}(y_{i+1}) = \mathbf{y}_{i+1} - \mathbf{y}_i - h\mathbf{f}(\mathbf{y}_{i+1}) \qquad (4.6.37)$$

In the algorithm, we have an outer loop that is updating the values of $\mathbf{y}_i$. At each step i, we have an inner loop (with counter k) that involves first solving

$$\mathbf{J}(\mathbf{y}^{(k)}_{i+1})\boldsymbol{\delta}^{(k+1)} = -\mathbf{R}(\mathbf{y}^{(k)}_{i+1}) \qquad (4.6.38)$$

and then updating the value for $\mathbf{y}$,

$$\mathbf{y}^{(k+1)}_{i+1} = \mathbf{y}^{(k)}_{i+1} + \boldsymbol{\delta}^{(k+1)} \qquad (4.6.39)$$

Similar to the case with one equation, we usually have a pretty good initial guess $\mathbf{y}_{i+1}^{(0)}$ by using the converged solution for $\mathbf{y}_i$. Since the accuracy of the methods is unchanged when we move from one equation to a system of equations, we are still forced to use a relatively small value of h to ensure an accurate solution. The small value of h also ensures that $\mathbf{y}_{i+1}$ is not much different than $\mathbf{y}$, hence zero-order continuation usually leads to Newton–Raphson converging in a few iterations.

The Jacobian for this problem has a nice form that is worth mentioning. We can write the Jacobian out as

$$\mathbf{J} = \begin{bmatrix} \dfrac{\partial R_1}{\partial y_{1,i+1}} & \dfrac{\partial R_1}{\partial y_{2,i+1}} & \cdots & \dfrac{\partial R_1}{\partial y_{n,i+1}} \\[2ex] \dfrac{\partial R_2}{\partial y_{1,i+1}} & \dfrac{\partial R_2}{\partial y_{2,i+1}} & \cdots & \dfrac{\partial R_2}{\partial y_{n,i+1}} \\[2ex] \vdots & \vdots & & \vdots \\[2ex] \dfrac{\partial R_n}{\partial y_{1,i+1}} & \dfrac{\partial R_n}{\partial y_{2,i+1}} & \cdots & \dfrac{\partial R_n}{\partial y_{n,i+1}} \end{bmatrix} \tag{4.6.40}$$

Given the form of R_i in Eq. (4.6.37), this can be written as

$$\mathbf{J} = \mathbf{I} - h \begin{bmatrix} \dfrac{\partial f_1}{\partial y_{1,i+1}} & \dfrac{\partial f_1}{\partial y_{2,i+1}} & \cdots & \dfrac{\partial f_1}{\partial y_{n,i+1}} \\[2ex] \dfrac{\partial f_2}{\partial y_{1,i+1}} & \dfrac{\partial f_2}{\partial y_{2,i+1}} & \cdots & \dfrac{\partial f_2}{\partial y_{n,i+1}} \\[2ex] \vdots & \vdots & & \vdots \\[2ex] \dfrac{\partial f_n}{\partial y_{1,i+1}} & \dfrac{\partial f_n}{\partial y_{2,i+1}} & \cdots & \dfrac{\partial f_n}{\partial y_{n,i+1}} \end{bmatrix} \tag{4.6.41}$$

or

$$\mathbf{J} = \mathbf{I} - h\mathbf{J}_f \tag{4.6.42}$$

where $\mathbf{J}_f$ is the Jacobian of the vector function $\mathbf{f}$. In other words, the Jacobian $\mathbf{J}_f$ corresponds to the steady state of the system of equations.

PROGRAM 4.5   *To solve a system of equations, we need to modify Program 4.2 for a single IVP to implement Newton–Raphson.*

```
1  function [xout,yout] = ivp_implicit_euler_sys(x0,y0,h,n_out,i_out)
2  %  x0 = initial value for x
3  %  y0 = initial value for y
4  %  h = step size
5  %  i_out = frequency to output x and y to [xout, yout]
6  %  n_out = number of output points
7  %  requires the function getR(y,yold,h) and getJ(y,h)
```

```
8   xout = zeros(n_out+1,1); xout(1) = x0;
9   yout = zeros(n_out+1,length(y0)); yout(1,:) = y0;
10  x = x0; y = y0; tol = 1e-8;
11  for j = 2:n_out+1
12      for i = 1:i_out
13          yold = y;
14          R = getR(y,yold,h);
15          count = 1;
16          while norm(R) > tol
17              J = getJ(y,h);
18              del = -J\R;
19              y = y + del;
20              R = getR(y,yold,h);
21              count = count + 1;
22              if count > 20
23                  fprintf('Failed to converge.\n')
24                  return
25              end
26          end
27          x = x + h;
28      end
29      xout(j) = x;
30      yout(j,:) = y;
31  end
```

This program now takes a vector of inputs $\mathbf{y}_0$ as the initial conditions. Since we do not know the number of equations a priori, we modified line 9 of the single equation problem to make the output `yout` into a matrix with `n_out+1` rows for each time point (including the initial condition, which goes in the first spot) and `length(y_0)` columns. The MATLAB command `length` computes the length of a vector, which gives us the flexibility to allow our program to take any input of unknowns. This means that the output of the program will be an $n \times 1$ vector `xout` with the value of the independent vector at each output point and an $n \times m$ matrix `yout` of the m independent variables at each of the n output points. At the end of line 9, we set the first row of the output `yout` to be the initial condition. The syntax `yout(1,:)` puts the vector `y0` on the first row of `yout` using all of the columns. MATLAB has a flexible syntax for inserting vectors into sections of matrices; this is the simplest case where the vector fills the entire row. MATLAB is also intelligent and it will transpose the vector if it is required by the colon operator, which is the case in our program.

In the inner **for** loop, we have now replaced the old **while** loop that we used for Newton's method of a single equation with Newton–Raphson for a system of equations. The call to the residual subfunction `getR` sends the current value of $\mathbf{y}_{i+1}$, the value of $\mathbf{y}_i$ (which we call `yold`), and the step size. Note that the call to the Jacobian subfunction `getJ` does not require the old value $\mathbf{y}_i$ since this term drops out during the differentiation.

Note that we have included a counter to check to see if the Newton–Raphson routine has converged. You may think this is unnecessary, since we said earlier that Newton–Raphson usually converges in a few iterations using zero-order continuation. This is usually true, but the Newton–Raphson scheme will fail if the solution is changing very *quickly. So it is still worth making sure that it is converging to avoid getting stuck in an infinite loop.*

Example 4.9   Write a program to solve the system of equations

$$\frac{dy_1}{dt} = y_1 y_2^2 \tag{4.6.43}$$

$$\frac{dy_2}{dt} = (y_1 - y_2)^2 \tag{4.6.44}$$

using implicit Euler. The initial conditions are $y_1(0) = 1$ and $y_2(0) = 2$.

Solution

This is an autonomous system of equations, so we do not need to add a third equation. If we write out the equations for implicit Euler, we have

$$y_{1,i+1} = y_{1,i} + h y_{1,i+1} y_{2,i+1}^2 \tag{4.6.45}$$

$$y_{2,i+1} = y_{2,i} + h \left(y_{1,i+1} - y_{2,i+1} \right)^2 \tag{4.6.46}$$

We can thus write out the residual

$$\mathbf{R} = \left[\begin{array}{c} y_{1,i+1} - y_{1,i} - h y_{1,i+1} y_{2,i+1}^2 \\ y_{2,i+1} - y_{2,i} - h \left(y_{1,i+1} - y_{2,i+1} \right)^2 \end{array} \right] \tag{4.6.47}$$

If we put our unknowns in the logical order

$$\mathbf{y} = \left[\begin{array}{c} y_{1,i+1} \\ y_{2,i+1} \end{array} \right] \tag{4.6.48}$$

then the Jacobian we need is

$$\mathbf{J} = \left[\begin{array}{cc} 1 - h y_{2,i+1}^2 & -2h y_{1,i+1} y_{2,i+1} \\ -2h \left(y_{1,i+1} - y_{2,i+1} \right) & 1 + 2h \left(y_{1,i+1} - y_{2,i+1} \right) \end{array} \right] \tag{4.6.49}$$

We have solved these equations using Program 4.5 with the residual function

```
function R = getR(y,yold,h)
R = zeros(2,1);
R(1) = y(1) - yold(1) - h*y(1)*y(2)^2;
R(2) = y(2) - yold(2) - h*(y(1) - y(2))^2;
```

and the Jacobian function

```
function J = getJ(y,h)
J = zeros(2);
J(1,1) = 1 - h*y(2)^2;
J(1,2) = -2*h*y(1)*y(2);
J(2,1) = -2*h*(y(1)-y(2));
J(2,2) = 1 + 2*h*(y(1)-y(2));
```

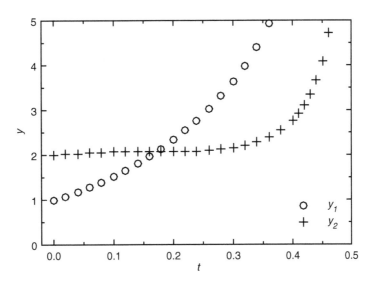

Figure 4.16 Solution to Eqs. (4.6.43) and (4.6.44) with the initial conditions $y_1(0) = 1$ and $y_2(0) = 2$ using implicit Euler.

The results are reported in Fig. 4.16. Notice that the solution is changing very rapidly, and this system diverges around $t = 0.48$. As a result, it is good that we put in a check for the convergence of Newton–Raphson into our program; otherwise, the solution gets stuck in an infinite loop near the point of divergence.

4.7   Higher-Order ODE IVPs

So far we have focused exclusively on first-order ODE IVPs. These are obviously important problems, as they arise frequently in kinetics. However, we also routinely encounter higher-order ODE IVPs in chemical engineering, most notably in process control. Let's see how we can solve a higher-order equation of the form

$$\frac{d^n y}{dx^n} = f\left(y, \frac{dy}{dx}, \frac{d^2 y}{dx^2}, \dots, \frac{d^{n-1} y}{dx^{n-1}}\right) \tag{4.7.1}$$

subject to the initial conditions

$$y(x_0) = b_0 \tag{4.7.2}$$

$$y'(x_0) = b_1 \tag{4.7.3}$$

$$\vdots \quad \vdots$$

$$y^{(n-1)}(x_0) = b_{n-1} \tag{4.7.4}$$

In the latter, we are using the superscript notation to indicate derivatives with respect to x.

It turns out that such problems can be transformed into a system of first-order ODEs, which can then be solved using the methods described in the previous section. This procedure involves defining a system of auxiliary variables

$$z_1 = y \tag{4.7.5}$$

$$z_2 = y' \tag{4.7.6}$$

$$\vdots \quad \vdots$$

$$z_{n-1} = y^{(n-2)} \tag{4.7.7}$$

$$z_n = y^{(n-1)} \tag{4.7.8}$$

To determine the differential equations governing each of the y_i, we simply differentiate each equation in the above definition with respect to x. The first one is straightforward and gives

$$\frac{dz_1}{dx} = \frac{dy}{dx} = z_2 \tag{4.7.9}$$

and the second one gives

$$\frac{dz_2}{dx} = \frac{d^2y}{dx^2} = z_3 \tag{4.7.10}$$

This pattern continues out to the $(n-1)$th equation,

$$\frac{dz_{n-1}}{dx} = \frac{d^{n-1}y}{dx^{n-1}} = z_n \tag{4.7.11}$$

For the nth equation, we end up with

$$\frac{dz_n}{dx} = \frac{d^ny}{dx^n} = f\left(y, \frac{dy}{dx}, \frac{d^2y}{dx^2}, \ldots, \frac{d^{n-1}y}{dx^{n-1}}\right) \tag{4.7.12}$$

which we can rewrite as

$$\frac{dz_n}{dx} = f(z_1, z_2, z_3, \ldots, z_n) \tag{4.7.13}$$

Putting these equations in vector form, we obtain

$$\frac{d}{dx}\begin{bmatrix} z_1 \\ z_2 \\ \vdots \\ z_{n-1} \\ z_n \end{bmatrix} = \begin{bmatrix} z_2 \\ z_3 \\ \vdots \\ z_n \\ f(z_1, z_2, \ldots, z_n) \end{bmatrix} \tag{4.7.14}$$

with an initial condition

$$\mathbf{z}(x_0) = \begin{bmatrix} b_0 \\ b_1 \\ \vdots \\ b_{n-1} \end{bmatrix} \tag{4.7.15}$$

Example 4.10   Write the system of first-order equations we need to solve

$$\frac{d^3y}{dx^3} + \frac{dy}{dx} = 0 \qquad (4.7.16)$$

subject to

$$y(0) = 1 \qquad (4.7.17)$$
$$y'(0) = 2 \qquad (4.7.18)$$
$$y''(0) = 0 \qquad (4.7.19)$$

Solution
Defining

$$z_1 = y \qquad (4.7.20)$$
$$z_2 = y' \qquad (4.7.21)$$
$$z_3 = y'' \qquad (4.7.22)$$

we obtain the following system of equations,

$$\frac{d}{dx}\begin{bmatrix} z_1 \\ z_2 \\ z_3 \end{bmatrix} = \begin{bmatrix} z_2 \\ z_3 \\ -z_2 \end{bmatrix} \qquad (4.7.23)$$

with the initial conditions

$$\mathbf{z}(0) = \begin{bmatrix} 1 \\ 2 \\ 0 \end{bmatrix} \qquad (4.7.24)$$

We now proceed to solve just like any other system of ordinary differential equations.

Example 4.11   Convert the second-order ODE

$$y'' - y^2 = x \quad y(0) = 2, y'(0) = -1$$

into an autonomous system of ODEs.

Solution
We first define

$$z_1 = y, \quad z_2 = y' \qquad (4.7.25)$$

Since the system is non-autonomous, we also need a variable

$$z_3 = x \qquad (4.7.26)$$

We thus have the vector

$$\mathbf{z} = \begin{bmatrix} y \\ y' \\ x \end{bmatrix} \qquad (4.7.27)$$

If we differentiate,

$$\frac{d\mathbf{z}}{dx} = \begin{bmatrix} y' \\ y'' \\ 1 \end{bmatrix} \tag{4.7.28}$$

Writing in terms of the z_i gives

$$\frac{d}{dx}\begin{bmatrix} z_1 \\ z_2 \\ z_3 \end{bmatrix} = \begin{bmatrix} z_2 \\ z_1^2 + z_3 \\ 1 \end{bmatrix} \tag{4.7.29}$$

The initial condition is

$$\mathbf{z}(x=0) = \begin{bmatrix} 2 \\ -1 \\ 0 \end{bmatrix} \tag{4.7.30}$$

4.8   Numerical Stability for Systems of ODEs

4.8.1   Analysis by Matrix Diagonalization

In Section 4.3, we discussed the stability of a single IVP. Now let's see what happens when we have a system of IVPs. Similar to what we did for a single ODE, we will consider first the model linear system

$$\frac{d\mathbf{y}}{dx} = \mathbf{A}\mathbf{y} \tag{4.8.1}$$

where $\mathbf{A}$ is an $n \times n$ matrix, with the initial condition

$$\mathbf{y}(0) = \mathbf{y_0} \tag{4.8.2}$$

We will initially address the analytical solution of this system using matrix diagonalization. This will on the one hand allow us to derive conditions under which the actual solution is bounded/convergent, and on the other hand to set the stage for deriving numerical stability limits for the different methods.

Let $\lambda_1, \lambda_2, \cdots, \lambda_n$ denote the eigenvalues of the matrix $\mathbf{A}$ and $\mathbf{x}_1, \mathbf{x}_2, \cdots, \mathbf{x}_n$ the corresponding eigenvectors. We will assume that these eigenvectors are linearly independent (although the analysis could also be extended to the case when the eigenvectors are not independent). This will always be the case if the eigenvalues are distinct, but multiple eigenvalues can also yield independent eigenvectors. Let us also form the $n \times n$ matrix

$$\mathbf{X} = [\mathbf{x}_1, \mathbf{x}_2, \dots, \mathbf{x}_n] \tag{4.8.3}$$

whose columns are the eigenvectors of $\mathbf{A}$, and the diagonal matrix

$$\mathbf{\Lambda} = \begin{bmatrix} \lambda_1 & 0 & \cdots & 0 \\ 0 & \lambda_2 & \cdots & 0 \\ \vdots & \vdots & \vdots & \vdots \\ 0 & 0 & \cdots & \lambda_n \end{bmatrix} \tag{4.8.4}$$

With these definitions, the corresponding eigenvalue problem can be written as

$$\mathbf{AX} = \mathbf{X\Lambda} \tag{4.8.5}$$

Note that, while we do not worry about the order of the right-hand side for a single eigen-equation, we need to be more careful here because we are doing matrix multiplication.

Since the columns of $\mathbf{X}$ are linearly independent, we know that $\mathbf{X}^{-1}$ exists. If we right multiply Eq. (4.8.5) by $\mathbf{X}^{-1}$, we can express $\mathbf{A}$ in terms of its eigenvalues and eigenvectors,

$$\mathbf{A} = \mathbf{X\Lambda X}^{-1} \tag{4.8.6}$$

Let's now rewrite the original linear problem (4.8.1) with the decomposition of $\mathbf{A}$ in Eq. (4.8.6),

$$\frac{d\mathbf{y}}{dx} = \mathbf{X\Lambda X}^{-1}\mathbf{y} \tag{4.8.7}$$

We can now left multiply on both sides by $\mathbf{X}^{-1}$,

$$\mathbf{X}^{-1}\frac{d\mathbf{y}}{dx} = \mathbf{X}^{-1}\mathbf{X\Lambda X}^{-1}\mathbf{y} \tag{4.8.8}$$

Since the eigenvectors are independent of x, we can bring $\mathbf{X}^{-1}$ inside the derivative and simplify the right-hand side,

$$\frac{d}{dx}(\mathbf{X}^{-1}\mathbf{y}) = \mathbf{\Lambda X}^{-1}\mathbf{y} \tag{4.8.9}$$

If we make a change of variables to

$$\hat{\mathbf{y}} = \mathbf{X}^{-1}\mathbf{y} \tag{4.8.10}$$

then we see that Eq. (4.8.9) takes the form

$$\frac{d\hat{\mathbf{y}}}{dx} = \mathbf{\Lambda}\hat{\mathbf{y}} \tag{4.8.11}$$

with a suitably modified set of initial conditions. Given the diagonal form of the matrix $\mathbf{\Lambda}$, this is a fully decoupled set of scalar ODEs,

$$\frac{d\hat{y}_1}{dx} = \lambda_1\hat{y}_1 \tag{4.8.12}$$

$$\frac{d\hat{y}_2}{dx} = \lambda_2\hat{y}_2 \tag{4.8.13}$$

$$\vdots \quad \vdots$$

$$\frac{d\hat{y}_n}{dx} = \lambda_n\hat{y}_n \tag{4.8.14}$$

The solution of this system of equations for $\hat{\mathbf{y}}$ can be obtained in a straightforward fashion, and from that result the solutions for y can also be obtained from Eq. (4.8.10). For these solutions to be bounded we do require that the eigenvalues have a non-positive real part, $\text{Re}(\lambda_i) \leq 0$. This is a pre-requisite to be able to talk about numerical stability.

Following a similar approach to the one that we followed for single ODEs, we can now derive analogous conditions for numerical stability. For explicit Euler, we now require that

$$h \leq \frac{2}{|\lambda^{\max}|} \tag{4.8.15}$$

where $\lambda^{\max}$ is the eigenvalue with the largest magnitude. Note that the eigenvalues can be complex,

$$\lambda = a + \iota b \tag{4.8.16}$$

in which case

$$|\lambda| = \sqrt{a^2 + b^2} \tag{4.8.17}$$

Implicit Euler was unconditionally stable for a single equation, and it remains unconditionally stable for a system of equations. For the predictor–corrector method, we have the same result as for the explicit Euler (4.8.15). For RK4, we require

$$h \leq \frac{2.785}{|\lambda^{\max}|} \tag{4.8.18}$$

Similar to what we saw in Section 4.3.2 for a single nonlinear IVP, the stability limits we just derived for a system also apply for nonlinear systems of ODEs, with the exact values of the eigenvalues depending on the specific point of the solution trajectory.

4.8.2   Stiff Systems

There is a particular class of systems called stiff, for which numerical stability is a critical issue. Stiff systems are those for which there is a big difference in the order of magnitude of their eigenvalues. This is common whenever fast and slow phenomena co-exist (e.g. in chemical kinetics, when fast and slow reactions exist).

Let us consider a simple example of a stiff system that is easily analyzed,

$$\frac{dy_1}{dt} = 998y_1 + 1998y_2; \quad y_1(0) = 1 \tag{4.8.19}$$

$$\frac{dy_2}{dt} = -999y_1 - 1999y_2; \quad y_2(0) = 0 \tag{4.8.20}$$

We can use matrix diagonalization to find the exact solution. Here,

$$\mathbf{A} = \begin{bmatrix} 998 & 1998 \\ -999 & -1999 \end{bmatrix} \tag{4.8.21}$$

and its eigenvalues are $\lambda_1 = -1$ and $\lambda_2 = -1000$. For λ_1, the eigenvector equation is

$$\mathbf{A}\mathbf{x}_1 = -\mathbf{x}_1 \tag{4.8.22}$$

which gives an eigenvector of

$$\mathbf{x}_1 = \begin{bmatrix} 2 \\ -1 \end{bmatrix} \tag{4.8.23}$$

For λ_2, the eigenvector equation is

$$\mathbf{A}\mathbf{x}_2 = -1000\mathbf{x}_2 \tag{4.8.24}$$

which gives the eigenvector

$$\mathbf{x}_2 = \begin{bmatrix} 1 \\ -1 \end{bmatrix} \tag{4.8.25}$$

The analytical solution to this problem is then

$$\begin{bmatrix} y_1 \\ y_2 \end{bmatrix} = c_1 \begin{bmatrix} 2 \\ -1 \end{bmatrix} e^{-t} + c_2 \begin{bmatrix} 1 \\ -1 \end{bmatrix} e^{-1000t} \tag{4.8.26}$$

The coefficients are obtained from the initial condition,

$$\begin{bmatrix} 1 \\ 0 \end{bmatrix} = c_1 \begin{bmatrix} 2 \\ -1 \end{bmatrix} + c_2 \begin{bmatrix} 1 \\ -1 \end{bmatrix} \tag{4.8.27}$$

which gives us $c_1 = 1$ and $c_2 = -1$. Thus, the solution of the problem is

$$y_1 = 2e^{-t} - e^{-1000t} \tag{4.8.28}$$
$$y_2 = -e^{-t} + e^{-1000t} \tag{4.8.29}$$

The key feature in this solution is the co-existence of fast and slow components, as seen in Fig. 4.17. The numerical solution for short times is in Fig. 4.17a, and the long-time solution is in Fig. 4.17b. If you only looked at the long-time solution, you might

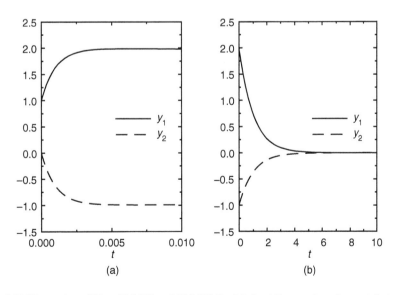

Figure 4.17 Illustration of Eqs. (4.8.28) and (4.8.29) for (a) short times, where λ_2 contributes and (b) long times, where this fast transient has decayed.

think the initial conditions for this problem were $y_1 = 2$ and $y_2 = -1$. This emphasizes the effect of the large separation between the eigenvalues – you cannot even see the short-time behavior with the time axis used in Fig. 4.17b.

The eigenvalue with the largest magnitude is responsible for the fast changes, and also sets the limit for numerical stability initially. For example, if we use the explicit Euler method, we have

$$h \leq \frac{2}{1000} = 0.002 \qquad (4.8.30)$$

However, at longer t, the large eigenvalue is no longer important because the fast exponential has decayed and the solution is dominated by the smallest (in magnitude) eigenvalue,

$$y_1 \approx 2e^{-t} \qquad (4.8.31)$$

$$y_2 \approx -e^{-t} \qquad (4.8.32)$$

If we were to maintain the same h as before, we would incur significant computational cost to march through the entire domain.

There are two ways to handle stiff systems. The most common approach is to use an implicit method, which is unconditionally stable. Another option is to adjust the step size "on-the-fly," depending on how fast the solution is changing. In the example above, once the fast changes decay, we could take much larger time steps and still maintain stability even with the explicit Euler method, using $h \leq 2$.

4.9   Case Study: Non-Isothermal PFR

4.9.1   Problem Statement

In the case study of Section 4.4, we used our knowledge of numerical methods to investigate the behavior of a continuous stirred tank reactor (CSTR) with a fluctuating inlet composition. The CSTR is one of two classic models for chemical reactors, with the other being the plug-flow reactor (PFR). In a plug-flow reactor, illustrated schematically in Fig. 4.18, the reactants flow through the reactor like a plug. In theory, we can imagine that each differential axial distance Δz is a little CSTR, with the PRF thus being the continuum limit.

Let's consider a PFR where a reaction $A \rightarrow B$ takes place with reaction rate constant

$$k = k_0 e^{-E/RT} \qquad (4.9.1)$$

where k_0 is the pre-exponential factor, E is the activation energy, R is the ideal gas constant, and T is the absolute temperature. The reactant A enters the reactor at a concentration c_0 at a superficial velocity v and temperature T_F. We take the reactor to be tubular, with an internal radius R. It will be clear from the dimensions of the terms whether R refers to the radius or the ideal gas constant. The terms in the shell balance on species A are

$$0 = \pi R^2 v \, [c(z) - c(z + \Delta z)] - [k_0 e^{-E/RT(z)} c(z)](\pi R^2 \Delta z) \qquad (4.9.2)$$

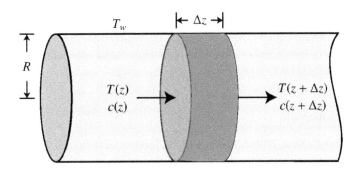

Figure 4.18 Schematic illustration of a plug-flow reactor.

Taking the limit as $\Delta z \to 0$ gives the differential equation

$$v\frac{dc}{dz} = -k_0 e^{-E/RT} c \qquad (4.9.3)$$

We further assume that the stream has a fixed density ρ and heat capacity $\hat{C}_p$. The reaction will either generate or consume heat with a heat of reaction ΔH_{rxn} that is positive for endothermic reactions and negative for exothermic reactions. The reactor is also surrounded by a heating jacket maintained at a temperature T_w that has an overall heat transfer coefficient U. The shell balance for energy is then

$$0 = (\pi R^2)\rho\hat{C}_p v\left[T(z) - T(z+\Delta z)\right] - \Delta H_{\mathrm{rxn}}\left[k_0 e^{-E/RT(z)}c(z)\right](\pi R^2 \Delta z)$$
$$+ U[T_w - T(z)]\,(2\pi R\Delta z) \qquad (4.9.4)$$

In the latter, note that the sign of the heat of reaction term is negative because endothermic reactions will reduce the energy inside the shell. Also, the factor $2\pi R\Delta z$ is the surface area available for heat exchange. Again taking the limit as $\Delta z \to 0$ we have

$$\rho\hat{C}_p v\frac{dT}{dz} = -\Delta H_{\mathrm{rxn}}k_0 e^{-E/RT}c + \frac{2U}{R}(T_w - T) \qquad (4.9.5)$$

There are quite a lot of symbols in these balance equations, so it proves useful to introduce some dimensionless notation. Let's define the dimensionless position $\zeta \equiv z/R$, concentration $\xi \equiv c/c_0$, and temperature $\theta \equiv T/T_F$. If we introduce these into the mass balance, we ultimately get

$$\frac{d\xi}{d\zeta} = -\alpha e^{-\beta/\theta}\xi, \quad \xi(0) = 1 \qquad (4.9.6)$$

where $\alpha = k_0 R/v$ and $\beta = E/RT_F$. Likewise, the dimensionless energy balance is

$$\frac{d\theta}{d\zeta} = -\gamma e^{-\beta/\theta} + \delta(\theta_w - \theta) \qquad (4.9.7)$$

where $\gamma = \Delta H_{\mathrm{rxn}}k_0 c_0 R/\rho\hat{C}_p vT_F$, $\delta = 2U/\rho\hat{C}_p v$, and $\theta_w = T_w/T_F$.

The goal of this problem is to determine the length of the reactor, in terms of the number of radii, required to achieve a conversion of 90% as a function of heating provided by the walls for an endothermic reaction. In other words, we will fix the parameters α,

$\beta, \gamma > 0$ and δ and figure out the value of ζ where $\xi = 0.1$ for different values of θ_w. Along the way, we will also get information about how the temperature and concentration evolve through the reactor and talk about the stability of the numerical solution using RK4.

4.9.2   Solution

To solve this problem, we modified Program 4.4 for RK4 to handle multiple equations while passing the parameters for this problem:

```
1  function [xout,yout] = ivp_RK4(x0,y0,h,n_out,i_out,greek,Tw)
2  %   x0 = initial value for x
3  %   y0 = initial value for y
4  %   h = step size
5  %   i_out = frequency to output x and y to [xout, yout]
6  %   n_out = number of output points
7  %   greek = vector with [alpha,beta,gamma,del]
8  %   Tw = dimensionless wall temperature
9  %   requires the function getf(x,y)
10 xout = zeros(n_out+1,1); xout(1) = x0;
11 yout = zeros(n_out+1,length(y0)); yout(1,:) = y0;
12 x = x0; y = y0;
13 for j = 2:n_out+1
14     for i = 1:i_out
15         k1 = getf(x,y,greek,Tw);
16         k2 = getf(x + 0.5*h,y + 0.5*h*k1,greek,Tw);
17         k3 = getf(x + 0.5*h,y + 0.5*h*k2,greek,Tw);
18         k4 = getf(x + h, y + h*k3,greek,Tw);
19         y = y + (h/6)*(k1 + 2*k2 + 2*k3 + k4);
20         x = x + h;
21     end
22     xout(j) = x;
23     yout(j,:) = y;
24 end
```

Compared to Program 4.4, the input of this new program has two new variables, a vector greek that has the parameters α, β, γ, and δ, and a scalar Tw that has the dimensionless wall temperature, θ_w. In line 11, we modified the yout variable in the same way we did for systems of equations in Program 4.5 except that since we know there are only two equations, the size of the matrix is hard coded. Line 23 has also been modified to save both the concentration (first entry) and temperature (second entry) in y.

For the function evaluation, we use the program

```
1  function f = getf(x,y,greek,Tw)
2  alpha = greek(1);
3  beta = greek(2);
4  gamma = greek(3);
5  del = greek(4);
6  f = zeros(2,1);
7  f(1) = -alpha*exp(-beta/y(2))*y(1);
8  f(2) = -gamma*exp(-beta/y(2))*y(1) + del*(Tw-y(2));
```

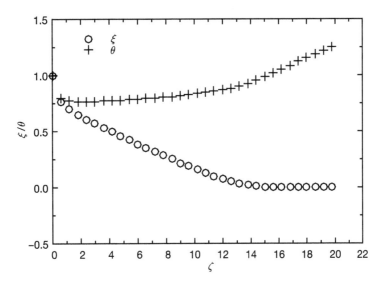

Figure 4.19 Evolution of the temperature and concentration in the non-isothermal PFR.

For convenience, we unpacked the vector `greek` into its components. This is not required, but it makes the program easier to read and debug. Indeed, we might have made the program even easier to read and debug by unpacking the vector `y` as well. In writing programs, you need to decide how much you want to balance the size of the code versus its readability.

Let's see what happens as a function of distance in the PFR for the parameter values $\alpha = 50\,000$, $\beta = 10$, $\gamma = 50\,000$, and $\delta = 0.3$ with the dimensionless wall temperature $\theta_w = 3$. The results in Fig. 4.19 are quite sensible. As the reaction proceeds down the PRF, the concentration decays and the temperature increases because the heat delivered by the heating jacket exceeds the heat removed by the endothermic reaction. Note that the temperature continues to increase after the reaction ceases due to the heating jacket; if we integrated out to $\xi \to \infty$, the temperature in the reactor would equilibrate with the heating jacket at $\theta = 3$.

Now let's see how the heating jacket affects the reaction rate by varying δ while keeping the other parameters fixed. Since we have an endothermic reaction, a good reactor design will constantly feed energy to the PFR to prevent the reaction from slowing to a negligible rate. Figure 4.20 shows how the distance to reach 90% conversion is affected by the heat transferred through the walls from the heating jacket. As expected, increasing the heat transfer through the heating jacket (by increasing δ) decreases the axial distance ζ required to reach high conversion.

Similar to what we did in the CSTR case study in Section 4.4, our example of the non-isothermal PRF is an excellent opportunity to study the numerical stability of a system of nonlinear IVPs in the context of a chemical engineering problem. Indeed, the exponential term involving the temperature is a much stronger nonlinearity than we had in the CSTR, which makes ensuring the numerical stability much more challenging. As we see in Fig. 4.21, the solution can easily become unstable. This problem especially

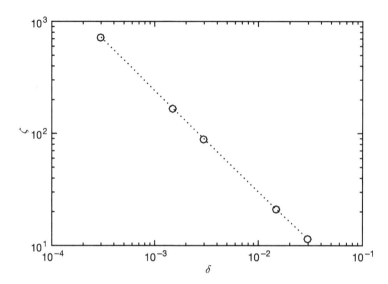

Figure 4.20 Number of tube radii required to reach a conversion of 90% in the non-isothermal PFR.

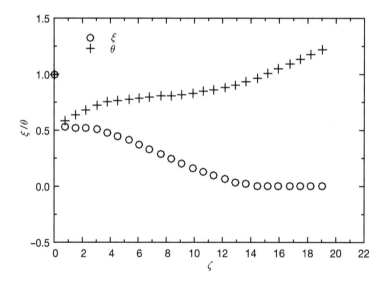

Figure 4.21 Example of an unstable numerical solution for the non-isothermal PFR using RK4.

illustrates the importance of understanding numerical stability. For a short period after the entrance of the reactor, the trajectory in Fig. 4.21 mimics the solution in Fig. 4.19, with both the concentration and temperature decreasing. Since Fig. 4.21 uses a larger step size, we would expect to see some truncation error. However, the point at which the truncation error is overtaken by the instability in the numerical integration is very difficult to detect. Indeed, if we only integrated out to $\zeta = 15$, we might think that the integration was stable since (i) the reactants were depleted and (ii) the temperature increases as the reaction ceases. However, by integrating further, we see that the reactants suddenly reappear. This is clearly impossible, since the reaction is irreversible and the concentrations ξ must be between $[0, 1]$.

4.10   Implementation in MATLAB

MATLAB has a host of different solvers for handling initial value problems, in particular for stiff systems. The recommended solver is ode45, which they claim is suitable for non-stiff problems that require medium accuracy for their solution. The solvers can handle both single equations and systems of equations.

Example 4.12   Solve the IVP

$$\frac{dy}{dt} = e^{-yt} \qquad (4.10.1)$$

with $y(0) = 0$ using ode45.

Solution
A MATLAB script to solve this problem is

```
1  function matlab_ode45
2  [t,y] = ode45(@getf,[0,1],0)
3
4  function f = getf(t,y)
5  f = exp(-t*y);
```

Similar to what we saw in Section 3.11 for solving nonlinear problems, the MATLAB IVP solver requires the use of anonymous functions. The second entry in the function call is the range of the integration, and the last entry is the initial value $y(0)$. The function returns two output vectors, one being the time and the other the value of the function.

Example 4.13   Solve the system of IVPs

$$\frac{dy_1}{dt} = y_1^2 y_2 \qquad (4.10.2)$$

$$\frac{dy_2}{dt} = y_1^2 - y_2^2 \qquad (4.10.3)$$

with $y_1(0) = y_2(0) = 1$ using ode45.

Solution
A MATLAB script to solve this problem is

```
1  function matlab_ode45_system
2  [t,y] = ode45(@getf,[0,0.5],[1,1])
3
4  function f = getf(t,y)
5  f(1,1) = y(1)^2*y(2);
6  f(2,1) = y(1)^2-y(2)^2
```

The syntax here is very similar to what we saw for a single equation. The only difference now is that we need to start with a vector of initial conditions as the third argument in the function `ode45`, rather than a single initial condition. The output is an $n \times 1$ vector of the time points and a $n \times 2$ vector $y = [y1, y2]$, where the first column is y_1 and the second column is y_2.

4.11   Further Reading

Numerical solution of IVPs is covered well in standard numerical methods books, often going beyond the material covered in this book to include advanced multistep methods which build on the interpolation methods to be covered later in the book. Good sources include

- C. W. Gear, *Numerical Initial Value Problems in Ordinary Differential Equations*, Prentice Hall, 1971
- L. Lapidus and J. H. Seinfeld, *Numerical Solution of Ordinary Differential Equations*, Academic Press, 1971
- G. H. Golub and J. M. Ortega, *Scientific Computing and Differential Equations*, Academic Press, 1992
- C. F. Gerald and P. O. Wheatley, *Applied Numerical Analysis*, McGraw Hill, 1992
- U. M. Ascher and L. R. Petzold, *Computer Methods for Ordinary Differential Equations and Differential-Algebraic Equations*, SIAM, 1998
- J. D. Hoffman and S. Frankel, *Numerical Methods for Engineers and Scientists*, McGraw Hill, 2001
- K. Atkinson, W. Han, and D. Stewart, *Numerical Solution of Ordinary Differential Equations*, Wiley, 2009
- J. C. Butcher, *Numerical Methods for Ordinary Differential Equations*, Wiley, 2009.

The solution of IVPs in systems of coupled differential and algebraic equations (DAEs) may present unique challenges and generally requires specialized techniques. Key references on this topic are

- E. Hairer, C. Lubich, and M. Roche, *The Numerical Solution of Differential-Algebraic Systems by Runge-Kutta Methods*, Springer-Verlag, 1989
- K. E. Brenan, S. L. Campbell, and L. R. Petzold, *Numerical Solution of Initial-Value Problems in Differential-Algebraic Equations*, SIAM, 1996.

The reactor problems discussed in the case studies in this chapter are covered in many reaction textbooks such as

- H. S. Fogler, *Elements of Chemical Reaction Engineering*, Prentice Hall, 2005
- L. D. Schmidt, *The Engineering of Chemical Reactions*, Oxford University Press, 2004

and more details about control problems can be found in control textbooks such as

- G. Stephanopoulos, *Chemical Process Control: An Introduction to Theory and Practice*, Prentice Hall, 1984
- B. Ogunnaike and W. H. Ray, *Process Dynamics, Modeling and Control*, Oxford University Press, 1994
- D. E. Seborg, D. A. Mellichamp, T. F. Edgar, and F. J. Doyle III, *Process Dynamics and Control*, Wiley, 2010.

Problems

4.1 Consider the implicit Euler integration of the unsteady mass balance equations with time step h corresponding to the batch reaction

$$A + 2B \rightarrow C$$

which we can write as

$$y_1' = -ky_1y_2^2$$
$$y_2' = -2ky_1y_2^2$$
$$y_3' = ky_1y_2^2$$

with a reaction rate k. What are the equations required for implicit Euler, the residual and the Jacobian to compute the values at the next time step?

4.2 Compare the exact solution, the result from one time step of forward Euler, and the result of one time step for backward Euler for a step size of $h = 0.01$, $h = 0.1$, and $h = 1$ for the differential equation

$$\frac{dy}{dx} = x^2 - x^3; \quad y(2) = 1$$

4.3 Perform one time step of the predictor–corrector method with $h = 0.01$ for the differential equation

$$\frac{dy}{dx} = \cos(\pi x^2) - y^2 \sin(\pi x); \quad y(1) = 10$$

4.4 Perform one time step of RK4 with $h = 0.1$ for the differential equation

$$\frac{dy}{dx} = \ln(xy); \quad y(1) = 2$$

4.5 Imagine that I have integrated the autonomous IVP

$$\frac{dy}{dt} = f(y), \quad y(t = 0) = y_0$$

using some method and at step n I have a value y_n. Under what conditions will explicit Euler and predictor–corrector yield the same result for y_{n+1}?

4.6   Consider the differential equation $y' = y^2$ subject to $y(0) = 2$.

(a) What is the value of y from the predictor–corrector scheme after a single step $h = 1/4$?

(b) It is easy to show that the exact value at $x = 1/4$ is $y = 4$. Based on your result for a step size $h = 1/4$, what step size do we need to reduce the error (defined as $|y_{numerical} - y_{exact}|$) to a value of $1/8$?

4.7   Consider the following nonlinear ordinary differential equation

$$\frac{dy}{dx} = y^2 \exp(-2x)$$

with $y(0) = 1$. Show the exact form of the equations needed to calculate y_{i+1}, with an integration step equal to h, using the:

(a) explicit Euler method

(b) implicit Euler method

(c) predictor–corrector method.

4.8   Consider the following nonlinear ordinary differential equation

$$\frac{dy}{dx} = -y^3(1 - x^2)$$

with $y(0) = 2$. Show the exact form of the equations needed to calculate y_{i+1}, with an integration step equal to h, using the:

(a) explicit Euler method

(b) predictor–corrector method

(c) implicit Euler method.

4.9   Consider the nonlinear ordinary differential equation

$$\frac{dy}{dx} = \frac{x + y}{x}$$

with $y(2) = 2$. Estimate $y(2.5)$

(a) using the explicit Euler's method with $h = 0.1$

(b) using the fourth-order Runge–Kutta method with $h = 0.5$.

4.10   Consider the system of ordinary differential equations

$$\frac{dy_1}{dx} = -0.5y_1$$
$$\frac{dy_2}{dx} = 4 - 0.3y_2 - 0.1y_1$$

with $y_1(0) = 4$ and $y_2(0) = 6$ and for step size $h = 0.5$. Find

(a) $y_1(2)$ and $y_2(2)$ using the explicit Euler method

(b) $y_1(0.5)$ and $y_2(0.5)$ using the fourth-order Runge–Kutta method.

4.11   Consider the nonlinear ordinary differential equation

$$\frac{dy}{dx} = \frac{x - y}{2}$$

with $y(0) = 1$. Estimate $y(1)$

(a) using the explicit Euler's method with $h = 0.2$
(b) using the fourth-order Runge–Kutta method with $h = 1$.

4.12   Answer the following questions about this program:

```
1   function problem4_12
2   y = 1;
3   t = 0;
4   dt = 1e-4;
5   for i = 1:8e3
6       t = t + dt;
7       a = y;
8       b = getb(y,a,t,dt);
9       k = 1;
10      while abs(b) > 1e-4
11          c = getc(y,t,dt);
12          y = y - b/c;
13          b = getb(y,a,t,dt);
14          k = k + 1;
15          if k > 20
16              break
17          end
18      end
19  end
20
21  function b = getb(y,a,t,dt)
22  b = y - a - dt*(y^2 + cos(y*t));
23
24  function c = getc(y,t,dt)
25  c = 1 - dt*(2*y - t*sin(y*t));
```

(a) What differential equation is solved by this program?
(b) What is the initial condition?
(c) What does the variable a represent?
(d) What does the variable b represent?
(e) What does the variable c represent?
(f) What numerical method is implemented by the **for** loop in lines 5–19?
(g) What numerical method is implemented by the **while** loop in lines 10–18?
(h) What is the purpose of the **if** block in lines 15–17?
(i) What is the value of t at the end of the program?
(j) What type of continuation method is used?

4.13   Perform a local stability analysis for the differential equation

$$\frac{dy}{dt} = t\cos(y) + \exp(-t); \quad y(0) = 1$$

What is the step size that you will need at a particular time t and corresponding value y_s if you try to integrate this problem with a forward Euler method? What happens to the step size at large times?

4.14 Consider the system of equations

$$\frac{dy_1}{dt} = y_2$$
$$\frac{dy_2}{dt} = -2y_1 - 3y_2$$

(a) Calculate the eigenvalues and eigenvectors.
(b) Using matrix diagonalization, obtain (analytically) the solution of the equations for $y_1(0) = y_2(0) = 1$.
(c) How does the solution behave for large t? Is this consistent with your eigenvalue analysis? Explain.
(d) Suppose you were using explicit Euler to integrate the system numerically. What would be the maximum integration step size to guarantee numerical stability? Explain.

4.15 Use a linear stability analysis to determine the maximum step size that you could use for the forward Euler integration of the differential equation $y'' + 2(y')^2 - y = 0$ as a function of the local slope, $y'(x)$.

4.16 Consider the following system of ordinary differential equations

$$\frac{dy_1}{dt} = y_1(1 - y_1) - y_1y_2$$
$$\frac{dy_2}{dt} = 2y_1y_2 - y_2$$

with $y_1(0) = 0.5$ and $y_2(0) = 5$. We consider the explicit Euler method for numerical integration. What is the maximum value for the time step h that guarantees numerically stable integration over a single time step? Do you expect the integration to be stiff in the vicinity of $t = 0$? Explain.

4.17 Consider the following MATLAB program:

```
1  function problem4_17
2  clc, close all
3  y = 1; x = 1; h = 0.05;
4  figure, hold on
5  plot(x,y,'o')
6  for i = 1:10
7      y = y + h*(x*y^3+sin(pi*y));
8      x = x+h;
9      plot(x,y,'o')
10 end
11 hold off
```

Show that the first step of the integration satisfies the stability criterion.

4.18 How many coupled first-order IVPs are required to solve the problem

$$\frac{d^3y}{dt^3} = \cos(t)$$

using RK4 with appropriate initial conditions?

4.19 Convert the following higher-order differential equations into a system of autonomous first-order differential equations

$$y''' + 3y'' - 2y' + y = 0; \quad y(0) = 2; \quad y'(0) = 1; \quad y''(0) = 3$$
$$y'''' - y = 0; \quad y(0) = 3; \quad y'(0) = 2; \quad y''(0) = 3; \quad y'''(0) = 0$$
$$y''' = \sin(x); \quad y(0) = 1; \quad y'(0) = 0; \quad y''(0) = 4$$
$$(yy')' + y = 0; \quad y(0) = 2; \quad y'(0) = 1$$

4.20 Consider nonlinear ordinary differential equation

$$y'' + (x^2y)'' + \cos(x)y = y$$

subject to initial conditions

$$y(0) = 3 \quad y'(0) = 1$$

Convert this problem into a system of autonomous equations and initial conditions.

4.21 Consider the solution of the differential equation

$$y'' + yy' + y^3 = \sin(x); \quad y(0) = 1; \quad y'(0) = 2$$

by an implicit Euler scheme. Determine the form of the residual and Jacobian that you would need to compute the values for time step $i + 1$.

4.22 Determine the Jacobian required to integrate the equation

$$\ddot{x} + \dot{x} - 2x + 3x^3 = 0$$

using implicit Euler.

4.23 What ODE is solved by the following MATLAB code and what method is used? Include the initial condition in your answer.

```
1  function problem4_23
2  nsteps = 100;
3  y = 1;
4  t = 0;
5  h = 0.01;
6  g=figure;
7  plot(t,y,'ob')
8  xlabel('t'), ylabel('y')
9  hold on
10 for i = 1:nsteps
11     a = feval(y);
12     b = feval(y+h/2*a);
13     c = feval(y+h/2*b);
14     d = feval(y+h*c);
```

```
15        t = t + h;
16        y = y + h/6*(a + 2*b + 2*c + d);
17        plot(t,y,'ob')
18    end
19    hold off
20
21    function out = feval(y)
22    out = sin(y)*cos(sqrt(y));
```

4.24   Consider the nonlinear differential equation

$$\frac{d^2x}{dt^2} + x^2 = t$$

subject to the conditions $x = 3$ and $dx/dt = 2$ at the time $t = 1$. Let's solve this problem using implicit Euler with a time step of $h = 1/2$. You may recall that a continuation method is often useful in this problem; you should use $x_{i+1}^{(0)} = x_i$ as the initial guess for the Newton–Raphson steps in the solution. Determine the *numerical* values for the residual and Jacobian corresponding to the first iteration of Newton–Raphson during the first time step of the implicit Euler integration of the differential equation.

4.25   Determine the stability condition for the step size h for the midpoint method, which has the form

$$y_{i+1} = y_i + hf\left(x_i + \frac{h}{2}, y_i + \frac{h}{2}f(x_i, y_i)\right)$$

4.26   Determine the stability condition for the step size h for the trapezoidal integration method, which has the form

$$y_{i+1} = y_i + \frac{h}{2}\left[f(y_i) + f(y_{i+1})\right]$$

4.27   Determine the stability condition for the step size h for Ralston's method, which has the form

$$y_{i+1} = y_i + h\left(\frac{1}{3}k_1 + \frac{2}{3}k_2\right)$$

where

$$k_1 = f(x_i, y_i)$$
$$k_2 = f\left(x_i + \frac{3}{4}h, y_i + \frac{3}{4}k_1 h\right)$$

4.28   One possible third-order Runge–Kutta method is given by

$$y_{i+1} = y_i + \frac{h}{6}(k_1 + 4k_2 + k_3)$$

where

$$k_1 = f(x_i, y_i)$$
$$k_2 = f(x_i + 0.5h, y_i + 0.5hk_1)$$
$$k_3 = f(x_i + h, y_i - hk_1 + 2hk_2)$$

By analyzing the linear problem

$$\frac{dy}{dt} = \lambda y, \quad \lambda < 0$$

determine the function $f(h\lambda)$ appearing in the criterion

$$|f(h\lambda)| \leq 1$$

for stable integration. You do not need to actually compute the maximum step size, as this would be best done using a computer. What numerical method would you use to compute the largest step size?

4.29 Answer the following seven questions about the program listed below that solves a system of two ODEs for the unknowns y_1 and y_2.

```
1   function problem4_29
2   yold = [2;-1];
3   h = 0.01;
4   nsteps = 10;
5   ynew = yold;
6   for i = 1:nsteps
7       R = residual(ynew,yold,h);
8       while norm(R) > 10^-8
9           J = jacobian(ynew,h);
10          delt = -J\R;
11          ynew = ynew + delt;
12          R = residual(ynew,yold,h);
13      end
14      yold = ynew;
15  end
16
17  function out = residual(ynew,yold,h)
18  out(1,1) = ynew(1)-yold(1)-h*(ynew(1)^2+ynew(1)*ynew(2));
19  out(2,1) = ynew(2)-yold(2)-h*(ynew(1)*ynew(2));
20
21  function out = jacobian(ynew,h)
22  out(1,1) = 1 - h*(2*ynew(1)+ynew(2));
23  out(1,2) = -h*ynew(1);
24  out(2,1) = ?
25  out(2,2) = 1-h*ynew(1);
```

(a) What *numerical* method is implemented to solve these ordinary differential equations?

(b) What system of ordinary differential equations is this program solving?

(c) Are these linear or nonlinear ODEs?

(d) What is the initial condition for the problem?

(e) What order continuation method is used for the Newton–Raphson part of the program?

(f) What is the entry that should replace the ? in the out(2,1) entry of the subfunction jacobian(ynew,h)?

Computer Problems

4.30   Write a MATLAB program that uses implicit Euler to integrate

$$\frac{dy}{dx} = y^x - x \ln y$$

Use an initial condition $y(0) = 1$ and integrate until $x = 2$. What equations are needed for Newton's method? Is this solution becoming unstable?

4.31   Write a MATLAB program that uses RK4 to solve the ODE

$$\frac{d^2y}{dx^2} + y^2 \frac{dy}{dx} - \sin(x)y = \exp(-x)$$

subject to the initial conditions $y(0) = 1$ and $y'(0) = 0$. You should compute the solution for the step size $h = 0.01$. Make a plot of the function $y(x)$ over the range $x = [0, 10]$.

4.32   Write a MATLAB program that does an RK4 integration of the differential equation

$$\frac{dy}{dx} = xy^{-1/2} + \sin(\pi xy)$$

for the initial condition $y(1) = 2$ up to a maximum value $x = 5$. Your program should automatically generate a single plot with a legend that has the evolution of the function $y(x)$ for the step sizes $h = 0.2$, $h = 0.08$ and $h = 0.01$.

4.33   This problem set deals with the Schlögel model for a bistable biochemical reaction. You can read more about it in the paper "On thermodynamics near a steady state" (*Z. Physik* **248**, 446–458, 1971). For our purposes, you can think of the model as being the solution to the ordinary differential equation

$$\frac{dx}{dt} = -k_1 x^3 + k_2 x^2 - k_3 x + k_4$$

where x is the concentration of some species and the k_i are the reaction rates. We'll work with a particular choice of these parameters: $k_1 = 0.0015$, $k_2 = 0.15$, $k_3 = 3.5$, and $k_4 = 20$.

(a) Write a MATLAB program that determines the steady states for this system. For each steady state, you should make a linear stability analysis to determine if the steady state is stable or unstable. In this type of analysis, you consider a steady state x_{ss} and a small perturbation Δx around the steady state. You then derive a *linear* differential equation for Δx by ignoring terms involving Δx^2 and higher-order powers. The solution of this equation tells you if the perturbation grows in time (unstable) or decays (stable). Provide the values of the steady states and their stability, along with a derivation of the stability condition.

(b) Write a MATLAB program that uses RK4 to integrate the differential equation until it reaches a value near one of the steady states. Your program should automatically generate a plot of $x(t)$ for initial conditions $x = 0$, $x = 15$, $x = 30$, and $x = 50$. It may be helpful (but not required) to also have the steady states on your plot. While it is not required for part (b), it is best that you have your program automatically detect if you are near the steady state so that you can use it for part (c).

(c) Use your program from part (b) (with some changes, if necessary), to determine the value of the steady state x_{ss} corresponding to initial conditions $x = 0, 1, \ldots, 100$. Your program should make plots of (i) the steady state concentration x_{ss} versus the initial concentration x_0 and (ii) the time (i.e., the number of time steps multiplied by the step size) to get to some fixed distance from the steady state versus the initial concentration. Note that the time depends on the time step and what you consider "close" to the steady state, so choose reasonable values for both quantities. Explain the behavior in the plots.

4.34 Write a MATLAB program that uses the predictor–corrector method to solve the ODE

$$\frac{dy}{dx} = x^2 y$$

subject to the initial condition $y(0) = 1$. You should compute the solution for the step sizes $h = 0.1, 0.01, \ldots, 10^{-7}$. Make a log–log plot of the absolute error in the numerical solution at the value $x = 2$ as a function of the step size. You should define the absolute error as

$$E = |y_{numerical} - y_{exact}|$$

4.35 Consider the ordinary differential equation from Problem 4.32 but with a new initial condition, $y(1) = 1$. What is the largest step size, h^*, in RK4 such that the system is stable for the first step? Rename your program from Problem 4.32 and modify the parameters so that it produces a solution for $h = 0.01$, $h = h^*$, and $h = h^* + 0.2$. Explain the result in the context of numerical stability. The results for these particular values are not so obvious, so you need to be careful in your explanation.

4.36 In this problem, you are going to look at how the accuracy of the solution of an ODE depends on the time step. Consider the integration of the differential equation

$$\frac{dy}{dx} = \sin(\pi x)\cos(\pi y)\exp(-[x^2 + y^2])$$

over the domain $x \in [0, 1]$ with the initial condition $y(0) = 0$. Write a program that uses implicit Euler to compute the value of $y(1)$ when the number of time steps $n = 10, 20, \ldots, 1000$. In other words, the first calculation divides the domain from $x = 0$ to $x = 1$ into ten intervals and so forth. As you are marching the solution forward in time, use a zero-order continuation method to produce the initial guess for Newton's method. Within Newton's method, your convergence criteria should be $|f(y_{i+1}^{(k)})| < 10^{-6}$. Your program should make a semilog-x plot of the value of $y(1)$ as a function of the number of time steps.

4.37 This problem deals with the numerical analysis of the solution to

$$\frac{dy}{dx} = y\sin x, \quad y(0) = 1$$

This is a separable ordinary differential equation with the solution

$$y^* = e^{1-\cos x}$$

The exact solution is denoted as y^* for future reference.

(a) We will begin by looking at the solutions produced by using either explicit Euler or RK4 and comparing them to the exact solution over the interval $x \in [0, 10]$. Write a MATLAB program that solves the ODE for $\Delta x = 10^{-3}, 10^{-2}, 10^{-1}, 0.2, 1$, and 2. For each value of Δx, your program should automatically produce a plot of $y(x)$ for the exact result, explicit Euler, and RK4.

(b) Now that we have a working code that (hopefully) produces sensible results, let's do some analysis of the accuracy of the solution and the time. Modify your program from part (a) so that it determines the error in the solution, which we will define as

$$\epsilon \equiv \frac{\|\mathbf{y}^* - \mathbf{y}\|}{n}$$

where $\mathbf{y}$ is the numerical solution, $\mathbf{y}^*$ is the exact solution evaluate at the same points in x, and n is the number of points in each vector. You should also keep track of the time required for the solution. MATLAB has built in functions for this purpose called `tic` and `toc`. The syntax is

```
1  tic %this will start the timer
2      action1 %this is where you put the stuff you want to do.
3      action2
4      action3
5      . . . .
6      action n
7  t = toc; %end the timer and stores result as t
```

This calculation should be done for 50 logarithmically spaced values of Δx between 10^{-5} and 1. MATLAB has a function `logspace` that you may find convenient for this purpose. Note that the integrals will not hit $x = 10$ exactly, so you can either stop at the point just before $x = 10$ or just after $x = 10$. Your program should automatically produce two plots: (i) the values of the error for explicit Euler and RK4 versus Δx, and (ii) the time required or explicit Euler and RK4 versus Δx. Discuss both plots with respect to errors in integration and the amount of work required for these methods.

(c) Let's now repeat part (b) using implicit Euler and compare the results for the error and the time with explicit Euler. Modify the program from part (b) to also do implicit Euler integration for the same range of Δx and produce the same two plots: (i) the values of the error for explicit and implicit Euler versus Δx, and (ii) the time required or explicit and implicit Euler versus Δx. Explain both the time for the solution and the error that you see in your plots. You may find it useful to make some plots of the solutions for representative values of Δx to help with your explanation. You should also include the details of how you implemented Newton's method in your implicit solution.

Important suggestion: The calculation of part (c) may take some time for your computer to complete. You should begin with a reasonably large step size, such as $\Delta x = 0.1$, to debug the program before you attempt to run all of the calculations. You can use the same idea in parts (a) and (b), but they should run much more quickly so it is less of an issue.

4.38 Use RK4 to find the solution to

$$\frac{d^5 y}{dt^5} - 3\left(\frac{d^3 y}{dt^3}\right)\left(\frac{d^2 y}{dt^2}\right) + \left(\frac{dy}{dt}\right)^{\frac{d^2 y}{dt^2}} = t^{1/2}$$

subject to the initial conditions $y(0) = -1$, $y'(0) = 1/2$, $y''(0) = -1/3$, $y''' = 1/4$, and $y'''' = -1/5$. Your program should make a plot of the fourth derivative of y versus the second derivative of y up to a time $t = 1$. What equations convert this into an autonomous system of first-order equations?

4.39 In this problem, you are going to use RK4 to find the minimum value of the solution of a third-order ordinary differential equation. Consider the solution of the differential equation

$$(yy'')' + y' = x^2$$

subject to the initial conditions

$$y(0) = 1, \quad y'(0) = -0.5, \quad y''(0) = 0.7$$

over the domain $x \in [0, 1]$. You should write a MATLAB program that uses a fourth-order Runge–Kutta integration scheme with step size $h = 0.001$ to integrate this equation over the domain. You should then compute the minimum value of the function from the information produced by your integration scheme. The minimum value will not occur exactly at one of the time steps, so you will need to use a lever rule or interpolation to find the position of the minimum. Your program should output the minimum value of the function, y_{min} and the corresponding x value. Make a plot of the function $y(x)$ as a solid line and add the minimum as an open circle. You can do this using the `plot` command in MATLAB.

4.40 Determine the solution to the kinetic system

$$A \rightarrow B \rightarrow C$$

$$B \rightarrow D$$

where the reaction rates are k_1, k_2, and k_3 (in the order written here). The corresponding ODEs that you need to solve are

$$\frac{dA}{dt} = -k_1 A$$
$$\frac{dB}{dt} = k_1 A - k_2 B - k_3 B$$
$$\frac{dC}{dt} = k_2 B$$

$$\frac{dD}{dt} = k_3 B$$

The initial condition is a concentration A_0 and no B, C, or D. Write a MATLAB program that plots the solution up to $t = 10$ for $k_1 = 2$, $k_2 = 0.5$, $k_3 = 0.3$, and $A_0 = 1$. Now write a MATLAB program that uses RK4 to integrate the system. Pick a time step so that the solution is stable and the error is small. Your program should plot the numerical solution and the analytical result together.

4.41 Consider the set of reactions and rate constants

$$A + B \rightarrow C \quad (k_1 = 2)$$
$$C \rightarrow A + B \quad (k_2 = 1)$$
$$B + C \rightarrow D + E \quad (k_3 = 5)$$
$$D + E \rightarrow B + C \quad (k_4 = 3)$$
$$E \rightarrow F \quad (k_5 = 1)$$

with an initial concentration of 0.75 moles of A and 1 mole of B. Write the system of ordinary differential equations describing the concentration of each species, along with appropriate initial conditions. Numerically integrate this system out to $t = 10$ using a stable method. Your program should automatically generate a plot of the concentrations of all species versus time. Explain the evolution of the concentrations using your knowledge of kinetics.

4.42 Consider two particles that are interacting via a Lennard–Jones potential,

$$V = r^{-12} - r^{-6}$$

The dynamic equation for the distance r between these particles is given by the force balance (in dimensionless form),

$$\frac{d^2 r}{dt^2} = 12r^{-13} - 6r^{-7}$$

where we used the relationship between potential energy and force, $f = -dV/dr$. Write a MATLAB program that uses implicit Euler to find the position r as a function of time t for an initial separation of $r = 1.3$ and no initial velocity. The program should produce a plot of the position versus time. Include the residual and Jacobian you used to solve this problem. There is something bizarre about the result, explain. *Hint*: Think about the minimum value of the potential energy and the dynamics of a particle in a potential field.

5   Dynamical Systems

5.1   Introduction

In the previous chapter, we considered in general how to solve a system of initial value problems. In this chapter we will focus on the specific class of systems of nonlinear IVPs of the form

$$\frac{d\mathbf{y}}{dt} = \mathbf{f}(\mathbf{y}) \tag{5.1.1}$$

where the independent variable t denotes time. These are called dynamical systems, and they arise from unsteady state conservation equations. The analysis of the behavior of such systems is the subject of nonlinear dynamics, a rich mathematical field which cuts across all engineering disciplines. Thus, we will first develop our understanding of dynamic systems in a generic context, and then see how these concepts apply to chemical engineering in the context of the non-isothermal continuous stirred tank reactor in Section 5.7.

For these systems, we are interested in answering the following three questions.

(1) What are the steady state (or equilibrium) solutions $\mathbf{y}_{ss}$ of the system? These are solutions that are time invariant, i.e. once the system is there, it never leaves. To determine them we simply set $\mathbf{f}(\mathbf{y}) = \mathbf{0}$ and solve the resulting system of nonlinear algebraic equations. Thus, we need to remember the methods we learned in Chapter 3 to answer this first question.

(2) What happens when a *small* perturbation from a steady state is applied to the system? Will the trajectories $\mathbf{y}(t)$ return to the steady state, or move away from it? This is a question of linear stability of the steady state, i.e. stability to infinitesimally small perturbations, assessed based on a linear approximation of the nonlinear system around the steady state. In this sense, there is a relationship with the stability analysis from Chapter 4. The difference here is that we are now talking about the stability of the steady state of the system, not the numerical stability of the integration method.

(3) What happens when a larger perturbation from a steady state is applied to the system? Answering this requires integrating the fully nonlinear IVP for different initial conditions (using the methods that we described in the previous chapter) to fully characterize the behavior of the system.

In what follows we will discuss the answers to these questions in more detail.

5.2   Equilibrium Points and Their Stability

For the dynamical system in Eq. (5.1.1), the steady states (or equilibrium points) are given by the solution of

$$\mathbf{f}(\mathbf{y}) = \mathbf{0} \qquad (5.2.1)$$

Let us denote a solution of this system of (nonlinear) algebraic equations as $\mathbf{y}_{ss}$. As we have seen in Chapter 3, such systems of nonlinear algebraic equations may indeed have multiple solutions. Newton–Raphson, with multiple initial guesses (whose choice is often aided by physical intuition) can be used to determine these solutions.

We now consider a new variable $\mathbf{\Delta}$ that represents a small (infinitesimal) perturbation to the steady state. We'd like to examine what happens to $\mathbf{y}$ when we apply such a perturbation, i.e. when we start integrating the system from

$$\mathbf{y}_0 = \mathbf{y}_{ss} + \mathbf{\Delta}_0 \qquad (5.2.2)$$

Of course, one can always solve Eq. (5.1.1) with this as the initial condition to answer this question. It turns out, however, that actually solving the problem is frequently unnecessary. This is a very satisfying approach, as it is almost like getting something for nothing.

To see how to figure out the behavior of $\mathbf{y}(t)$ without actually solving the IVP, let's start with a Taylor series expansion of $\mathbf{f}(\mathbf{y})$ about the steady state solution $\mathbf{y}_{ss}$. For a given component f_i of the vector function $\mathbf{f}$, this expansion will be

$$f_i(\mathbf{y}) = f_i(\mathbf{y}_{ss}) + \sum_{j=1}^{n} \left.\frac{\partial f_i}{\partial y_j}\right|_{\mathbf{y}_{ss}} (y_j - y_{j,ss}) + \cdots \qquad (5.2.3)$$

In a compact vector form, we can then write

$$\mathbf{f}(\mathbf{y}) = \mathbf{0} + \mathbf{J}_{ss}(\mathbf{y} - \mathbf{y}_{ss}) \qquad (5.2.4)$$

To arrive at this result, we (i) neglected the higher-order terms; (ii) used the fact that $f_i(\mathbf{y}_{ss}) = 0$, by the definition of the steady state; and (iii) collected all the partial derivatives in the Jacobian matrix, $\mathbf{J}_{ss}$. Evaluating the expansion in Eq. (5.2.4) at $\mathbf{y}_{ss} + \mathbf{\Delta}$, we get

$$\mathbf{f}(\mathbf{y}_{ss} + \mathbf{\Delta}) = \mathbf{J}_{ss}\mathbf{\Delta} \qquad (5.2.5)$$

We are interested in the time evolution of this perturbed variable

$$\frac{d}{dt}(\mathbf{y}_{ss} + \mathbf{\Delta}) = \frac{d\mathbf{\Delta}}{dt} \qquad (5.2.6)$$

Based on Eq. (5.1.1) and the linear approximation of Eq. (5.2.5), the perturbed variable evolves according to

$$\frac{d\mathbf{\Delta}}{dt} = \mathbf{J}_{ss}\mathbf{\Delta} \qquad (5.2.7)$$

We have already seen previously in Section 4.8.1 that if $\mathbf{J}_{ss}$ has n linearly independent eigenvectors, then the solution to this equation is

$$\boldsymbol{\Delta} = \mathbf{X}e^{\boldsymbol{\Lambda}t}\mathbf{X}^{-1}\boldsymbol{\Delta}_0 \tag{5.2.8}$$

where $\mathbf{X}$ is an $n \times n$ matrix containing the eigenvectors of $\mathbf{J}_{ss}$ and $\boldsymbol{\Delta}_0$ is the initial perturbation. In the latter, we have used the matrix exponential notation

$$e^{\boldsymbol{\Lambda}t} = \begin{bmatrix} e^{\lambda_1 t} & 0 & \cdots & 0 \\ 0 & e^{\lambda_2 t} & \cdots & 0 \\ \vdots & \vdots & \vdots & \vdots \\ 0 & 0 & \cdots & e^{\lambda_n t} \end{bmatrix} \tag{5.2.9}$$

Depending on the nature of the eigenvalues λ_i, the following cases emerge for the stability of the steady state.

(1) *Stable steady state:* If $\mathrm{Re}(\lambda_i) < 0$ for every eigenvalue λ_i, then the steady state $\mathbf{y}_{ss}$ is (asymptotically) stable; all of the exponential terms in Eq. (5.2.8) decay with time, and as a result,

$$\boldsymbol{\Delta} \to \mathbf{0} \text{ for } t \to \infty \tag{5.2.10}$$

and the trajectories converge back to the steady state asymptotically.
(2) *Unstable steady state:* If $\mathrm{Re}(\lambda_i) > 0$ for any eigenvalue, then the steady state is unstable because at least one of the exponential terms in Eq. (5.2.8) grows with time. As a result the solution of the nonlinear system $\mathbf{y}$ cannot be guaranteed to converge back to the steady state $\mathbf{y}_{ss}$ or even to remain bounded. Indeed, it may diverge or it may converge to a different steady state or even a more exotic equilibrium structure! We will talk about this in more detail later in the chapter.
(3) *Neutrally stable state:* If $\mathrm{Re}(\lambda_i) \le 0$ for all of the eigenvalues but there are some for which $\mathrm{Re}(\lambda_i) = 0$, then although stability of the linearized system describing the perturbation evolution can be assessed with some additional effort, no conclusions can be drawn for the stability of the steady state of the nonlinear system. In this case, nonlinear effects may become significant.

It is important to remember that this analysis is valid only for small perturbations from the steady state, as it was based on a Taylor series expansion that is valid only locally around the steady state. Also, if there are multiple steady states, then the stability of each steady state needs to be analyzed individually.

5.3 Dynamical Systems on the Plane

The nature of the eigenvalues of a dynamic system does not only provide information on the stability of steady state solutions, but also on the qualitative shape of the trajectories

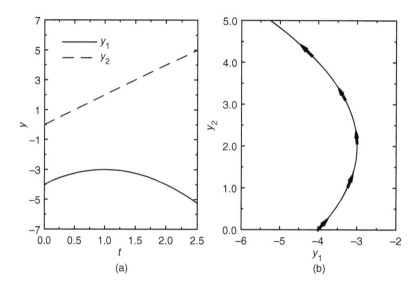

Figure 5.1 (a) Schematic illustration of the trajectories $y_1(t)$ and $y_2(t)$. (b) Phase plane representation of $y_2(t)$ as a function of $y_1(t)$. Time is implicit in this trajectory and indicated by the arrows.

close to the steady state. This is best illustrated by looking at two-dimensional systems (i.e. systems of two ODEs) of the form

$$\frac{dy_1}{dt} = f_1(y_1, y_2) \tag{5.3.1}$$

$$\frac{dy_2}{dt} = f_2(y_1, y_2) \tag{5.3.2}$$

Solving these equations for an arbitrary initial condition will result in trajectories, y_1 versus t and y_2 versus t, as seen in Fig. 5.1a. An alternative way to plot these trajectories would be to plot y_1 versus y_2, for each value of t. This approach is seen in Fig. 5.1b. In the resulting curve, time is implicit, and we capture the relative change between the dependent variables along the solution. This is called a phase plane (or phase space) plot and is used extensively in the analysis of two-dimensional systems, as it gives rise to distinct trajectory shapes, depending on the nature of eigenvalues. Let's consider this in more detail.

5.3.1   Linear Systems on the Plane – Classification of Steady States

We begin by analyzing linear systems, for which the trajectories can be obtained analytically. We specifically consider systems of the form of Eq. (4.8.1),

$$\frac{d\mathbf{y}}{dt} = \mathbf{A}\mathbf{y} \tag{5.3.3}$$

where $\mathbf{A}$ is a 2×2 matrix, and the steady state corresponds to $\mathbf{y} = \mathbf{0}$. We exclude the degenerate case of one or both eigenvalues being zero (i.e. the matrix $\mathbf{A}$ being singular), which leads to a continuum of steady states rather than an isolated one.

Distinct Real Eigenvalues

We consider first the case where the two eigenvalues λ_1, λ_2 are distinct and real. In this case, two linearly independent eigenvectors $\mathbf{x}_1, \mathbf{x}_2$ exist, the matrix

$$\mathbf{X} = [\mathbf{x}_1, \mathbf{x}_2] \tag{5.3.4}$$

is non-singular, and the change of variables

$$\hat{\mathbf{y}} = \mathbf{X}^{-1}\mathbf{y} \tag{5.3.5}$$

transforms the system above into the decoupled equations

$$\frac{d\hat{y}_1}{dt} = \lambda_1\hat{y}_1 \tag{5.3.6}$$

$$\frac{d\hat{y}_2}{dt} = \lambda_2\hat{y}_2 \tag{5.3.7}$$

Their solution is

$$\hat{y}_1 = \hat{y}_1(0)e^{\lambda_1 t} \tag{5.3.8}$$

$$\hat{y}_2 = \hat{y}_2(0)e^{\lambda_2 t} \tag{5.3.9}$$

Eliminating time, we obtain

$$\frac{\hat{y}_1}{\hat{y}_1(0)} = \left(\frac{\hat{y}_2}{\hat{y}_2(0)} \right)^{\frac{\lambda_1}{\lambda_2}} \tag{5.3.10}$$

which describes the trajectory in the phase plane.

The following classification of the steady states, illustrated in Fig. 5.2, now emerges.

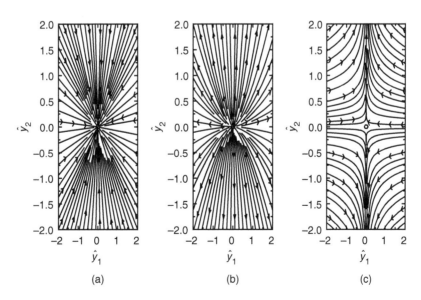

Figure 5.2 Illustration of (a) a stable node, (b) an unstable node, and (c) a saddle point.

(1) **Stable node.** This corresponds to the case where $\lambda_1 < 0$ and $\lambda_2 < 0$. In this case both $\hat{y}_1, \hat{y}_2$ approach zero asymptotically, and since $\lambda_1/\lambda_2 > 0$, they do so in a parabolic fashion. Figure 5.2a shows a stable node, where the arrows indicate the direction of time.

(2) **Unstable node.** This corresponds to the case where $\lambda_1 > 0$ and $\lambda_2 > 0$. The trajectories diverge to infinity, again in a parabolic fashion. They are identical to those for a stable node, but with the time arrows reversed. Figure 5.2b illustrates an unstable node.

(3) **Unstable saddle point.** This corresponds to the case where $\lambda_1 < 0$ and $\lambda_2 > 0$. The steady state is still unstable, and the trajectories diverge to infinity, but since $\lambda_1/\lambda_2 < 0$, they do so in the hyperbolic fashion seen in Fig. 5.2c. Furthermore, if the initial condition is exactly on the $\hat{y}_1$ axis (i.e., $\hat{y}_2(0) = 0$), the solution will asymptotically converge to 0. This does not contradict the fact that the steady state is unstable. To begin with, due to finite precision, it is impossible to have an initial condition exactly on this axis. Any infinitesimally small perturbation will activate divergent trajectories. Also, the notion of stability, by definition, requires that there exists an entire neighborhood (even a very small one) around a steady state such that, starting from this neighborhood, the trajectories remain bounded or converge to the steady state. Clearly this is not the case here – the stable initial conditions lie on a line.

Note that all of the above trajectories were drawn in the phase plane $(\hat{y}_1, \hat{y}_2)$. We can also visualize them in the original phase plane (y_1, y_2), by simply using the fact that

$$\mathbf{y} = \mathbf{X}\hat{\mathbf{y}} \qquad (5.3.11)$$

In this case the movement will be along the eigenvectors (which were simply the $\hat{y}_1, \hat{y}_2$ axes for the transformed, decoupled system).

Complex Eigenvalues

We now consider the case where the two eigenvalues are a complex conjugate pair,

$$\begin{aligned} \lambda_1 &= a + \iota b \\ \lambda_2 &= a - \iota b \end{aligned} \qquad (5.3.12)$$

In this case, although we can still form the eigenvector matrix, it will be a complex one, and the approach we followed before is a bit more involved. We can still though analyze the trajectories of the solutions and recognize that they will comprise combinations of complex exponential terms, i.e. terms of the form

$$e^{\lambda_1 t} = e^{at + b\iota t} = e^{at}(\cos bt + \iota \sin bt) \qquad (5.3.13)$$

and

$$e^{\lambda_2 t} = e^{at - b\iota t} = e^{at}(\cos bt - \iota \sin bt) \qquad (5.3.14)$$

Recall that the coefficients multiplying these solutions can also be complex numbers, so the final result can be complex, pure imaginary, or, as will be the case here, real. The first term on the right-hand sides above will converge to 0 or diverge to ∞, depending

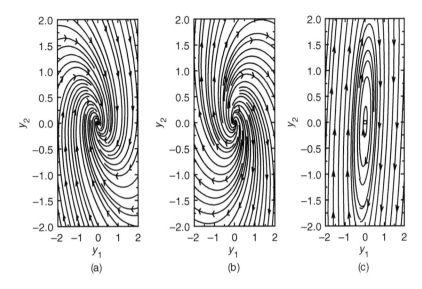

Figure 5.3 Illustration of (a) a stable focus, (b) an unstable focus, and (c) a center.

on the sign of the real part of the eigenvalue, a. The second term, corresponding to $e^{\iota b t}$, will make the trajectory oscillate.

As illustrated in Fig. 5.3, the following cases emerge:

(1) **Stable focus.** For the case where $a < 0$, the steady state is stable. The imaginary part leads to oscillations, and the net effect is an oscillatory trajectory with the amplitude of the oscillations decreasing. This produces a spiral in the phase plane, as shown in Fig. 5.3a.

(2) **Unstable focus.** This is the case where $a > 0$. The steady state is unstable, and the oscillations increase in amplitude. Trajectories starting near the steady state spiral away as shown in the phase plane in Fig. 5.3b.

(3) **Center.** This is the case of purely imaginary eigenvalues, where $a = 0$. In this case, the solution is a periodic trajectory with the sustained oscillations seen in Fig. 5.3c. Note that for different initial conditions, we will get a different oscillatory trajectory. Also note that when this case arises for the linearization of a nonlinear system, as we will see in Section 5.3.2, further analysis will be necessary to determine the behavior of the nonlinear system itself.

Repeated Eigenvalues

We finally consider the case of a double real eigenvalue λ. (Note that we cannot get repeated complex eigenvalues since they must occur as complex conjugate pairs.) In this case we cannot guarantee that two linearly independent eigenvectors exist. If they do, the procedure followed for real distinct eigenvalues can also be followed. In this case, the ratio $\lambda_1/\lambda_2 = 1$ in Eq. (5.3.10). The trajectories in the phase plane are lines passing through the origin, and depending on whether λ is negative or positive, we have

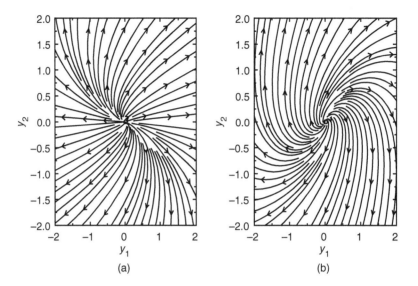

Figure 5.4 Illustration of the case of repeated eigenvalues for (a) two linearly independent eigenvectors and (b) one independent eigenvector.

a stable or unstable node. Figure 5.4a shows the case of an unstable node with two linearly independent eigenvectors.

If only one independent eigenvector can be found, then the analysis is a bit more complicated and we cannot fully decouple the equations. Still, depending on whether λ is negative or positive, we have a stable or unstable node, respectively, but the trajectories converge or diverge in the slightly more complicated fashion seen in Fig. 5.4b.

5.3.2   Nonlinear Systems on the Plane

We now return to analyzing the steady states of the nonlinear system

$$\frac{dy_1}{dt} = f_1(y_1, y_2) \tag{5.3.15}$$

$$\frac{dy_2}{dt} = f_2(y_1, y_2) \tag{5.3.16}$$

Considering again a Taylor series expansion of the nonlinear terms around the steady state $\mathbf{y}_{ss}$, and truncating after the linear terms, we obtain the approximate, linearized system of Eq. (5.2.7),

$$\frac{d\mathbf{\Delta}}{dt} = \mathbf{J}_{ss}\mathbf{\Delta} \tag{5.3.17}$$

where $\mathbf{\Delta} = \mathbf{y} - \mathbf{y}_{ss}$.

The basic result here is that as long as $\mathbf{J}_{ss}$ does not have purely imaginary eigenvalues, the stability properties and the phase portrait, *locally* around the steady state, will resemble those of the linearized system. Note that this excludes the case of a center; in this case the linear analysis is inconclusive, and a more detailed analysis of the effect of the nonlinear terms is needed to ascertain the qualitative properties of the solution.

An additional point here is that multiple steady states may be present in a nonlinear system. In this case trajectories that are repelled by a steady state may be attracted by another stable one. Of course in this case integration of the full ODE system is required to capture the exact trajectories away from the steady states. Finally, similar classifications of the steady states exist for higher-order nonlinear systems, although trajectory visualization is harder in these cases.

Example 5.1   Analyze the stability and the nature of the nearby trajectories for the steady states of

$$\frac{dy_1}{dt} = y_1 - y_1^3 + y_2 \qquad (5.3.18)$$

$$\frac{dy_2}{dt} = y_1 - y_1^2 \qquad (5.3.19)$$

Solution
We begin with the calculation of the steady states of the system. Setting the derivative to zero we get from the second equation

$$y_1(1 - y_1) = 0 \qquad (5.3.20)$$

implying that at steady state $y_1 = 1$ or $y_1 = 0$. From the first equation, it follows that the corresponding steady state value of y_2 is $y_2 = 0$ for both cases. Thus we have two steady states, $\mathbf{y}_{ss1} = (0,0)$ and $\mathbf{y}_{ss2} = (1,0)$.

The Jacobian of this system of equations is

$$\mathbf{J} = \begin{bmatrix} 1 - 3y_1^2 & 1 \\ 1 - 2y_1 & 0 \end{bmatrix} \qquad (5.3.21)$$

There is a nice formula for the eigenvalues of a 2×2 system that looks very similar to the quadratic equation,

$$\lambda = \frac{\tau \pm \sqrt{\tau^2 - 4\Delta}}{2} \qquad (5.3.22)$$

where τ is the trace of $\mathbf{J}$ and Δ is the determinant of $\mathbf{J}$. To prove this formula, consider the general case

$$\mathbf{A} = \begin{bmatrix} a & b \\ c & d \end{bmatrix} \qquad (5.3.23)$$

The trace of this matrix is $\tau = a + d$ and its determinant is $\Delta = ad - bc$. The eigenvalue equation is

$$\det \begin{bmatrix} a - \lambda & b \\ c & d - \lambda \end{bmatrix} = 0 \qquad (5.3.24)$$

which gives the polynomial equation

$$\lambda^2 - (a+d)\lambda + ad - bc = 0 \tag{5.3.25}$$

If we substitute for the trace and determinant, we have

$$\lambda^2 - \tau\lambda + \Delta = 0 \tag{5.3.26}$$

Using the quadratic formula gives Eq. (5.3.22).

For the steady state at $(0,0)$, we have a Jacobian

$$\mathbf{J}_{ss1} = \begin{bmatrix} 1 & 1 \\ 1 & 0 \end{bmatrix} \tag{5.3.27}$$

The trace is $\tau = 1$ and the determinant is $\Delta = -1$, giving eigenvalues

$$\lambda = \frac{1 \pm \sqrt{5}}{2} \tag{5.3.28}$$

Thus, this steady state is an unstable saddle point.

For the steady state at $(1,0)$, we have a Jacobian

$$\mathbf{J}_{ss2} = \begin{bmatrix} -2 & 1 \\ -1 & 0 \end{bmatrix} \tag{5.3.29}$$

The trace is $\tau = -2$ and the determinant is $\Delta = 1$, giving eigenvalues

$$\lambda = \frac{-2 \pm \sqrt{0}}{2} = -1 \tag{5.3.30}$$

This is stable node with a repeated eigenvalue.

To investigate the dynamics of this system more thoroughly, Fig. 5.5 shows the phase plane for this problem using RK4 to integrate from a number of different initial conditions. As we can see, the analysis based on the linearized problem is accurate.

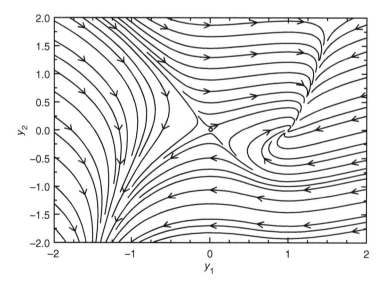

Figure 5.5 Phase plane for Example 5.1.

Example 5.2   Analyze the stability and the nature of the nearby trajectories for the steady states of

$$\frac{dx}{dt} = -y + ax(x^2 + y^2) \tag{5.3.31}$$

$$\frac{dy}{dt} = x + ay(x^2 + y^2) \tag{5.3.32}$$

where a is a constant.

Solution
We start with the calculation of the steady states for this system. Setting the derivatives to zero, multiplying the first equation by x, the second by y, and adding them, we obtain that at steady state $(x^2 + y^2) = 0$. Therefore, the only steady state for this system is $(0, 0)$. The Jacobian for this system is

$$\mathbf{J} = \begin{bmatrix} 3ax^2 & -1 + 2axy \\ 1 + 2axy & 3ay^2 \end{bmatrix} \tag{5.3.33}$$

At the steady state $(0, 0)$ we have

$$\mathbf{J}_{ss} = \begin{bmatrix} 0 & -1 \\ 1 & 0 \end{bmatrix} \tag{5.3.34}$$

The eigenvalue equation is

$$\det \begin{bmatrix} -\lambda & -1 \\ 1 & -\lambda \end{bmatrix} = 0 \tag{5.3.35}$$

which gives the polynomial equation

$$\lambda^2 + 1 = 0 \tag{5.3.36}$$

The eigenvalues are purely imaginary, $\lambda = \pm i$. This is a case of a neutrally stable steady state. Although the linearized analysis predicts a center, further analysis is required to determine the nature of the trajectories of the nonlinear system close to the steady state.

In some cases, important conclusions can be drawn by switching to polar coordinates,

$$x = r \cos \theta \tag{5.3.37}$$

$$y = r \sin \theta \tag{5.3.38}$$

Our aim is to determine the resulting differential equations in terms of θ and r. To get the radial position equation, we can take advantage of the identity

$$x^2 + y^2 = r^2 \tag{5.3.39}$$

which upon differentiation with respect to time becomes

$$x\dot{x} + y\dot{y} = r\dot{r} \tag{5.3.40}$$

In the latter, we use the dot to denote differentiation with respect to time. Substituting from the original differential equation gives

$$r\dot{r} = x\left[-y + ax(x^2 + y^2)\right] + y\left[x + ay(x^2 + y^2)\right]$$
$$= a(x^2 + y^2)^2$$
$$= ar^4 \tag{5.3.41}$$

As a result, the radial position differential equation is

$$\dot{r} = ar^3 \tag{5.3.42}$$

To get the equation for the angular position, we differentiate the relationships

$$x = r\cos\theta \tag{5.3.43}$$
$$y = r\sin\theta \tag{5.3.44}$$

with respect to time,

$$\dot{x} = \dot{r}\cos\theta - (r\sin\theta)\dot{\theta} = \dot{r}\cos\theta - y\dot{\theta} \tag{5.3.45}$$
$$\dot{y} = \dot{r}\sin\theta + (r\cos\theta)\dot{\theta} = \dot{r}\sin\theta + x\dot{\theta} \tag{5.3.46}$$

It then follows that

$$x\dot{y} - y\dot{x} = \left(x\dot{r}\sin\theta + x^2\dot{\theta}\right) - \left(y\dot{r}\cos\theta - y^2\dot{\theta}\right)$$
$$= r^2\dot{\theta} + \dot{r}(r\cos\theta\sin\theta - r\sin\theta\cos\theta)$$
$$= r^2\dot{\theta} \tag{5.3.47}$$

from which we can solve for $\dot{\theta}$

$$\dot{\theta} = \frac{x\dot{y} - y\dot{x}}{r^2} \tag{5.3.48}$$

Substituting $\dot{x}$ and $\dot{y}$ from the original differential equations, we get

$$\dot{\theta} = \frac{x(x + ayr^2) - y(-y + axr^2)}{r^2} = \frac{x^2 + y^2}{r^2} \tag{5.3.49}$$

As a result, the differential equation for the angular velocity becomes

$$\dot{\theta} = 1 \tag{5.3.50}$$

These equations confirm that the steady state of the system corresponds to $r = 0$, i.e. the origin $(0, 0)$ in the original coordinates. The stability of this steady state depends on a. When $a = 0$, the original system in Eqs. (5.3.31) and (5.3.32) is a linear one, and indeed the steady state is a center. When $a > 0$, the differential equation for the radial position shows that a small positive perturbation of r will grow, with constant angular velocity (increasing phase angle). Thus the solution will rotate counter-clockwise away from the origin, indicating an unstable focus. On the other hand, when $a < 0$, the differential equation for the radial position shows that a small positive perturbation of r will decay, again with constant angular velocity. In this case the steady state behaves like a stable focus. These features of the solution are shown in Fig. 5.6. The take home

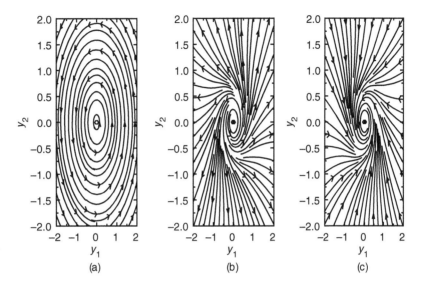

Figure 5.6 Solution to Example 5.2 leading to (a) a stable focus, (b) a center, and (c) an unstable focus.

message here is that in the case of a center in the linearized equations, nonlinear effects can change the stability of the steady state dramatically.

5.4 Nonlinear Phenomena

In this section we will discuss some qualitative features of the behavior of dynamical systems that are unique to nonlinear systems. There are of course a number of fundamental differences between the solution properties of linear versus nonlinear ODEs. For example, the existence, uniqueness, and continuity of solutions is significantly more difficult to characterize in nonlinear systems. Also, although divergence to infinity is possible in linear ODEs, it happens asymptotically, i.e. as $t \to \infty$. For nonlinear ODEs the solutions may escape to infinity in finite time. For example, the problem

$$\frac{dx}{dt} = x^2, \quad x(0) = 1 \tag{5.4.1}$$

has the solution

$$x(t) = \frac{1}{1 - t} \tag{5.4.2}$$

in which x approaches infinity as t approaches 1. Interestingly, the sign of x depends on whether t approaches 1 from the left or right.

In what follows we will discuss two types of solutions for two-dimensional dynamical systems for which the linear stability analysis that we presented before is inconclusive, and we need to expend more effort to determine the solution behavior around a steady state.

5.4.1   Periodic Solutions

As an example of a system that exhibits periodic solutions, consider the so-called predator–prey (or Volterra–Lottka) model

$$\frac{dy_1}{dt} = ay_1 - by_1y_2 \tag{5.4.3}$$

$$\frac{dy_2}{dt} = cy_1y_2 - dy_2 \tag{5.4.4}$$

Here y_1 denotes the prey population and y_2 denotes the predator population. All variables and parameters are positive. The model is based on the assumptions that

(1) in the absence of predator the prey population grows proportionally to its size,
(2) in the absence of prey the predator population declines in size again proportionally to its size,
(3) when both are present their interactions are proportional to the product of their populations.

The steady state solutions are obtained by setting the time derivatives to zero and solving to obtain

$$\mathbf{y}_{ss1} = (0, 0) \tag{5.4.5}$$

$$\mathbf{y}_{ss2} = \left(\frac{d}{c}, \frac{a}{b}\right) \tag{5.4.6}$$

with the corresponding Jacobian

$$\mathbf{J} = \begin{bmatrix} a - by_2 & -by_1 \\ cy_2 & cy_1 - d \end{bmatrix} \tag{5.4.7}$$

If we evaluate the Jacobian at the first steady state $\mathbf{y}_{ss1}$ we have

$$\mathbf{J}_{ss1} = \begin{bmatrix} a & 0 \\ 0 & -d \end{bmatrix} \tag{5.4.8}$$

The eigenvalues are therefore the diagonal elements and this steady state is an unstable saddle.

For $\mathbf{y}_{ss2}$, the Jacobian becomes

$$\mathbf{J}_{ss2} = \begin{bmatrix} 0 & -b\left(\frac{d}{c}\right) \\ c\left(\frac{a}{b}\right) & 0 \end{bmatrix} \tag{5.4.9}$$

The eigenvalues in this case are $\lambda_1 = \iota\sqrt{ad}$ and $\lambda_2 = -\iota\sqrt{ad}$. This corresponds to a center for the linearized system. Recall from Example 5.2 that we cannot draw any conclusion about a nonlinear system from a center.

Now that we know the predictions (or lack thereof) for the linear problem, let's analyze the motions of y_1, y_2 in the first quadrant (since both are positive). Dividing the two differential equations to eliminate time, we get

$$\frac{dy_2}{dy_1} = \frac{y_2(-d + cy_1)}{y_1(a - by_2)} \tag{5.4.10}$$

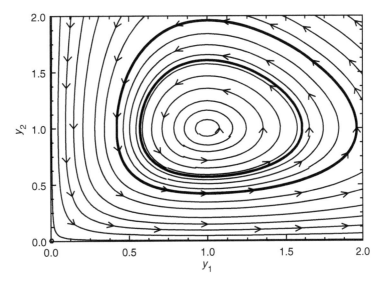

Figure 5.7 Phase plane for the predator–prey equations for $a = b = c = d = 1$. The thicker lines were meant to illustrate the trajectory for the two different initial conditions (0.5,1.5) and (0.75, 1.5).

Separating and integrating, we get

$$a \ln y_2 - by_2 = -d \ln y_1 + cy_1 + C \qquad (5.4.11)$$

where C is the constant of integration. It can be shown that for each C (i.e. different initial conditions), this is a closed curve surrounding the steady state $\mathbf{y}_{ss2}$.

Figure 5.7 shows one example. We observe that the trajectories look like a center but slightly distorted. These types of periodic solutions are observed in many ecological systems where two species compete. We stress that, as seen in Fig. 5.7, different initial conditions yield a different trajectory. This will contrast with another kind of periodic solutions that we discuss next.

5.4.2   Limit Cycles

In the present section, we consider a more complicated type of periodic solution that can occur in nonlinear systems. The idea is best illustrated through an example that, like our analysis in Example 5.2, can be solved using polar coordinates.

Example 5.3   Analyze the phase plane trajectories for the system

$$\frac{dx}{dt} = x - y - x(x^2 + y^2) \qquad (5.4.12)$$

$$\frac{dy}{dt} = x + y - y(x^2 + y^2) \qquad (5.4.13)$$

Solution

To find the steady state solutions, let's set the time derivatives to zero, multiply the first equation by y, the second by x, and subtract the resulting steady state equations to obtain $x^2 + y^2 = 0$. This suggests that the unique steady state is $(0, 0)$; see Fig. 5.8.

The Jacobian at this steady state is

$$\mathbf{J}_{ss} = \begin{bmatrix} 1 & -1 \\ 1 & 1 \end{bmatrix} \tag{5.4.14}$$

and its eigenvalues are $\lambda_1 = 1 + \iota$ and $\lambda_2 = 1 - \iota$. Therefore, the equilibrium point is an unstable focus. The question here is what happens as the trajectories spiral away from the steady state.

To determine this, it is helpful to use the polar transformations that we used in Example 5.2. Following a similar procedure like in that example, the original equations can be converted to

$$\frac{dr}{dt} = r(1 - r^2) \tag{5.4.15}$$

$$\frac{d\theta}{dt} = 1 \tag{5.4.16}$$

As was the case in Example 5.2, Eq. (5.4.16) suggests that the phase angle θ increases linearly with time. However, in contrast to what we saw in Example 5.2, Eq. (5.4.15) suggests two equilibrium states, one at $r = 0$ that corresponds to our steady state $\mathbf{y}_{ss} = (0, 0)$ and a *second one* at $r = 1$ that corresponds to a circle of radius 1, in the y_1, y_2 plane. This periodic (equilibrium) solution was hidden and we were able to uncover it using polar coordinates!

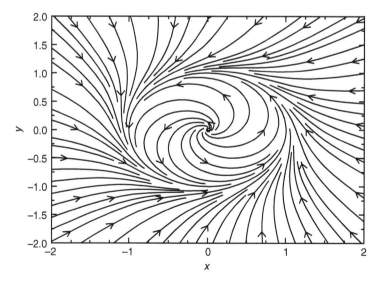

Figure 5.8 Phase plane for Eqs. (5.4.12) and (5.4.13).

Can we assess the stability of the steady state corresponding to $r = 1$? Well, if $r < 1$ in Eq. (5.4.15), then we have $\dot{r} > 0$ and the radial position increases with time. Conversely, if $r > 1$ in Eq. (5.4.15), then $\dot{r} < 0$ and the radial position decreases in time. Since $r = 1$ is an equilibrium state, this implies that initial conditions outside or inside this closed trajectory will converge to it asymptotically!

The type of periodic solution we saw in Example 5.3 is called a *limit cycle*. Since the particular limit cycle in this example attracts the trajectories, it is a *stable* limit cycle. We can also have the opposite case, where a periodic steady state repels trajectories and this is an *unstable* limit cycle.

Limit cycles are fundamentally different from the type of periodic solution we encountered before, such as a center. For one thing, a limit cycle is very robust; it does not depend on a pair of eigenvalues being zero, which could easily change if some parameter in our system changes. We will explore this point in more detail in Example 5.4. Moreover, a limit cycle is also a single, isolated solution, which attracts or repels trajectories starting from nearby initial conditions. In contrast, in the case of a center (or a periodic trajectory such as the one in the predator–prey example), there is a continuum of periodic solutions as the initial conditions are varied. It is also much more difficult to predict a limit cycle *a priori*; in Example 5.3, polar coordinates helped. In general, more sophisticated techniques are needed to either confirm or refute the presence of a limit cycle. In many problems, they are simply discovered by numerical integration.

5.5 Bifurcations

We now provide a brief introduction to one of the most fascinating features of nonlinear systems, i.e. the fact that equilibrium solutions may branch out as we vary one or more parameters, giving rise to multiple branches with very different stability properties.

5.5.1 One-Dimensional Systems

We will initially focus on a scalar equation

$$\frac{dy}{dt} = f(y, p) \tag{5.5.1}$$

where p is a parameter that can vary. We saw in Section 3.9 when we discussed continuation how to efficiently calculate the steady state solutions y_{ss} as p varies. These solutions will be smooth functions of p, as long as

$$\frac{\partial f}{\partial y} \equiv f_y \neq 0 \tag{5.5.2}$$

at the steady state solution. (This results from the implicit function theorem.) We choose to switch to subscript notation for the derivative, as defined in Eq. (5.5.2) simply to make the subsequent text easier to read. Note that in the vector case this would generalize to the Jacobian being non-singular. At a point where this condition fails, several branches of equilibria may come together. This is called a point of bifurcation.

Let's consider some of the different types of bifurcations that we can see in a one-dimensional system in the context of some concrete examples. This is not an exhaustive list of all possible bifurcations, but they will give you a basic understanding of the concept of a bifurcation and how complicated they can be – even for a single simple equation.

Pitchfork Bifurcation
Consider the ODE

$$\frac{dy}{dt} = y(p - y^2) \tag{5.5.3}$$

There are three steady state solutions to this problem: $y_{s1} = 0$, $y_{s2} = \sqrt{p}$, and $y_{s3} = -\sqrt{p}$. Also $f_y = p - 3y^2$.

For the case where $p < 0$, the steady states $\pm\sqrt{p}$ are not real valued. The only real valued steady state is $y_{s1} = 0$ and it is stable because $f_y = p < 0$. When $p = 0$ all three of the possible steady states are now real and converge onto a single one, $y = 0$. Finally, for $p > 0$, all three possible steady states are real and distinct. To characterize their stability, we need to look at f_y. This quantity is positive for $y_{s1} = 0$, meaning that this steady state which was stable for $p < 0$ has now become unstable for $p > 0$. For the other two steady states, $f_y = -2p < 0$, i.e. they are both stable. What we observe in this example is that as we vary the parameter p the number and stability properties of the steady states change, with a point of bifurcation at $p = 0$ giving rise to a branch of three steady state solutions.

Figure 5.9 shows the so-called bifurcation diagram for Eq. (5.5.3). In a sense, this is a continuation diagram where in addition to plotting the steady state solutions as the

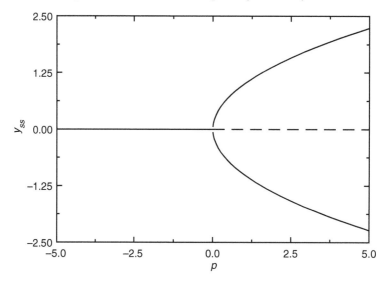

Figure 5.9 Bifurcation diagram for Eq. (5.5.3), which exhibits a pitchfork bifurcation.

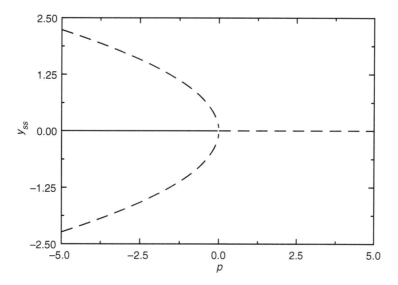

Figure 5.10 Bifurcation diagram for Eq. (5.5.4), which exhibits a subcritical pitchfork bifurcation.

parameter p is varied, we indicate their stability too. Stable steady state branches are usually denoted by solid lines, and unstable ones by dashed lines. This type of bifurcation is called a pitchfork bifurcation because of the characteristic shape of the bifurcation diagram. It can be further classified as a supercritical one to denote that the branches with the multiple equilibria occur for values of p for which the zero steady state is unstable.

Subcritical Pitchfork Bifurcation
Let's consider a very simple modification of Eq. (5.5.3)

$$\frac{dy}{dt} = y(p + y^2) \tag{5.5.4}$$

where we have just switched the sign of y^2 inside the parenthesis. The steady state solutions are $y_{s1} = 0$, $y_{s2} = \sqrt{-p}$, and $y_{s3} = -\sqrt{-p}$. Also $f_y = p + 3y^2$.

This is in a sense the reverse situation compared with the previous example. For $p > 0$ there is only one steady solution at $y_{s1} = 0$ and it is unstable because $f_y = p > 0$. At $p = 0$ and $y = 0$, we have again a point of bifurcation. As we keep decreasing p, we have all three steady states. For $y_{s1} = 0, f_y = p < 0$, i.e. the unstable steady state has now become stable. For the other two steady states $f_y = -2p > 0$, i.e. they are both unstable.

Figure 5.10 shows this bifurcation diagram. This type of bifurcation is called a subcritical pitchfork bifurcation.

Trans-Critical Bifurcation
Let's now reduce the power of the last term in Eq. (5.5.4),

$$\frac{dy}{dt} = yp + y^2 \tag{5.5.5}$$

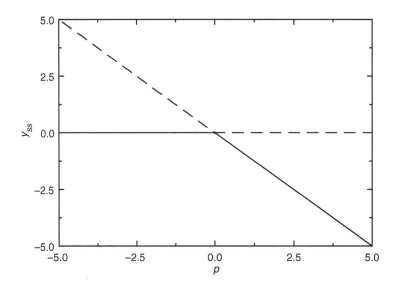

Figure 5.11 Bifurcation diagram for Eq. (5.5.5), which exhibits a trans-critical (exchange of stability) bifurcation.

The steady state equation is now quadratic with two real roots, $y_{s1} = 0$, $y_{s2} = -p$. Also $f_y = p + 2y$.

As was the case in our previous examples, the bifurcation again occurs at $p = 0$ and $y = 0$. However, in contrast to the pitchfork bifurcations, none of the steady states appears/disappears as we pass through the bifurcation point. Rather, for $p > 0$, $y_{s1} = 0$ is unstable while $y_{s2} = -p$ is stable. Conversely, for $p < 0$, $y_{s1} = 0$ is stable while $y_{s2} = -p$ is unstable. This type of bifurcation, called a trans-critical bifurcation or an exchange of stability bifurcation, is illustrated in Fig. 5.11.

Hysteresis Bifurcation

Let's go back to a problem with three possible steady states,

$$\frac{dy}{dt} = p + y - y^3 \tag{5.5.6}$$

The steady state equation is of the form $y^3 - y = p$, which is known as a depressed cubic equation. This problem can be solved analytically (indeed, it is the approach used in general to solve a cubic equation), but the algebra is a bit messy. Let's simply state here that this problem can have one, two or three real steady state solutions depending on the parameter p. Also, for this problem, $f_y = 1 - 3y^2$.

Figure 5.12 shows the bifurcation diagram for Eq. (5.5.6). As we can see, the number of steady states goes from one to two to three, and back to two and one as we increase p.

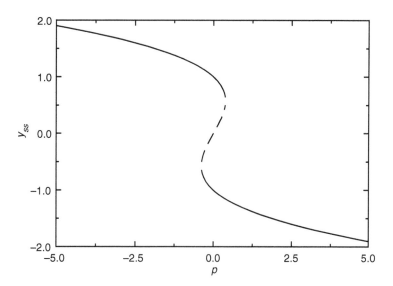

Figure 5.12 Bifurcation diagram for Eq. (5.5.6), which exhibits a hysteresis bifurcation.

5.5.2   Two-Dimensional Systems

For two-dimensional systems even more exotic possibilities than the ones just discussed arise. Remember that we can have periodic equilibrium solutions for such systems, and these can also bifurcate! One of the most celebrated such bifurcations is the so-called Hopf bifurcation, which we illustrate through the subsequent example.

Example 5.4   Analyze the phase plane trajectories for the system

$$\frac{dx}{dt} = -y + x(p - x^2 - y^2) \tag{5.5.7}$$

$$\frac{dy}{dt} = x + y(p - x^2 - y^2) \tag{5.5.8}$$

Solution

One can show that for all p, $(0,0)$ is the only equilibrium point. The Jacobian at this steady state is

$$\mathbf{J}_{ss} = \begin{bmatrix} p & -1 \\ 1 & p \end{bmatrix} \tag{5.5.9}$$

and its eigenvalues are $\lambda_1 = p + \iota$ and $\lambda_2 = p - \iota$. So for $p < 0$ we have a stable focus and for $p > 0$ we have an unstable focus. Using the polar coordinates that we

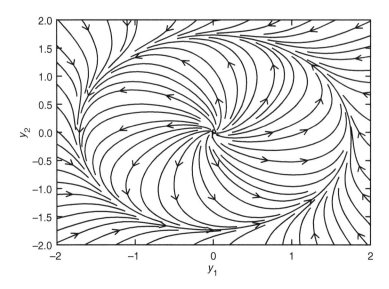

Figure 5.13 Hopf bifurcation for the system of Eqs. (5.5.7) and (5.5.8). For $p > 0$, there is a stable limit cycle surrounding an unstable focus.

introduced in Example 5.2, we get the following description of the system in terms of the radial distance and the phase angle

$$\frac{dr}{dt} = r(p - r^2) \tag{5.5.10}$$

$$\frac{d\theta}{dt} = 1 \tag{5.5.11}$$

Similar to what we saw in Example 5.3, the polar coordinates uncover the presence of an additional equilibrium solution. There is the steady state $r = 0$ which corresponds to $\mathbf{y}_{ss} = (0, 0)$, while an additional steady state emerges at $r = \sqrt{p}$ for $p > 0$. This latter solution is a limit cycle.

To analyze the stability of the limit cycle, we need to analyze how the radial position changes near the limit cycle. Starting inside the limit cycle, i.e. for $0 < r < \sqrt{p}$, $\dot{r}$ is positive and the radial position increases. Starting outside the limit c, i.e. for $r > \sqrt{p}$, we have that $\dot{r}$ is negative and the radial distance decreases. This implies that this limit cycle attracts the trajectories and is thus stable. As before, $p = 0$ corresponds to a bifurcation point, whereby a stable focus becomes unstable, and a stable limit cycle develops around it. At that bifurcation point the Jacobian has a pair of eigenvalues on the imaginary axis. As p increases, the amplitude of the oscillation of the limit cycle also increases as $\sqrt{p}$. Figure 5.13 shows the phase plane for Example 5.4.

A Hopf bifurcation can be further characterized as subcritical or supercritical. In more general systems, it occurs when, as we vary p, a pair of the Jacobian eigenvalues crosses the imaginary axis, with their real part changing sign. At the critical value of p where this happens, all of the remaining eigenvalues must have strictly negative real parts. Naturally, for a two-dimensional system there are no other eigenvalues, but we need this condition if we are looking for a Hopf bifurcation in higher dimensions.

5.6   Deterministic Chaos

As we go to higher-dimensional systems, even more complicated phenomena can take place, including chaotic-like behavior. Indeed, one of the most interesting and pervasive behaviors exhibited by dynamical systems is called deterministic chaos. Deterministic chaos is not random behavior, but very complicated deterministic behavior. It can occur when there are at least three coupled, differential equations that contain nonlinear terms. Perhaps the most famous equations that exhibit this behavior are known as the Lorenz equations,

$$\frac{dx}{dt} = -\sigma x + \sigma y \qquad (5.6.1)$$

$$\frac{dy}{dt} = -xz + rx - y \qquad (5.6.2)$$

$$\frac{dz}{dt} = xy - bz \qquad (5.6.3)$$

where σ, r, and b are parameters. These non-dimensional equations were derived to approximate the behavior of convective flows in the earth's atmosphere. The unusual behavior exhibited by these seemingly simple equations was reported by Lorenz in a seminal paper in 1963 that many feel marked the birth of the field of modern nonlinear dynamics.

Let's take a look at how these equations behave for $\sigma = 10$, $b = 8/3$, and $r = 28$. Figure 5.14 plots the trajectory of the three variables versus time. The behavior is very striking. It looks like it *might* be periodic, but that's not exactly the case. Lorenz was puzzled by these results, too. He called this behavior "aperiodic," i.e., behavior which is almost periodic, but not quite.

Lorenz was alerted to the unusual behavior of these equations when he stopped his computations for a coffee break. Rather than saving the solution to a file, he simply jotted down the first several digits of the solution, stopped the computer, and restarted the computation later by inputing the variables by hand.

We can re-create history here. Let's take our prior solution and restart it with a few digits missing. In Fig. 5.15, we have started the trajectory from one of the points in Fig. 5.14, except that we truncated the values of x, y, and z after the fourth decimal place. Like Lorenz, you probably think that this should make no difference. Indeed, you probably spent quite some time in your laboratory classes learning about significant digits and measurement accuracy. In lab classes, your instructors surely told you that there is no way that a result should require an infinite precision to be reproducible. However, notice how the new solution in Fig. 5.15 tracks the old one in Fig. 5.14 for about four time units, then dramatically departs from it. Lorenz couldn't believe that

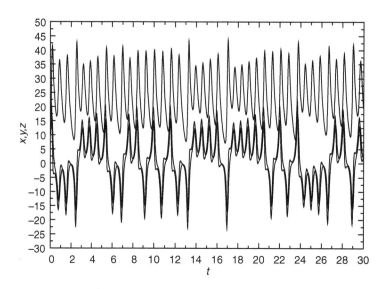

Figure 5.14 Integration of the Lorenz equations for an initial condition $x = y = z = 10$ out to a time of $t_f = 30$.

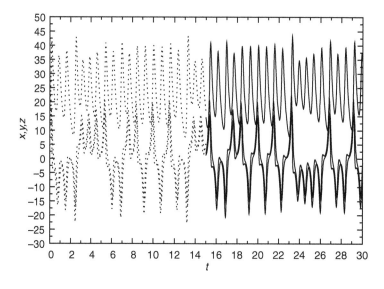

Figure 5.15 Integration of the Lorenz equations to $t = 15$, where $x = 4.0049$, $y = 5.9239$, $z = 16.8624$ (dashed lines). The data are rounded to two decimal places and then integrated out to a time of $t_f = 30$.

this solution was so sensitive to such a very small change to the initial conditions. He eventually discovered that this behavior is characteristic of chaotic systems, namely extreme sensitivity to initial conditions. Extremely small changes in initial conditions can lead, over time, to dramatically different outcomes, a phenomenon that has been called the "butterfly effect," after the notion that a butterfly flapping its wings half-way

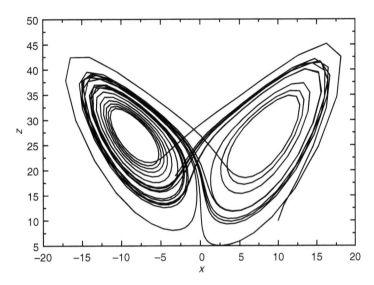

Figure 5.16 Phase portrait for the Lorenz equations.

around the globe could affect the weather we have locally. (The earth's atmosphere is a prime example of a nonlinear dynamical system.)

The Lorenz equations have three unknowns, so the phase portrait is a trajectory in three-dimensional space. As such, it is somewhat harder to visualize than the two-dimensional problems we have investigated so far. Figure 5.16 shows the phase behavior for the parameters that we have used so far. What you see in Fig. 5.16 is something called a "strange attractor," which is also one of the signatures of chaotic behavior. Nearby initial conditions are attracted to this region of phase space (hence the name "attractor"); however, once there, they never leave (the attractor is bounded in phase space) and never cross paths (the trajectory goes around and around phase space in an "aperiodic" manner). This attractor has been called the "Lorenz mask" and also, fittingly, the "Lorenz butterfly."

5.7   Case Study: Non-Isothermal CSTR

In Chapter 4, we presented two case studies of chemical reactors. In Section 4.4, we looked at a continuous stirred tank reactor with a fluctuating inlet. This problem was dynamic, since the independent variable was time, and one-dimensional since the only dependent variable in the problem was the concentration in the tank. In Section 4.9, we looked at the conversion in a non-isothermal plug flow reactor (PFR). This example was a system of two equations for the mass and energy balances, where the heat generated/consumed by the reaction leads to a change in the reaction rate, which thus couples it to the mass conservation. However, the system of equations for the non-isothermal PFR are not a dynamic system, *per se*, since the independent variable is the axial position in the reactor.

In this example, we will consider the non-isothermal CSTR, which is dynamic (independent variable being time) and a two-dimensional system (two dependent variables). While we have been discussing phase planes, limit cycles, and bifurcations in a rather generic way so far, this example will show you their relevance for a chemical engineering application.

5.7.1   Problem Statement

To get started, let's merge the analysis we needed for the CSTR in Section 4.4 with the non-isothermal reaction in Section 4.9. The overall process is illustrated in Fig. 5.17. The process consists of a tank with reaction volume V containing a reactant at concentration C_A and temperature T. Species A decomposes to produce B, with a reaction rate

$$r = k_0 e^{-E/RT} C_A \qquad (5.7.1)$$

where k_0 is the pre-exponential factor, E is the activation energy and R is the ideal gas constant. There is a volumetric flow rate q into the tank of a stream at a concentration $C_{A,F}$ and temperature T_F. There is an equal volumetric flow rate q coming out of the tank, thus keeping the volume constant.

We already discussed in Chapter 4 the derivation of the balance equations for a CSTR and the energy balance, so we simply state the final equations here. The mass balance on species A is

$$\frac{d}{dt}(C_A V) = q C_{A,F} - q C_A - k_0 e^{-E/RT} C_A V \qquad (5.7.2)$$

and the energy balance is

$$\frac{d}{dt}(\rho C_p V T) = (\rho C_p T_F) q - (\rho C_p T) q - k_0 e^{-E/RT} C_A V \Delta H_{rxn} \qquad (5.7.3)$$

where ρ and C_p are the density and heat capacity of the fluid in the reactor. The heat of reaction, ΔH_{rxn}, is defined to be negative for an exothermic reaction, so the negative sign in the energy balance causes an exothermic reaction to increase the energy inside the reactor. In our energy balance, we have assumed adiabatic operation with $Q = 0$.

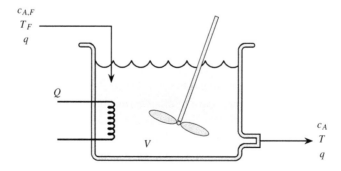

Figure 5.17 Schematic illustration of a non-isothermal, continuous stirred tank reactor (CSTR).

We will assume constant physical properties and volume such that

$$\frac{d}{dt}(C_A V) = V\frac{dC_A}{dt} \tag{5.7.4}$$

and

$$\frac{d}{dt}(\rho C_p V T) = \rho C_p V\frac{dT}{dt} \tag{5.7.5}$$

Let us introduce the dimensionless concentration,

$$x_1 = \frac{C_A}{C_{A,F}} \tag{5.7.6}$$

and dimensionless temperature,

$$x_2 = \frac{T}{T_F} \tag{5.7.7}$$

In the latter, we are using absolute zero as a reference temperature for convenience. It will also be convenient to take advantage of two additional dimensionless quantities, the dimensionless time

$$\theta = \frac{t}{V/q} \tag{5.7.8}$$

where we have used the residence time V/q as the characteristic time scale, and the dimensionless Arrhenius factor

$$\beta = \frac{E}{RT_F} \tag{5.7.9}$$

With these dimensionless variables, the mass balance becomes

$$\frac{V}{V/q}\frac{d}{d\theta}(C_{A,F}x_1) = qC_{A,F} - qC_{A,F}x_1 - k_0 e^{-\beta/x_2}C_{A,F}x_1 V \tag{5.7.10}$$

which simplifies to

$$\frac{dx_1}{d\theta} = 1 - x_1 - \left(\frac{k_0 V}{q}\right)e^{-\beta/x_2}x_1 \tag{5.7.11}$$

The quantity in brackets is the Dämkohler number,

$$\mathrm{Da} = \frac{k_0 V}{q} \tag{5.7.12}$$

measuring the relative rates of reaction and convention through the reactor. We thus can write our dimensionless mass balance in the form

$$\frac{dx_1}{d\theta} = 1 - x_1 - \mathrm{Da}\, e^{-\beta/x_2}x_1 \tag{5.7.13}$$

For the energy balance, the dimensionless variables lead to

$$\frac{\rho C_p V}{V/q}\frac{d}{d\theta}(T_F x_2) = \rho C_p T_F q - \rho C_p T_F x_2 q - k_0 e^{-\beta/x_2}C_{A,F}x_1 V\Delta H_{rxn} \tag{5.7.14}$$

This simplifies to

$$\frac{dx_2}{d\theta} = 1 - x_2 + \left(\frac{k_0 V}{q}\right)\left[\frac{C_{A,F}(-\Delta H_{rxn})}{\rho C_p T_f}\right]e^{-\beta/x_2}x_1 \tag{5.7.15}$$

We again see the presence of the Dämkohler number along with a second dimensionless quantity,

$$\gamma = \frac{C_{A,F}(-\Delta H_{rxn})}{\rho C_p T_f} \qquad (5.7.16)$$

known as the Prater number, which is a balance between the energy produced (consumed) by the reaction relative to the energy brought in by the incoming fluid. The dimensionless equation governing the temperature is thus

$$\frac{dx_2}{d\theta} = 1 - x_2 + \mathrm{Da}\,\gamma x_1 e^{-\beta/x_2} \qquad (5.7.17)$$

Equations (5.7.13) and (5.7.17) represent a system of coupled dynamic equations that can be solved for a particular initial reactor temperature and reactant concentration. We will now analyze this problem in the context of the dynamic systems concepts we learned in this chapter.

5.7.2   Solution

Steady States

If we want to look for the steady states for the reactor, we need to solve the coupled nonlinear algebraic equations

$$1 - x_1 - \mathrm{Da}\,e^{-\beta/x_2}x_1 = 0 \qquad (5.7.18)$$

$$1 - x_2 + \mathrm{Da}\,\gamma e^{-\beta/x_2}x_1 = 0 \qquad (5.7.19)$$

The most interesting question to ask is how the steady state concentration (x_1) and temperature (x_2) depend on the residence time $\tau = V/q$, since this is a property that we can control easily by changing either the size of the reaction vessel or the volumetric flow rate. We can thus use a continuation method where the parameter p is the Dämkohler number (5.7.12), which has the residence time as one of its parameters.

To proceed further, we use some reasonable numbers (from Lanny Schmidt's book):

$$\gamma = 0.1333, \quad \beta = 50.327 \qquad (5.7.20)$$

indicating an exothermic reaction, and

$$\mathrm{Da} = 2.6 \times 10^{20}\tau \qquad (5.7.21)$$

where τ is a dimensionless residence time. This choice of parameters makes τ a number of $O(1)$.

To find the steady states, we need to use Newton–Raphson. We already have the residuals in Eqs. (5.7.18) and (5.7.19), which can be computed from the MATLAB function:

```
1  function R = getR(x,b,g,D)
2  x1 = x(1);
3  x2 = x(2);
4  R(1,1) = 1 - x1 - D*x1*exp(-b/x2);
5  R(2,1) = 1 - x2 + D*g*x1*exp(-b/x2);
```

Note that this function takes as inputs not only the vector of unknowns **x**, but also the parameters β, γ, and Da. For convenience, we unpack the input vector into the concentration, x(1), and the temperature, x(2).

Since we put the unknown vector in the logical order $[x_1, x_2]$, the Jacobian is

$$J = \begin{bmatrix} -1 - Dae^{-\beta/x_2} & -Dax_1e^{-\beta/x_2}\beta x_2^{-2} \\ Da\gamma e^{-\beta/x_2} & -1 + Da\gamma x_1 e^{-\beta/x_2}\beta x_2^{-2} \end{bmatrix} \quad (5.7.22)$$

which is computed from a similar MATLAB function:

```
function J = getJ(x,b,g,D)
%this is the jacobian
x1 = x(1);
x2 = x(2);
J(1,1) = -1 - D*exp(-b/x2);
J(1,2) = -D*x1*exp(-b/x2)*b/x2^2;
J(2,1) = D*g*exp(-b/x2);
J(2,2) = -1 + D*g*x1*exp(-b/x2)*b/x2^2;
```

As was the case in getR, we are passing the problem parameters to getJ, and we have unpacked the unknown vector for convenience.

For a given value of the residence time τ, and thus a given Dämkohler number, Da, we need to use a slight modification of Program 3.5 to implement Newton–Raphson:

```
function [x,flag] = nonlinear_newton_raphson(x,tol,b,g,D)
%   x = initial guess for x
%   tol = criteria for convergence
%   requires function R = getR(x)
%   requires function J = getJ(x)
%   flag = 0 if failed, 1 if converged
%   parameters b,g,tau from CSTR problem
R = getR(x,b,g,D);
k = 0; %counter
kmax = 20; %max number of iterations allowed
while norm(R) > tol
    J = getJ(x,b,g,D); %compute Jacobian
    del = -J\R; %Newton Raphson
    x = x + del; %update x
    k = k + 1; %update counter
    R = getR(x,b,g,D); %f for while loop and counter
    if k > kmax
        fprintf('Did not converge.\n')
        break
    end
end
if k > kmax || max(x) == Inf || min(x) == -Inf
    flag = 0;
else
    flag = 1;
end
```

The obvious difference with the original program is that we are sending the values of the parameters β, γ, and Da. The more subtle difference is in line 10, where we cut off the maximum number of iterations at `kmax = 20`, whereas our original program allowed this number to go to 100. We will explain the reason for this choice in a moment.

This problem has multiple steady states including two turning points that make the continuation method fail. The strategy to find the steady states is as follows.

(1) Fill in the lower part up to the first turning point starting from a low temperature.
(2) Fill in the upper part down to the second turning point starting from a high temperature.
(3) Find an intermediate point (the trickiest part).
(4) Fill up to the second turning point from the intermediate point.
(5) Fill down to the first turning point from the intermediate point.

For each "fill" step, we wrote a program to perform the continuation method until it fails.

```
1  function [tau_out,x2_out] = continuation_tau(x,tau,g,b,dtau)
2  stop = 1;
3  count = 1;
4  while stop > 0
5      D = TauToD(tau);
6      tol = 1e-8;
7      [x,flag] = nonlinear_newton_raphson(x,tol,b,g,D);
8      if flag == 0
9          stop = -1; %stop the loop
10     else
11         x2_out(count) = x(2);
12         tau_out(count) = tau;
13         tau = tau + dtau;
14         count = count+1;
15     end
16 end
```

The inputs to this function are the initial guesses for the concentration and temperature, contained in the vector x, at the first value of the residence time τ. We also send the parameters γ and β, as well as the step size $\Delta\tau$ that we are going to use for the continuation method. Remember from Section 3.9 that we need to take a small step size if we use a zero-order continuation method, which is the case here.

The main part of `continuation_tau` is a **while** loop that executes so long as a new solution was found in the last iteration. Before we try to find this new solution, we first convert τ to Da with a simple program:

```
1  function D = TauToD(tau)
2  D = 2.6e20*tau;
```

You may be wondering why we bothered to write a separate program with such a simple calculation when we could have just put this on line 5 of `continuation_tau`. The reason is that we decided to work in terms of τ, rather than Da, for the continuation. If at some point we decide to change the problem parameters, in particular the conversion factor in Eq. (5.7.21), we would need to remember to make the change in `continuation_tau`. You may be very good at remembering the details of your programs, but we are getting old! By separating this calculation into its own subfunction, we are less likely to forget where we put Eq. (5.7.21).

Now let's go back to `continuation_tau`. Line 7 calls Newton–Raphson to compute the concentration and temperature at the current value of τ. Recall from our modified version of `nonlinear_newton_raphson.m` that we will stop the algorithm if we go past 20 iterations or the solution diverges. In either case, the value of `flag` will be zero. If this happens, we know that Newton–Raphson failed and line 9 stops the continuation method. If Newton–Raphson converged, then we record the value of the temperature and the residence time in the output vectors in lines 11 and 12. To get ready for the next iteration, we update the value of the residence time in line 13 and update our counter for the output vectors in line 14 so that we can record the new values in the right spot. Our algorithm is a zero-order continuation method since, when the **while** loop executes again, we use the value of x from the last iteration as the initial guess for the next iteration.

Let's see how well our program works. For the lower branch, we start with a very small residence time of $\tau = 0.1$ and step in increments of $\Delta\tau = 0.005$. If the residence time is small, then there should be hardly any reaction at all. We thus selected a dimensionless concentration of $x_1^{(0)} = 0.9$ and temperature of $x_2^{(0)} = 1$, corresponding to a small decrease in concentration with no accompanying temperature rise. As we see in Fig. 5.18, the continuation method works well until $\tau = 1.870$. When we try to solve the problem for $\tau = 1.875$ using the converged solution at $\tau = 1.870$, we cannot get convergence. Thus, it is reasonable to assume that this is the turning point on the lower branch and we need a better initial guess to find a solution.

For the upper branch, we now need to consider a residence time that exceeds the end of the lower branch. We chose to use an initial residence time $\tau = 2$ and the same

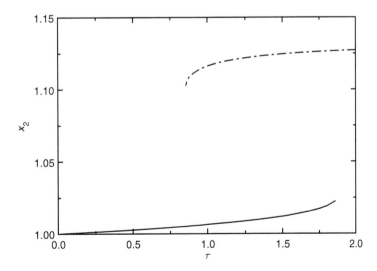

Figure 5.18 Upper and lower branches of steady states for the non-isothermal CSTR with the parameters in Eq. (5.7.20).

increment $\Delta \tau = -0.005$, except that for the upper branch we are trying to march down in residence time. If you think about this case for a moment, you would expect that a large residence time should lead to substantial reaction. Thus, we would expect a high temperature and low outlet concentration. We decided to use $x_1^{(0)} = 0.04$, i.e. 96% conversion, and $x_2^{(0)} = 1.14$ as the initial guess. You might not think that a 14% increase in the temperature relative to the feed is very hot, but remember that this is in absolute temperature. So if we feed the reactor near room temperature, say 300 K, then the outlet of the reactor (342 K = 156 °F) exceeds the hottest weather temperature ever recorded on Earth (134 °F in Death Valley, CA on July 10, 2013). Figure 5.18 shows that this was indeed a good initial guess. The continuation method marches down to $\tau = 0.8550$ before failing at $\tau = 0.8500$. We thus assume this is the region for the turning point for the upper branch.

There is a rather peculiar result in our continuation method for the upper branch. As we mentioned above, we set the maximum number of iterations in Newton–Raphson to kmax = 20, and this indeed is a value that works. If we increase the number of iterations to kmax = 100 and let Newton–Raphson wander around the phase space this long near the turning point, we do eventually find a solution. The problem is that the solution is not physical! This highlights an important point about nonlinear equations governing physical processes; they may have many real solutions, so we need to use our knowledge of chemistry and physics to decide whether or not the solution is correct.

We are now ready for the hardest step of this turning point problem, namely finding a middle point to fill in the branch between the two turning points. Fortunately, our solutions for the upper and lower branch give us some very useful insights. First, we know that the two turning points are near $\tau = 0.85$ and $\tau = 1.87$. So we should pick an intermediate value, say $\tau = 1$, that should be on the branch connecting the two turning points. Second, we know the values of the concentration and temperatures at these turning points. We would expect that the other steady state at $\tau = 1$ should have a temperature and concentration intermediate between the upper branch and lower branch. We thus guessed $x_1^{(0)} = 0.4$ and $x_2^{(0)} = 1.08$ and fortunately Newton–Raphson converged to the solution $x_1 = 0.401\,627$ and $x_2 = 1.079\,781$.

We are now in great shape to complete the set of steady states. We can just march up and down in τ from the converged solution, using the continuation method until we reach the turning points we found in Fig. 5.18. Figure 5.19 shows the final result for the steady states of the CSTR. Below the turning point near $\tau = 0.85$ there is a single steady state at a low temperature. Above the turning point near $\tau = 1.87$ there is also a single steady state, but this time we have a hot temperature. The interesting behavior occurs between the two turning points. Here, for a given value of the residence time τ, there appear to be three steady states. Thus, it seems that the reactor could operate at three different temperatures for exactly the same residence time!

We should also try to figure out whether these steady states are stable or not; if the steady state is stable, then when we start up the reactor we know that it will not eventually reach that temperature and stay there. To make this calculation, we made a small modification to our continuation method program to also determine the stability:

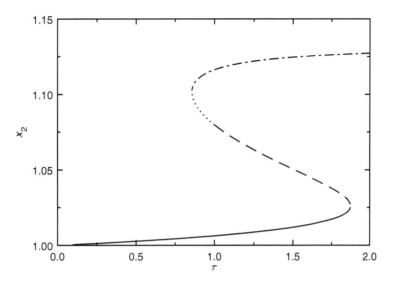

Figure 5.19 Complete set of steady states for the non-isothermal CSTR with the parameters in Eq. (5.7.20). The different line styles indicate the different parts of the solution scheme.

```
1   function [tau_out,x2_out,stable] = cont_tau_stab(x,tau,g,b,dtau)
2   stop = 1;
3   count = 1;
4   while stop > 0
5       D = TauToD(tau);
6       tol = 1e-8;
7       [x,flag] = nonlinear_newton_raphson(x,tol,b,g,D);
8       if flag == 0
9           stop = -1; %stop the loop
10      else
11          J = getJ(x,b,g,D);
12          stable(count) = check_stability(J);
13          x2_out(count) = x(2);
14          tau_out(count) = tau;
15          tau = tau + dtau;
16          count = count+1;
17      end
18  end
```

The new parts of the program begin on line 11. We first compute the Jacobian from the steady state values of x returned from line 7, and we then call another function to determine whether this Jacobian corresponds to a stable steady state:

```
1   function stable = check_stability(J)
2   %0 = unstable, 1 = stable
3   a = J(1,1);
4   b = J(1,2);
5   c = J(2,1);
6   d = J(2,2);
7
8   discr = (a+d)^2 - 4*(a*d-b*c);
```

```
 9   if discr < 0
10       %the eigenvalues are imaginary
11       if a + d > 0
12           stable = 0; %the real part is positive
13       else
14           stable = 1; %the real part is negative
15       end
16   else
17       %the eigenvalues are real
18       if a + d + sqrt(discr) > 0
19           stable = 0; %the biggest eigenvalue is positive
20       else
21           stable = 1; %the biggest eigenvalue is negative
22       end
23   end
```

For convenience, this function starts by unpacking the entries of the Jacobian into the form

$$\mathbf{J} = \begin{bmatrix} a & b \\ c & d \end{bmatrix} \tag{5.7.23}$$

We then check to see if the largest real part of the eigenvalues is positive. This is broken up into two steps. First, we see if the eigenvalues are imaginary. If so, then we just have to look at the trace of the matrix to find the real part. If the eigenvalues are real, then the term in the square root of Eq. (5.3.22) is also important. Notice that we only have to consider the positive term, since if the larger real eigenvalue is negative, then the smaller real eigenvalue also must be negative. The program `cont_tau_stab` returns a vector called `stable` that has a value of zero if the steady state is unstable and a value of 1 if the steady state is stable. Note that we are not considering the case of neutral stability. From a computational standpoint, round-off error makes it impossible to actually get a value of identically zero from the eigenvalue calculation in `check_stability`. (It is worth pointing out that MATLAB may return a value of zero because it "thinks" you wanted to work with integers instead of floating point.)

Figure 5.20 reports the results of this stability calculation. As we can see, only the upper and lower branches of the system are stable; the branch connecting the two turning point consists of a number of unstable steady states. Thus, for residence times between the two turning points, the reactor is either hot (the upper branch) or cold (the lower branch). Comparing Fig. 5.20 with Fig. 5.12, we see that this problem exhibits a hysteresis bifurcation.

Phase Plane

If there are two stable steady states between the turning points, how do we know which one is the observed state? After all, a well mixed reactor can only be operating at one temperature at a given point in time. The answer is that the steady state depends on the initial conditions. Some initial conditions will lead to the hot steady state, others will lead to the cold steady state. Since the problem is highly nonlinear, due to the exponential in the reaction rate, it is not obvious that a reactor that starts hot will stay hot and vice versa. Moreover, since there is an unstable steady state at a temperature

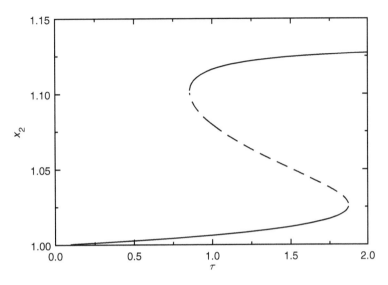

Figure 5.20 Complete set of steady states for the non-isothermal CSTR with the parameters in Eq. (5.7.20). The solid lines are stable steady states and the dashed line is unstable steady states.

between the two stable steady states, any trajectory that passes near the unstable state could be deflected.

To look at this more concretely, let's consider the case where $\tau = 1$. We already computed the unstable steady state at $x_1 = 0.401\,627$ and $x_2 = 1.079\,781$ as the seed point to create the branch connecting the turning points. For this residence time, the cold steady state corresponds to $x_1 = 0.952\,803$ and $x_2 = 1.006\,293$ and the hot steady state is $x_1 = 0.127\,103$ and $x_2 = 1.116\,383$. These three steady states are seen in Fig. 5.20.

The phase plane for these parameters is in Fig. 5.21. We can clearly see the so-called "basin of attraction" for each of the stable steady states, namely the region of the phase space initial concentration and temperature that leads to a particular steady state. While it is true that lower initial temperatures tend to lead to the cold steady state, the dividing line between the two basins of attraction is not at a fixed temperature. Rather, the system selects the hot or cold steady state depending on both the initial temperature of the reactor and the amount of reactant inside the reactor when it starts. Recall that we are looking at an exothermic reaction. Thus, even if the reactor is somewhat cold when it starts, if there is a large amount of reactant inside at the start then the reaction rate will be somewhat fast (due to the concentration), which will increase the temperature. The increased temperature further increases the reaction rate, reducing the concentration. These two effects compete – if the temperature does not increase fast enough relative to the rate at which the reactants are depleted, then the reaction falls into the low temperature steady state. However, if the temperature can go up fast enough, then it can be maintained at a high temperature even as the reactant concentration approaches zero. The reaction must eventually slow down, since the reactant flow into the reactor is very slow compared to the rate of reaction at these temperatures, and eventually the system settles into the hot steady state.

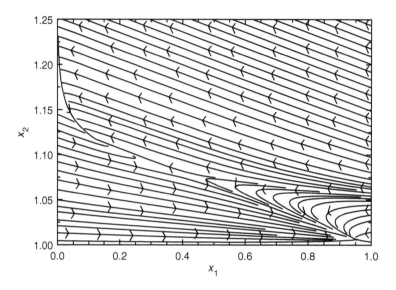

Figure 5.21 Phase plane for the non-isothermal CSTR with the parameters in Eq. (5.7.20) with $\tau = 1$.

The phase plane in Fig. 5.21 raises some important chemical engineering issues that are discussed at length in reaction engineering textbooks. First, we should note that the high temperature state corresponds to a high conversion. (Recall that the conversion is given by $1 - x_1$, since we scaled concentration with the feed concentration.) Thus, if we want the reaction to be efficient, we need to use care to make sure that we actually end up operating at the higher temperature. This is a delicate control problem, since the way we set up the initial conditions here is artificial – it would be extremely difficult to impose a fixed concentration and temperature at $t = 0$ and then instantaneously start the reaction at $t > 0$ unless the reaction vessel is empty at the start of the process. Second, some of the initial conditions at high initial concentrations inside the reactor lead to very high maximum temperatures, much larger than the steady state temperature. The dynamics towards the steady state are important if, for example, the reaction temperature or pressure exceed the operating limits of the equipment. Moreover, very high temperatures could activate alternate reaction pathways with higher Gibbs free energies that we could safely ignore at a lower temperature. In short, understanding the steady state is obviously important, but ignoring the transients can literally kill you.

Limit Cycles

Depending on the various physical parameters, the non-isothermal CSTR can exhibit limit cycles of the type seen in Fig. 5.22. Qualitatively, you may be able to rationalize the existence of a limit cycle from the coupling between the reaction rate and the temperature. As the reactor gets hot, the reactants are consumed more quickly. While an exothermic reaction will continue to release heat, eventually there are very few reactants inside the CSTR to generate the heat. The reactants thus need to be replenished by the feed stream. If the feed stream is cold, the fluid flowing into the reactor will

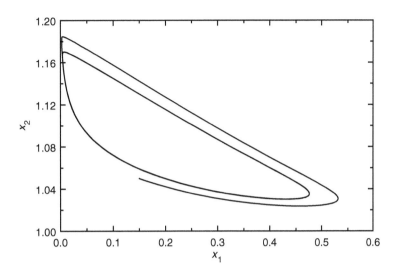

Figure 5.22 Limit cycle for the non-isothermal CSTR. This limit cycle corresponds to $\beta = 50$, Da $= 1.04 \times 10^{21}$, $\gamma = 0.324$, $U = 3$, and $x_c = 1$.

lower the temperature. (The temperature can also be lowered by having heat transfer at the boundaries, which we have not included in our model.) Eventually, there may be enough reactants to again send the temperature surging upward, and the cycle repeats again. While this logic sounds like it might produce a limit cycle, it is not necessarily a limit cycle or even cyclic behavior in the first place. For example, the trajectories in Fig. 5.21 seem to show part of the logic (a quick rise in temperature) but there is no cyclic behavior – the trajectories simply cool down to a steady (ignited) state. We could just as easily imagine that the scenario we just described would lead to chaotic behavior where the ups and downs of concentration and temperature are similar on each iteration but do not exactly overlap.

It turns out that Uppal, Ray, and Poore, in a classic paper in 1974, *discovered* limit cycles by numerical analysis of the non-isothermal CSTR when they also included heat transfer at the surface of the reactor. This additional heat transfer mechanism requires modifying Eq. (5.7.3) to include an adddicional term

$$\frac{d}{dt}(\rho C_p V T) = (\rho C_p T_F)q - (\rho C_p T)q - k_0 e^{-E/RT} C_A V \Delta H_{rxn} - hA(T - T_c) \quad (5.7.24)$$

where h is the heat transfer coefficient over the area A of the reactor and T_c is the temperature of the cooling fluid (where the notation "cooling" assumes that $T_c < T$). When we convert this to dimensionless form, we have

$$\frac{dx_2}{d\theta} = 1 - x_2 + \text{Da}\,\gamma x_1 e^{-\beta/x_2} - U(x_2 - x_c) \quad (5.7.25)$$

where the dimensionless heat transfer coefficient is

$$U = \frac{hA}{\rho C_p q} \quad (5.7.26)$$

and the cooling fluid temperature is measured relative to the inlet fluid

$$x_c = T_c/T_f \qquad (5.7.27)$$

Figure 5.22 shows one particular limit cycle for the non-isothermal CSTR corresponding to the case where the cooling fluid is at the same temperature of the feed temperature. We will give you the chance to explore limit cycles in the non-isothermal CSTR in more detail in Problem 5.20. The analysis by Uppal, Ray, and Poore, which includes the conditions we used to produce Fig. 5.22, was a major advance in reaction engineering.

5.8   Implementation in MATLAB

Dynamical systems require the analysis of systems of initial value problems. As a result, the MATLAB solver `fsolve` for finding the roots of systems of equations (Section 3.11) and the integrators such as `ode45` (Section 4.10) are the main tools that you need to analyze dynamic systems.

You may also find the built-in eigenvalue solver useful for determining the stability of steady states. The function is called as `eigs(A)` and returns the eigenvalues of a matrix **A**.

Finally, you may want to produce flows in the phase plane, for example what we saw in Fig. 5.21, from your own data. A very nice tool for creating these types of plots is called `directedplot`. This is not available as part of the general release of MATLAB. Rather, it is one of many files at the "File Exchange" on the MATLAB Central website at http://www.mathworks.com/matlabcentral/fileexchange. This website is a repository of functions that people have written that go beyond what is available by default in MATLAB.

5.9   Further Reading

There is an extensive literature on the analysis of nonlinear dynamical systems. Excellent, albeit rather advanced sources from the mathematics literature include

- J. M. Guckenheimer and P. Holmes, *Nonlinear Oscillations, Dynamical Systems and Bifurcations of Vector Fields*, Springer-Verlag, 1983
- F. Verhulst, *Nonlinear Differential Equations and Dynamical Systems*, Springer-Verlag, 1985
- S. Wiggins, *Introduction to Applied Nonlinear Dynamical Systems and Chaos*, Springer-Verlag, 1990
- J. Hale and H. Kocak, *Dynamics and Bifurcations*, Springer-Verlag, 1991
- Y. A. Kuznetsov, *Elements of Applied Bifurcation Theory*, Springer, 1995.

A particularly easy to read book with lots of good problems is

- S. Strogatz, *Nonlinear Dynamics and Chaos with Applications in Physics, Biology, Chemistry and Engineering*, Perseus Publishing, 1994.

A study of the Lorenz equations specifically can be found in

- E. N. Lorenz, "Deterministic nonperiodic flow," *Journal of Atmospheric Sciences* **20**, 130, 1963
- C. Sparrow, *The Lorenz Equations: Bifurcations, Chaos, and Strange Attractors*, Springer-Verlag, 1982.

A broad coverage of dynamical systems (which has influenced that in the book) with emphasis on chemical engineering applications can be found in

- A. Varma and M. Morbidelli, *Mathematical Methods in Chemical Engineering*, Oxford University Press, 1997.

A classic overview paper on chaotic behavior from the chemical engineering literature is

- M. F. Dougherty and J. M. Ottino, "Chaos in deterministic systems: strange attractors, turbulence, and applications in chemical engineering," *Chemical Engineering Science* **43**, 139, 1988.

A comprehensive study of the dynamic behavior of the CSTR can be found in

- A. Uppal, W. H. Ray, and A. B. Poore, "On the dynamic behavior of continuous stirred tank reactors," *Chemical Engineering Science* **29**, 967, 1974

and the particular parameters that we used here are from the discussion in

- L. D. Schmidt, *The Engineering of Chemical Reactions*, Oxford University Press, 2004.

Problems

5.1 Perform a linear stability analysis of the system

$$x_1' = \sin(\pi x_1)\cos(\pi x_2) + x_2^2 - 4x_1$$
$$x_2' = x_2 \ln x_1$$

about the steady state $x_1 = 1$ and $x_2 = 2$. What type of steady state do you find?

5.2 Consider the dynamics of the damped, forced oscillator

$$\ddot{x} + \dot{x} - 2x + 3x^3 = 0$$

Classify the steady states in the position/velocity phase space $(x, \dot{x})$.

5.3 Consider the dynamical system

$$\dot{x} = -2\cos x - \cos y$$
$$\dot{y} = -2\cos y - \cos x$$

What type of steady state (or steady states) are nearest to the origin?

5.4   Classify the two steady states of the dynamic system and make a sketch of the phase plane near the steady states for the system

$$\frac{dy_1}{dt} = y_1^2 - 3y_1y_2 + 2$$

$$\frac{dy_2}{dt} = y_1 - y_2$$

Your phase plane only needs to show the basic idea and you do not need to worry about the directions of the eigenvectors or the sense of rotation.

5.5   Use linear stability analysis to make a sketch of the phase plane for

$$\frac{d}{dt} \begin{bmatrix} x \\ y \end{bmatrix} = \begin{bmatrix} e^{x^2 - y^2} - 1 \\ x^2 y^{-1} + 2 \end{bmatrix}$$

5.6   Consider the force balance for the position x of a particle subject to the following force balance

$$\frac{d^2x}{dt^2} - x\frac{dx}{dt} + x^2 = 1$$

Sketch the phase plane where the x-axis is the position and the y-axis is the velocity.

5.7   Is the steady state at $(x_1 = 0, x_2 = 0)$ for

$$\dot{x}_1 = x_1 x_2 - x_1$$

$$\dot{x}_2 = x_2^2 - x_1$$

stable, neutrally stable, or unstable?

5.8   Determine the stability of the steady state solution of

$$\frac{dy_1}{dt} = 2y_1 + 2y_2 + y_3$$

$$\frac{dy_2}{dt} = 3y_1 + 2y_2 + y_3$$

$$\frac{dy_3}{dt} = y_1 + 2y_2 + y_3$$

5.9   Consider the system of equations

$$\frac{dy_1}{dt} = y_1(1 + y_2)$$

$$\frac{dy_2}{dt} = y_1^3 - y_2$$

(a) Is the steady state at $(y_1 = 0, y_2 = 0)$ stable, unstable, or neutrally stable?
(b) Solve this system corresponding to the initial conditions $y_1(0) = 0$ and $y_2(0) = 1$.
(c) Explain why the steady state solution to part (b) does not agree with the answer to part (a).

5.10 Consider the dynamic system

$$\frac{d}{dt}\begin{bmatrix} x \\ y \end{bmatrix} = \begin{bmatrix} 3 & -2 \\ -3 & 2 \end{bmatrix}\begin{bmatrix} x \\ y \end{bmatrix}$$

subject to the initial conditions $x(0) = 1$ and $y(0) = 1$. What are the steady state values of x and y for this system of equations? (b) What are the eigenvalues for this system of equations? Consider the numerical integration of this system of equations where you start somewhere away from the steady state. Will this system reach the steady state?

5.11 For the dynamical system

$$\frac{dy_1}{dt} = y_2(1 + y_1)$$
$$\frac{dy_2}{dt} = y_1(y_2 + 3)$$

(a) Calculate the steady state values of y_1, y_2.
(b) For each steady state, calculate the eigenvalues and assess its stability.
(c) Provide sketches of the corresponding phase plane plots, indicating clearly their key features.

5.12 How do the location and classifications of the steady states for the system

$$\frac{d}{dt}\begin{bmatrix} x_1 \\ x_2 \end{bmatrix} = \begin{bmatrix} x_1^2 + x_2 \\ x_1 + x_2 \end{bmatrix}$$

change if the second equation becomes

$$\frac{dx_2}{dt} = x_1 - x_2$$

5.13 Perform a linear stability analysis of the system

$$x_1' = ax_1 + bx_2$$
$$x_2' = bx_1 + ax_2$$

around the steady state solution. How does the approach to the steady state depend on the parameters a and b?

5.14 Consider a system of equations with a steady state Jacobian of

$$\mathbf{J} = \begin{bmatrix} 1 & 3 \\ a & 0 \end{bmatrix}$$

where a is some rational number. For what values of a is the steady state a saddle point?

5.15 Consider the nonlinear system of ODEs

$$\dot{x}_1 = x_1 x_2 - x_1$$
$$\dot{x}_2 = x_2^2 - x_1$$

If we want to perform a linear stability analysis at some point in the solution corresponding to x_1 and x_2, what is the Jacobian that we need for this system of equations?

5.16   Consider the system of ordinary differential equations

$$\frac{dx}{dt} = xy + x^2$$
$$\frac{dy}{dt} = x^2 y - y$$

(a) Confirm that $(x,y) = (0,0)$, $(1,-1)$, and $(-1,1)$ are steady states of this system.
(b) What is the 2×2 steady state Jacobian for this system in terms of x and y?
(c) Determine the stability and the classification (node, spiral, saddle point, center, or unknown) for each of the steady states in part (a).

5.17   The system

$$\dot{x}_1 = x_1 x_2 - x_1$$
$$\dot{x}_2 = x_2^2 - x_1$$

has a steady state at $(x_1 = 1, x_2 = 1)$. You can analyze this steady state using linear stability analysis, or just consider the phase plane in Fig. 5.23. What can you say about the real and imaginary parts of the eigenvalues of the steady state Jacobian for this problem? You can provide a qualitative answer based on the phase plane or a numerical value based on linear stability analysis, whichever you prefer.

5.18   Consider the force balance for a damped harmonic oscillator with mass $m = 1$ kg, a spring constant $k > 0$ in N/m, and a friction coefficient $\xi > 0$ in kg/s,

$$\frac{d^2 x}{dt^2} = -kx - \xi \frac{dx}{dt}$$

subject to some initial position x_0 and initial velocity v_0.

(a) Define the right-hand-side functions f_1 and f_2 and state the initial conditions if this is converted into a system of ordinary differential equations of the form

$$\frac{dx}{dt} = f_1(x, v)$$
$$\frac{dv}{dt} = f_2(x, v)$$

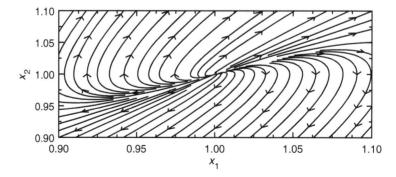

Figure 5.23 Phase plane for Problem 5.17.

(b) What is the steady state for the system?

(c) Is the steady state stable or unstable? You do not need to show any work here if you can answer this from the physics. If you do not know the answer, do part (d) and then come back to this part.

(d) How does the classification of the steady state depend on the parameters ξ, m, and k? Report your result in terms of a friction coefficient $\xi^* = \xi^*(k, m)$ that governs the crossover. (*Hint*: There are two different classifications that you can figure out if you just think about the physics.)

(e) Sketch the phase plane for the two cases from part (d).

(f) Let's assume that I integrated this system of equations using RK4. What is the maximum step size that I could take? Think carefully here with respect to the previous parts of the problem. You can report your answer as a fraction – you do not need to do the arithmetic.

(g) How does the maximum step size depend on the current values of $x(t)$ and $v(t)$?

Computer Problems

5.19 Consider the system of equations

$$\dot{x}_1 = 2x_1 - x_1 x_2$$

$$\dot{x}_2 = 2x_1^2 - x_2$$

Determine the stability of the steady states of these equations and sketch the phase plane. Confirm that your sketch is qualitatively correct by integrating these equations using RK4 with initial guesses $x_1 = 0.01$, $x_2 = -5$ and $x_1 = -0.01$, $x_2 = -5$. Do you feel that this phase plane reflects your feelings towards this book?

5.20 Uppal, Ray, and Poore provide the following system of differential equations describing irreversible, first-order, exothermic reactions in a CSTR:

$$\frac{dx_1}{dt} = -x_1 + \mathrm{Da}(1 - x_1)\exp\left(\frac{x_2}{1 + \dfrac{x_2}{\gamma}}\right)$$

$$\frac{dx_2}{dt} = -x_2 + B\mathrm{Da}(1 - x_1)\exp\left(\frac{x_2}{1 + \dfrac{x_2}{\gamma}}\right) - \beta x_2$$

where x_1 and x_2 are dimensionless concentration and temperature variables respectively, and Da, γ, B, and β are dimensionless parameters that describe the dynamic behavior of the CSTR. Write a MATLAB function that will create phase planes for each of the following cases. Briefly explain the result of each phase plane.

(a) Use the parameters $\mathrm{Da} = 0.085$, $B = 22.0$, and $\beta = 3.0$ with an initial condition of $x_1 = 0.8$, $x_2 = 4.25$.

(b) Use the parameters Da $= 0.150$, $B = 19.0$, and $\beta = 3.0$ with an initial condition of $x_1 = 0.9$, $x_2 = 6.25$.

(c) Use the parameters Da $= 0.320$, $B = 11.2$, and $\beta = 3.0$ with initial conditions of $x_1 = 0.95$, $x_2 = 2.5$, and $x_1 = 0.7$, $x_2 = 1.8$.

5.21   The van der Pol equation is

$$\ddot{x} + \mu(x^2 - 1)\dot{x} + x = 0$$

where μ is a parameter. This system exhibits a limit cycle in the plane $(x, \dot{x})$ (i.e., in momentum-position phase space). For the parameter $\mu = 1.1$, write a MATLAB program that uses numerical integration to determine whether this is a stable limit cycle or an unstable limit cycle. You may need to consider a number of initial guesses in order to answer this question. Your program should automatically generate a plot that demonstrates the stability/instability of the limit cycle.

5.22   Consider the dynamical system

$$\frac{d}{dt}\begin{bmatrix} x \\ y \end{bmatrix} = \begin{bmatrix} xy + \sin \pi y \\ x(y^2 - 1) \end{bmatrix}$$

(a) Use linear stability analysis to characterize the steady states in the interval $x \in [-3.5, 3.5]$ and $y \in [-3.5, 3.5]$. You are *not* allowed to use the eigenvalue functions in MATLAB to compute the eigenvalues but you can use this function to check your answer.

(b) Numerically construct a phase plane portrait for the system in part (a) on the interval $x \in [-3.5, 3.5]$ and $y \in [-3.5, 3.5]$. Your program should automatically generate the overall phase plane as well as additional figures that zoom in on the different steady states. Use your knowledge of numerical integration and MATLAB to figure out a good way to solve the problem. For each plot for the zooms near the steady state, discuss how the numerical solution compares to the linear stability analysis.

5.23   Consider a bioreactor described by the set of balance equations

$$\frac{dx_1}{dt} = (\mu - D)x_1$$

$$\frac{dx_2}{dt} = D(x_{2f} - x_2) - \frac{\mu x_1}{Y}$$

where x_1 and x_2 represent the biomass and substrate concentrations in the reactor. The growth rate depends on the substrate concentration. Different models have been proposed to represent this dependance.

- Monod kinetics:

$$\mu = \frac{\mu_{max} x_2}{k_m + x_2}$$

- Substrate inhibition:

$$\mu = \frac{\mu_{max} x_2}{k_m + x_2 + k_1 x_2^2}$$

(a) Determine the steady states of the system and assess their stability properties using both the models for growth rate. The values of various parameters are: $\mu_{max} = 0.53$, $k_m = 0.12$, $k_1 = 0.4545$, $Y = 0.4$, $x_{2f} = 4$, and $D = 0.3$.

(b) Generate the phase plot for each case in part (a).

5.24 Consider a simple predator–prey model

$$\frac{dx}{dt} = ax - bxy$$

$$\frac{dy}{dt} = -cy + dxy$$

with parameters $a = 1.2$, $b = 0.6$, $c = 0.8$, and $d = 0.3$. Employ initial conditions of $x(0) = 2$ and $y(0) = 1$ and integrate numerically from $t = 0$ to 30. Compare the performance of explicit Euler and Runge–Kutta fourth-order method for a time step of $\Delta t = 0.0625$. Plot time trajectories as well as phase plane plots for each method.

5.25 Consider the Lorenz equations

$$\frac{dx}{dt} = -\sigma x + \sigma y$$

$$\frac{dy}{dt} = rx - y - xz$$

$$\frac{dz}{dt} = xy - bz$$

with parameters $\sigma = 10$, $b = 8/3$, and $r = 28$. Employ initial conditions of $x(0) = y(0) = z(0) = 5$ and integrate from $t = 0$ to 20. Use the Runge–Kutta fourth-order method for the integration with a time step of $\Delta t = 0.03125$.

(a) For the above initial condition generate all three two-dimensional phase plots as well as the three-dimensional phase plot.

(b) Now consider a small perturbation in the initial condition for x (from 5 to 5.001). Compare the time trajectories of x for the two initial conditions. Is the system chaotic?

5.26 Consider the equation

$$\ddot{x} + \dot{x} - 2x + 3x^3 = 0$$

(a) Write a MATLAB program that produces the phase plane. You are welcome to use any integration method you would like. The phase plane portrait that you make should be wide enough to show all of the steady states. Explain why you do (or do not) see the behavior predicted by a linear stability analysis.

(b) This problem gets very interesting if you also include a time-dependent forcing term. Let's modify the problem to

$$\ddot{x} + \dot{x} - 2x + 3x^3 = 1.5 \cos(0.5t)$$

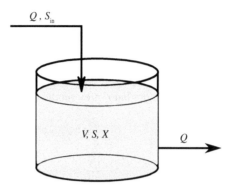

Figure 5.24 Schematic of a CSTR bioreactor for Problem 5.27.

Modify your MATLAB program from part (a) to integrate this modified problem from the initial condition $x_0 = -1$ and $\dot{x}_0 = 1$ out to a very long time – at least up to a total time of $t = 10^3$. Can you explain what is going on here? (It is complicated, but try to explain it.)

5.27   In this problem, we will analyze a dynamic model of the two-species CSTR bioreactor shown in Fig. 5.24. On the left, fresh media with a substrate concentration S_{in} enters at a volumetric flow rate Q. Inside the reactor volume V, bacteria at a concentration X consume the substrate at concentration S which allows them to multiply with a specific growth rate given by Monod kinetics

$$\mu = \frac{\mu_{max}S}{k_m + S}$$

On the right of the diagram, bacteria, substrate, and water exit the bioreactor with flowrate Q.

A mass balance on X and S yields the following system of equations

$$V\frac{dX}{dT} = -QX + \mu VX$$
$$V\frac{dS}{dT} = Q(S_{in} - S) - \frac{\mu VX}{Y}$$

where T is time. Note the constant Y, which is the yield of biomass produced per substrate consumed. To simplify the mathematical problem we define dimensionless variables $x = X/S_{in}$, $s = S/S_{in}$, $t = TQ/V$. The non-dimensional equations are

$$\frac{dx}{dt} = \alpha\frac{sx}{\gamma + s} - x$$
$$\frac{ds}{dt} = 1 - s - \frac{\alpha}{Y}\frac{sx}{\gamma + s}$$

where the dimensionless parameters $\alpha = \mu_{max}V/Q$ and $\gamma = k_m/S_{in}$ appear. Note that to be physically meaningful x, s, and $Y \in [0, 1]$ and t, α, and $\gamma \in [0, \infty)$.

(a) Perform a linear stability analysis of the dimensionless dynamic system when $\alpha = 2$, $Y = 1/2$, and $\gamma = 1/2$. Do this in the following steps.

   (1)   Explain the physical meaning of the dimensionless parameters α and γ.

   (2)   Find the steady states of the dynamic system. What do these correspond to physically?

   (3)   Find the Jacobian *before* substituting the numerical values of the parameters.

   (4)   Find the eigenvalues of the Jacobian for each steady state. Are they stable states? What class of behavior do they show?

(b) Suppose you wish to innoculate the bioreactor with bacteria. You let the bioreactor run for a long time until $S = S_{in}$ and then innoculate with enough microbes so that $X = 0.05S_{in}$. Use RK4 to find the trajectory that your innoculation takes in phase space. What happens if you are lazy and decide not to wait until the reactor reaches $S = S_{in}$ before innoculating? Instead you innoculate when $S = 0.3S_{in}$, but you increase the innoculation concentration to $X = 0.1S_{in}$ (just to be safe). Use RK4 to find this trajectory as well. In a few words explain what is physically happening as a function of time in each of the two trajectories. To get an idea of the entire phase space, plot the velocity field for $x \in [0, 1]$ and $s \in [0, 1]$. As a hint, a velocity field may be plotted using the following MATLAB syntax. In the code example, f is a vector function that holds the "right-hand side" of the equation for the dynamical system.

```
1   [x,s] = meshgrid(linspace(0, 1),linspace(0, 1));
2   RHS = zeros(length(x), length(x), 2);
3   for i = 1:length(x)
4     for j = 1:length(x)
5       RHS(i, j, :) = f([x(i, j), s(i, j)]);
6     end
7   end
    ⋮
20  streamslice(x, s, RHS(:, :, 1), RHS(:, :, 2))
```

Combine your trajectories and the points for the steady state solution (from part (a)) to the plot of the velocity field. You may find the MATLAB commands `hold on` and `hold off` useful for overlaying multiple functions in a plot.

(c) The steady states of the bioreactor are sensitive to the operation parameters, such as the flow rate. Changing the flow rate changes the dimensionless parameter α. Make a plot of both x and s for both steady states as α varies from 1 (high flow rate) to 2 (low flow rate). This type of behavior in your plot is an example of a phenomena called a *transcritical bifurcation*. Qualitatively answer the following questions regarding the plot you just made: What happens to the location of the steady states? Does the stability of the steady states change or stay the same as α changes? What makes the point $\alpha = \gamma + 1$ special; what is physically happening here?

5.28 In this problem, you will use your knowledge of solving systems of ordinary differential equations and phase planes to look at the classic predator–prey problem in biology. The model itself is very simple and covered in numerous textbooks on biology, population dynamics, and nonlinear analysis. Let x be the population of prey and y be the population of predators. The rate of change of these populations is modeled by the coupled set of ordinary differential equations

$$\frac{dx}{dt} = (\alpha - \beta y - \lambda x)x$$

$$\frac{dy}{dt} = (\delta x - \gamma - \mu y)y$$

The parameters α, β, λ, δ, γ, and μ are positive. Naturally, the predator population y and prey population x cannot be negative.

If we think about the left-hand side as the growth of each population, then we can interpret the different terms as follows:

- the terms $\alpha x - \lambda x^2$ ensure that the prey population grows if it is small but decays when it gets large
- the term $-\beta xy$ decreases the prey population as both the predator and the prey become abundant
- the terms $\delta x - \gamma$ increases the predator population if there is more prey to eat and decreases it when there is not enough food
- the term $-\mu y^2$ decreases the predator population as it gets large.

The problem can be put in a "dimensionless" form through the transformations $\tau = \alpha t$, $u = \delta x/\alpha$, and $v = \beta y/\alpha$. This is not really a dimensional analysis of the type you see in transport phenomena, for example, but it does reduce the number of parameters to

$$\frac{du}{d\tau} = (1 - v - Au)u$$

$$\frac{dv}{d\tau} = (u - C - Bv)v$$

where A, B, and C are combinations of the original parameter set.

(a) Determine how the parameters A, B, and C are related to the original parameter set α, β, λ, δ, γ, and μ.
(b) One possible steady state for this system is

$$u^* = \frac{C + B}{1 + AB} \qquad v^* = \frac{1 - AC}{1 + AB}$$

Confirm that this is a steady state solution. Do we need to put any restrictions on these parameter values to ensure that the stable points are non-negative?
(c) For the following values of A, B, and C, calculate the corresponding steady state values of u^* and v^*.

	A	B	C
(i)	0.50	0.10	0.200
(ii)	0.05	0.01	0.200
(iii)	10.00	0.50	0.001
(iv)	0.01	100.00	50.000
(v)	0.70	0.40	1.200
(vi)	0.70	1.20	1.100

You can either make the calculation by hand or use a MATLAB program to make the calculations using the result from part (b). Use MATLAB to create a plot (v vs. u) using a circle to mark each steady state. Are there bounds on the possible steady state values of u^* and v^*?

(d) Determine the entries of the Jacobian as a function of A, B, C, u, and v. Using the values of A, B, and C from the table above, perform a linear stability analysis from this Jacobian and determine the eigenvalues for each case near the steady state. What do the eigenvalues for each of these cases tell you (qualitatively) about the stability of the steady states?

(e) Write a MATLAB program that integrates the governing equations using a fourth order Runge–Kutta method. Produce a phase plot for each of the above sets of A, B, and C with initial conditions $u(0) = 2$ and $v(0) = 2$. Use 0.01 as the dimensionless time step and integrate until you reach a distance of $r = 10^{-4}$ from the steady state solution. If the solution becomes unstable, repeat the calculation by sequentially halving the time step until the time integration is stable. Include in your code a way to count the number of oscillations for those solutions which are oscillatory. Note the time step required to have a stable integration for each of the phase planes.

5.29 In this problem, you will use your knowledge of the phase plane and stability to analyze a simple model of glycolysis. The glycolysis pathway that first convert glucose into fructose-6-phosphate (F6P), and later lead to two molecules of pyruvate, which then enters the TCA cycle. In the course of the reaction, there are reactions which convert ATP to ADP, as well as the reverse reactions.

A simplified model of glycolysis considers only the concentrations of ADP, which we will call x, and the concentration of F6P, which we will call y. The model equations are

$$\dot{x} = -x + ay + x^2y$$
$$\dot{y} = b - ay - x^2y$$

where a and b are positive coefficients. The goal of this problem is to analyze the dynamics of this simplified model of glycolysis.

This problem consists of two parts. The first one is primarily analytical with some programming tasks to assist in your analysis. The second one is primarily computational. While both parts refer to the same problem, you can probably do most of the second one without doing the first one. So if you get stuck early on, don't neglect the second part.

Part I

In the first part of the problem, we will analyze the stability of the steady states of this reaction system. The final result will be a detailed output file, along with two plots if you do the bonus activity. Your final program will be a function called `stability`, which calls a subfunction called `eigenval`.

(a) Determine the steady state of this system, the Jacobian at steady state, and the eigenvalues at steady state. Include your calculation in the written submission.

(b) Determine the boundary between the stable and unstable steady states in the problem. Include your calculation in the written submission. *Hint: If you are having trouble with this calculation, you should do the numerical calculations below. If you can do them successfully, it may help you to see the answer to this part of the problem.*

(c) Write a MATLAB function called `eigenval(a,b)` that takes as its inputs the parameters a and b. The function should return a vector containing the following outputs (in this order).

   (1) The real part of the larger eigenvalue.
   (2) The imaginary part of the larger eigenvalue.
   (3) The real part of the smaller eigenvalue.
   (4) The imaginary part of the smaller eigenvalue.
   (5) A number indicating the type of steady state: 1 = stable node, 2 = unstable node, 3 = saddle point, 4 = stable spiral, 5 = unstable spiral, 6 = center.

   Since you will never actually get a center in a numerical calculation, if the real part of the eigenvalues is less than 10^{-6}, you can assume that the steady state is a center. In this calculation, you are *not* allowed to use the imaginary number capabilities of MATLAB. You should construct your function with conditional `if-else` type statements to determine the real and imaginary parts, as well as the type of eigenvalue.

(d) Write a MATLAB function called `stability` for analyzing the steady states. This function must call your `eigenval` function for the relevant calculations. You are to consider all combinations of the values $a = 0.005, 0.010, \ldots, 0.14$ and $b = 0.05, 0.1, \ldots, 1.2$. For each (a, b) pair, this function should compute the steady state (x_{ss}, y_{ss}), the real and imaginary parts of each eigenvalue, and the type of steady state using the numbering system defined above. Compile all of these data into a matrix whose columns correspond to the following:

   (1) a.
   (2) b.
   (3) The steady state value x_{ss}.
   (4) The steady state value y_{ss}.
   (5) The real part of the larger eigenvalue.
   (6) The imaginary part of the larger eigenvalue.
   (7) The real part of the smaller eigenvalue.

(8) The imaginary part of the smaller eigenvalue.

(9) A number indicating the type of steady state: 1 = stable node, 2 = unstable node, 3 = saddle point, 4 = stable spiral, 5 = unstable spiral, 6 = center.

The matrix should be saved to a file using `dlmwrite`.

Bonus: Have your program create the following plots.

(1) A plot of the type of steady state in the x, y plane. Use blue color for stable states, red color for unstable states, circles for nodes, x for spirals, a green triangle for a saddle point, and a black pentagram for a center. Label the axes, provide a title, and save the plot as a jpg. *Hint: Do not worry if you have trouble seeing all of the steady states in your plot. If you look back at your answers for the values of the steady state as a function of a and b, it may help you understand this plot.*

(2) A plot of the type of steady state in the a, b plane. Use blue color for stable states, red color for unstable states, circles for nodes, x for spirals, a green triangle for a saddle point, and a black pentagram for a center. You should also plot the boundary of the stability envelope in the second figure. Label the axes, provide a title, and save the plot as a jpg.

Hint: If you have many different conditions, it is easier to use a `switch` *statement than a bunch of* `if-else` *statements. In order to make plots using different types of symbols, you should use the* `hold` *on and* `hold` *off commands.*

You can get a very deep understanding of the stability of this problem from these plots, which is why you should try to make them. However, the plotting is included as extra credit because it involves a reasonably good understanding of MATLAB plotting commands.

Part II

In the second part of the problem, you will construct three phase diagrams for particular conditions of a, b, and initial conditions. The various conditions are in Table 5.1. You will need to construct phase planes for these conditions using RK4 and explain the results. Your final program will a function called `makeplots` that calls the subfunctions `RK4`, `RK`, and `feval`.

(a) Write a MATLAB function called `feval(z,a,b)` that takes in a vector of positions $z = [x, y]$ and the parameters a, b and returns the values of $\dot{x}$ and $\dot{y}$.

(b) Write a MATLAB function called `RK(z,h,a,b)` that takes in a vector of the current values z, the time step h, and the parameters a and b and returns the new values of z. Your function must call `feval(z,a,b)` to compute the various k values.

(c) Write a MATLAB function called `RK4(a,b,z)` that takes in a vector initial positions $z = [x_0, y_0]$ and the parameters a and b and performs RK4 for 10^4 time steps with a step size $h = 0.01$. The function should plot the starting point as a blue circle

Table 5.1 Conditions for phase plane plots in Problem 5.29.

a	b	x_0	y_0
0.02	0.1	0.1	4
		0.4	1
		1	1.5
		0.4	3.5
		0.3	5
0.015	0.4	0.5	2.4
		0.4	1
		1.5	3.5
		4	1.5
		4	3
		3	5
0.025	0.15	0.5	2.4
		0.2	2.5
		0.3	3.5
		0	3.5

and the trajectory as a blue line. Your function must call `RK(z,h,a,b)` for the integration.

(d) Write a MATLAB function called `makeplots` that will create the phase planes. For each case, you should plot the steady state as a red x and then have each trajectory plotted from the `RK4(a,b,z)` function. These extra points make it easier to see the direction of the trajectory. Label the axes, provide a title and save each plot as a jpg file. *Hint: In order to make plots using different types of symbols, you should use the* `hold on` *and* `hold off` *commands.*

(e) Explain the result of each phase plane.

5.30   In this problem, we will look at the phase plane dynamics of a particle in a double-well potential given by the dynamic equation

$$\ddot{x} = x - x^3$$

The term on the left-hand side is the acceleration, and the term on the right-hand side is the force due to a potential

$$V = -\frac{x^2}{2} + \frac{x^4}{4}$$

If you plot the potential, you will see that it has a familiar form from physics. This problem is easily converted into a dynamic phase plane by considering the position, x, as the x-axis and the velocity, $v = \dot{x}$, as the y-axis.

You should write a MATLAB program that uses a fourth-order Runge–Kutta integration scheme with step size $h = 0.01$ to integrate the dynamical system. Consider the following initial positions: $(x = 1.4142, v = 0)$ and $(x = 1.4143, v = 0)$. Plot the trajectory for the first position as a blue line and the trajectory for the second position as a red line. To complete at least one cycle, use 10^4 time steps. Use your knowledge of the phase plane and stability to explain what is happening with these two initial positions that differ only by $x = 10^{-4}$. It may be useful for you to zoom your plot to look at where the trajectories begin to diverge but this is not necessary to solve the problem.

6 Boundary Value Problems

6.1 ODE Boundary Value Problems

In the previous two chapters we discussed the solution of ODE IVPs, and in particular dynamical systems where the independent variable is time. In this chapter we turn our attention back to the general ODE form introduced in Chapter 4,

$$F\left(x, y, \frac{dy}{dx}, \frac{d^2y}{dx^2}, \ldots, \frac{d^ny}{dx^n}\right) = 0 \tag{6.1.1}$$

but with the specification of n auxiliary conditions at *different* values of the independent variable x. Such conditions are termed *boundary conditions* and the resulting problems are called boundary value problems (BVPs).

In most practical problems of interest in chemical engineering the ODE will be of the form

$$\frac{d^2y}{dx^2} = f\left(x, y, \frac{dy}{dx}\right) \tag{6.1.2}$$

where x will belong to an interval $[x_l, x_r]$. Common sources of such problems are conservation equations of mass and energy in one spatial dimension at steady state, leading to equations of the form

$$v(x)\frac{dy}{dx} = D(x)\frac{d^2y}{dx^2} + r(x, y) \tag{6.1.3}$$

where y would be the temperature or concentration, v is the entraining fluid velocity, and D is the diffusivity, either thermal or mass. The second-order derivative terms arise from diffusive transport, while the first-order derivative terms arise from convective transport. This equation, in the absence of convection, will be the focus of our discussion in this chapter. Reaction terms, $r(x, y)$, are typical sources of nonlinearities, although we can also get nonlinearities when $v(x)$ and/or $D(x)$, which Eq. (6.1.3) treats as functions of position, become functions of y as well. The positions $x = x_l$ and $x = x_r$ define the boundaries of the spatial domain of interest. To guarantee a unique solution to this equation, we need to specify two boundary conditions, typically one at each boundary (although mathematically, one can also consider more general formulations).

The most common types of boundary conditions are the following.

(1) **Dirichlet (boundary condition of the first kind)**: In this type, we specify the value of y on the boundary, e.g.

$$y(x_l) = \alpha \tag{6.1.4}$$

In physical systems, this may correspond to fixing the concentration or temperature at a boundary.

(2) **Neumann (boundary condition of the second kind)**: In this type, we specify the derivative of y on the boundary, e.g.

$$\left. \frac{dy}{dx} \right|_{x_l} = \alpha \tag{6.1.5}$$

This typically corresponds to fixing the flux of mass or energy. The constant α is 0 when we impose a no-flux condition, for example when we have insulated or impermeable boundaries.

(3) **Robin (boundary condition of the third kind)**: This type involves a combination of y and the derivative of y on the boundary. For example, we may have an expression of the form

$$\alpha_1 \left. \frac{dy}{dx} \right|_{x_r} + \alpha_2 y(x_r) = \alpha_3 \tag{6.1.6}$$

This may correspond to a heat or mass transport correlation for the flux at the boundary.

There are many ways to solve ODE BVPs. We will consider only the finite difference method, which works fine for simple geometries. This is often the first approach that students learn to solve BVPs, since its theoretical basis is easily grasped and it is easy to program. Other more sophisticated approaches, such as finite elements or spectral methods, are more powerful for many applications but beyond the scope of the book. As was the case for IVPs, the idea is that since we cannot find the exact solution function for the ODE BVP analytically, we seek an approximation of the solution, after discretizing the ODE by approximating the derivatives with finite differences. For the single ODE in Eq. (6.1.1), we seek approximate values of $y(x)$, y_i, at discrete values of the independent variable, x_i.

Note that for the IVPs discussed in Chapters 4 and 5, we had an open domain (for most of the problems we considered the independent variable was time) and we generated approximations of the solution successively at different points as we marched along the domain up to a desired point. In BVPs, the boundaries of the domain are usually specified *a priori*, and so are conditions on these boundaries. The solution strategy will again involve approximating the derivatives with finite differences at each discrete point on the domain, but now we will be solving for all of the unknowns y_i simultaneously. This leads to systems of linear equations (for linear differential equations) or nonlinear equations (for nonlinear differential equations). As a result, we will draw heavily from the work in Chapters 2 and 3 to solve these problems. Moreover, we will not be able to easily write "generic" BVP programs analogous to the programs we saw in Chapter 4 for IVPs.

Rather, we will set up each individual problem to generate the appropriate system of equations, and then solve those equations using methods for solving algebraic equations.

6.2   Finite Difference Approximations

6.2.1   Discretization in Space

The first step to applying the finite difference method is to discretize the spatial domain. Usually, this is done by defining n equidistant nodes $x_1, \ldots, x_n$, where $x_1 = x_l$, $x_n = x_r$ and the discretization step Δx is

$$\Delta x = \frac{x_r - x_l}{n - 1} \qquad (6.2.1)$$

It is very common for students to forget the term -1 in this equation. Note that we started our numbering with $i = 1$ instead of $i = 0$, to be consistent with MATLAB's indexing of vectors. In other languages, such as C, it is easier to use $i = 0, 1, \ldots, n - 1$. Equation (6.2.1) is very simple but arguably the most important equation in this entire chapter! We can also use this equation to get the value of x at node i,

$$x_i = x_l + (i - 1)\Delta x = x_l + (i - 1)\frac{x_r - x_l}{n - 1}, \quad (i = 1, 2, \ldots, n) \qquad (6.2.2)$$

Sometimes it is useful to use non-equal spacing between the nodes. For example, if we have a reacting system with sharp concentration gradients near the reacting boundary, then we would want many nodes close to that boundary to capture the solution more accurately. In the bulk, where the concentration changes slowly, we can use fewer nodes and still get an accurate solution. Creating good non-uniform grids is a bit of an art, and outside the scope of our discussion. We will focus exclusively on uniform grids here, which can be inefficient but are easy to implement.

6.2.2   Derivation of Finite Difference Equations

We now proceed with the derivation of the finite difference approximation formulas used in the discretization of the ODE. Consider a Taylor series expansion of $y(x)$ around one of the nodes x_i in Fig. 6.1,

$$y(x) = y(x_i) + \left.\frac{dy}{dx}\right|_{x_i} (x - x_i) + \frac{1}{2} \left.\frac{d^2y}{dx^2}\right|_{x_i} (x - x_i)^2 + \cdots \qquad (6.2.3)$$

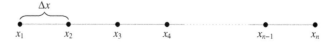

Figure 6.1 Discretization of spatial domain into n equidistant nodes.

If we evaluate the Taylor series at the point $x = x_{i+1}$ and note that, for equally spaced nodes, Eq. (6.2.1) can also be written as $\Delta x = x_{i+1} - x_i$, we have

$$y(x_{i+1}) = y(x_i) + \Delta x \left.\frac{dy}{dx}\right|_{x_i} + \frac{\Delta x^2}{2} \left.\frac{d^2y}{dx^2}\right|_{x_i} + \mathcal{O}(\Delta x^3) \qquad (6.2.4)$$

Solving for the first derivative, we obtain

$$\left.\frac{dy}{dx}\right|_{x_i} = \frac{y(x_{i+1}) - y(x_i)}{\Delta x} - \frac{\Delta x}{2} \left.\frac{d^2y}{dx^2}\right|_{x_i} + \mathcal{O}(\Delta x^2) \qquad (6.2.5)$$

Keeping only the first term on the right-hand side and using the approximate values of y_i instead, we get an approximation for the first derivative,

$$\left.\frac{dy}{dx}\right|_{x_i} = \frac{y_{i+1} - y_i}{\Delta x} \qquad (6.2.6)$$

where the truncation error is of $\mathcal{O}(\Delta x)$. Equation (6.2.6) is called the (first-order) forward difference approximation.

We could also have used the Taylor series (6.2.3) and evaluated the function at $x = x_{i-1}$,

$$y(x_{i-1}) = y(x_i) + \left.\frac{dy}{dx}\right|_{x_i} (-\Delta x) + \frac{\Delta x^2}{2} \left.\frac{d^2y}{dx^2}\right|_{x_i} + \mathcal{O}(\Delta x^3) \qquad (6.2.7)$$

Solving for the first derivative, we now obtain

$$\left.\frac{dy}{dx}\right|_{x_i} = \frac{y(x_i) - y(x_{i-1})}{\Delta x} + \frac{\Delta x}{2} \left.\frac{d^2y}{dx^2}\right|_{x_i} + \mathcal{O}(\Delta x^2) \qquad (6.2.8)$$

Keeping only the first term on the right-hand side and using the approximate values of $y(x_i)$, we now get

$$\left.\frac{dy}{dx}\right|_{x_i} = \frac{y_i - y_{i-1}}{\Delta x} \qquad (6.2.9)$$

whose accuracy is again of $\mathcal{O}(\Delta x)$. This is called the (first-order) backward difference approximation.

Both the forward difference (6.2.6) and backward difference (6.2.9) formulas have errors that are $\mathcal{O}(\Delta x)$. We can actually do better by adding Eqs. (6.2.5) and (6.2.8) to get

$$\left.\frac{dy}{dx}\right|_{x_i} = \frac{y(x_{i+1}) - y(x_{i-1})}{2\Delta x} + \mathcal{O}(\Delta x^2) \qquad (6.2.10)$$

where the $\mathcal{O}(\Delta x)$ terms have cancelled out. Keeping only the first term on the right-hand side and using the approximate values of $y(x_i)$, we get

$$\left.\frac{dy}{dx}\right|_{x_i} = \frac{y_{i+1} - y_{i-1}}{2\Delta x} \qquad (6.2.11)$$

This is called a centered finite difference approximation. Since the truncation error is of $\mathcal{O}(\Delta x^2)$, it is a more accurate approximation than either the forward or the backward approximation.

If we now add the initial Taylor series expansions, Eqs. (6.2.4) and (6.2.7), we get

$$y(x_{i+1}) + y(x_{i-1}) = 2y(x_i) + \Delta x^2 \left. \frac{d^2 y}{dx^2} \right|_{x_i} + \mathcal{O}(\Delta x^4) \qquad (6.2.12)$$

Solving for the second derivative on the right-hand side, and using the approximate values of $y(x)$, we get

$$\left. \frac{d^2 y}{dx^2} \right|_{x_i} = \frac{y_{i+1} - 2y_i + y_{i-1}}{\Delta x^2} \qquad (6.2.13)$$

This is the centered finite difference approximation for the second derivative which has an accuracy of $\mathcal{O}(\Delta x^2)$ too.

The centered finite difference approximations above will be used for the remainder of the book, with only a few rare exceptions. The procedure to obtain finite difference equations for higher-order derivatives, or more accurate formulas for first and second derivatives is similar to the one we used here. However, it is helpful to be a bit more formal in the derivation. In both of these cases, we will want to use additional points to evaluate the derivative. For a centered formula, we use an equal number of points on either side of x_i. For example, we can derive formulas for the third or fourth derivative by using the points x_{i-2} and x_{i+2} in addition to the points x_{i-1} and x_{i+1}. At each evaluation point, we need to apply a Taylor series of the form of Eq. (6.2.3) except that we retain terms involving derivatives up to the order of the number of evaluation points. We then add these equations together with appropriate weights to remove all terms except the derivative of interest. Determining the weights leads to an algebraic system of equations, which we can readily solve using the concepts in Chapter 2. Let's see how to do this in an example.

Example 6.1   Derive a centered finite difference formula for the second derivative using four evaluation points.

Solution
For each evaluation point, we need to use the Taylor series

$$y(x) = y(x_i) + y'(x - x_i) + \frac{1}{2}y''(x - x_i)^2 + \frac{1}{6}y'''(x - x_i)^3 + \frac{1}{24}y''''(x - x_i)^4 + \cdots \quad (6.2.14)$$

to produce the equations

$$y_{i+2} = y_i + y'(2\Delta x) + \frac{1}{2}y''(2\Delta x)^2 + \frac{1}{6}y'''(2\Delta x)^3 + \frac{1}{24}y''''(2\Delta x)^4 + \cdots \qquad (6.2.15)$$

$$y_{i+1} = y_i + y'(\Delta x) + \frac{1}{2}y''(\Delta x)^2 + \frac{1}{6}y'''(\Delta x)^3 + \frac{1}{24}y''''(\Delta x)^4 + \cdots \qquad (6.2.16)$$

$$y_{i-1} = y_i + y'(-\Delta x) + \frac{1}{2}y''(-\Delta x)^2 + \frac{1}{6}y'''(-\Delta x)^3 + \frac{1}{24}y''''(-\Delta x)^4 + \cdots \qquad (6.2.17)$$

$$y_{i-2} = y_i + y'(-2\Delta x) + \frac{1}{2}y''(-2\Delta x)^2 + \frac{1}{6}y'''(-2\Delta x)^3 + \frac{1}{24}y''''(-2\Delta x)^4 + \cdots \qquad (6.2.18)$$

The idea is to multiply each equation by some weight c_i and add them together to remove all the derivatives except for y''. This leads to the system of equations

$$\begin{bmatrix} 2 & 1 & -1 & -2 \\ 4 & 1 & 1 & 4 \\ 8 & 1 & -1 & -8 \\ 16 & 1 & 1 & 16 \end{bmatrix} \begin{bmatrix} c_{i+2} \\ c_{i+1} \\ c_{i-1} \\ c_{i-2} \end{bmatrix} = \begin{bmatrix} 0 \\ 2 \\ 0 \\ 0 \end{bmatrix} \tag{6.2.19}$$

where we chose the value of 2 on the right-hand side to cancel the factor of 1/2 in the second derivative. We can solve by Gauss elimination to get

$$\begin{bmatrix} c_{i+2} \\ c_{i+1} \\ c_{i-1} \\ c_{i-2} \end{bmatrix} = \frac{1}{12} \begin{bmatrix} -1 \\ 16 \\ 16 \\ -1 \end{bmatrix} \tag{6.2.20}$$

Adding up the equations with these coefficients gives

$$\frac{-y_{i+2} + 16y_{i+1} + 16y_{i-1} - y_{i-2}}{12} = \frac{30}{12}y_i + y''\Delta x^2 \tag{6.2.21}$$

Solving for the derivative gives

$$y'' = \frac{-y_{i+2} + 16y_{i+1} - 30y_i + 16y_{i-1} - y_{i-2}}{12\Delta x^2} \tag{6.2.22}$$

What is the advantage of the higher-order formula? Let's look at the error. If we continued our expansion out to the fifth derivatives, the weights and coefficients give

$$\frac{-(2\Delta x)^5 + 16(\Delta x)^5 + 16(-\Delta x)^5 - (-2\Delta x)^5}{12} = 0 \tag{6.2.23}$$

so the terms cancel out. When we take the expansion out to sixth derivatives we get a non-zero result,

$$\frac{-(2\Delta x)^6 + 16(\Delta x)^6 + 16(-\Delta x)^6 - (-2\Delta x)^6}{12} = -8\Delta x^6 \tag{6.2.24}$$

Since we divided by Δx^2, the error is $\mathcal{O}(\Delta x^4)$.

This is one possible approach to derive higher-order finite difference approximations. We will discuss an alternate approach in the context of interpolation in Section 8.2.4.

6.3 Solution Procedure

After the spatial discretization and the derivation of the finite difference approximations [Eqs. (6.2.11) and (6.2.13)], we can proceed with the transformation of the ODE BVP into a set of algebraic equations. At each interior node $i = 2, 3, \ldots, n-1$, we use the centered finite difference approximations to convert the ODE

$$\frac{d^2y}{dx^2} = f\left(x, y, \frac{dy}{dx}\right) \tag{6.3.1}$$

into the discretized form

$$\frac{y_{i+1} - 2y_i + y_{i-1}}{\Delta x^2} = f\left(x_i, y_i, \frac{y_{i+1} - y_{i-1}}{2\Delta x}\right) \tag{6.3.2}$$

The result is a set of $n - 2$ algebraic equations in n unknowns. The remaining two equations will come from the boundary conditions. The resulting set of (in general, nonlinear) algebraic equations will have to be solved simultaneously to obtain the values of the unknowns $y_1, \ldots, y_n$.

Before we get into the details of solving several different types of problems, let's first outline the general approach to solving BVPs. This is a useful list that you can refer to when it is time to solve your own problems.

(1) Discretize the domain into n nodes.
(2) For the interior nodes ($i = 2, 3, \ldots, n-1$) discretize the ODE using finite difference approximations of the derivatives. You need to be careful in the presence of convection to respect the direction of information propagation to get a stable solution. We will focus solely on reaction–diffusion here, so centered finite differences will be fine for all of our problems.
(3) For the boundary nodes ($i = 1$ and $i = n$):

   (1) If there is a Dirichlet boundary, set the boundary node to the specified value.
   (2) Otherwise, make a fictitious node and discretize both the boundary condition and the ODE.

(4) For the solution:

   (1) If the problem is linear, solve using any method the resulting system $\mathbf{Ax} = \mathbf{b}$.
   (2) If the problem is nonlinear, form the Jacobian and solve by Newton–Raphson. Be careful about the existence of multiple solutions. If the nonlinearity is strong, it may be helpful to start from a linearized version of the problem (which has a unique solution) and then use a continuation method to develop good initial guesses as the size of the nonlinear term increases.

(5) Check for mesh refinement.

Some of the topics in this list may not be familiar to you yet. Don't worry – we will address all of these issues in turn as we work our way through the examples in this section.

6.3.1   Dirichlet Boundary Conditions

When we have Dirichlet boundary conditions, the solution method is very straightforward. Let's illustrate this procedure via two examples. The first example is linear and analytically solvable, so it will let us look a bit more closely at the error in finite differences. The second example is nonlinear and highlights the challenges in such solutions.

Example 6.2   Show how the error in the finite difference approximation depends on the number of nodes n for the reaction–diffusion equation

$$\frac{d^2c}{dx^2} = \phi^2 c \tag{6.3.3}$$

with $c(0) = 0$ and $c(1) = 1$. In chemical engineering lingo, the quantity ϕ is the Thiele modulus for this problem.

Solution

This problem is solved analytically using the characteristic polynomial,

$$r^2 - \phi^2 = 0 \tag{6.3.4}$$

which has roots $r = \pm\phi$. The solution is then

$$c(x) = a \sinh \phi x + b \cosh \phi x \tag{6.3.5}$$

From the left boundary condition we get $b = 0$ since $\sinh 0 = 0$. From the right boundary condition we get $a^{-1} = \sinh \phi$. Thus, the exact solution is

$$c(x) = \frac{\sinh \phi x}{\sinh \phi} \tag{6.3.6}$$

To solve this problem by finite differences, we approximate the second derivative with finite differences,

$$\frac{c_{i+1} - 2c_i + c_{i-1}}{\Delta x^2} - \phi^2 c_i = 0 \tag{6.3.7}$$

For the boundary conditions, we have $c_1 = 0$ and $c_n = 1$.

This leads to a problem of the form

$$\mathbf{Ac} = \mathbf{b} \tag{6.3.8}$$

or, more specifically,

$$\begin{bmatrix} 1 & 0 & 0 & 0 & 0 & \cdots & 0 & 0 & 0 \\ 1 & \alpha & 1 & 0 & 0 & \cdots & 0 & 0 & 0 \\ 0 & 1 & \alpha & 1 & 0 & \cdots & 0 & 0 & 0 \\ \vdots & \vdots & \vdots & \vdots & \vdots & \vdots & \vdots & \vdots & \vdots \\ 0 & 0 & 0 & 0 & 0 & \cdots & 1 & \alpha & 1 \\ 0 & 0 & 0 & 0 & 0 & \cdots & 0 & 0 & 1 \end{bmatrix} \begin{bmatrix} c_1 \\ c_2 \\ c_3 \\ \vdots \\ c_{n-1} \\ c_n \end{bmatrix} = \begin{bmatrix} 0 \\ 0 \\ 0 \\ \vdots \\ 0 \\ 1 \end{bmatrix} \tag{6.3.9}$$

where $\alpha = -2 - \phi^2 \Delta x^2$. We can then solve this system using any of the methods from Chapter 2. Since the finite difference method leads to a tridiagonal band structure ($p = q = 1$) for Dirichlet boundary conditions, we should take advantage of a banded solver.

Let's take a look at the MATLAB program to solve this problem.

```
1  function c = bvp_linear(n,L)
2  dx = 1/(n-1);
3  alpha = -2 - (L*dx)^2;
4  A = zeros(n);
5  b = zeros(n,1);
6  A(1,1) = 1;
7  for i = 2:n-1
8      A(i,i-1) = 1;
9      A(i,i) = alpha;
10     A(i,i+1) = 1;
11 end
12 A(n,n) = 1;
13 b(n) = 1;
14 A = sparse(A);
15 c = A\b;
```

The program is just setting up the system of linear equations produced by the finite difference method and then solving them using a banded solver. Note that lines 4 and 5 declare the matrix $\mathbf{A}$ and the vector $\mathbf{b}$ to be all zeros. We thus only have to fill in the entries that are non-zero. For the left boundary condition, we have

$$A_{1,1} = 1, \quad b_1 = 0 \tag{6.3.10}$$

since the corresponding equation is $c_1 = 0$. Since we already set all the entries in $\mathbf{b}$ to be zero at the start, we do not need to reassign them to zero here. For the interior nodes, we have the repeating pattern $[1, \alpha, 1]$. The **for** loop in lines 7–11 writes these parts of the matrix. The differential equation is homogeneous, so again we do not need to do anything about the vector $\mathbf{b}$ for the interior nodes. The only time we have to address the vector $\mathbf{b}$ is in the nth row, where we set it equal to 1 in line 13. Note that line 14 declares the matrix $\mathbf{A}$ to be sparse. This will force MATLAB to use the banded solver.

Figure 6.2 compares the exact solution in Eq. (6.3.6) to the numerical solution. Even with just a few nodes, we get a reasonably good prediction of the solution.

It is worthwhile spending a bit more time to consider the solution of a linear problem, such as the one in Example 6.2, to understand how the accuracy of the solution changes with the number of nodes. Clearly, if we increase the number of nodes, the accuracy of the solution should increase. In Fig. 6.3, we plot

$$\text{err} \equiv \frac{|c(x_i) - c_i|}{|c_i|} \tag{6.3.11}$$

where $c(x_i)$ is the exact solution evaluated at x_i and c_i is the numerical solution. As we see in Fig. 6.3, the size of this norm decreases linearly with the number of nodes. This might seem surprising to you, since we saw in the Taylor series expansion leading to Eq. (6.2.13) that centered finite differences have an error of $\mathcal{O}(\Delta x^2)$, which implies an error of $\mathcal{O}(n^{-2})$. The reason for the linear decrease is that *each* finite difference approximation introduces an error of $\mathcal{O}(n^{-2})$. Since we need n nodes, the total error is

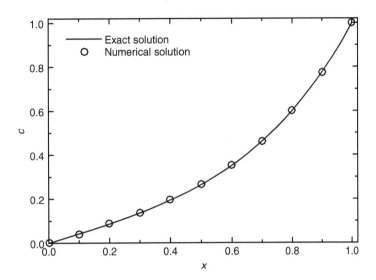

Figure 6.2 Comparison of the exact solution and the numerical solution of the reaction–diffusion equation with $n = 11$ nodes for $\phi^2 = 2.5$.

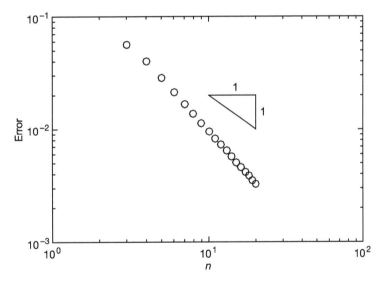

Figure 6.3 Comparison of the exact solution and the numerical solution of the reaction–diffusion equation as a function of the number of nodes.

of order $n\mathcal{O}(n^{-2}) = \mathcal{O}(n^{-1})$. This result is quite similar to what we saw in the global truncation error for IVPs in Chapter 4. Although the truncation error in the derivative had a certain order in terms of the step size, we needed n steps to get to the end of the integration. As a result, the global truncation error for IVPs was also n times the local truncation error.

This discussion of truncation error brings up an interesting point. What happens if we just solve the diffusion equation in the absence of reaction,

$$\frac{d^2c}{dx^2} = 0 \qquad (6.3.12)$$

with Dirichlet boundary conditions. If we again put $c(0) = 0$ and $c(1) = 1$, we have the solution

$$c(x) = x \qquad (6.3.13)$$

The finite difference approximation to Eq. (6.3.12) is

$$\frac{c_{i+1} - 2c_i + c_{i-1}}{\Delta x^2} = 0 \qquad (6.3.14)$$

with $c_1 = 0$ and $c_n = 1$. The solution to this system of equations gives back the analytical result in Eq. (6.3.13), modulo any round-off errors from the numerical solution of the linear algebraic equation. Centered finite differences are exact in this case because the truncation error in Eq. (6.2.13) involves second derivatives and higher, but these are identically zero.

Now that we know a bit about what goes on in the solution of a linear problem, including the error in the solution, let's consider a nonlinear problem where we do not know the answer in advance.

Example 6.3   Develop a MATLAB code to solve

$$\frac{d^2y}{dx^2} = -2y(1 - 2y^2), \quad y(-1) = -1, \quad y(1) = 1 \qquad (6.3.15)$$

for an arbitrary number of nodes.

Solution
Following the procedure above, for the interior nodes we have the following algebraic equations

$$\frac{y_{i+1} - 2y_i + y_{i-1}}{\Delta x^2} + 2y_i(1 - 2y_i^2) = 0 \qquad (6.3.16)$$

For the boundary nodes, in this case of Dirichlet boundary conditions, we can directly fix the values of the variables y_1 and y_n,

$$y_1 = -1 \qquad (6.3.17)$$

and

$$y_n = 1 \qquad (6.3.18)$$

The above can now be put in a residual form $\mathbf{R}(\mathbf{y}) = \mathbf{0}$ with

$$\begin{aligned} R_1 &= y_1 + 1 \\ R_i &= \frac{y_{i+1} - 2y_i + y_{i-1}}{\Delta x^2} + 2y_i(1 - 2y_i^2) \\ R_n &= y_n - 1 \end{aligned} \qquad (6.3.19)$$

for $i = 2, \ldots, n - 1$. To solve this problem with Newton–Raphson, we need to be able to generate the residual vector for an arbitrary number of nodes:

```
1  function R = getR(y,dx,n)
2  R = zeros(n,1);
3  R(1,1) = y(1) + 1;
4  for i = 2:n-1
5      R(i,1) = (y(i+1)-2*y(i)+y(i-1))/(dx^2)+2*y(i)*(1-2*y(i)^2);
6  end
7  R(n,1) = y(n) - 1;
```

This function takes in the values of the unknowns, the spacing between nodes, and the number of nodes. While the latter two variables are not independent, we chose to send them both in this example rather than compute, for example, Δx from n. On line 2, we make space in the memory for the residual. If we use many nodes, we can get a major slowdown in MATLAB if we forget to initialize matrices and vectors to the correct size before filling them in! (In other structured programming languages, you generally have to declare all of your variables at the outset anyway, so this is a problem that really only arises when you program in MATLAB.) After making space in the memory, we fill in the vector with the residual equations. Line 3 handles the left boundary condition, lines 4–6 loop over the interior of the domain, and line 7 writes the entry for the right boundary.

We also need to specify the non-zero entries in the Jacobian matrix. For the first line, there is only one non-zero entry,

$$J_{1,1} = 1 \qquad (6.3.20)$$

Likewise, for the last row, there is also only one non-zero entry,

$$J_{n,n} = 1 \qquad (6.3.21)$$

Note that we could reduce the problem from an $n \times n$ system to an $(n - 2) \times (n - 2)$ system by substituting for the first and last nodes. This is a matter of preference, since for $n \gg 1$ the time saved by this substitution is not really significant. We will generally include the boundary conditions in all of the problems that we solve here, since the code is easier to relate back to the original differential equation. For the interior rows of the Jacobian, corresponding to the index $i = 2, 3, \ldots, n - 1$, there are three non-zero entries

$$J_{i,i-1} = \frac{1}{\Delta x^2} \qquad (6.3.22)$$

$$J_{i,i} = \frac{-2}{\Delta x^2} + 2 - 12y_i^2 \qquad (6.3.23)$$

$$J_{i,i+1} = \frac{1}{\Delta x^2} \qquad (6.3.24)$$

The MATLAB program to write the Jacobian is then:

```
1  function J=getJ(y,dx,n)
2  J = zeros(n);
3  J(1,1) = 1;
```

```
4   denom = dx^(-2);
5   for i = 2:n-1
6       J(i,i-1) = denom;
7       J(i,i) = -2*denom + 2 - 12*y(i)^2;
8       J(i,i+1) = denom;
9   end
10  J(n,n) = 1;
```

The structure of the program is quite similar to getR; the function takes in the vector of unknowns and the discretization parameters, makes space in the memory to write the Jacobian, and then fills in the non-zero entries. Notice in line 4 that we have defined a new variable denom to be equal to $(\Delta x)^{-2}$. This step removes quite a lot of computation, since we calculate this term $3(n-2)$ times when we write the Jacobian. There are often many opportunities to speed up your codes, but we generally suggest that you write the simplest code possible as you are learning the methods and worry about speed later. The reason for this advice is that most increases in speed will reduce the readability of the code and make it harder to find mistakes.

To solve this problem, we need to make an initial guess and then iterate using Newton–Raphson. Let's go back to Program 3.5 and make a few small changes:

```
1   function [x,flag] = nonlinear_newton_raphson(x,tol,dx,n)
2   %  x = initial guess for x
3   %  tol = criteria for convergence
4   %  requires function R = getR(x)
5   %  requires function J = getJ(x)
6   %  flag = 0 if failed, 1 if converged
7   %  dx, n are parameters from finite differences
8   R = getR(x,dx,n);
9   k = 0; %counter
10  kmax = 100; %max number of iterations allowed
11  while norm(R) > tol
12      J = getJ(x,dx,n); %compute Jacobian
13      J = sparse(J); %use banded solver in Matlab
14      del = -J\R; %Newton Raphson
15      x = x + del; %update x
16      k = k + 1; %update counter
17      R = getR(x,dx,n); %f for while loop and counter
18      if k > kmax
19          fprintf('Did not converge.\n')
20          break
21      end
22  end
23  if k > kmax || max(x) == Inf || min(x) == -Inf
24      flag = 0;
25  else
26      flag = 1;
27  end
```

Obviously, we need to send the values of Δx and n to this program, since they are needed by getR and getJ. We again are using a banded solver due to the tridiagonal structure of the Jacobian. Note that the banded solver can be even more important for nonlinear

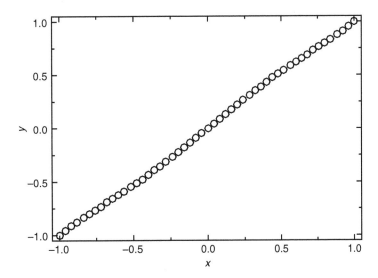

Figure 6.4 Solution to Eq. (6.3.15) using finite differences with 51 nodes.

problems than linear problems, since we may need to iterate a number of times before we converge on a solution to a nonlinear problem.

The only remaining item is to decide on an initial guess. Based on the boundary conditions, the choice is $y_i^{(0)} = x_i$ since it satisfies the boundary conditions and the homogeneous problem for Eq. (6.3.15). As we see in Fig. 6.4, this guess is adequate to get the solution to converge. Indeed, the nonlinearity is relatively weak and the end result is quite similar to a straight line with some small fluctuations.

We want to point out that nonlinear BVPs, just like nonlinear algebraic equations, can possess multiple solutions. This should be clear in the context of the last example, since our solution of the differential equation involves converting it to a system of algebraic equations. These can be identified by using different initial guesses in the Newton–Raphson iteration.

6.3.2   Neumann and Robin Boundary Conditions

When we have a derivative at one (or both) of the boundary conditions, we need to be more careful. For example, let's consider a Neumann boundary condition at the left boundary,

$$\left.\frac{dy}{dx}\right|_{x=x_l} = \alpha \tag{6.3.25}$$

where α is a constant. We could try to use a forward finite difference approximation for this boundary condition, which would give us

$$\frac{y_2 - y_1}{\Delta x} = \alpha \tag{6.3.26}$$

Figure 6.5 Introduction of a fictitious node for Neumann and Robin boundary conditions.

While this approach would work, it leads to an error at the boundary of $\mathcal{O}(\Delta x)$, which is larger than the error for the interior nodes of $\mathcal{O}(\Delta x^2)$. It would be preferable to use the centered finite difference approximation of the first derivative at the boundary so that the equation at node $i = 1$ has the same error as the other nodes.

We can keep the same level of accuracy at the boundary by introducing an auxiliary (or fictitious) node called x_0, on the left of x_1, with a corresponding value of $y(x = x_0) = y_0$. This idea is illustrated in Fig. 6.5 for the left boundary. The same idea applies for a derivative boundary condition on the right boundary; we now introduce a node x_{n+1} to the right of x_n. In the case here, neither x_0 nor y_0 has physical meaning. They exist only to allow us to use a centered finite difference approximation in the implementation of the boundary condition,

$$\frac{y_2 - y_0}{2\Delta x} = \alpha \qquad (6.3.27)$$

which we can solve for the value of the fictitious node,

$$y_0 = y_2 - 2\alpha\Delta x \qquad (6.3.28)$$

Since we have added an additional unknown y_0, we need one more equation to complete the system of equations. This can be obtained by discretizing the differential equation at x_1. As a result, we can solve for y_0 from Eq. (6.3.27) and substitute the result into the differential equation at x_1. We thus have a $\mathcal{O}(\Delta x^2)$ accurate approximation at the boundary. Note that, for some simple boundary conditions and differential equations, the result from a fictitious node is the same as using a forward or backward difference, but this is not true in general.

Example 6.4   Consider the same ODE as in Example 6.3 but use a Neumann boundary condition on the left boundary, e.g.,

$$\left.\frac{dy}{dx}\right|_{x=-1} = 10 \qquad (6.3.29)$$

Solution
In this case, the form of the equations for the interior nodes and the right boundary are the same as before, given by Eqs. (6.3.16) and (6.3.18). For the boundary condition with the fictitious node, we get

$$\frac{y_2 - y_0}{2\Delta x} = 10 \qquad (6.3.30)$$

which we solve to get

$$y_0 = y_2 - 20\Delta x \qquad (6.3.31)$$

We now discretize the differential equation at x_1, to get

$$\frac{y_2 - 2y_1 + y_0}{\Delta x^2} + 2y_1(1 - 2y_1^2) = 0 \qquad (6.3.32)$$

Substituting for y_0 gives a new first row for the residual

$$R_1 = \frac{2y_2 - 2y_1 - 20\Delta x}{\Delta x^2} + 2y_1(1 - 2y_1^2) = 0 \qquad (6.3.33)$$

We thus modify the program getR from the last example:

```
1  function R = getR(y,dx,n)
2  R = zeros(n,1);
3  R(1,1) = (2*y(2)-2*y(1)-20*dx)/dx^2 + 2*y(1)*(1-2*y(1)^2);
4  for i = 2:n-1
5      R(i,1) = (y(i+1)-2*y(i)+y(i-1))/(dx^2)+2*y(i)*(1-2*y(i)^2);
6  end
7  R(n,1) = y(n) - 1;
```

The only change is in line 3, where we now write Eq. (6.3.33) in place of the Dirichlet condition.

Since we have a new R_1, the non-zero entries in the first row of the Jacobian are now

$$J_{1,1} = \frac{-2}{\Delta x^2} + 2 - 12y_1^2 \qquad (6.3.34)$$

$$J_{1,2} = \frac{2}{\Delta x^2} \qquad (6.3.35)$$

We thus need to make small changes to the program to compute the Jacobian as well:

```
1  function J=getJ(y,dx,n)
2  J = zeros(n);
3  denom = dx^(-2);
4  J(1,1) = -2*denom + 2 - 12*y(1)^2;
5  J(1,2) = 2*denom;
6  for i = 2:n-1
7      J(i,i-1) = denom;
8      J(i,i) = -2*denom + 2 - 12*y(i)^2;
9      J(i,i+1) = denom;
10 end
11 J(n,n) = 1;
```

Lines 4 and 5 of this program compute the new entries in the Jacobian for the Neumann boundary condition.

These changes to the programs for the residual and Jacobian are all we need; the rest of the solution is identical to Example 6.3. If we also continue to use the same initial guess, it takes eight iterations for Newton–Raphson to converge to the solution in Fig. 6.6. The solution $y(x)$ is now much more nonlinear due to the very steep boundary condition at $x = -1$.

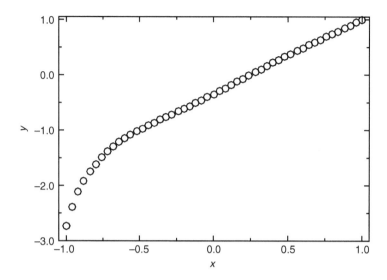

Figure 6.6 Solution to Eq. (6.3.15) with the Neumann boundary condition (6.3.29) on the left boundary.

In the last example, we showed how to use a fictitious node for a Neumann condition. The same approach can be used with the Robin boundary conditions in Eq. (6.1.6). For example, if we are dealing with a Robin boundary condition on the left boundary, we introduce the fictitious node x_0 and discretize the boundary condition to get

$$\alpha_1 \left(\frac{y_2 - y_0}{2\Delta x} \right) + \alpha_2 y_1 = \alpha_3 \tag{6.3.36}$$

We can solve this equation for y_0,

$$y_0 = y_2 - \frac{2\Delta x}{\alpha_1} (\alpha_3 - \alpha_2 y_1) \tag{6.3.37}$$

We then substitute this value of y_0 into the discretized differential equation at node x_1.

6.3.3   Position-Dependent Terms

The previous two examples involved a nonlinear differential equation with constant coefficients. Let's now consider a linear problem with position-dependent terms. These problems can often be solved, but their solutions can be a mess if the functional form of these terms is not so simple.

Example 6.5   Consider the solid bar in Figure 6.7. The temperature of the bar is maintained at the constant value T_0 at the boundaries $x = -1$ and $x = 1$. In the middle of the bar, from a region $-\epsilon < x < \epsilon$, there is a heating element that delivers heat at a rate q per unit volume. The steady state energy balance for this system is

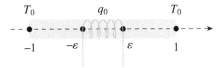

Figure 6.7 Schematic illustration of the heating problem in Example 6.5.

$$0 = k\frac{d^2T}{dx^2} + q(x) \tag{6.3.38}$$

where k is the thermal conductivity and $q(x)$ is the position-dependent heat generation rate term

$$q(x) = \begin{cases} 0 & \text{for} \quad -1 \leq x < -\epsilon \\ q_0 & \text{for} \quad -\epsilon \leq x \leq \epsilon \\ 0 & \text{for} \quad \epsilon < x \leq 1 \end{cases} \tag{6.3.39}$$

Compute the temperature profile in this bar as a function of position using finite differences.

Solution

We first discretize the spatial domain into n nodes. The corresponding boundary conditions are

$$T_1 = T_0 \quad \text{and} \quad T_n = T_0 \tag{6.3.40}$$

In the interior, we apply centered finite differences to get the discretized form of the ODE

$$k\left[\frac{T_{i+1} - 2T_i + T_{i-1}}{\Delta x^2}\right] = -q(x_i) \tag{6.3.41}$$

We can rewrite this equation in the form

$$T_{i+1} - 2T_i + T_{i-1} = Q_i \tag{6.3.42}$$

where the entries in the vector $\mathbf{Q}$ have the value

$$Q_i = \begin{cases} -\dfrac{q_0 \Delta x^2}{k} & \text{for } i \in [i_{\min}, i_{max}] \\ 0 & \text{otherwise} \end{cases} \tag{6.3.43}$$

The lower bound for the index i where we have non-zero Q_i values is

$$-\epsilon = x_l + (i_{\min} - 1)\Delta x \tag{6.3.44}$$

which reduces to

$$i_{\min} = \frac{(1 - \epsilon)(n - 1)}{2} + 1 \tag{6.3.45}$$

given the definition of Δx and the boundary $x_l = -1$. Likewise, the maximum index for the heater is determined from

$$\epsilon = x_l + (i_{\max} - 1)\Delta x \tag{6.3.46}$$

Figure 6.8 Approach for handling the case where the heater start/end does not align with the location of the nodes.

which gives us

$$i_{\max} = \frac{(1 + \epsilon)(n - 1)}{2} + 1 \qquad (6.3.47)$$

The problem we run into in the computer is that Eq. (6.3.45) generally will not produce an integer value. A simple approximation is to round off $i_{\min}$. The question is whether to round up or down. Consider a small problem with $\epsilon = 1/3$ and $n = 5$. Equation (6.3.45) gives $i_{\min} = 2.33$. The value of $x_2 = -1/2$ and the value of $x_3 = 0$. So we should round up,

$$i_{\min} = \left\lceil \frac{(1 - \epsilon)(n - 1)}{2} + 1 \right\rceil \qquad (6.3.48)$$

where $\lceil \cdot \rceil$ is the "ceiling" operator that rounds up to the nearest integer. For the same reason that we rounded up for the start of the heater, we now need to round down to get an integer,

$$i_{\max} = \left\lfloor \frac{(1 + \epsilon)(n - 1)}{2} + 1 \right\rfloor \qquad (6.3.49)$$

where $\lfloor \cdot \rfloor$ is the "floor" operator that rounds down to the nearest integer.

The original differential equation is linear, so we end up with a linear discretized system of the form

$$\mathbf{AT} = \mathbf{Q} \qquad (6.3.50)$$

where $\mathbf{A}$ is a tri-diagonal matrix and $\mathbf{T}$ is the vector of unknown temperatures. Recall that the effort to solve this equation scales with np^2, where n is equal to the number of nodes and $p = 1$ is the half bandwidth of the matrix. Recall also that the truncation error for the centered finite differences approximation scales with Δx^2, where Δx is inversely proportional to the number of nodes n. It is therefore clear that there is a trade-off between the computational effort and the error in the solution.

To solve this problem, we can write a MATLAB file to generate the matrix $\mathbf{A}$ and forcing vector $\mathbf{b}$:

```
1  function T = bvp_heater(n,epsilon,q,k)
2  A = zeros(n);
3  b = zeros(n,1);
4  dx = 2/(n-1);
5  imin = ceil((1-epsilon)*(n-1)/2 + 1);
6  imax = floor((1+epsilon)*(n-1)/2+1);
```

```
7   Q = -q*dx^2/k;
8   A(1,1) = 1;
9   for i = 2:n-1
10      A(i,i-1) = 1;
11      A(i,i) = -2;
12      A(i,i+1) = 1;
13  end
14  A(n,n) = 1;
15  A = sparse(A);
16  b(1) = 1;
17  for i = imin:imax
18      b(i) = Q;
19  end
20  b(n) = 1;
21  T = A\b;
```

In this program, lines 8–14 write the coefficient matrix. These terms are independent of the location of the heater, but we do need to be sure to handle the boundaries (Dirichlet conditions) differently than the interior nodes. As was the case in Example 6.2, line 15 declares the matrix to be sparse so that we can take advantage of the tri-diagonal structure. We then use lines 16–20 to write the forcing vector. In line 3, we previously set all the values to be equal to zero, so we just need to write the non-zero entries. Finally, line 21 solves the BVP.

We can see that this program is very flexible, allowing us to handle a wide range of different scenarios just by changing the inputs of ϵ and q_0. (Changing k is unnecessary since it always appears with q_0 in the single dimensionless parameter $q_0 \Delta x^2 / k$.) Figure 6.9a shows how the temperature profile depends on the heating rate q, while Fig. 6.9b shows how the temperature profile depends on the size of the heater. Note that we can

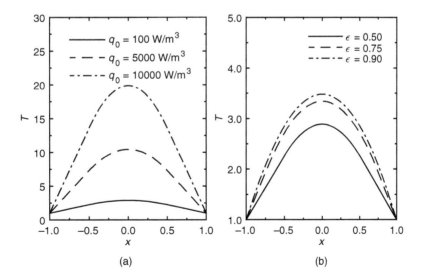

(a)   (b)

Figure 6.9 Temperature profile for Example 6.5 for (a) different heating rates at $\epsilon = 0.5$ and (b) different heating areas at $q_0 = 1000$ W/m^3.

run into some issues in (b) when we make ϵ very small or close to unity. In both cases, either the spatial domain with the heater (small ϵ) or the one without the heater ($\epsilon \approx 1$) become small. In order to resolve the temperature gradients in these regions, we need to have a sufficient number of nodes. Since we are using evenly spaced nodes, this can increase the problem size. Fortunately, we know how to use a banded solver!

This example showed how to modify the solution procedure when the terms in the differential equation depend on the independent variable x. The procedure for handling transport coefficients that depend on both variables is similar. Consider for example a second derivative whose coefficient depends on both x and y. If we apply the finite difference approximation, we would have

$$D(x, y) \frac{d^2 y}{dx^2}\bigg|_{x_i, y_i} = D(x_i, y_i) \left(\frac{y_{i+1} - y_i + y_{i-1}}{\Delta x^2} \right) \qquad (6.3.51)$$

In this case the resulting algebraic equation system would be nonlinear due to the product terms involving y, and would require us to use Newton–Raphson for the solution. You should also keep in mind your transport phenomena for these types of equations, as the derivations often assume constant transport parameters and the differential equations may change if the transport parameters are functions of position.

6.3.4   Mesh Refinement

The last item in our recipe for solving BVPs by finite difference at the start of this section refers to a process known as "mesh refinement." Since the accuracy of the solution depends on the number of nodes n, it is important to consider a discretization of the domain which is fine enough to capture the essential features of the solution. It is customary to start with relatively few nodes (i.e., a coarse discretization) and reduce the discretization step until no significant changes in the solution are observed. We always want to use a solution that is accurate enough, but we do not want to waste time making an ultra-high resolution calculation if a coarse calculation would be sufficient.

You need to use your judgement with respect to "how fine is fine." Figure 6.10 illustrates a relatively simple approach to test the accuracy of the solution. We first solve the

Figure 6.10 Schematic illustration of mesh refinement from $n = 3$ to $n = 5$ to $n = 9$. Each time, the new grid has an additional node between each of the old grid points.

problem on a grid with n points. We then solve the problem on a new grid where we have put an additional point between each of the old grid points, i.e., we are now using $n_{\text{new}} = 2n_{\text{old}} - 1$ points. We then evaluate the difference in the solution between the two grids at their common grid points. If the difference is below our tolerance, then our solution is converged. We should report the results from the old grid, since we know that those values have not changed when the mesh is refined. If the difference is not below our tolerance, then we again add grid points between nodes and repeat the procedure.

Let's see how mesh refinement works for a small, linear problem that we can solve easily.

Example 6.6 Use the problem in Example 6.2 with $\phi = 1$ to illustrate mesh refinement from $n = 3$ to $n = 7$.

Solution
Recalling the notation of Eq. (6.3.9), we have $\Delta x = 1/2$ so $\alpha = -2.25$. The problem we need to solve is

$$\begin{bmatrix} 1 & 0 & 0 \\ 1 & -2.25 & 1 \\ 0 & 0 & 1 \end{bmatrix} \begin{bmatrix} c_1^{(3)} \\ c_2^{(3)} \\ c_3^{(3)} \end{bmatrix} = \begin{bmatrix} 0 \\ 0 \\ 1 \end{bmatrix} \tag{6.3.52}$$

We use the superscript notation to denote the number of nodes in order to keep the numbering convention the same as we change n. The solution to this problem is

$$\begin{bmatrix} c_1^{(3)} \\ c_2^{(3)} \\ c_3^{(3)} \end{bmatrix} = \begin{bmatrix} 0 \\ 0.4444 \\ 1 \end{bmatrix} \tag{6.3.53}$$

The only value that is of interest is $c_2^{(3)} = c(1/2)$, since we get the boundary conditions exactly.

For $n = 5$, we have $\Delta x = 1/4$ so

$$\begin{bmatrix} 1 & 0 & 0 & 0 & 0 \\ 1 & -2.0625 & 1 & 0 & 0 \\ 0 & 1 & -2.0625 & 1 & 0 \\ 0 & 0 & 1 & -2.0625 & 1 \\ 0 & 0 & 0 & 0 & 1 \end{bmatrix} \begin{bmatrix} c_1^{(5)} \\ c_2^{(5)} \\ c_3^{(5)} \\ c_4^{(5)} \\ c_5^{(5)} \end{bmatrix} = \begin{bmatrix} 0 \\ 0 \\ 0 \\ 0 \\ 1 \end{bmatrix} \tag{6.3.54}$$

Notice that the diagonal entry for the interior nodes has changed, since the value of Δx depends on n. The solution is

$$\begin{bmatrix} c_1^{(5)} \\ c_2^{(5)} \\ c_3^{(5)} \\ c_4^{(5)} \\ c_5^{(5)} \end{bmatrix} = \begin{bmatrix} 0 \\ 0.2151 \\ 0.4437 \\ 0.7000 \\ 1 \end{bmatrix} \tag{6.3.55}$$

To see if we have a sufficiently refined mesh, we want to compare the interior nodes that are the same. For this case we want to look at $c_2^{(3)}$ and $c_3^{(5)}$, which are at the same point $x = 1/2$. To evaluate the error, we could use

$$\epsilon = \frac{|c_2^{(3)} - c_3^{(5)}|}{|c_2^{(3)}|} = 0.0017 \tag{6.3.56}$$

So the change in the solution is only 0.2% going from three nodes to five nodes.

What if this is not accurate enough for us? After all, we frequently solve problems down to machine precision, which is 10^{-16} or thereabouts. If we want to do another stage of mesh refinement, we should go to $n = 9$ nodes. The matrix problem becomes

$$\begin{bmatrix} 1 & 0 & 0 & 0 & 0 & 0 & 0 & 0 & 0 \\ 1 & \alpha & 1 & 0 & 0 & 0 & 0 & 0 & 0 \\ 0 & 1 & \alpha & 1 & 0 & 0 & 0 & 0 & 0 \\ 0 & 0 & 1 & \alpha & 1 & 0 & 0 & 0 & 0 \\ 0 & 0 & 0 & 1 & \alpha & 1 & 0 & 0 & 0 \\ 0 & 0 & 0 & 0 & 1 & \alpha & 1 & 0 & 0 \\ 0 & 0 & 0 & 0 & 0 & 1 & \alpha & 1 & 0 \\ 0 & 0 & 0 & 0 & 0 & 0 & 1 & \alpha & 1 \\ 0 & 0 & 0 & 0 & 0 & 0 & 0 & 0 & 1 \end{bmatrix} \begin{bmatrix} c_1^{(9)} \\ c_2^{(9)} \\ c_3^{(9)} \\ c_4^{(9)} \\ c_5^{(9)} \\ c_6^{(9)} \\ c_7^{(9)} \\ c_8^{(9)} \\ c_9^{(9)} \end{bmatrix} = \begin{bmatrix} 0 \\ 0 \\ 0 \\ 0 \\ 0 \\ 0 \\ 0 \\ 0 \\ 1 \end{bmatrix} \tag{6.3.57}$$

where $\alpha = -2.0156$ for $n = 9$. The solution is now

$$\begin{bmatrix} c_1^{(9)} \\ c_2^{(9)} \\ c_3^{(9)} \\ c_4^{(9)} \\ c_5^{(9)} \\ c_6^{(9)} \\ c_7^{(9)} \\ c_8^{(9)} \\ c_9^{(9)} \end{bmatrix} = \begin{bmatrix} 0 \\ 0.1067 \\ 0.2150 \\ 0.3267 \\ 0.4435 \\ 0.5672 \\ 0.6998 \\ 0.8433 \\ 1 \end{bmatrix} \tag{6.3.58}$$

The nodes that we now need to compare are at $x = 1/4$, $x = 1/2$ and $x = 3/4$. Let's define these as the vectors

$$\mathbf{c}_{\text{interior}}^{(5)} = \begin{bmatrix} 0.2151 \\ 0.4437 \\ 0.7000 \end{bmatrix} \tag{6.3.59}$$

and

$$\mathbf{c}_{\text{interior}}^{(9)} = \begin{bmatrix} 0.2150 \\ 0.4435 \\ 0.6998 \end{bmatrix} \tag{6.3.60}$$

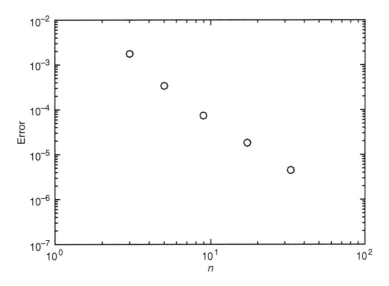

Figure 6.11 Result of mesh refinement for the reaction–diffusion equation with a Thiele modulus of unity.

for the two solutions. We can thus assess the improvement in mesh refinement by computing

$$\epsilon = \frac{||\mathbf{c}_{\text{interior}}^{(5)} - \mathbf{c}_{\text{interior}}^{(9)}||}{||\mathbf{c}_{\text{interior}}^{(5)}||} = 3.4261 \times 10^{-4} \qquad (6.3.61)$$

We can continue on in this vein indefinitely, but it becomes tedious to write out the equations. Figure 6.11 shows the results of the mesh refinement where we define the error between the grids as

$$\epsilon(n^{(\text{old})}) = \frac{||\mathbf{c}_{\text{interior}}^{(\text{new})} - \mathbf{c}_{\text{interior}}^{(\text{old})}||}{||\mathbf{c}_{\text{interior}}^{(\text{old})}||} \qquad (6.3.62)$$

We can thus use a plot of this type to dial in the desired amount of error in the solution.

While the previous example illustrates the idea of mesh refinement, this linear problem has the exact result given in Eq. (6.3.6). Indeed, we used this exact result to study the error in finite differences in Example 6.2. So why are we so worried about having a way to evaluate the accuracy of the solution by mesh refinement? In general, we use numerical methods to solve problems where we do not have an analytical result. It is important to remember that, while you can keep decreasing the error caused by the truncation error in finite differences by increasing the number of mesh points, you cannot arbitrarily increase the accuracy of the solution. The idea in finite differences is to reduce the ordinary differential equation to a system of linear or nonlinear differential equations. In Chapters 2 and 3, we saw that there are additional errors that arise when we solve these systems of algebraic equations. If we use an iterative method like Newton–Raphson (for a nonlinear problem) or SOR (for a linear problem), the accuracy of the solution is set by the tolerance for converging the iterations. If we use a direct solver for a linear problem, we still have accuracy issues related to round-off errors during the elimination steps. In all of these cases, the error is ultimately limited by machine error.

6.4   Case Study: Reaction–Diffusion in a Packed Bed

6.4.1   Problem Statement

We consider a packed bed reactor, shown in Fig. 6.12. At the left edge of the bed, $x = 0$, the concentration of the reactant is equal to the bulk concentration, $c = c_0$. There is a wall at the right edge of the bed, $x = l$, with the corresponding no-flux boundary condition,

$$\left.\frac{dc}{dx}\right|_{x=l} = 0 \tag{6.4.1}$$

For a first-order reaction in the bed, the mass balance takes the form

$$0 = D\frac{\partial^2 c}{\partial x^2} - k(x)c \tag{6.4.2}$$

We assume that, due to the size of the particles, the reaction rate is spatially dependent. We adopt a simple model, where the reaction is at full strength in the center half of the particle and zero otherwise. For example, in the first particle of the bed, the reaction rate constant is

$$k(x) = \begin{cases} 0 & \text{for} & x \in [0, 0.05l] \\ k_0 & \text{for} & x \in (0.05l, 0.15l) \\ 0 & \text{for} & x \in [0.15l, 0.20l] \end{cases} \tag{6.4.3}$$

This periodic function repeats five times over the length l of the array.

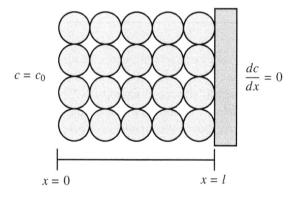

Figure 6.12 Schematic illustration of a packed bed with a spatially dependent reaction rate.

The goal of this case study is to use finite differences to see how the concentration profile inside the reactor depends on the Thiele modulus.

6.4.2   Solution

If we are going to work in terms of a Thiele modulus, we should convert the problem to dimensionless form,

$$\tilde{x} = x/l, \quad \tilde{c} = c/c_0, \quad \tilde{k} = k/k_0 \tag{6.4.4}$$

With this choice of dimensionless variables, you can convert the unsteady equations into

$$0 = \frac{d^2 \tilde{c}}{d\tilde{x}^2} - \tilde{k}(\tilde{x})\phi^2 \tilde{c}. \tag{6.4.5}$$

The dimensionless reactivity has the same spatial dependence as eq. (6.4.3) in each particle, but now the maximum value is $\tilde{k}(\tilde{x}) = 1$ instead of k_0. For a first-order reaction, the Thiele modulus is

$$\phi = \sqrt{\frac{k_0 l^2}{D}} \tag{6.4.6}$$

The equation for the interior nodes is

$$c_{i+1} + c_{i-1} - \left[2 + \phi^2 \Delta x^2 k_i\right] c_i = 0 \tag{6.4.7}$$

where

$$\Delta x = \frac{1}{n-1} \tag{6.4.8}$$

and k_i is the reaction rate constant evaluated at node i. The left node equation is

$$c_1 = 1 \tag{6.4.9}$$

The fictitious node equation at the right boundary gives

$$\frac{c_{n+1} - c_{n-1}}{2\Delta x} = 0 \tag{6.4.10}$$

so $c_{n+1} = c_{n-1}$. Using this result in the discretized form of the differential equation gives

$$2c_{n-1} - \left[2 + \phi^2 \Delta x^2 k_n\right] c_n = 0 \tag{6.4.11}$$

Since the reaction rate constant k_n is zero at the boundary, this result simplifies to

$$c_{n-1} - c_n = 0 \tag{6.4.12}$$

Let's write first a function that solves the differential equation for a given value of the Thiele modulus and the number of nodes n, in case we want to do some mesh refinement later on.

```
1   function c = linear_react(n,phi2)
2   dx = 1/(n-1);
3   k = zeros(n,1);
4   for i = 1:n
5       x = mod((i-1)*dx,0.2);
6       if x >= 0.05 && x <= 0.15
7           k(i) = 1;
8       end
9   end
10  A = zeros(n);
11  b = zeros(n,1);
12  A(1,1) = 1;
13  b(1,1) = 1;
14  for i = 2:n-1
15      A(i,i-1) = 1;
16      A(i,i) = -2 - dx^2*phi2*k(i);
17      A(i,i+1) = 1;
18  end
19  A(n,n-1) = 1;
20  A(n,n) = -1;
21  A = sparse(A);
22  c = A\b;
```

The beginning of this program sets up the reaction rate vector using the modulo operator. We have already seen the modulo operator when we set the output frequency for integrating IVPs, for example in Program 4.1. Here, we are using the modulo operator to map the position in the "global" domain $x \in [0, 1]$ back onto a "local" domain $x \in [0, 0.2]$. The **for** loop starting in line 4 loops through the n nodes. The value of x for that node is $x_i = (i - 1)\Delta x$. When we compute the modulus of x_i with 0.2 in line 5, we get the remainder after dividing by 0.2, which maps the result back to the first catalyst particle. For example, if $x_i = 0.55$, the operation x=mod(0.55,0.2) returns $x = 0.15$, since you can divide 0.2 into 0.55 twice with a remainder of 0.15. Line 6 then checks to see if this position is inside the catalyst particle. We need to have an "and" operator && to be sure that we are to the right of the start of the particle and to the left of the end of the particle. If this is the case, the reaction rate constant is set to 1. If not, then the reaction rate constant is zero because we initialized the vector **k** to be all zeros in line 3.

Aside from setting the reaction rate vector, this program should look remarkably similar to the one we used for Example 6.2, since the underlying differential equations are the same. Lines 12 and 13 of the present program implement the Dirichlet boundary condition on the left, while lines 19 and 20 implement the Neumann boundary condition on the right. For the interior nodes, we need the local reaction rate k_i, which is called in line 16. Finally, we solve this tridiagonal system using a banded solver in lines 21 and 22.

Figure 6.13 shows the concentration profile in this model of a packed bed for different values of the Thiele modulus. Clearly, as the Thiele modulus increases and reaction becomes dominant over diffusion, there is a sharp reduction in concentration away from the reservoir at $x = 1$. If you look closely, you may also be able to see the discrete nature of the reaction, since it only takes place inside the particles. You may have thought that

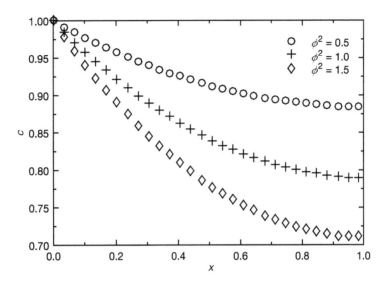

Figure 6.13 Concentration profile in a packed bed with discrete reaction sites as a function of the Thiele modulus.

the concentration profile would be more step-like with this reactivity profile. However, remember that diffusion takes place in both the catalyst pellets and the interstitial space. Although the reaction is only occurring inside the particles, diffusion both inside the particles and in the space between particles is smearing out the discreteness of the reaction rate.

6.5 Coupled BVPs

We will now consider systems of two ODEs that are coupled. For simplicity, let's consider the following system of two equations

$$\frac{d^2c}{dx^2} = f_1\left(c, T\right) \tag{6.5.1}$$

$$\frac{d^2T}{dx^2} = f_2\left(c, T\right) \tag{6.5.2}$$

with appropriate boundary conditions on c and T. You may recognize this as a coupled heat and mass transfer problem at steady state, for example in a reactive system where the reaction term acts as a source (or sink) of heat too. For simplicity, we are assuming that the functions f_1, f_2 do not depend on spatial derivatives of the dependent variables. For such a problem the goal is to determine approximate solutions for both c and T.

The starting point is the same as before. We discretize the domain into n nodes with

$$\Delta x = \frac{x_r - x_l}{n - 1} \tag{6.5.3}$$

Figure 6.14 shows that, at each node x_i, there is a corresponding value of c_i and T_i, to be determined. In other words, there are $2n$ unknowns.

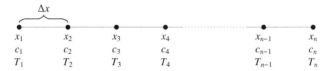

Figure 6.14 Discretization of space for a system of two coupled boundary value problems.

We will consider two different approaches to solve this problem.

6.5.1   Approach #1: Appending the Unknown Vectors

In this approach, we append the unknown vectors to obtain the following single vector:

$$\mathbf{y} = \begin{bmatrix} c_1 \\ c_2 \\ \vdots \\ c_{n-1} \\ c_n \\ \hline T_1 \\ T_2 \\ \vdots \\ T_{n-1} \\ T_n \end{bmatrix} \tag{6.5.4}$$

In other words, the indexing for $\mathbf{y}$ is

$$y_k = \begin{cases} c_k & \text{for} \quad k \in [1, n] \\ T_{k-n} & \text{for} \quad k \in [n+1, 2n] \end{cases} \tag{6.5.5}$$

where k is a global index.

We proceed with discretization of the ODEs at each interior node $i = 2, 3, \ldots, n-1$. The discretized form of the equations for concentration is

$$\frac{c_{i+1} - 2c_i + c_{i-1}}{\Delta x^2} = f_1\,(c_i, T_i) \tag{6.5.6}$$

while the discretized equation for the temperature is

$$\frac{T_{i+1} - 2T_i + T_{i-1}}{\Delta x^2} = f_2\,(c_i, T_i) \tag{6.5.7}$$

Rewriting these equations in terms of the variable $\mathbf{y}$ and in residual form, Eq. (6.5.6) becomes

$$\frac{y_{i+1} - 2y_i + y_{i-1}}{\Delta x^2} - f_1\,(y_i, y_{i+n}) = 0 \tag{6.5.8}$$

and Eq. (6.5.7) becomes

$$\frac{y_{i+n+1} - 2y_{i+n} + y_{i+n-1}}{\Delta x^2} - f_2\,(y_i, y_{i+n}) = 0 \tag{6.5.9}$$

The problem formulation is completed by implementing the boundary conditions and forming the nonlinear algebraic system $\mathbf{R}(\mathbf{y}) = \mathbf{0}$. This will be solved using Newton–Raphson,

$$\mathbf{J}|_{\mathbf{y}^{(k)}}\,\boldsymbol{\delta}^{(k+1)} = -\mathbf{R}(\mathbf{y}^{(k)}) \tag{6.5.10}$$

Let us consider the structure of the Jacobian, assuming Dirichlet boundary conditions at the left,

$$\mathbf{J} = \begin{bmatrix}
1 & 0 & 0 & \cdots & \cdots & \cdots & \cdots & \cdots & \cdots & \cdots & \cdots & \cdots & 0 \\
* & * & * & 0 & 0 & \cdots & \cdots & 0 & * & 0 & 0 & 0 & \cdots & 0 \\
0 & * & * & * & 0 & & \cdots & \cdots & 0 & * & 0 & 0 & \cdots & 0 \\
\vdots & \vdots & \vdots & \vdots & \vdots & \vdots & \vdots & \vdots & \vdots & \vdots & \vdots & \vdots & \vdots \\
0 & 0 & 0 & 0 & \cdots & \cdots & 0 & 1 & 0 & 0 & 0 & 0 & \cdots & 0 \\
0 & * & 0 & 0 & \cdots & \cdots & 0 & * & * & * & 0 & 0 & \cdots & 0 \\
0 & 0 & * & 0 & \cdots & \cdots & 0 & 0 & * & * & * & 0 & \cdots & 0 \\
\vdots & \vdots & \vdots & \vdots & \vdots & \vdots & \vdots & \vdots & \vdots & \vdots & \vdots & \vdots & \vdots
\end{bmatrix} \tag{6.5.11}$$

where $*$ denotes a non-zero entry. Note that because of the dependence of f_1 on y_i, y_{i+n} in Eq. (6.5.8), the ith row of the Jacobian has non-zero elements that extend from the $i-1$ column to the $i+n$ column. Similarly, because of the dependence of f_2 on y_i, y_{i+n} in Eq. (6.5.9), the $(i+n)$th row of the Jacobian has non-zero elements that extend from the i column to the $i+n+1$ column. Thus, although the Jacobian is extremely sparse, it has half bandwidths $p = n$ and $q = n$. This approach is very easy to implement, but the computational cost scales with n^3 in this case.

6.5.2   Approach #2: Interlacing the Variables

In this approach we interlace the two variable sets,

$$\mathbf{y} = \begin{bmatrix} c_1 \\ T_1 \\ c_2 \\ T_2 \\ \vdots \\ c_n \\ T_n \end{bmatrix} \tag{6.5.12}$$

In other words, we now have

$$c_i = y_{2i-1} \tag{6.5.13}$$

and

$$T_i = y_{2i} \tag{6.5.14}$$

for $i = 1, \ldots, n$. In terms of the global index k, we have

$$y_k = \begin{cases} c_{(k+1)/2} & \text{for } k \text{ odd} \\ T_{k/2} & \text{for } k \text{ even} \end{cases} \qquad (6.5.15)$$

If we rewrite Eqs. (6.5.6) and (6.5.7) in terms of y_k, we now have for R_{2i-1}

$$\frac{y_{2i+1} - 2y_{2i-1} + y_{2i-3}}{\Delta x^2} = f_1(y_{2i-1}, y_{2i}) \qquad (6.5.16)$$

and for R_{2i} we have

$$\frac{y_{2i+2} - 2y_{2i} + y_{2i-2}}{\Delta x^2} = f_2(y_{2i-1}, y_{2i}) \qquad (6.5.17)$$

The corresponding Jacobian, again assuming Dirichlet boundary conditions at the left, now has the form

$$\mathbf{J} = \begin{bmatrix} 1 & 0 & 0 & 0 & 0 & 0 & 0 & 0 & 0 & \cdots \\ 0 & 1 & 0 & 0 & 0 & 0 & 0 & 0 & 0 & \cdots \\ * & 0 & * & * & * & 0 & 0 & 0 & 0 & \cdots \\ 0 & * & * & * & 0 & * & 0 & 0 & 0 & \cdots \\ 0 & 0 & * & 0 & * & * & * & 0 & 0 & \cdots \\ 0 & 0 & 0 & * & * & * & 0 & * & 0 & \cdots \\ \vdots & \vdots & \vdots & \vdots & \vdots & \vdots & \vdots & \vdots & \vdots & \vdots \end{bmatrix} \qquad (6.5.18)$$

where again $*$ denotes a non-zero entry. The half bandwidths in the Jacobian are now only $p = 2$ and $q = 2$, leading to a bandwidth of 5. As a result, the cost for solving the problem scales with n. For a large problem, this is huge savings over the first approach. In general, the problems are large because the truncation error of finite differences is $\mathcal{O}(\Delta x^2)$. The problem is the bookkeeping; but it is worth the effort!

6.6   Case Study: Coupled Heat and Mass Transfer

6.6.1   Problem Statement

Let's now consider the problem of calculating the steady state concentration and temperature profile for the system in Fig. 6.15. The concentration of reactant on the left is $c(0) = 0$, while the concentration on the right is fixed at $c(L) = c_0$. The steady state mass balance inside the domain is given by

$$0 = D\frac{d^2c}{dx^2} - r(c, T) \qquad (6.6.1)$$

where

$$r(c, T) = k_0 e^{-E/RT} c \qquad (6.6.2)$$

is the first-order rate of consumption of the reactant. In the latter, k_0 is the pre-exponential factor, E is the activation energy, R is the ideal gas constant, and T is the temperature. The temperature field also obeys a reaction–diffusion type equation

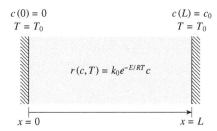

$$c(0) = 0 \qquad\qquad\qquad c(L) = c_0$$
$$T = T_0 \qquad\qquad\qquad T = T_0$$

$$r(c,T) = k_0 e^{-E/RT} c$$

$$x = 0 \qquad\qquad\qquad\qquad x = L$$

Figure 6.15 Non-isothermal reaction–diffusion between an empty reservoir on the left and a full reservoir on the right.

$$0 = k\frac{d^2 T}{dx^2} + q(c, T) \tag{6.6.3}$$

where

$$q(c, T) = r(c, T)\left(\frac{-\Delta H_{rxn}}{V}\right) \tag{6.6.4}$$

is the rate of heat generation inside volume V. We'll fix the temperature to be at the same value T_0 on both sides of the reactor.

The goal of this problem is to see how the concentration and temperature profiles depend on the Thiele modulus and the activation energy. To keep things interesting, let's consider endothermic reactions where $\Delta H_{rxn} > 0$.

6.6.2 Solution

Dimensionless Formulation
Let's first convert the problem to a dimensionless form with $\tilde{c} = c/c_0$, $\tilde{T} = T/T_0$, $\tilde{x} = x/L$, and the parameter $\beta = E/RT_0$. Using these dimensionless variables for the concentration gives us

$$\frac{Dc_0}{L^2}\frac{d^2\tilde{c}}{d\tilde{x}^2} = k_0 e^{-\beta/\tilde{T}} c_0 \tilde{c} \tag{6.6.5}$$

Rearranging the terms gives

$$\frac{d^2\tilde{c}}{d\tilde{x}^2} = \left(\frac{k_0 L^2}{D}\right) e^{-\beta/\tilde{T}} \tilde{c} \tag{6.6.6}$$

The term in brackets is the square of the Thiele modulus,

$$\phi = \sqrt{\frac{k_0 L^2}{D}} \tag{6.6.7}$$

giving us

$$\frac{d^2\tilde{c}}{d\tilde{x}^2} = \phi^2 e^{-\beta/\tilde{T}} \tilde{c} \tag{6.6.8}$$

The boundary conditions for this equation are

$$\tilde{c}(\tilde{x} = 0) = 0, \quad \tilde{c}(\tilde{x} = 1) = 1 \tag{6.6.9}$$

Applying the same procedure for the temperature gives

$$\left(\frac{kT_0}{L^2}\right)\frac{d^2\tilde{T}}{d\tilde{x}^2} = k_0 e^{-\beta/\tilde{T}}\tilde{c}c_0\frac{\Delta H_{rxn}}{V} \tag{6.6.10}$$

Rearranging the terms gives

$$\frac{d^2\tilde{T}}{d\tilde{x}^2} = \left(\frac{k_0 c_0 \Delta H_{rxn}L^2}{Vk_0 T_0}\right)\tilde{c}e^{-\beta/\tilde{T}} \tag{6.6.11}$$

The term in brackets is a second dimensionless number,

$$\gamma = \frac{k_0 c_0 \Delta H_{rxn}L^2}{VkT_0} \tag{6.6.12}$$

which gives us

$$\frac{d^2\tilde{T}}{d\tilde{x}^2} = \gamma e^{-\beta/\tilde{T}}\tilde{c} \tag{6.6.13}$$

The boundary conditions for this equation are

$$\tilde{T}(\tilde{x}=0) = \tilde{T}(\tilde{x}=1) = 1 \tag{6.6.14}$$

Numerical Solution

We thus need to solve Eqs. (6.6.8) and (6.6.13) subject to Eqs. (6.6.9) and (6.6.14). To discretize the domain, we use

$$\Delta x = \frac{1}{n-1} \tag{6.6.15}$$

where n is the number of nodes. We also interlace the variables such that

$$c_i = y_{2i-1}, \quad T_i = y_{2i} \tag{6.6.16}$$

For the interior nodes, Eq. (6.6.8) gives

$$\frac{y_{2i+1} - 2y_{2i-1} + y_{2i-3}}{\Delta x^2} - \phi^2 y_{2i-1}e^{-\beta/y_{2i}} = 0 \tag{6.6.17}$$

for $i = 2, \ldots, n-1$. For these same nodes, Eq. (6.6.13) gives

$$\frac{y_{2i+2} - 2y_{2i} + y_{2i-2}}{\Delta x^2} - \gamma y_{2i-1}e^{-\beta/y_{2i}} = 0 \tag{6.6.18}$$

The boundary conditions at node $i = 1$ correspond to

$$y_1 = 0 \tag{6.6.19}$$

for concentration and

$$y_2 - 1 = 0 \tag{6.6.20}$$

for temperature. Likewise, the boundary conditions at node $i = n$ correspond to

$$y_{2n-1} - 1 = 0 \tag{6.6.21}$$

for concentration and

$$y_{2n} - 1 = 0 \tag{6.6.22}$$

for temperature.

Equations (6.6.17)–(6.6.22) constitute a system of $2n$ nonlinear algebraic equations, where we have written all of the equations in the form of the residual. A MATLAB file to calculate the residual is:

```
1   function R = getR(y,alpha,beta,gamma,n)
2   dx = 1/(n-1);
3   R = zeros(2*n,1);
4   R(1,1) = y(1) - 0;
5   R(2,1) = y(2) - 1;
6   for i = 2:n-1
7       conc_L = (y(2*i+1) -2*y(2*i-1) +y(2*i-3))/dx^2;
8       conc_nL = -alpha*y(2*i-1)*exp(-beta/y(2*i));
9       R(2*i-1,1) =  conc_L + conc_nL;
10      T_L = (y(2*i+2) -2*y(2*i) +y(2*i-2))/dx^2;
11      T_nL = -gamma*y(2*i-1)*exp(-beta/y(2*i));
12      R(2*i,1)= T_L + T_nL;
13  end
14  R(2*n-1,1) = y(2*n-1) - 1;
15  R(2*n,1) = y(2*n) - 1;
```

In the loop for computing the interior node equations, we have broken up the operators into the linear part and the nonlinear part. We use `alpha` for the square of the Thiele modulus.

We also need the non-zero entries in the Jacobian. For the first row, we have

$$J_{1,1} = 1 \tag{6.6.23}$$

from the left boundary condition on concentration (6.6.19). We get the same result for the second row,

$$J_{2,2} = 1 \tag{6.6.24}$$

from the left boundary condition on temperature (6.6.20). For the interior nodes, using the local counter $i = 2, 3, \ldots, n - 1$, the non-zero entries in the Jacobian are

$$J_{2i-1,2i-3} = \Delta x^{-2} \tag{6.6.25}$$

$$J_{2i-1,2i-1} = -2\Delta x^{-2} - \phi^2 e^{-\beta/y_{2i}} \tag{6.6.26}$$

$$J_{2i-1,2i} = -\phi^2 y_{2i-1} e^{-\beta/y_{2i}} \beta y_{2i}^{-2} \tag{6.6.27}$$

$$J_{2i-1,2i+1} = \Delta x^{-2} \tag{6.6.28}$$

for the entries corresponding to the concentration equations (6.6.17) and

$$J_{2i,2i-2} = \Delta x^{-2} \tag{6.6.29}$$

$$J_{2i,2i-1} = -\gamma e^{-\beta/y_{2i}} \tag{6.6.30}$$

$$J_{2i,2i} = -2\Delta x^{-2} - \gamma y_{2i-1} e^{-\beta/y_{2i}} \beta y_{2i}^{-2} \tag{6.6.31}$$

$$J_{2i,2i+2} = \Delta x^{-2} \tag{6.6.32}$$

for the entries corresponding to the temperature equations (6.6.18). For the last two rows, we again get

$$J_{2n-1,2n-1} = 1 \qquad (6.6.33)$$

from the right boundary condition on concentration (6.6.21) and

$$J_{2n,2n} = 1 \qquad (6.6.34)$$

from the right boundary condition on temperature (6.6.22).

The Jacobian also requires a MATLAB function:

```
function J = getJ(y,alpha,beta,gamma,n)
dx = 1/(n-1);
dx2 = 1/dx^2;
J(1,1) = 1;
J(2,2) = 1;
for i = 2:n-1
    r = exp(-beta/y(2*i));
    J(2*i-1,2*i-3) = dx2;
    J(2*i-1,2*i-1) = -2*dx2 - alpha*r;
    J(2*i-1,2*i) = -alpha*y(2*i-1)*r*beta/y(2*i)^2;
    J(2*i-1,2*i+1) = dx2;
    J(2*i,2*i-2) = dx2;
    J(2*i,2*i-1) = - gamma*r;
    J(2*i,2*i) = -2*dx2 -gamma*y(2*i-1)*r*beta/y(2*i)^2;
    J(2*i,2*i+2) = dx2;
end
J(2*n-1,2*n-1) = 1;
J(2*n,2*n) = 1;
```

Since the quantity $e^{-\beta/y_{2i}}$ appears frequently in the Jacobian, we compute it once on line 7 and use it for the rest of the terms at a given interior node. Also, note that we are doing the counting using the local counter i, which simplifies the bookkeeping. Lines 8–11 implement the entries in the Jacobian for concentration, while lines 12–15 do the same for temperature. Lines 4 and 5 are the left boundary condition, and lines 17 and 18 are the right boundary condition.

Since we are passing parameters for the BVP through the program, we also need to modify the Newton–Raphson program (3.5) to take the inputs:

```
function [x,flag] = nonlinear_newton_raphson(x,tol,paramet)
%  x = initial guess for x
%  tol = criteria for convergence
%  requires function R = getR(x)
%  requires function J = getJ(x)
%  flag = 0 if failed, 1 if converged
%  alpha,beta,gamma,n are parameters from the BVP
alpha = paramet(1);
beta = paramet(2);
gamma = paramet(3)
n = paramet(4);
R = getR(x,alpha,beta,gamma,n);
k = 0; %counter
kmax = 100; %max number of iterations allowed
while norm(R) > tol
    J = getJ(x,alpha,beta,gamma,n); %compute Jacobian
```

```
17        J = sparse(J);
18        del = -J\R; %Newton Raphson
19        x = x + del; %update x
20        k = k + 1; %update counter
21        R = getR(x,alpha,beta,gamma,n);
22        if k > kmax
23            fprintf('Did not converge.\n')
24            break
25        end
26    end
27    if k > kmax || max(x) == Inf || min(x) == -Inf
28        flag = 0;
29    else
30        flag = 1;
31    end
```

There are a number of parameters that get passed through the program, so for convenience in our main program they are passed as a vector and then unpacked in lines 8–11 to send to the subfunctions.

To solve the problem, we also need to initialize the vector **y**. A choice $y_i = 1$ for this problem is normally good enough to get the solution to converge in three iterations of the Newton–Raphson. When we are done with the calculation, we will normally want to unpack the unknowns into two separate vectors for plotting. We can do this with the following program:

```
1    function [x,c,T] = extract_coupled_vars(y,n)
2    c = zeros(n,1);
3    T = zeros(n,1);
4    x = linspace(0,1,n);
5    for i = 1:n
6        c(i) = y(2*i-1);
7        T(i) = y(2*i);
8    end
```

This simple program just loops over the unknown vector **y** and extracts the odd entries into a new vector **c** and the even entries into a vector **T**. For convenience, the program also returns the corresponding entries of the independent variable in the vector **x**.

Parametric Analysis

We now have a set of MATLAB programs that will solve this nonlinear reaction–diffusion problem given an initial condition and a set of parameters. To look at how the concentration and temperature profiles evolve as a function of the parameters, let's start with the simplest one. The parameter β is a dimensionless activation temperature relative to the temperature at the boundaries. In other words, it sets the bar for the difficulty in achieving a substantial reaction. It is useful then to consider what happens as β becomes large. In this case, there should be no reaction. Indeed, as we see in Fig. 6.16, the temperature profile becomes flatter and flatter as we increase β, while the concentration profile is approaching the linear one that we would expect in the absence of reaction. From a physical standpoint, the dependence on β is not particularly interesting. From a

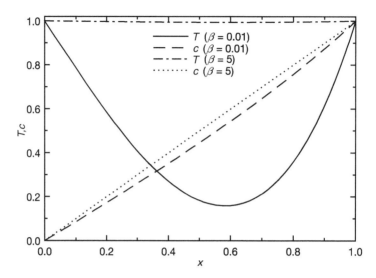

Figure 6.16 Impact of the activation energy on the concentration and temperature profile for $\beta = 0.01$ and $\beta = 5$ with $\alpha = 1$ and $\gamma = 15$.

numerical standpoint, however, this serves as an important check on the accuracy of our solution and whether or not we have any bugs in the code. When you are writing more and more complicated programs, it becomes very useful to test them on cases where you know the answer. Indeed, our next test will bring up a problem with our work so far!

If we make the reaction endothermic and pick a relatively small activation temperature, say $\beta = 0.01$, we see in Fig. 6.16 that the temperature gets very low without much reaction. This particular example shows a very strong coupling between the transport of heat and species. There is also an analogy with the case of a stiff system of IVPs that we discussed in Section 4.8.2. For stiff IVPs, we have a wide separation of eigenvalues and we need to keep the time step small with respect to the largest eigenvalue so that we have numerical stability. In the case of coupled BVPs, it is possible to have one of the dependent variables change much more rapidly than the other dependent variables. The most rapidly changing dependent variable, in this case the temperature, sets the spatial discretization size Δx that we need to avoid large errors in the solution. Since the equations are coupled, errors in the solution for the temperature will be transmitted to the concentration through the reaction term. The difference between BVPs and IVPs is that there is no equivalent of the stability condition for a BVP – we will obtain a solution to the BVP for large Δx, but it will just be inaccurate. It is unlikely that the solution will diverge, although it is always possible that a nonlinear system will find an unphysical solution due to discretization error.

Let's consider a more interesting case where we consider the balance between reaction and diffusion embodied in the Thiele modulus. If $\phi = 0$, then there should be no reaction. Thus, we would expect a small Thiele modulus to be the same as the large β case in Fig. 6.16. This is obviously not the case in Fig. 6.17!

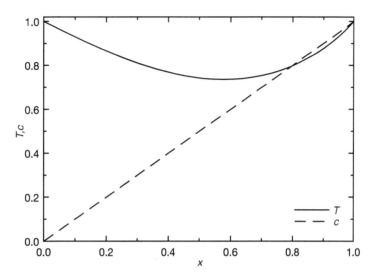

Figure 6.17 A problem when $\phi = 0$ for an endothermic reaction.

The question we need to answer is whether this is due to a bug in our program, an unphysical solution caused by a bad guess, or some other problem. In general, this is a very difficult question to answer. We normally look at known cases to test out the code before considering other problems where we don't know the answer. In the case of no reaction, the coupling in Eqs. (6.6.1) and (6.6.3) should disappear – the concentration should be a linear profile between the reservoirs and the temperature should be uniform. The solution we see in Fig. 6.17 gives us a clue to the source of the problem. The concentration profile looks correct, with a perfectly linear solution between the two reservoirs. So the problem must be in the temperature equations. If we look back at the definition of the parameter γ in Eq. (6.6.12), we see that it could be written in the form

$$\gamma = \left(\frac{k_0 L^2}{D}\right)\left(\frac{Dc_0 \Delta H_{rxn}}{VkT_0}\right) \tag{6.6.35}$$

where the first term in brackets is the square of the Thiele modulus. Thus, we are not actually free to set the parameters ϕ and γ independently – if we wanted to set the Thiele modulus to zero and eliminate the reaction, we would have needed the second term to be infinitely large to lead to a non-zero limiting value. This is physically unrealistic.

We put in a deliberate trap in this problem in the definition of γ to bring up an important point about non-dimensionalization of a problem, especially a complicated one that requires numerical methods. This is a powerful approach, but you should be careful that your dimensionless parameters are indeed independent. If you look back to the non-isothermal CSTR problem in Section 5.7, you will see that we were careful to separate out the coefficient for the reactive contribution to the temperature equation into the Thiele modulus and Prater number, which avoids mixing different effects. Fortunately, we were able to use our numerical analysis to find the shortcoming in the problem formulation that we used here.

6.7   Implementation in MATLAB

There is no native implementation of finite differences for boundary value problems in MATLAB. Rather, you need to use the tools in this chapter to develop the system of equations, which then can be solved by using methods in this book or by built-in MATLAB solvers. If your system is linear, such as we saw in Example 6.2, then you can use the `slash` solver to solve the system of equations. Since the system can become quite large if you use a small discretization, and because finite differences lead to nice banded matrices, you should be sure to declare the matrix as `sparse` before solving it. If the system is nonlinear, such as Example 6.3, you could use `fsolve` to solve the resulting system of equations. As you can see from that example, we prefer to use our Newton–Raphson programs.

6.8   Further Reading

A good perspective on the solution of BVPs using finite differences but also other approaches can be found in

- G. H. Golub and J. M. Ortega, *Scientific Computing and Differential Equations*, Academic Press, 1992
- U. M. Ascher, R. M. M. Matheij, and R. D. Russell, *Numerical Solution of Boundary Value Problems for Ordinary Differential Equations*, SIAM, 1995
- R. J. LeVeque, *Finite Difference Methods for Ordinary and Partial Differential Equations*, SIAM, 2007

and the details of the transport phenomena used in our case studies are covered in more detail in

- R. B. Bird, W. E. Stewart, and E. N. Lightfoot, *Transport Phenomena*, Wiley, 2007.

Problems

6.1   What is the grid spacing Δx if you discretize the space $x \in [-1, 1]$ with 51 nodes?

6.2   What is the equation for c_1 for the solution of the diffusion equation $d^2c/dx^2 = 0$ with a no-flux boundary condition at the left boundary?

6.3   In a one-dimensional boundary value problem with a no-flux condition on the left boundary, what is the relationship between the fictitious node and the nodes in the domain?

6.4   Consider a coupled system of boundary value problems for x and y. If the unknowns are in a vector z in an order that preserves the band structure, what are the first five entries in z?

6.5   Consider the differential equation

$$\frac{d^2y}{dx^2} = xy$$

subject to $y(0) = 0$ and $y(1) = 1$. If this equation is solved using centered finite differences and three nodes, what is the value of interior node, y_2?

6.6 Use Taylor series expansions to derive the centered finite difference formula for the third derivative,

$$\frac{d^3y}{dx^3} = \frac{y_{i+2} - 2y_{i+1} + 2y_{i-1} - y_{i-2}}{2(\Delta x)^3}$$

6.7 The centered finite difference formulas can be written in the form

$$\frac{d^ky}{dx^k} = \frac{1}{(\Delta x)^k} \sum_{j=j_{min}}^{j_{max}} w_j y_{i+j}$$

where k is the order of the derivative and the values of j denote locations relative to node i. For this problem, we will consider the lowest-order form for the fourth derivative using the Taylor series approach.

(a) Write out the Taylor series expansion for an arbitrary number of nodes j away from node i.
(b) What is the smallest range $j \in [j_{min}, j_{max}]$ that can be used to compute the centered fourth derivative?
(c) How many equations are required to determine the weights w_j?
(d) Using the results from parts (a)–(c), develop a system of algebraic equations that would allow you to solve for the weights w_j.

6.8 Determine the residual and Jacobian needed to solve the equation $(yy')' = \sqrt{y}$ subject to $y(0) = 1$ and $y(1) = 0$ using centered finite differences and $n = 5$ nodes.

6.9 Determine the residual and Jacobian needed to solve the diffusion equation,

$$D\frac{d^2c}{dx^2} = 0$$

subject to $c(0) = 1$ on the left boundary and a second-order depletion reaction,

$$D\frac{dc}{dx} = kc^2$$

on the right boundary using centered finite differences and $n = 4$ nodes.

6.10 Consider the centered finite difference solution of the coupled system of equations

$$\frac{d^2w}{dz^2} = \frac{dx}{dz} + w$$
$$\frac{d^2x}{dz^2} = x^2w$$

using four nodes. Discretize the differential equations for w_3 and x_3 and then rewrite these equations in terms of a single variable y that interleaves the variables w and x.

6.11   Consider the coupled system of second-order differential equations

$$a'' = a^2 b$$
$$b'' = x^2 a - b^2$$

subject to $a(0) = 1$, $a'(1) = 2$, $b'(0) = 0$ and $b(1) = 1$. Define a single system of unknowns $\mathbf{y} = [a_1, b_1, a_2, b_2, \ldots]$ to solve this system by finite differences. You do not need to convert this into an autonomous system of equations. Determine the residuals (in terms of the unknowns y_i) for $n = 4$ nodes.

6.12   We can get an accurate idea of the error in centered finite differences by studying the solution to an exactly solvable problem and directly calculating the error in the finite difference representation. Consider the simple reaction–diffusion problem

$$\frac{d^2 c}{dx^2} = \mathrm{Da}\, c$$

where Da is the Dämkohler number. (We have already made the problem dimensionless for you.) The system is subject to the Dirichlet boundary conditions at $x = 0$ and $x = 1$.

Use the exact solution to this problem to calculate the error in the finite difference approximation at node i as a function of the exact concentration at that position, $c_i = c(x = x_i)$, the grid spacing, Δx, and the Dämkohler number. For which values of the Dämkohler number is the finite difference approximation exact for this problem?

6.13   In the course of solving a boundary value problem for some function $c(x)$ using centered finite differences, there is a subfunction for computing the residuals that has the following structure:

```
1  function out = problem6_13
2  n = length(c);
3  dx = 2/(n-1);
4  R(1,1) = c(1) - 1;
5  for i = 2:n-1
6      R(i,1) = c(i+1) - 2*c(i) + c(i-1) - (dx)^2*c(i)^2;
7  end
8  R(n,1) = 2*c(n-1) + 2*(dx-1)*c(n) -c(n)^2*dx^2;
9  out = R;
```

Answer the following questions about this problem.

(a)   If $x_1 = 0$, what is x_n?
(b)   What is the ordinary differential equation solved by this program?
(c)   What are the boundary conditions?
(d)   Write a subfunction that takes the vector c as its input and returns the Jacobian needed for the Newton–Raphson solution of the problem.

Computer Problems

6.14   Write a MATLAB program that solves the diffusion equation

$$\frac{d^2 y}{dx^2} = 0$$

subject to the boundary conditions $y(0) = 1$ and $dy/dx = -y^{3/2}$ at $x = 1$ using finite centered differences. Show that your numerical result agrees with the analytical solution to this nonlinear problem by plotting both results together.

6.15   Write a MATLAB program that uses centered finite differences to solve the one-dimensional diffusion equation

$$\frac{d}{dx}\left(D(x)\frac{dc}{dx}\right) = 0$$

for a spatially dependent diffusion coefficient $D(x) = D_0(1 - ax)$ and the boundary conditions $c(0) = 1$ and $c(1) = 0$. Compute the solution for $a = 0, 0.2, 0.4, 0.6$, and 0.8 using 11 nodes. Plot the concentration profiles versus position for each value of a on a single plot.

6.16   Consider the solution to the reaction diffusion equation

$$D\frac{d^2c}{dx^2} = kc$$

with boundary conditions $c(0) = 1$ and $c(L) = 0$. Write a MATLAB program that uses centered finite differences to solve this equation for $Da = kL^2/D$ values of 0.01, 0.1, 1, 10, 100, 1000. Your program should be written so that you specify an arbitrary number of nodes. This problem has an exact solution that you can find analytically. Figure out how the error in your solution, defined as

$$\text{err} = \frac{||c_{\text{numerical}} - c_{\text{exact}}||}{||c_{\text{exact}}||}$$

depends on the number of nodes, n, for $n = 5$ to $n = 200$. You should evaluate the exact solution on the finite difference grid points to compute the error. Your program should automatically make a log–log plot of the error as a function of n for each value of Da. Include an explanation of the results in your plot. It may be helpful if you plot the exact solution at the different values of Da to figure out what is going on at the higher values of Da.

6.17   Write a MATLAB code to solve the diffusion equation

$$\frac{d^2c}{dx^2} = 0$$

over the domain $x \in [0, 1]$ subject to

$$c(0) = 1, \quad \left.\frac{dc}{dx}\right|_{x=0} = kc(x = 0)$$

for the (dimensionless) reaction rate $k = 2$. Your program should plot the solution and compare the result to the exact solution of this linear problem. Include the derivation of the exact result along with a discussion of how you made the numerical solution and the plot.

6.18   In this problem, we will look at the convergence of the solution to a boundary value problem as a function of the grid spacing. Consider the reaction–diffusion equation

$$\frac{d^2c}{dx^2} = k(x)c$$

where the spatially dependent reaction term is

$$k(x) = 10[1 + \sin(\pi x)]$$

The boundary conditions for the problem are $c(-1) = 1$ and $c(1) = 0.5$.

  You should write a MATLAB code that solves this problem using centered finite differences for grid spacings $\Delta x = 1/2, 1/4, \ldots, 1/1024$. For the smallest value of $\Delta x = \Delta x_{\min}$, make a plot of the concentration versus position. We will estimate the error in the solution by comparing the value of the solution at $x = 0$ for the different values of Δx. Define the error of the solution as

$$\epsilon(\Delta x) = \frac{c(0, \Delta x)}{c(0, \Delta x_{\min})} - 1$$

In other words, lets assume that the finest grid spacing corresponds to a "perfect" solution and assess the fractional error at the other values of Δx. Make a semilog-x plot of ϵ versus Δx.

6.19   Consider the equation

$$\frac{d^2c}{dx^2} - c = 0$$

(a) Write a MATLAB program that solves the diffusion–reaction problem on the domain $x \in [0, 0.5]$ subject to $c(0) = 1$ and $c(0.5) = 0$. Your program should generate two plots: (i) the numerical solution and the exact solution for $\Delta x = 0.01$ and (ii) the norm of the difference between the numerical and exact solution for values of Δx between 0.001 and 0.5. Explain the behavior in the plot of the error.

(b) Now let's see how well the program from part (a) can solve the same problem for the unbounded domain $x \in [0, \infty)$ with $c(0) = 1$ and $c \to 0$ as $x \to \infty$. Modify your program so that you can change the upper bound. The new program should produce three plots: (i) the exact solution and numerical solution using an upper bound of $x = 0.5$ for the numerical solution, which is the same as part (a); (ii) the same plot with an upper bound $x = 15$; and (iii) a comparison of the error between the exact and numerical solution as a function of the upper bound for values between these two limits. Explain the behavior in the plot of the error. *Hint:* It may be useful to look at the results for $c(x)$ in semilogarithmic plots to understand the error.

6.20   Write a MATLAB program that uses finite differences to solve

$$\frac{d^2y}{dx^2} = \sin(yz)$$

$$\frac{d^2z}{dx^2} = \cos(yz)$$

over the interval $x \in [0, \pi]$. The boundary conditions at $x = 0$ are $y = z = 1$ and the boundary conditions at $x = \pi$ are $y = z = 0$. Your program should automatically generate a plot of y versus x and z versus x.

6.21   Write a MATLAB code to solve the coupled system

$$\frac{d^2y}{dx^2} = z; \quad \frac{d^2z}{dx^2} = y$$

on the domain $x \in [0, 1]$ subject to $y(0) = 1$, $y(1) = 2$, $z(0) = 0$, and $z(1) = 2$. Your program should use the optimal band structure for this problem and plot the solution. Include the finite difference equations used in your program.

6.22   Write a MATLAB program that solves the coupled system of equations

$$\frac{d^2x}{dz^2} = xe^{-y}$$

$$\frac{d^2y}{dz^2} = x + y^2$$

subject to $x(1) = 0.8$, $y(1) = 1$, $dx/dz = 0$ at $z = 0$, and $dy/dz = x$ at $z = 0$ using interleaved variables. Your program should generate a plot of $x(z)$ and $y(z)$.

6.23   Write a MATLAB program that uses finite difference to solve

$$\frac{d^2y}{dx^2} = x^3 y$$

on $x \in [0, 1]$ subject to $y(0) = 1$ and $y'(1) = 0$ and

$$\frac{d^2y}{dx^2} = x^3 y^2$$

on $x \in [0, 1]$ subject to $y(0) = 1$ and $y'(1) = 0$. Use the solution to the linear problem, $y'' = x^3 y$ as the initial guess. Your program should plot the solutions to both the linear and the nonlinear problems on the same graph.

6.24   This problem consists of three numerical calculations that involve the coupled reaction–diffusion system

$$D_A \frac{d^2 c_A}{dx^2} = k_1 c_A - k_2 c_B$$

$$D_B \frac{d^2 c_B}{dx^2} = -k_1 c_A + k_2 c_B$$

where D_A and D_B are the diffusion coefficients for species A and B, respectively, and k_1 and k_2 are reaction rates. The boundary conditions for these equations are

$$c_A(x = 0) = 0$$

$$c_B(x = 0) = 1$$

$$\frac{dc_A}{dx} = 0 \quad \text{at} \quad (x = 1)$$

$$\frac{dc_B}{dx} = 0 \quad \text{at} \quad (x = 1)$$

You will write three different programs for this problem, but each one builds on the previous one. It may be helpful to copy-and-paste between programs.

(a) Write a MATLAB program that computes the composition as a function of position using centered finite differences with $D_A = 2$, $D_B = 1$, $k_1 = 9$, and $k_2 = 10$. What are all of the equations that you will need in the numerical solution? For the numerical solution, use a sparse matrix, especially if you want to use a large number of grid points. Your program should automatically plot the concentrations of A and B as a function of position.

(b) While the concentrations are fixed at the left boundary, the concentrations at the right boundary are harder to predict. In this problem, develop a bracketing method that determines the value of $k_2 = k_2^*$ such that $c_A(x = 1) = c_B(x = 1)$ for $D_A = D_B = 1$ and $k_1 = 10$. In this bracketing method, you should pick some value of k_2 and use the integration method from part (a) to compute the values of c_A and c_B at the right boundary. If the ratio of concentrations c_A/c_B is not within to 1 ± 10^{-6}, your program should update the value of k_2 in some intelligent manner until it converges to within the latter tolerance. There are many ways to design this bracketing method, so this is an opportunity for you to demonstrate your creativity in the numerical method. Explain how your bracketing method works and report the value of k_2 that gives equal concentrations at $x = 1$.

(c) Use the programs from the previous two problems to explore how the value of k_2^* depends on the diffusion coefficient D_B with $D_A = 1$ and $k_1 = 10$. Copying from the programs you have already written, make a new MATLAB file that loops through the values $D_B = 0.1, 0.2, \ldots, 10$ and automatically determines the corresponding value of k_2^*. Your program should also plot k_2^* versus D_B.

6.25   This problem concerns the solution of heat transfer from a radial fin. Although this problem has a solution in terms of Bessel functions, we are going to see how to compute it numerically directly from the boundary value problem.

(a) Consider a radial fin of thickness t and length L. The fin has a thermal conductivity k and a heat transfer coefficient (from the sides) of h to a far-field temperature T_∞. The shell balance for the fin is

$$-kA_c \left. \frac{dT}{dr} \right|_r = -kA_c \left. \frac{dT}{dr} \right|_{r+dr} + hA_s(T - T_\infty)$$

where $A_c = 2\pi rt$ is the cross-sectional area of the fin and $A_s = 4\pi r \Delta r$ is the surface area on the top and the bottom of the fin in the shell volume. Complete the derivation of the differential equation and find an appropriate change of variables to cast the problem as the modified Bessel equation

$$z^2 \frac{d^2 w}{dz^2} + z \frac{dw}{dz} - z^2 w = 0$$

(b) Write a MATLAB program that uses centered finite differences to compute the fin temperature profile for a temperature $T=100\,°C$ at the hot end of the fin and a far-field temperature of $T_\infty = 25\,°C$. You can treat the cold end of the fin as insulating. For the physical parameters, you can use $k = 210\ \mathrm{W/mK}$, $h = 20\ \mathrm{W/m^2K}$, $t = 5$ cm, and $L = 25$ cm. Your program should automatically produce a plot of the temperature (in celsius) versus the distance along the fin (in centimeters).

(c) Write a MATLAB program that finds the fin size (to a tolerance of 1 mm) that produces a temperature within 1 degree of T_∞ using the physical parameters from part (b). Your program should provide the size of this fin to the screen. (It would be helpful to display more information to the screen so that we can understand your program and give you full credit!) Does this seem like a practical design?

7   Partial Differential Equations

7.1   Introduction

In this chapter we turn our attention to Partial Differential Equations (PDEs), i.e. differential equations with more than one independent variable that therefore contain partial derivatives. PDEs come in a variety of forms, and their solution properties may differ greatly depending on the form of the equation. We will focus on some specific forms that arise commonly in conservation equations and the particular methods required to solve these equations. We will begin with PDEs of the form

$$\frac{\partial y}{\partial t} = f\left(x, y, \frac{\partial y}{\partial x}, \frac{\partial^2 y}{\partial x^2}\right) \qquad (7.1.1)$$

which are a direct generalization of the ODE BVPs that we studied in the previous chapter. Here, t denotes time, $x \in [x_l, x_r]$ is a spatial domain, y is the dependent variable (typically temperature, T, or concentration, c), and f is a nonlinear function of its arguments. Specific examples of this equation are

$$\frac{\partial T}{\partial t} = \alpha \frac{\partial^2 T}{\partial x^2} - \frac{\partial(v_x T)}{\partial x} + H(t, x) \qquad (7.1.2)$$

and

$$\frac{\partial c}{\partial t} = D \frac{\partial^2 c}{\partial x^2} - \frac{\partial(v_x c)}{\partial x} + R(c, x) \qquad (7.1.3)$$

which describe unsteady state energy and mass balances in the presence of: (i) conduction/diffusion, which gives rise to the second spatial derivatives, with α and D denoting the thermal and mass diffusivitites, respectively; (ii) convection, which gives rise to the first spatial derivatives, with v_x denoting the convective velocity; and (iii) a source term H (or a reaction term R in the case of the mass balance). For an incompressible, unidirectional flow, the convective terms reduce to

$$\frac{\partial(v_x T)}{\partial x} = v_x \frac{\partial T}{\partial x} \qquad (7.1.4)$$

with an analogous simplification for Eq. (7.1.3) for concentration. These equations require the specification of two boundary conditions, i.e. conditions on y at the two boundary values of x that hold for all times. The boundary conditions could be of Dirichlet, Neumann, or Robin type, as discussed in Chapter 6. In addition, we need to specify an initial condition, i.e. the value of y at a time t_0 for all values of x. The resulting problem is called an Initial Boundary Value Problem (IBVP). While these are the general forms of the conservation equations, similar to Chapter 6 we will only consider reaction–diffusion problems here, setting the convection terms to zero.

The next class of PDEs that we will consider will be of the form

$$\nabla^2 u = \frac{\partial^2 u}{\partial x^2} + \frac{\partial^2 u}{\partial y^2} = 0 \tag{7.1.5}$$

or, more generally,

$$\frac{\partial^2 u}{\partial x^2} + \frac{\partial^2 u}{\partial y^2} = H(x, y) \tag{7.1.6}$$

These two equations are called Laplace's and Poisson's equations, respectively. Here we have switched the notation for the dependent variable to u, with x, y denoting spatial dimensions. These equations arise from steady state mass and energy balances in two spatial dimensions, where $H(x, y)$ represents a chemical or energy source. For this type of PDE, the natural auxiliary conditions are conditions for u at all four boundary points (two for x and two for y), hence we have again a (two-dimensional) BVP.

In this chapter, we will restrict ourselves to two-dimensional (2D) problems, where the independent variables are either time and one spatial dimension or two spatial dimensions. We put this restriction in place simply to reduce the size of the problems that we consider; you will see shortly that solving PDEs numerically can be a very computationally intensive task. That being said, the methods that we will discuss for the solutions of these problems can be generalized directly to more general equations, such as

$$\frac{\partial u}{\partial t} = \alpha \left(\frac{\partial^2 u}{\partial x^2} + \frac{\partial^2 u}{\partial y^2} \right) + f(x, y) \tag{7.1.7}$$

which gives rise to an IBVP in two spatial dimensions, or

$$\frac{\partial^2 u}{\partial x^2} + \frac{\partial^2 u}{\partial y^2} + \frac{\partial^2 u}{\partial z^2} = H(x, y, z) \tag{7.1.8}$$

giving a BVP in three spatial dimensions. The computational cost, as well as the bookkeeping required to construct the solution, goes up quickly with the number of independent variables.

7.2   Solution of IBVPs – Method of Lines

Let's begin with the case where the independent variables are time and a single spatial coordinate x in Eq. (7.1.1) with $x \in [x_l, x_r]$, subject to the appropriate boundary and initial conditions. Our goal is to determine the solution function $y(x, t)$. Physically, this solution captures the evolution in time of spatial temperature or concentration profiles, as they (hopefully!) approach the steady state profiles that are determined from the solution of the corresponding steady state BVP.

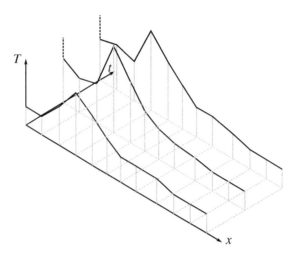

Figure 7.1 Schematic illustration of the method of lines solution to an IBVP.

7.2.1 Solution Method

The approach that we will follow to generate numerical approximations of this function will entail the following.

(1) Discretize the spatial domain in n nodes $x_1, \cdots, x_n$; this will result in n unknowns $y_1, \cdots, y_n$ which are now functions of time.
(2) Discretize the PDE in each nodal point using finite difference approximations of the spatial derivatives; this will result in a set of ODEs (in time) for the nodal variables $y_i(t)$.
(3) Integrate the resulting set of ODEs to determine $y_i(t)$.

This approach, illustrated schematically in Fig. 7.1, is called the method of lines because, in a sense, we determine the time trajectories along lines starting from the different spatial nodes.

Example 7.1 Use the method of lines to construct a numerical solution to the one-dimensional unsteady heat conduction equation

$$\frac{\partial T}{\partial t} = \alpha \frac{\partial^2 T}{\partial x^2} \tag{7.2.1}$$

subject to the boundary conditions

$$T(x_l, t) = 0, \quad T(x_r, t) = 1 \tag{7.2.2}$$

as well as an initial condition,

$$T(x, 0) = 2; x \in (x_l, x_r) \tag{7.2.3}$$

Solution

The first step is the definition of the spatial discretization step

$$\Delta x = \frac{x_r - x_l}{n - 1} \tag{7.2.4}$$

At each node i, there is now a time-dependent variable $T_i(t)$ that we seek to determine.

Using a centered finite difference approximation of the spatial derivative for an interior node, the PDE in Eq. (7.2.1) becomes

$$\frac{dT_i}{dt} = \alpha \left[\frac{T_{i+1}(t) - 2T_i(t) + T_{i-1}(t)}{\Delta x^2} \right] \tag{7.2.5}$$

Note that this is now a system of coupled ODEs for the n variables T_i, since these discrete variables no longer depend on position. The initial condition at the interior nodes is the initial condition for the original PDE (7.2.3),

$$T_i(t = 0) = 2 \tag{7.2.6}$$

The Dirichlet conditions imposed in Eqs. (7.2.2) suggest that the boundary nodes have a constant value for all times, i.e. $T_1 = 0$ and $T_n = 1$. There are two ways that we can handle this situation. One option is to substitute these fixed values into the ODEs for the $i = 2$ and $i = n - 1$ nodes. As a result, Eq. (7.2.5) for node $i = 2$ becomes

$$\frac{dT_2}{dt} = \alpha \left[\frac{T_3(t) - 2T_2(t)}{\Delta x^2} \right] \tag{7.2.7}$$

and for node $i = n - 1$ we have

$$\frac{dT_{n-1}}{dt} = \alpha \left[\frac{1 - 2T_{n-1}(t) + T_{n-2}(t)}{\Delta x^2} \right] \tag{7.2.8}$$

We then proceed by integrating the system of $n - 2$ ODEs embodied by Eqs. (7.2.5)–(7.2.8). Alternatively, and perhaps more conveniently, the following ODEs can be directly postulated for the boundary nodes,

$$\frac{dT_1}{dt} = 0, \quad T_1(0) = 0 \tag{7.2.9}$$

$$\frac{dT_n}{dt} = 0, \quad T_n(0) = 1 \tag{7.2.10}$$

which together with the ODEs for the interior nodes lead to a system of n coupled ODEs in n unknown functions $T_i(t)$. In compact form these will be

$$\frac{d\mathbf{T}}{dt} = \mathbf{f}(\mathbf{T}), \quad \mathbf{T}(0) = \mathbf{T}_0 \tag{7.2.11}$$

where $\mathbf{T}$ is the vector of the unknown variables

$$\mathbf{T} = \begin{bmatrix} T_1 \\ T_2 \\ \vdots \\ T_n \end{bmatrix}, \tag{7.2.12}$$

$$\mathbf{f} = \frac{\alpha}{\Delta x^2} \begin{bmatrix} 0 \\ (T_3 - 2T_2 + T_1) \\ \vdots \\ (T_{i+1} - 2T_i + T_{i-1}) \\ \vdots \\ (T_n - 2T_{n-1} + T_{n-2}) \\ 0 \end{bmatrix} \qquad (7.2.13)$$

and the initial condition vector is

$$\mathbf{T}_0 = \begin{bmatrix} 0 \\ 2 \\ \vdots \\ 2 \\ 1 \end{bmatrix} \qquad (7.2.14)$$

We can then integrate Eq. (7.2.11) using any of the methods we studied for systems of ordinary differential equations (e.g., RK4).

Recall that the case of Neumann or Robin boundary conditions required some special care in the solution of BVPs; specifically, we used a fictitious node beyond the actual domain to be able to use a centered finite difference approximation of the boundary condition; we also used the discretized PDE at the boundary to account for the extra unknown variable corresponding to the fictitious node. A similar approach can be used for IBVPs. For example, let's change the left boundary condition in Example 7.1 to that of an insulating wall,

$$\left. \frac{\partial T}{\partial x} \right|_{x_l} = 0 \qquad (7.2.15)$$

We introduce the fictitious node T_0 which allows us to obtain the discretized form of the boundary condition,

$$\frac{T_2(t) - T_0(t)}{2\Delta x} = 0 \qquad (7.2.16)$$

The only difference between what we saw for BVPs in Chapter 6 and here is that we have retained the explicit time dependence of the $T_i(t)$ for the PDE. The PDE in Eq. (7.2.5) at node 1 yields

$$\frac{dT_1}{dt} = \alpha \left[\frac{T_2(t) - 2T_1(t) + T_0(t)}{\Delta x^2} \right] \qquad (7.2.17)$$

The discretized boundary condition requires $T_0 = T_2$, which upon substituting into the discretized PDE yields

$$\frac{dT_1}{dt} = \alpha \left[\frac{2T_2(t) - 2T_1(t)}{\Delta x^2} \right] \qquad (7.2.18)$$

as the ODE for node 1. The initial condition for this node can be set to be the same as the rest of the interior nodes,

$$T_1(0) = 2 \qquad (7.2.19)$$

7.2.2   Numerical Stability

We now turn our attention to the last step of the method of lines that we described above, namely the integration of the ODE system that results from the spatial discretization. This can be performed using any of the methods that we described in Chapter 4. Recall though that whenever we use an explicit method, numerical stability is a concern and the time step has to be chosen carefully to guarantee that the numerical approximation will be bounded (provided that the actual solution in time is bounded). It turns out that there is an interesting twist to this requirement in the case of IBVPs solved by the method of lines.

To illustrate this, let's return to Example 7.1 with the original Dirichlet boundary conditions. The system of ODEs (7.2.11) in this case has the form

$$\frac{d\mathbf{T}}{dt} = \frac{\alpha}{\Delta x^2}
\begin{bmatrix}
0 & 0 & 0 & 0 & 0 & \cdots & 0 & 0 & 0 & 0 & 0 \\
1 & -2 & 1 & 0 & 0 & \cdots & 0 & 0 & 0 & 0 & 0 \\
0 & 1 & -2 & 1 & 0 & \cdots & 0 & 0 & 0 & 0 & 0 \\
\vdots & \vdots & \vdots & \vdots & \vdots & \vdots & \vdots & \vdots & \vdots & \vdots & \vdots \\
0 & 0 & 0 & 0 & 0 & \cdots & 0 & 1 & -2 & 1 & 0 \\
0 & 0 & 0 & 0 & 0 & \cdots & 0 & 0 & 1 & -2 & 1 \\
0 & 0 & 0 & 0 & 0 & \cdots & 0 & 0 & 0 & 0 & 0
\end{bmatrix}
\begin{bmatrix}
T_1 \\ T_2 \\ T_3 \\ \vdots \\ T_{n-2} \\ T_{n-1} \\ T_n
\end{bmatrix} \qquad (7.2.20)$$

or, in compact form,

$$\frac{d\mathbf{T}}{dt} = \frac{\alpha}{\Delta x^2} \mathbf{A}\mathbf{T} \qquad (7.2.21)$$

with the matrix $\mathbf{A}$ clearly defined in Eq. (7.2.20). Note that this is a linear system of ODEs, for which an analytical solution can be derived, along with a concrete set of conditions for numerical stability when using an explicit numerical method. The analysis below follows closely the analysis in Section 4.8.1.

Let $\lambda_1, \lambda_2, \ldots, \lambda_n$ denote the eigenvalues of the matrix $\mathbf{A}$ and $\mathbf{x}_1, \mathbf{x}_2, \ldots, \mathbf{x}_n$ the corresponding eigenvectors. (To do this carefully, we should substitute for T_1 and T_n to avoid null rows in $\mathbf{A}$, which would then have $n - 2$ eigenvalues.) Assuming that these eigenvectors are linearly independent, the $n \times n$ matrix,

$$\mathbf{X} = [\mathbf{x}_1, \mathbf{x}_2, \ldots, \mathbf{x}_n] \qquad (7.2.22)$$

is non-singular. Defining

$$\tilde{\mathbf{T}} = \mathbf{X}^{-1}\mathbf{T} \qquad (7.2.23)$$

the ODE system in Eq. (7.2.20) can then be transformed into

$$\frac{d\tilde{\mathbf{T}}}{dt} = \frac{\alpha}{\Delta x^2} \mathbf{\Lambda} \tilde{\mathbf{T}} \tag{7.2.24}$$

where

$$\mathbf{\Lambda} = \begin{bmatrix} \lambda_1 & 0 & \cdots & 0 \\ 0 & \lambda_2 & \cdots & 0 \\ \vdots & \vdots & \vdots & \vdots \\ 0 & 0 & \cdots & \lambda_n \end{bmatrix} \tag{7.2.25}$$

In other words, we obtain a decoupled set of ODEs of the form

$$\frac{d\tilde{T}_i}{dt} = \frac{\alpha \lambda_i}{\Delta x^2} \tilde{T}_i \tag{7.2.26}$$

where λ_i is the eigenvalue corresponding to node i.

As we discussed in Section 4.8.1, for these solutions to be bounded the eigenvalues must have a non-negative real part, $\mathrm{Re}(\lambda_i) \leq 0$. Furthermore, for numerical stability there is an upper positive limit in the time step h that we can use with an explicit method. For the system above, when using explicit Euler for example, this condition becomes

$$h \leq \frac{2\Delta x^2}{\alpha |\lambda^{\max}|} \tag{7.2.27}$$

or

$$\frac{h}{\Delta x^2} \leq \frac{2}{\alpha |\lambda^{\max}|} \tag{7.2.28}$$

where $\lambda^{\max}$ is the eigenvalue with the largest magnitude. Similar conditions will hold for other explicit methods.

Note that for unsteady PDEs, it is the combination of the time discretization step h and the spatial discretization step Δx which determines the stability limit. For explicit Euler, different choices of these steps should obey

$$\frac{h_2}{h_1} \approx \left(\frac{\Delta x_2}{\Delta x_1} \right)^2 \tag{7.2.29}$$

to maintain numerical stability. This means that if we decrease the spatial discretization step by half (to improve the accuracy of the finite difference approximations), the time step has to be reduced by four times to maintain numerical stability. This will of course increase the computational effort significantly. For this reason, implicit methods are often a good choice for these problems since they are unconditionally stable. As you may recall, however, their implementation is more tedious, requiring the iterative solution of a nonlinear (in general) algebraic equation system in each step of the time integration. We will illustrate these points through the following two examples.

Example 7.2 Let's return to Example 7.1 and use explicit Euler to integrate the system of equations with $\alpha = 1$ on the domain $x \in [0, 1]$.

Solution

If we want to solve Eq. (7.2.11) using explicit Euler, the formula for propagating the solution in time becomes

$$\mathbf{T}^{(k+1)} = \mathbf{T}^{(k)} + h\mathbf{f}(\mathbf{T}^{(k)}) \tag{7.2.30}$$

We first need to modify Program 4.1 to solve this problem and output a vector of values T_i at each desired time step:

```
function [xout,yout] = ivp_explicit_euler(x0,y0,h,n_out,i_out)
%   x0 = initial value for x
%   y0 = initial value for y
%   h = step size
%   i_out = frequency to output x and y to [xout, yout]
%   n_out = number of output points
%   requires the function getf(x,y)
xout = zeros(n_out+1,1); xout(1) = x0;
yout = zeros(n_out+1,length(y0)); yout(1,:) = y0;
x = x0; y = y0;
for j = 2:n_out+1
    for i = 1:i_out
        y = y + h*getf(y);
        x = x + h;
    end
    xout(j) = x;
    yout(j,:) = y;
end
```

The changes required for a system of variables are in lines 8 and 9, where we set up the output, and lines 16 and 17, where we put in each output vector. We also need to write the program for computing the function $\mathbf{f}(\mathbf{T}^{(k)})$:

```
function f = getf(y)
n = length(y);
dx = 1/(n-1);
f = zeros(n,1);
f(1) = 0;
for i = 2:n-1
    f(i) = y(i+1) - 2*y(i) + y(i-1);
end
f(n) = 0;
f = f/dx^2;
```

Since we are not passing the number of nodes to the function, we determine the value of n in line 2 by looking at the number of entries in the vector $\mathbf{y}$. We then compute the value of Δx on line 3. When we fill in the entries in $\mathbf{f}(\mathbf{T}^{(k)})$, notice that we do not multiply each time by $\alpha/\Delta x^2$. Rather, we do this multiplication once at the end of the function in line 10.

Figure 7.2 shows the trajectory of the solution for a small value of h and a larger value of h. For the small value in Fig. 7.2a, we see that the solution is stable and we get the expected path to the steady state. However, as we increase h in Fig. 7.2b, the system

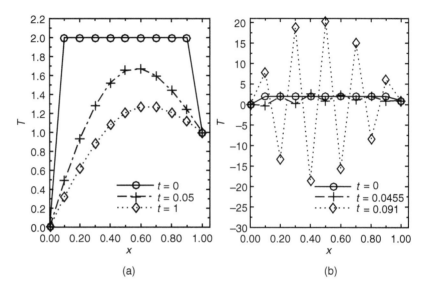

Figure 7.2 Integration of Eq. (7.2.1) by explicit Euler with (a) $h = 0.001$ and (b) $h = 0.0065$.

eventually becomes numerically unstable. For this particular value of h the instability is gradual. If we make h even larger, the solution blows up very quickly.

Example 7.3 Write a MATLAB program that uses implicit Euler and centered finite differences to solve the diffusion–reaction equation

$$\frac{\partial c}{\partial t} = D\frac{\partial^2 c}{\partial x^2} - Kc^2 \tag{7.2.31}$$

subject to $c(0, t) = 1$, $c(1, t) = 1$, and $c(x, 0) = 1$.

Solution

Recall from our discussion in Section 4.6.2 that the steady state residual for a system of equations leads to a compact way to write the implicit Euler algorithm. It is thus useful to start with the steady state solution of this problem, i.e. the solution to the equation

$$0 = D\frac{\partial^2 c}{\partial x^2} - Kc^2 \tag{7.2.32}$$

subject to the boundary conditions above. We first define n equidistant nodes in the interval $[0, 1]$. For the first node, from the boundary condition we have

$$c_1 = 1 \tag{7.2.33}$$

For the interior nodes, using a centered finite difference approximation we get

$$D\left(\frac{c_{i+1} - 2c_i + c_{i-1}}{\Delta x^2}\right) - Kc_i^2 = 0 \tag{7.2.34}$$

For the last node, again from the boundary condition, we have

$$c_n = 1 \tag{7.2.35}$$

These equations in residual form become

$$R_1^{ss} = c_1 - 1 \tag{7.2.36}$$

$$R_i^{ss} = \frac{D}{\Delta x^2}c_{i+1} - \frac{2D}{\Delta x^2}c_i - Kc_i^2 + \frac{D}{\Delta x^2}c_{i-1} \tag{7.2.37}$$

$$R_n^{ss} = c_n - 1 \tag{7.2.38}$$

A MATLAB file to compute this residual is:

```
1  function R = getR(y,paramet)
2  dx = paramet(1);
3  D = paramet(2);
4  K = paramet(3);
5  n = length(y);
6  R = zeros(n,1);
7  R(1) = y(1) - 1;
8  for i = 2:n-1
9      R(i) = D/dx^2*(y(i+1) - 2*y(i) + y(i-1)) - K*y(i)^2;
10 end
11 R(n) = y(n) - 1;
```

As usual, we also need the steady state Jacobian. The non-zero Jacobian entries on the first row are

$$J_{1,1}^{ss} = 1 \tag{7.2.39}$$

and on the last row are

$$J_{n,n}^{ss} = 1 \tag{7.2.40}$$

For the interior rows, the non-zero entries are

$$J_{i,i-1}^{ss} = \frac{D}{\Delta x^2} \tag{7.2.41}$$

$$J_{i,i}^{ss} = -\frac{2D}{\Delta x^2} - 2Kc_i \tag{7.2.42}$$

$$J_{i,i+1}^{ss} = \frac{D}{\Delta x^2} \tag{7.2.43}$$

The Jacobian is evaluated from the MATLAB function:

```
1  function J = getJ(y,paramet)
2  dx = paramet(1);
3  D = paramet(2);
4  K = paramet(3);
5  n = length(y);
6  J = zeros(n);
7  J(1,1) = 1;
8  for i = 2:n-1
9      J(i,i-1) = D/dx^2;
10     J(i,i) = -2*D/dx^2 - 2*K*y(i);
11     J(i,i+1) = D/dx^2;
12 end
13 J(n,n) = 1;
```

If we want to compute the steady state solution, we can use Newton–Raphson:

```
function [x,flag] = nonlinear_newton_raphson(x,tol,paramet)
%   x = initial guess for x
%   tol = criteria for convergence
%   requires function R = getR(x)
%   requires function J = getJ(x)
%   flag = 0 if failed, 1 if converged
%   paramet = [dx, D, K]
R = getR(x,paramet);
k = 0; %counter
kmax = 100; %max number of iterations allowed
while norm(R) > tol
    J = getJ(x,paramet); %compute Jacobian
    J = sparse(J);
    del = -J\R; %Newton Raphson
    x = x + del; %update x
    k = k + 1; %update counter
    R = getR(x,paramet);
    if k > kmax
        fprintf('Did not converge.\n')
        break
    end
end
if k > kmax || max(x) == Inf || min(x) == -Inf
    flag = 0;
else
    flag = 1;
end
```

Let's move on to the unsteady problem. For each node, we derive the ODE that describes the evolution of the variable in time. For the first node this is

$$\frac{dc_1}{dt} = 0 \tag{7.2.44}$$

For an interior node i, it is

$$\frac{dc_i}{dt} = \frac{D}{\Delta x^2}c_{i+1} - \frac{2D}{\Delta x^2}c_i - Kc_i^2 + \frac{D}{\Delta x^2}c_{i-1} \tag{7.2.45}$$

For the last node it is

$$\frac{dc_n}{dt} = 0 \tag{7.2.46}$$

The initial condition is $c_i(0) = 1$ for all i.

The implicit Euler iterative scheme for propagating the solution in time takes the form

$$R_1 = c_1^{(k+1)} - c_1^{(k)} \tag{7.2.47}$$

$$R_i = c_i^{(k+1)} - c_i^{(k)}$$
$$\qquad -h\left[\frac{D}{\Delta x^2}(c_{i+1}^{(k+1)} - 2c_i^{(k+1)} + c_{i-1}^{(k+1)}) - K(c_i^{(k+1)})^2\right] \tag{7.2.48}$$

$$R_n = c_n^{(k+1)} - c_n^{(k)} \tag{7.2.49}$$

If you look at these equations, you will see that they look remarkably similar to the steady state solution. The residual for the interior nodes can also be written in the shorthand notation as

$$\mathbf{c}^{(k+1)} - \mathbf{c}^{(k)} - h\mathbf{R}_{ss}^{(k+1)} = 0 \tag{7.2.50}$$

where $\mathbf{R}_{ss}^{(k+1)}$ refers to the steady state residual. Likewise, we can reuse our Jacobian for the steady state problem since the partial derivatives of Eq. (7.2.50) with respect to $c_i^{(k+1)}$ are simply

$$\mathbf{J} = \mathbf{I} - h\mathbf{J}_{ss}^{(k+1)} \tag{7.2.51}$$

The easiest way to combine the time integration with what we wrote for the steady state is to modify our program for implicit Euler for a system of equations (Program 4.5) to mesh with the functions we already wrote for getR and getJ:

```
1  function [xout,yout] = ivp_implicit_euler_sys(x0,y0,h,n_out,i_out,p)
2  %   x0 = initial value for x
3  %   y0 = initial value for y
4  %   h = step size
5  %   i_out = frequency to output x and y to [xout, yout]
6  %   n_out = number of output points
7  %   requires the function getR(y,yold,h) and getJ(y,h)
8  %   p = [dx, D, K]
9  xout = zeros(n_out+1,1); xout(1) = x0;
10 yout = zeros(n_out+1,length(y0)); yout(1,:) = y0;
11 x = x0; y = y0; tol = 1e-8;
12 for j = 2:n_out+1
13     for i = 1:i_out
14         yold = y;
15         R = y - yold - h*getR(y,p);
16         while norm(R) > tol
17             J = eye(length(y)) - h*getJ(y,p);
18             del = -J\R;
19             y = y + del;
20             R = y - yold - h*getR(y,p);
21         end
22         x = x + h;
23     end
24     xout(j) = x;
25     yout(j,:) = y;
26 end
```

Notice that Eq. (7.2.50) is implemented on lines 15 and 20, while Eq. (7.2.51) is implemented on line 17. Figure 7.3 shows the integration of this problem for a Thiele modulus of $\sqrt{2}$.

We close this section by noting a nice physical interpretation of the stability condition (7.2.27) in terms of transport phenomena. Recall that the quantity $\Delta x^2/\alpha$ is the time required for thermal diffusion across some distance Δx. As you reduce Δx, you are resolving finer and finer scale features in the thermal diffusion. In order to deal with how these features change in time, you need to make a smaller and smaller time step.

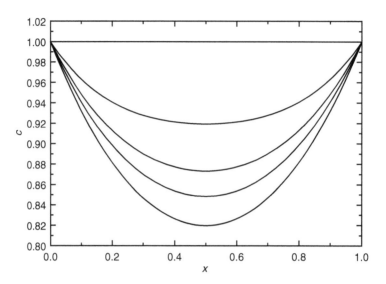

Figure 7.3 Solution to Eq. (7.2.31) using implicit Euler for $D = 1$ and $K = 2$.

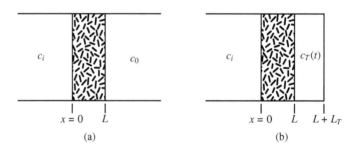

(a)   (b)

Figure 7.4 Diffusion through a membrane of thickness L connecting two reservoirs at initial concentrations c_i and c_0. (a) Both reservoirs are infinite and maintain constant concentration for all time. (b) The right reservoir is a tank of length L_T. The tank is well mixed, so the concentration inside, c_T, is a function only of time.

7.3   Case Study: Unsteady Diffusion Through a Membrane

7.3.1   Problem Statement

Consider the membrane in Fig. 7.4 separating two well-mixed fluids, each containing solute A at a fixed concentration. On the left side of the membrane the solute has a concentration c_i, while on the right side the concentration is c_0. Within the membrane itself, the initial concentration is also c_0. We want to study how the concentration changes as time progresses, and solute diffuses from the left side of the membrane to the right.

The diffusion equation for a binary, dilute system with a constant diffusion coefficient is

$$\frac{\partial c}{\partial t} = D\frac{\partial^2 c}{\partial x^2} \qquad (7.3.1)$$

At this point you might notice that the formulation of this problem is identical to determining the temperature profile due to conduction through a solid wall, or to determining the velocity profile for Couette flow between two flat plates. Although the physical mechanisms governing these three problems are different, the mathematics describing the processes are the same.

It is convenient before going further to put this problem in dimensionless form. If we choose $\theta = (c - c_o)/(c_i - c_o)$, $\xi = x/L$, and $\tau = tD/L^2$, then we get

$$\frac{\partial \theta}{\partial \tau} = \frac{\partial^2 \theta}{\partial \xi^2} \tag{7.3.2}$$

with

$$\theta(\xi, 0) = 0, \quad \theta(0, \tau) = 1, \quad \theta(1, \tau) = 0. \tag{7.3.3}$$

We are going to consider two cases. In the first case, both reservoirs are of infinite size. As a result, transport of solute from one reservoir to the other does not change the boundary concentration. This is a solvable problem via separation of variables. We just need to be a bit careful about making the boundary conditions homogeneous. To solve the problem, we define another function

$$y = \theta - (1 - \xi) \tag{7.3.4}$$

which is just subtracting off the steady state solution for θ. The transformed variable also obeys the diffusion equation

$$\frac{\partial y}{\partial \tau} = \frac{\partial^2 y}{\partial \xi^2} \tag{7.3.5}$$

with the homogeneous boundary conditions $y(0, \tau) = 0$ and $y(1, \tau) = 0$. The initial condition is now different,

$$y(\xi, 0) = -1 + \xi \tag{7.3.6}$$

The differential equation for y is solved by separation of variables to give

$$y = \sum_{n=1}^{\infty} c_n \sin(n\pi \xi) e^{-(n\pi)^2 \tau} \tag{7.3.7}$$

which you can readily verify satisfies the homogeneous Dirichlet boundary conditions at $\xi = 0$ and $\xi = 1$. From the initial condition, we have

$$\xi - 1 = \sum_{n=1}^{\infty} c_n \sin(n\pi \xi) \tag{7.3.8}$$

We can compute the coefficients

$$c_n = \frac{2}{n\pi} \tag{7.3.9}$$

following the approach we used in Section 1.2.6, integrating by parts to handle the ξ term on the left-hand side. The solution for the concentration is then

$$\theta = 1 - \xi - \sum_{n=1}^{\infty} \frac{2}{n\pi} \sin(n\pi\xi)e^{-(n\pi)^2\tau} \qquad (7.3.10)$$

We will use this solution to test out the accuracy of the numerical solution.

In the second case, we will consider a finite size reservoir on the right boundary. In this case, the transport from the left boundary will lead to an increase in the concentration on the right boundary. The steady state for this finite reservoir problem is $\theta = 1$, since eventually the reservoir on the right will be filled up from diffusion from the left one.

7.3.2 Solution

Infinite Reservoir

Let's begin with the infinite reservoir case, using a second-order centered finite difference approximation for the spatial derivative, and forward Euler integration for the time derivative. We will use n equally spaced nodes. For an interior node i, we get

$$\frac{d\theta_i}{d\tau} = \frac{\theta_{i+1}^{(k)} - 2\theta_i^{(k)} + \theta_{i-1}^{(k)}}{\Delta x^2} \qquad (7.3.11)$$

Solving the system of ODEs using explicit Euler leads to

$$\theta_i^{(k+1)} = \theta_i^{(k)} + \frac{h}{\Delta x^2}\left(\theta_{i+1}^{(k)} - 2\theta_i^{(k)} + \theta_{i-1}^{(k)}\right) \qquad (7.3.12)$$

The left and right nodes are unchanged in time, so their explicit Euler form becomes

$$\theta_1^{(k+1)} = \theta_1^{(k)} \qquad (7.3.13)$$

and

$$\theta_n^{(k+1)} = \theta_n^{(k)} \qquad (7.3.14)$$

This is the problem we solved in Example 7.2, just with different boundary conditions and initial conditions. As a result, we don't need to write any new MATLAB programs yet.

Figure 7.5 compares the result from explicit Euler to the exact solution in Eq. (7.5). For the finite differences solution, we used 51 nodes and a time step size of 10^{-4}. For very short times, the finite difference solution seems to be more accurate than evaluating the Fourier series because we need many terms in the Fourier series to accurately capture the boundary layer near the left reservoir. (There are more sophisticated analytical techniques, known as perturbation methods, that circumvent this shortcoming with Fourier series solutions.) The numerical solution does not oscillate because it is not actually a sum of sine waves. Moreover, there is an interesting issue related to the rate of propagation of information in finite differences. Recall that the concentrations at each node are connected to the previous node, so it takes a finite number of time steps before the concentration at a node can propagate to a distant node. While it seems like the numerical integration is superior to the exact solution, remember that neither approach is perfect.

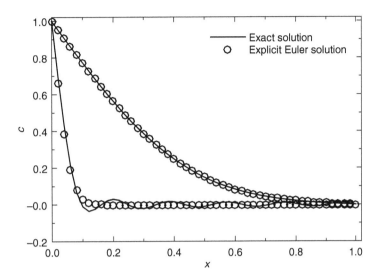

Figure 7.5 Comparison of the explicit Euler solution and the exact solution in Eq. (7.3.10) for two different times $\tau = 0.001$ and $\tau = 0.06$.

The numerical solution has multiple sources of error, including the truncation errors of finite differences and explicit Euler.

If we look at the two solutions for a longer time point, we see that the agreement is quite good. The Fourier series becomes accurate because we need only a few modes to capture most of the features of the solution; the exponential terms in Eq. (7.3.10) decay quickly with n when τ is large. So long as we use a sufficient number of nodes and a small enough time step, the truncation errors in the numerical solution are also small with respect to the resolution of the plot.

While Eq. (7.3.10) is an exact result, it requires an infinite number of terms until it gives a result with infinite precision. Obviously, when we produced Fig. 7.5, we had to truncate the sum at a finite number of terms. To see the relative difference between the two solutions, Fig. 7.6 presents the quantity $||\theta_{analytical} - \theta_{numerical}||$ as a function of the K terms that we included to compute the sum. The result is interesting. The plot of the error versus K steadily decreases for small K, but levels out for large K. This is because, when we reach large enough K, the exponential term in Eq. (7.3.10) becomes very small. As a result, when we add another term, we are below machine precision and nothing happens. The total error is then given by the error in the numerical solution. Note that if you use a smaller grid and time step size, then the plateau is reached at a lower value of error and a higher value of K.

Finite Reservoir

Now that we understand the infinite reservoir case, let's consider the case where the reservoir on the right has a finite size L_T. Returning briefly to dimensional variables, the mass flux at the right boundary of the membrane is

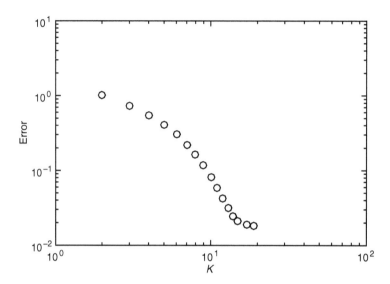

Figure 7.6 Comparison of the explicit Euler solution and the exact solution in Eq. (7.3.10) as a function of the number of terms in the Fourier series.

$$j = -D \frac{\partial c}{\partial x}\bigg|_{x=L} \tag{7.3.15}$$

Over a time step of duration Δt, we then expect the amount of mass inside the tank to increase by an amount

$$\Delta m = jA \Delta t \tag{7.3.16}$$

where A is the area of the membrane. To convert this into concentration in the tank, c_T, we should divide by the volume $V = AL_T$ of the tank,

$$\Delta c_T = \frac{j \Delta t}{L_T} \tag{7.3.17}$$

If we assume that the tank is well mixed and that there is no resistance to mass transfer to the tank (i.e., the partition coefficient is unity), then the concentration of the tank is equal to the concentration $c(1, t)$.

With the finite reservoir model, all that has changed in our system of ODEs is the concentration at node n. Putting everything together, using an explicit Euler approximation for all of the changes in time, and a backward difference equation for the flux, we have

$$c_n^{(k+1)} - c_n^{(k)} = -D \left(\frac{c_n^{(k)} - c_{n-1}^{(k)}}{\Delta x} \right) \frac{\Delta t}{L_T} \tag{7.3.18}$$

Converting to dimensionless notation gives

$$\theta_n^{(k+1)} = \theta_n^{(k)} - \epsilon \frac{h}{\Delta x} \left(\theta_n^{(k)} - \theta_{n-1}^{(k)} \right) \tag{7.3.19}$$

where $\epsilon = L/L_T$ is the size of the membrane relative to the size of the tank on the right. This is generally a small number since we think of the membrane as being thin.

To solve the finite tank problem, we just need to make small changes to our original programs. First, we need to alter the explicit Euler program to take the value of ϵ:

```
function [xout,yout] = ...
    ivp_explicit_euler_sys(x0,y0,h,n_out,i_out,eps)
% x0 = initial value for x
% y0 = initial value for y
% h = step size
% i_out = frequency to output x and y to [xout, yout]
% n_out = number of output points
% requires the function getf(x,y)
xout = zeros(n_out+1,1); xout(1) = x0;
yout = zeros(n_out+1,length(y0)); yout(1,:) = y0;
x = x0; y = y0;
for j = 2:n_out + 1
    for i = 1:i_out
        y = y + h*getf(y,eps);
        x = x + h;
    end
    xout(j,1) = x;
    yout(j,:) = y;
end
```

We then need to modify the function `getf` to account for the finite reservoir at node n:

```
function f = getf(y,eps)
n = length(y);
dx = 1/(n-1);
f = zeros(n,1);
f(1) = 0;
for i = 2:n-1
    f(i) = y(i+1) - 2*y(i) + y(i-1);
end
f = f/dx^2;
f(n) = -eps/dx*(y(n)-y(n-1));
```

The changes are in lines 9 and 10; instead of setting $f(n) = 0$, we now use Eq. (7.3.19). For simplicity, we moved the division by Δx^2 before we set the value of node n so that line 10 in our program corresponds exactly to the notation in Eq. (7.3.19).

Figure 7.7 shows how the finite size of the tank affects the solution. As we see in Fig. 7.7a, the concentration profile inside the membrane looks the same as that for an infinite reservoir at short times. This makes sense, since for short times there is no accumulation in the right tank. However, as we see in Fig. 7.7b, we start to see substantial accumulation in the right reservoir as material finally reaches the tank. This raises the concentration of the right boundary of the membrane, reducing the driving force for mass transfer. As a result, the mass transfer slows down as a function of time.

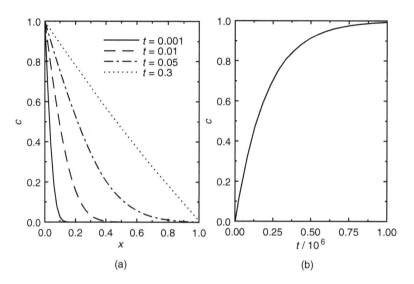

Figure 7.7 Diffusion through a membrane into a finite tank. (a) Concentration profile in the membrane as a function of time. (b) Concentration in the tank as a function of time.

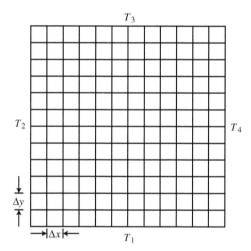

Figure 7.8 Schematic illustration of the problem of heat transfer in a square domain. The space is discretized into nodes with equal spacing Δx and Δy.

7.4   Two-Dimensional BVPs

Having considered the time-dependent solutions for one-dimensional problems, let's now turn our attention to two-dimensional (2D) BVPs of the form of the Poisson equation (7.1.6). For illustration purposes, we will focus for now on a specific example of heat conduction in a square plate in Fig. 7.8, modeled by the PDE

$$\frac{\partial^2 T}{\partial x^2} + \frac{\partial^2 T}{\partial y^2} = 0 \qquad (7.4.1)$$

subject to fixed temperature conditions on each edge,

$$T(x,0) = T_1 \tag{7.4.2}$$

$$T(0,y) = T_2 \tag{7.4.3}$$

$$T(x,1) = T_3 \tag{7.4.4}$$

$$T(1,y) = T_4 \tag{7.4.5}$$

The approach that we follow for such problems is similar to the one we followed for one-dimensional BVPs in the previous chapter. It involves the following.

(1) Discretize the spatial domain.
(2) Discretize the PDE.
(3) Write the equations for the boundary conditions.
(4) Solve the resulting algebraic equation system.

The new challenge is the required bookkeeping, which is much more tedious than that in a one-dimensional problem. Let's examine each step in more detail.

We begin with the discretization of the spatial domain, which in this case is a plane. We define the discrete points $x_1, \ldots, x_n$ and $y_1, \ldots, y_n$, with the discretization steps

$$\Delta x \equiv x_{i+1} - x_i \tag{7.4.6}$$

and

$$\Delta y \equiv y_{i+1} - y_i \tag{7.4.7}$$

Most often, for the simple problems we will consider here, we can use the same number of discretization points in both the x and the y axes, with a constant and uniform discretization step $\Delta x = \Delta y = \Delta$. This approach is illustrated in Fig. 7.8. Non-uniform discretizations become useful when there are sharp gradients in the function that need to be resolved accurately. With the above discretization we have created a grid of n^2 points (x_i, y_j) with $1 \leq i, j \leq n$. The goal is to determine the approximate temperature values at these points, $T(x_i, y_j) = T_{i,j}$, i.e. there are n^2 unknowns.

We proceed with the discretization of the PDE, starting as before from the interior nodes. Figure 7.9 shows such a node, along with the nodes in its immediate neighborhood. For the specific example of the heat conduction problem, using the centered

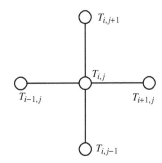

Figure 7.9 Discretization of the PDE at node i, j.

finite differences approximations for the second derivatives in T and considering that $\Delta x = \Delta y = \Delta$, we have

$$\left.\frac{\partial^2 T}{\partial x^2}\right|_{(i,j)} = \frac{T_{i+1,j} - 2T_{i,j} + T_{i-1,j}}{\Delta^2} \tag{7.4.8}$$

$$\left.\frac{\partial^2 T}{\partial y^2}\right|_{(i,j)} = \frac{T_{i,j+1} - 2T_{i,j} + T_{i,j-1}}{\Delta^2} \tag{7.4.9}$$

The discretized PDE at the node (i,j) will therefore yield the algebraic equation

$$\frac{T_{i+1,j} - 2T_{i,j} + T_{i-1,j}}{\Delta^2} + \frac{T_{i,j+1} - 2T_{i,j} + T_{i,j-1}}{\Delta^2} = 0 \tag{7.4.10}$$

This equation is valid for $2 \leq i,j \leq n - 1$, so we will have a total of $(n - 2)^2$ such equations.

Let's turn now to the boundaries of the domain. For the edges (excluding the corner nodes), the boundary conditions in Eqs. (7.4.2)–(7.4.5) translate directly in the following equations

$$T_{i,1} = T_1 \quad \text{for } 2 \leq i \leq n - 1 \tag{7.4.11}$$

$$T_{1,j} = T_2 \quad \text{for } 2 \leq j \leq n - 1 \tag{7.4.12}$$

$$T_{i,n} = T_3 \quad \text{for } 2 \leq i \leq n - 1 \tag{7.4.13}$$

$$T_{n,j} = T_4 \quad \text{for } 2 \leq j \leq n - 1 \tag{7.4.14}$$

For the corners, there is an apparent discontinuity as the two adjacent edges have different boundary conditions. This, by convention, is dealt with by using the average value of the variable for these nodes,

$$T_{1,1} = \frac{T_1 + T_2}{2} \tag{7.4.15}$$

$$T_{1,n} = \frac{T_2 + T_3}{2} \tag{7.4.16}$$

$$T_{n,1} = \frac{T_1 + T_4}{2} \tag{7.4.17}$$

$$T_{n,n} = \frac{T_3 + T_4}{2} \tag{7.4.18}$$

Note that, for this case of Dirichlet boundary conditions, the corner nodes are not involved in any of the other equations, so their specification does not affect the solution of the remaining variables. This is not the case when we have Neumann boundary conditions, since the use of the fictitious nodes will couple the corner node equations to those of the edges, which themselves are coupled to those of the interior nodes. Also, in the case where one side of the corner has Dirichlet boundary conditions and the other side has Neumann or Robin boundary conditions, the Dirichlet boundary condition takes precedence.

We have now generated n^2 equations needed to determine the n^2 unknowns. The resulting linear system of equations (or nonlinear system in the more general case) will have to be solved using the methods that we have discussed previously.

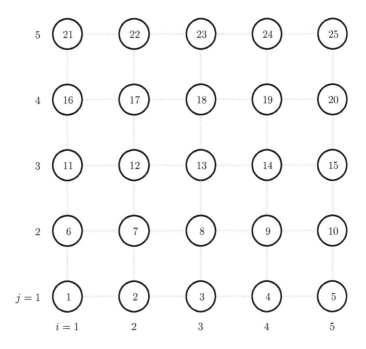

Figure 7.10 Conversion from local indices to a global index.

We are now in a situation similar to what we saw in Section 6.5 for coupled boundary value problems, since we have two indices i and j that are analogous to the indices for the two independent variables (e.g., c_i and T_i) for a coupled system. Similar to what we did for the coupled systems, we want to pick a good global variable that endows the Jacobian with a nice, banded structure. To accomplish this goal we use a global index for the unknowns $k = 1, 2, \ldots, n^2$ defined as

$$k = i + (j - 1)n \qquad (7.4.19)$$

As we can see in Fig. 7.10, the number of adjacent nodes to a point k now involves $k - n$, $k - 1$, $k + 1$ and $k + n$. The discretized PDE at the interior node k becomes

$$\frac{T_{k+1} - 2T_k + T_{k-1}}{\Delta^2} + \frac{T_{k+n} - 2T_k + T_{k-n}}{\Delta^2} = 0 \qquad (7.4.20)$$

The equations on the boundary nodes can also be converted to this global index, which we will see shortly in the context of an example. In the problem we are discussing here, the original PDE is linear, so the end result of the discretization is a linear system of n^2 equations of the form

$$\mathbf{AT} = \mathbf{b} \qquad (7.4.21)$$

where $\mathbf{A}$ is a banded matrix with $p = n$, $q = n$, and a bandwidth of $2n + 1$.

If we use Gauss elimination with a banded matrix solver, we know that the number of operations scale with Np^2, where N is the size of the matrix. In our case, this gives a scaling of n^4. Without accounting for the banded structure of the matrix, the operations scale with n^6, which is even worse! Clearly, there is a heavy computational penalty that we have to pay as we increase the spatial resolution of the domain. This makes the use of iterative solution methods preferable for these problems, although we could also improve the solution by adopting a linear solver that is designed for our particular band structure.

One additional thing to note about the matrix $\mathbf{A}$ is that it is sparse. Although there are n^4 entries, only around n^2 of them are non-zero – the "density" of non-zero entries in the matrix goes to zero as $n \to \infty$. In practice, such matrices are usually stored in a sparse form to save memory, which can be very important. For example, if you wanted to use a 100×100 grid and double precision (8 bytes per number), you would have to store 8×10^8 bytes = 762 Mb of data. (Note that the conversion to megabytes may not be what you would expect naively from the units.) In contrast, a sparse matrix would have to store 8×10^4 bytes = 78 kb of data. Not only is the storage much lower, but the time to write the matrix is much faster.

Let's see how all of this comes together in a non-trivial (but still linear) example problem.

Example 7.4   Write a MATLAB program to solve

$$\frac{\partial^2 T}{\partial x^2} + \frac{\partial^2 T}{\partial y^2} = 0 \tag{7.4.22}$$

in a square boundary subject to $\mathbf{n} \cdot \nabla T = 0$ on all of the sides except the top, where $T = 0$. There is a hot spot in the center corresponding to $T(0.5, 0.5) = 1$. Note that we have already converted this problem into dimensionless form.

Solution
We have already worked out the interior node equation in Eq. (7.4.20). This equation will be valid for all of the nodes $2 \leq i, j \leq n - 1$ except for the central node. We can write a short MATLAB program to handle the interior:

```
1  function A = interior(A,n)
2  for i = 2:n-1
3      for j = 2:n-1
4          k = i + (j-1)*n;
5          A(k,k) = -4;
6          A(k,k+1) = 1;
7          A(k,k-1) = 1;
8          A(k,k+n) = 1;
9          A(k,k-n) = 1;
10     end
11 end
```

This program, like several that follow, modifies the matrix $\mathbf{A}$ to account for a new part of the problem. Thus, it takes $\mathbf{A}$ as an input and returns it as an output. Note that we use nested **for** loops to handle the global counter k.

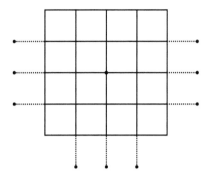

Figure 7.11 Discretization with the location of the central node and examples of the fictitious nodes on the boundaries.

When we use the above program for the interior nodes, we are writing the wrong equation for the hot spot, so we need to fix this row in **A**. As we can see in Fig. 7.11, we need to use an odd number of grid points n in each direction if we want to have a node at the center. This node will be at the middle of the local index i,

$$i = \frac{n+1}{2} \tag{7.4.23}$$

and the local index j,

$$j = \frac{n+1}{2} \tag{7.4.24}$$

The global index is

$$k = i + (j-1)n \tag{7.4.25}$$

so the center node is at

$$k_{center} = \frac{n+1}{2} + \left(\frac{n+1}{2} - 1\right)n \tag{7.4.26}$$

and we have

$$T_{k,center} = 1 \tag{7.4.27}$$

We can implement these equations in a second MATLAB file:

```
1  function [A,b] = hotspot(A,b,n)
2  i = (n+1)/2; j = i;
3  midpt = i + (j-1)*n;
4  insert_eqn = zeros(1,n^2);
5  insert_eqn(1,midpt) = 1;
6  A(midpt,:) = insert_eqn;
7  b(midpt,1) = 1;
```

This program computes the location of the mid-point, and then sets up a new row for **A** called insert_eqn that is all zeros except for the diagonal entry. We also modify the vector **b** to have the intensity of the hot spot.

For the top nodes, $k = n(n-1)+2,\ldots,n^2-1$, we have

$$T_k = 0 \tag{7.4.28}$$

which is identical to what we saw in our representative example. For the remaining boundaries, we need to make fictitious nodes as indicated in Fig. 7.11 For the bottom nodes $k = 2,\ldots,n-1$, we need to use a fictitious node T_{k-n}

$$\frac{T_{k+n} - T_{k-n}}{2\Delta x} = 0 \tag{7.4.29}$$

which gives us

$$T_{k+n} = T_{k-n} \tag{7.4.30}$$

We then rewrite Eq. (7.4.20) as

$$2T_{k+n} + T_{k+1} - 4T_k + T_{k-1} = 0 \tag{7.4.31}$$

For the left nodes, the fictitious node condition is very similar,

$$T_{k+1} = T_{k-1} \tag{7.4.32}$$

where T_{k-1} is the fictitious node, which gives us

$$T_{k+n} + 2T_{k+1} - 4T_k + T_{k-n} = 0 \tag{7.4.33}$$

For the right nodes, the fictitious node condition is the same as Eq. (7.4.32), except that T_{k+1} is now the fictitious node. As a result, we get

$$T_{k+n} + -4T_k + 2T_{k-1} + T_{k-n} = 0 \tag{7.4.34}$$

All of the boundary nodes are implemented in the following program:

```
1  function A = edges(A,n)
2  for k = 2:n-1 %bottom
3      A(k,k)   = -4;
4      A(k,k+1) = 1;
5      A(k,k-1) = 1;
6      A(k,k+n) = 2;
7  end
8  for k = n*(n-1)+2:n^2-1 %top
9      A(k,k)   = 1;
10 end
11 for i = 1:n-2 %left
12     k = n*i+1;
13     A(k,k)   = -4;
14     A(k,k+1) = 2;
15     A(k,k+n) = 1;
16     A(k,k-n) = 1;
17 end
18 for i = 2:n-1 %right
19     k = i*n;
20     A(k,k)   = -4;
21     A(k,k-1) = 2;
22     A(k,k+n) = 1;
23     A(k,k-n) = 1;
24 end
```

While it is possible to write the left nodes in terms of the global index k, it is simpler to just increment over the interior rows $i = 1, \ldots, n-2$ and then use $k = ni + 1$. You could also use $i = 2, \ldots, n-1$ and $k = n(i-1) + 1$. Similar to the left nodes, it is easiest to count through the right nodes using two counters, $i = 2, \ldots, n-1$ and then $k = ni$.

We now need to handle the corners. For the upper corners, the Dirichlet condition takes precedence and we have

$$T_{n(n-1)+1} = 0 \tag{7.4.35}$$

and

$$T_{n^2} = 0 \tag{7.4.36}$$

For the lower corners, we have no-flux conditions. If you look at the direction of the normal vector to the lower left corner ($k = 1$) in Fig. 7.11, you can see that we need to consider both

$$\frac{\partial T}{\partial x} = \frac{T_2 - T_1}{\Delta x} \tag{7.4.37}$$

and

$$\frac{\partial T}{\partial y} = \frac{T_{n+1} - T_1}{\Delta y} \tag{7.4.38}$$

In the latter, we have used forward finite differences for convenience; a centered finite difference solution will give us the same result in the end. The no-flux condition is

$$\mathbf{n} \cdot \nabla T = n_x \left(\frac{T_2 - T_1}{\Delta x} \right) + n_y \left(\frac{T_{n+1} - T_1}{\Delta y} \right) = 0 \tag{7.4.39}$$

Since the normal vectors $n_x = n_y$ for our problem, we simply get

$$T_{n+1} + T_2 - 2T_1 = 0 \tag{7.4.40}$$

The same logic applies to the lower right corner ($k = n$), where we have

$$T_{2n} + T_{n-1} - 2T_n = 0 \tag{7.4.41}$$

The corners are thus implemented by again modifying $\mathbf{A}$:

```
function A = corners(A,n)
A(1,1) = -2; A(1,2) = 1; A(1,n+1) = 1; %lower left
A(n,n) = -2; A(n,n-1) = 1; A(n,2*n) = 1; %lower right
A(n*(n-1)+1,n*(n-1)+1) = 1; %upper left
A(n^2,n^2) = 1; %upper right
```

You may have noticed that the programs `interior`, `edges`, and `corner` only modify the matrix $\mathbf{A}$, while the program `hotspot` modifies $\mathbf{A}$ and $\mathbf{b}$. The reason we only modify $\mathbf{A}$ in the first three programs is because these equations are homogeneous. If we have inhomogeneous boundary conditions, such as a non-zero temperature at one of the boundaries, then we would need to modify both $\mathbf{A}$ and $\mathbf{b}$ there as well. We will see this change in Section 7.5.

Once we have all of the equations, we just need to use a technique from Chapter 2 to solve the resulting system. The program below solves the problem using a banded solver in MATLAB:

```
function T = pde_heat_source(n)
A = zeros(n^2);
b = zeros(n^2,1);
A = interior(A,n);
[A,b] = hotspot(A,b,n);
A = edges(A,n);
A = corners(A,n);
A = sparse(A);
T = A\b;
```

This is a very compact program – it sets up the space for **A** and **b**, fills them in using the programs that we already wrote, and then solves the problem. This modular approach to the program makes modifications very easy, since we have compartmentalized all of the boundary conditions and the governing PDE into their own subfunctions. One very important point to note is that the program takes in the number of nodes on an edge, n, so the matrix and vector are of size n^2.

The output of our main program is a vector of values of the temperature at each node. To make plots, we should unpack the vector in a manner similar to what we did for coupled BVPs in Section 6.6:

```
function Tplot = unpack(T,n)
Tplot = zeros(n);
for i = 1:n
    row_start = n*(i-1)+1;
    row_end = n*i;
    Tplot(i,:) = T(row_start:row_end);
end
```

Figure 7.12 shows the solution to the problem. Visualizing these solutions is sometimes tricky, since they are functions of two variables. This particular projection was chosen so that you can easily see the Dirichlet boundary condition at the upper boundary and how the heat diffuses from the hot spot down to the sink on the upper boundary. On the other walls, the no-flux condition leads to a relatively boring result, where the temperature profile is relatively flat in the rest of the domain. We encourage you to play around with the solution to this problem, in particular to see how the location of the hot spot affects the solution. It is easy to move the hot spot around by just changing the variable k_hot in the program. You can also add additional hot spots relatively easily by adding additional subfunctions like hotspot that change other nodes from normal interior nodes to hotspots. For example, Fig. 7.13 shows two hot spots, one at half the intensity of the second. Clearly, we can construct some very interesting temperature profiles from this program!

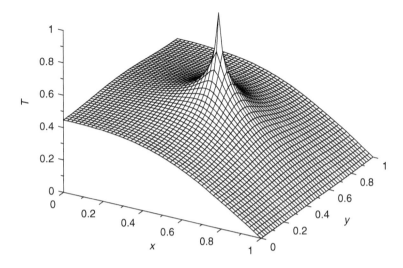

Figure 7.12 Solution to the heat equation with a source at the center.

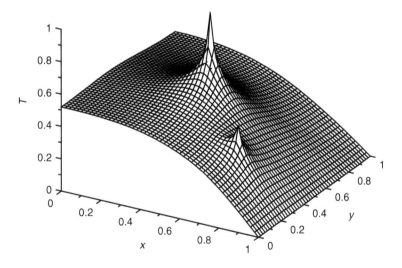

Figure 7.13 Solution to the heat equation with two sources, one at full intensity and the second at half intensity.

7.5 Case Study: 2D Reaction–Diffusion

7.5.1 Problem Statement

In this last case study, we will consider a basic question about reaction–diffusion problems. We will look at the solution to the steady state reaction–diffusion equation

$$D\nabla^2 c = R(c) \tag{7.5.1}$$

in a box of size $L \times L$, where $R(c)$ is the rate of consumption of the reactant. We will consider either a first-order reaction,

$$R = k_1 c \tag{7.5.2}$$

or a second-order reaction

$$R = k_2 c^2 \tag{7.5.3}$$

The boundaries are fixed at the bulk concentration c_0.

Our goal is to determine how the minimum concentration inside the box depends on the Thiele modulus for these two reaction models – which reaction type gives us the smaller value of the concentration?

To answer this question, we make the equations dimensionless with a length scale L and concentration c_0. For the linear problem, this gives

$$\frac{d^2 c}{dx^2} + \frac{d^2 c}{dy^2} - \phi^2 c = 0 \tag{7.5.4}$$

with the Thiele modulus

$$\phi = \sqrt{\frac{k_1 L^2}{D}} \tag{7.5.5}$$

For the nonlinear case, the same dimensionless form gives

$$\frac{d^2 c}{dx^2} + \frac{d^2 c}{dy^2} - \phi^2 c^2 = 0 \tag{7.5.6}$$

with

$$\phi = \sqrt{\frac{k_2 c_0 L^2}{D}} \tag{7.5.7}$$

In both cases, the Thiele modulus measures the relative rate of reaction to diffusion.

7.5.2   Solution

We will discretize the domain into an $n \times n$ grid, using an equally spaced grid where $\Delta x = \Delta y = \Delta$. For the linear problem, the interior nodes are of the form

$$\frac{c_{i+1,j} - 2c_{i,j} + c_{i-1,j}}{\Delta^2} + \frac{c_{i,j+1} - 2c_{i,j} + c_{i,j-1}}{\Delta^2} - \phi^2 c_{i,j} = 0 \tag{7.5.8}$$

while the residual for the nonlinear problem is

$$\frac{c_{i+1,j} - 2c_{i,j} + c_{i-1,j}}{\Delta^2} + \frac{c_{i,j+1} - 2c_{i,j} + c_{i,j-1}}{\Delta^2} - \phi^2 c_{i,j}^2 = 0 \tag{7.5.9}$$

In both cases, it proves convenient to multiply through by the grid spacing. If we also switch to the global variable $k = i + (j - 1)n$ the linear problem becomes

$$c_{k+1} - (4 + \phi^2 \Delta^2)c_k + c_{k-1} + c_{k+n} + c_{k-n} = 0 \tag{7.5.10}$$

while the nonlinear problem becomes

$$c_{k+1} - 4c_k + c_{k-1} + c_{k+n} + c_{k-n} - \phi^2 \Delta^2 c_k^2 = 0 \qquad (7.5.11)$$

For both the linear and nonlinear case, the boundary conditions are $c_i = 1$ for all of the nodes. In the linear case, we want to write these equations as

$$c_k = 1, \quad k \in \text{boundary} \qquad (7.5.12)$$

since the problem we are solving is of the form

$$\mathbf{Ac} = \mathbf{b} \qquad (7.5.13)$$

In contrast, for the nonlinear problem, we are solving a nonlinear algebraic problem of the form

$$\mathbf{R(c)} = \mathbf{0} \qquad (7.5.14)$$

so we want to write the boundary equations as

$$c_k - 1 = 0, \quad k \in \text{boundary} \qquad (7.5.15)$$

We already have enough information to write the program to solve the linear reaction case for a given value of the Thiele modulus. Let's follow the approach we used in Example 7.4 and write a simple main file:

```
1  function c = pde_reaction_linear(n,phi2)
2  A = zeros(n^2);
3  b = zeros(n^2,1);
4  A = interior(A,n,phi2);
5  [A,b] = edges(A,b,n);
6  [A,b] = corners(A,b,n);
7  A = sparse(A);
8  c = A\b;
```

We then have subfunctions for the interior nodes:

```
1   function A = interior(A,n,phi2)
2   dx = 1/(n-1);
3   for i = 2:n-1
4       for j = 2:n-1
5           k = i + (j-1)*n;
6           A(k,k) = -4 - phi2*dx^2;
7           A(k,k+1) = 1;
8           A(k,k-1) = 1;
9           A(k,k+n) = 1;
10          A(k,k-n) = 1;
11      end
12  end
```

the edges:

```
1   function [A,b] = edges(A,b,n)
2   for k = 2:n-1 %bottom
3       A(k,k) = 1;
```

```
4        b(k) = 1;
5    end
6    for k = n*(n-1)+2:n^2-1 %top
7        A(k,k) = 1;
8        b(k,1) = 1;
9    end
10   for i = 1:n-2 %left
11       k = n*i+1;
12       A(k,k) = 1;
13       b(k) = 1;
14   end
15   for i = 2:n-1 %right
16       k = i*n;
17       A(k,k) = 1;
18       b(k) = 1;
19   end
```

and the corners:

```
1    function [A,b] = corners(A,b,n)
2    A(1,1) = 1; b(1) = 1; %lower left
3    A(n,n) = 1; b(n) = 1; %lower right
4    A(n*(n-1)+1,n*(n-1)+1) = 1; b(n*(n-1)+1) = 1; %upper left
5    A(n^2,n^2) = 1; b(n^2) = 1; %upper right
```

Notice that our boundary conditions are inhomogeneous, so we have to modify both the vector **b** and the matrix **A** in the boundary condition functions. The Thiele modulus appears only in the interior node equations, so we only send it to `interior`.

Since this is a linear problem, there is nothing interesting that we need to do – just use a direct solver and get the result. The problem can get larger very quickly. For example, if we use $\Delta x = 0.025$, we get the nice smooth profile in Fig. 7.14. This is already a

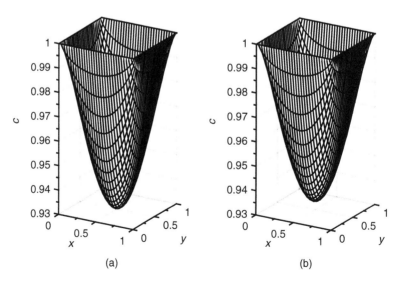

(a)   (b)

Figure 7.14 Comparison of the concentration profile for the (a) linear and (b) nonlinear reaction.

reasonably large problem with $41^2 = 1681$ unknowns. It is already time to start using a banded solver! However, if we wanted to go to a very large system with an even more fine discretization, it might make sense to switch to an iterative solver and use a continuation method for the initial guess, where we use the converged solution for the previous value of the Thiele modulus as the guess for the next value.

For the nonlinear problem, we first write down the residual equations:

```
1   function R = getR(c,paramet)
2   n = paramet(1);
3   phi2 = paramet(2);
4   R = zeros(n^2,1);
5   dx = 1/(n-1);
6   d = phi2*dx^2;
7   for i = 2:n-1 %interior nodes
8       for j = 2:n-1
9           k = (j-1)*n+i;
10          R(k) = c(k+1)+c(k-1)+c(k+n)+c(k-n)-4*c(k)-d*c(k)^2;
11      end
12  end
13  for k = 1:n
14      R(k) = c(k)-1; %bottom
15  end
16  for k = (n-1)*n+1:n^2
17      R(k) = c(k)-1;   %top
18  end
19  for j = 2:n-1
20      k = n*(j-1)+1;
21      R(k) = c(k)-1;   %left
22  end
23  for j = 2:n-1
24      k = j*n;
25      R(k) = c(k)-1;   %right
26  end
```

Note that we also could have used the procedure from the linear problem, breaking up the residual calculation into the interior and the boundary nodes. We chose here to keep all of the calculations in a single file. Since each row of the residual is a single equation, the file is still relatively compact.

We also need the entries in the Jacobian. For the boundary nodes, we simply get

$$J_{k,k} = 1, \quad k \in \text{boundary} \tag{7.5.16}$$

For the interior nodes, there are five non-zero entries from the differentiation of Eq. (7.5.11),

$$J_{k,k-n} = 1 \tag{7.5.17}$$

$$J_{k,k-1} = 1 \tag{7.5.18}$$

$$J_{k,k} = -4 - 2\phi^2 \Delta^2 c_k \tag{7.5.19}$$

$$J_{k,k+1} = 1 \tag{7.5.20}$$

$$J_{k,k+n} = 1 \tag{7.5.21}$$

We can code the Jacobian in a manner similar to the residual:

```
1   function J = getJ(c,paramet)
2   n = paramet(1);
3   phi2 = paramet(2);
4   dx = 1/(n-1);
5   d = 2*phi2*dx^2;
6   J = zeros(n^2);
7   for i = 2:n-1 %interior nodes
8       for j = 2:n-1
9           k = (j-1)*n+i;
10          J(k,k+1) = 1;
11          J(k,k-1) = 1;
12          J(k,k+n) = 1;
13          J(k,k-n) = 1;
14          J(k,k) = -4 -d*c(k);
15      end
16  end
17  for k = 1:n
18      J(k,k) = 1; %bottom
19  end
20  for k = (n-1)*n+1:n^2
21      J(k,k) = 1;   %top
22  end
23  for j = 2:n-1
24      k = n*(j-1)+1;
25      J(k,k) = 1;   %left
26  end
27  for j = 2:n-1
28      k = j*n;
29      J(k,k) = 1;   %right
30  end
```

As was the case in the residual, we chose to write the Jacobian as a single program rather than breaking it up into interior nodes and boundary nodes.

For a given value of the Thiele modulus, we can solve the problem using Newton–Raphson:

```
1   function [x,flag] = nonlinear_newton_raphson(x,tol,paramet)
2   %  x = initial guess for x
3   %  tol = criteria for convergence
4   %  requires function R = getR(x)
5   %  requires function J = getJ(x)
6   %  flag = 0 if failed, 1 if converged
7   %  n and Da are the paramet
8   R = getR(x,paramet);
9   k = 0; %counter
10  kmax = 100; %max number of iterations allowed
11  while norm(R) > tol
12      J = getJ(x,paramet); %compute Jacobian
13      J = sparse(J);
14      del = -J\R; %Newton Raphson
15      x = x + del; %update x
16      k = k + 1; %update counter
```

```
17        R = getR(x,paramet);
18        if k > kmax
19            fprintf('Did not converge.\n')
20            break
21        end
22    end
23    if k > kmax || max(x) == Inf || min(x) == -Inf
24        flag = 0;
25    else
26        flag = 1;
27    end
```

Since we are using an iterative method, we need to have an initial guess. The easiest approach is to use a zero-order continuation method that starts from the case $\phi = 0$, which is the no-reaction case. Here, we know that the solution is $c = 1$ everywhere. We can then make a small step in ϕ, corresponding to a weak reaction, and use the no-reaction solution as the initial guess. After converging that solution, we will use it as the initial guess for the next iteration.

If we just want to get a solution at a single value of ϕ^2, we can use a program like the following:

```
1   function [x,flag] = pde_continue(phi2_max,n,nsteps)
2   del_phi2 = phi2_max/(nsteps+1);
3   x = ones(n^2,1);
4   phi2 = 0;
5   tol = 1e-8;
6   for i = 1:nsteps+1 %continue up to the desired answer
7       [x,flag] = nonlinear_newton_raphson(x,tol,[n,phi2]);
8       phi2 = phi2 + del_phi2;
9   end
10  [x,flag] = nonlinear_newton_raphson(x,tol,[n,phi2]);
```

The **for** loop of this program does zero-order continuation up to a distance $\Delta\phi^2$ away from the desired value of the square of the Thiele modulus. We then compute the concentration field at the desired value of the square of the Thiele modulus in line 10. While this program will work to compute a single solution for the concentration field, it is not particularly efficient if we want to build up a plot of the concentration versus Thiele modulus since we throw away all of the intermediate calculations.

Before we proceed to answer our original question, it is worthwhile to take a look at the form of the solution for the linear and nonlinear reaction. We can do this by using the programs listed above, along with the program `unpack` from Example 7.4 to convert the vector solution into a matrix for easy plotting. Figure 7.14 shows the concentration profiles for $\phi = 1$ in both cases. If we did not tell you which profile was the linear or nonlinear reaction, would you be able to figure it out from the plots? This seems rather unlikely.

To determine which type of reaction leads to a lower minimum concentration, Figure 7.15 plots the concentration at the center of the domain as a function of the Thiele modulus. From the symmetry of the problem, the minimum value of the concentration will

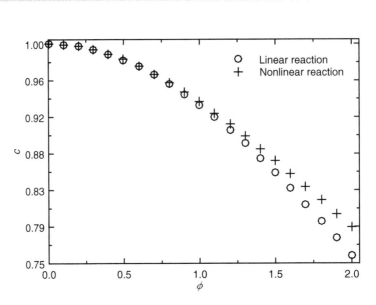

Figure 7.15 Concentration at the center of the domain as a function of Thiele modulus.

be at the center of the domain, so this value is a good metric for the extent of reaction. We already saw how to find the mid-point of the domain in Example 7.4, so we just took advantage of the algorithm in `hotspot` from that example to output the minimum concentration. For small values of the Thiele modulus, the difference between the linear and nonlinear reactions is negligible. As we increase the rate of reaction, it is clear that the linear reaction leads to the lowest concentration in the center of the domain. The reason for this behavior is that we have selected equal values of the Thiele modulus. Thus, we set the reaction rate in the linear case, k_1, equal to the maximum possible reaction rate $k_2 c_0$ in the nonlinear case. Since the concentration inside the domain is always smaller than that at the boundary, the second-order reaction rate $k_2 c^2 = k_1 c(c/c_0)$ is always smaller than the first-order reaction rate $k_1 c$. As a result, the concentration is lower in the first-order reaction case.

7.6   Implementation in MATLAB

From the standpoint of solving PDEs with finite differences, the built-in tools that are available in MATLAB are similar to what we saw in previous chapters. For two-dimensional BVPs, such as the case study in Section 7.5, you can use the `slash` or `fsolve` functions to solve the resulting system of linear or nonlinear algebraic equations. For initial-boundary value problems, such as the case study in Section 7.3, you could use `ode45` or a similar solver to integrate the system of equations produced by finite difference discretization of the spatial domain.

MATLAB also has a graphical user interface called `pdetool` for solving partial differential equations using the finite element method.

7.7   Further Reading

A very good text on the analysis of PDEs is

- R. Haberman, *Elementary Applied Differential Equations*, Prentice-Hall, 1998.

Numerical solution of PDEs using finite differences is covered well in

- G. D. Smith, *Numerical Solution of Partial Differential Equations: Finite Difference Methods*, Oxford University Press, 1978
- M. A. Celia and W. G. Gray, *Numerical Methods for Differential Equations*, Prentice-Hall, 1992
- C. F. Gerald and P. O. Wheatley, *Applied Numerical Analysis*, McGraw Hill, 1992
- J. D. Hoffman and S. Frankel, *Numerical Methods for Engineers and Scientists*, McGraw Hill, 2001
- R. J. LeVeque, *Finite Difference Methods for Ordinary and Partial Differential Equations*, SIAM, 2007.

There are numerous books on more advanced method for solving PDEs, e.g. spectral methods and finite element methods. Two good examples, one from the mathematics literature and the other from the chemical engineering one, are

- G. Strang and G. Fix, *An Analysis of the Finite Element Method*, SIAM, 2008
- C. Pozrikidis, *Introduction to Finite and Spectral Element Methods Using MATLAB*, CRC Press, 2014

and an excellent book on solving PDEs via spectral methods in MATLAB is

- L. N. Trefethen, *Spectral Methods in MATLAB*, SIAM, 2000.

Problems

7.1   Write the ordinary differential equation for an interior node produced by the method of lines for the PDE

$$\frac{\partial y}{\partial t} = \frac{\partial^2 y}{\partial x^2} - xy$$

7.2   How many non-zero entries are in the matrix required to solve the two-dimensional diffusion equation $\nabla^2 T = 0$ with n nodes in each direction and Dirichlet conditions on the boundaries? In the solution, include all of the boundary nodes in the system of equations.

7.3   If it takes 2 seconds to solve a two-dimensional, diffusion equation $\nabla^2 T = 0$ with 20 nodes in x and y using banded Gauss elimination, how long does it take for 40 nodes in x and y?

7.4   Consider the solution to the unsteady diffusion equation

$$\frac{\partial c}{\partial t} = D\frac{\partial^2 c}{\partial x^2}$$

by the method of lines and centered finite differences with a spacing Δx. The initial condition is $c(x,0) = 0$ and the left boundary condition satisfies no-flux. At the start of the process, concentration at the boundary at $x = L$ is instantaneously increased to some value c_A. Write the ODEs and initial conditions for (i) the left node, (ii) the interior nodes, and (iii) the right node. If you solve this problem with implicit Euler, what numerical method should you use to compute the concentrations $c_i^{(k+1)}$ from the values of $c_i^{(k)}$?

7.5 Consider the method of lines solutions of the unsteady reaction–diffusion equation with a second-order bulk reaction,

$$\frac{\partial c}{\partial t} = D\frac{\partial^2 c}{\partial x^2} - kc^2$$

with an initial concentration $c = 0$ and boundary conditions $c(0) = 1$ and $\partial c/\partial x = 0$ at $x = L$. Answer the following questions.

(a) If we discretize with centered finite differences and use an explicit Euler equation, what are the equations we need to solve to compute $c_i^{(k)}$ at each node i and time step k?

(b) If we instead use implicit Euler, what are the residuals and Jacobian for the nonlinear system?

7.6 Consider the method of lines solution to the unsteady diffusion equation

$$\frac{\partial c}{\partial t} = D\frac{\partial^2 c}{\partial x^2}$$

subject to an initial condition $c = 1$ and first-order depletion reactions on each boundary, $\mathbf{n} \cdot \mathbf{J} = -kc$, where $\mathbf{n}$ is an outward pointing normal vector and $\mathbf{J} = -D\nabla c$ is the flux. Determine the system of ordinary differential equations that you would need to solve using the method of lines and centered finite differences.

7.7 We would like to use centered finite differences and the method of lines to solve the unsteady diffusion–reaction problem

$$\frac{\partial c}{\partial t} = \frac{\partial^2 c}{\partial x^2} - kc$$

subject to no-flux on the left boundaries, $\partial c/\partial x = 0$ at $x = 0$, a constant concentration on the right boundary, $c(1, t) = 1$, and an initial concentration $c(x, 0) = 1$. We will use $n = 3$ nodes for this problem and a reaction rate $k = 2$.

(a) Convert the PDE into a system of coupled ODEs governing the concentrations at each node, c_1, c_2, and c_3. Remember to state the initial conditions.

(b) Let us do one time step of size $h = 1/4$ using implicit Euler. Write down the system of two equations you need to solve to compute c_1 and c_2 after the first time step. You do not need to compute c_1 or c_2.

7.8 We want to use the method of lines to solve the linear reaction–diffusion equation

$$\frac{\partial c}{\partial t} = D\frac{\partial^2 c}{\partial x^2} - kc$$

subject to the boundary condition

$$\frac{\partial c}{\partial x} = 0$$

at $x = 0$ and $c(x = 1, t) = 1$. The initial condition is $c(x, 0) = 2$. We will set $D = 4$ and $k = 3$. Using a grid with six nodes, determine the initial values for c_i and the values of c_i after one time step of explicit Euler using a time step $h = 0.1$.

7.9 Consider the solution to the steady state concentration profile $\nabla^2 c = 0$ using centered finite differences. The domain is discretized into nine nodes with a spacing Δx and Δy. The concentration at the top and right sides of the slab is $c = 2$, and the concentration on the bottom and left sides of the slab is $c = 1$. Answer the following questions:

(a) If $\Delta x = \Delta y$, what is the temperature at the center node?
(b) If $\Delta x \neq \Delta y$, how does the temperature at the center node depend on the spacing? This is equivalent to discretizing a rectangular domain with an equal number of nodes on each side.
(c) If $\Delta x \neq \Delta y$ and the top side boundary condition is changed to $c = 3$, how does the temperature at the center node depend on the spacing?
(d) If $\Delta x = \Delta y$ and the top side boundary condition is replaced with a no-flux condition $\partial c/\partial y = 0$, what is the temperature at each node in the domain?

7.10 Consider the solution of a 2D finite difference solution of the diffusion equation $\nabla^2 T = 0$ where the boundary conditions correspond to fixed temperatures. What fraction of the entries in the matrix needed to solve this problem are non-zero if there are n_x nodes in x and n_y nodes in y? Assume that n_x and n_y are large enough so that you can neglect the role of the boundary conditions in your answer. What is the bandwidth of the matrix in this problem? If we wanted to now solve the transient problem, $\partial T/\partial t = \nabla^2 T$, by the method of lines using n_x nodes in x and n_y nodes in y, how many coupled ODEs will we have?

7.11 Consider the solution to the diffusion equation

$$\nabla^2 c = 0$$

on a 4×4 rectangular grid with equal spacing in the x and y directions. The boundary conditions for the problem are

$$c(x = 0, y) = 1$$
$$c(x, y = 0) = 1$$
$$c(x, y = 1) = 1/2$$
$$D\frac{\partial c}{\partial x} = kc^2 \quad \text{on} \quad (x = 1, y)$$

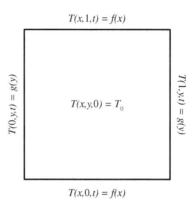

Figure 7.16 Boundary conditions for Problem 7.12.

Using the global index k and the ordering of equations and unknowns in order of the index k, determine the non-zero entries on the seventh, eighth, and fourteenth rows of the Jacobian required to solve this problem. You should clearly state the corresponding residual equations, since the form of the Jacobian will depend on how you write the residual.

7.12 Suppose you are interested in the steady state temperature profile $T(x, y)$ of a square surface with a known spatially dependent thermal diffusivity $\alpha(x, y)$ subject to Dirichlet boundary conditions. The time-dependent PDE governing this system is

$$\frac{\partial T}{\partial t} = \frac{\partial}{\partial x}\left[\alpha(x, y)\frac{\partial T}{\partial x}\right] + \frac{\partial}{\partial y}\left[\alpha(x, y)\frac{\partial T}{\partial y}\right]$$

and the boundary conditions are given by Fig. 7.16.

(a) List the most efficient combination of methods discussed in this book that you should use to solve the *steady* problem. In addition, using scaling arguments, estimate the time it would take to solve the problem with $n_x = n_y = 100$ nodes. Assume that if you discretize the x-domain using $n_x = 10$ nodes and the y-domain using $n_y = 10$ nodes, you get a program that runs in 10^{-1} seconds.

(b) Suppose the unsteady problem must reach a time $t = 10$ in order to reach equilibrium. List the combination of methods you would use to solve the unsteady problem. If one time step requires 10 ms, estimate the minimum number of seconds it will take to reach the steady state using a stable, explicit time integration method. Assume that $\max(\lambda_i) = 10^2$, $\max[\alpha(x, y)] = 1$ and that the discretization is $n_x = n_y = 101$.

7.13 Consider the case of unsteady diffusion of heat and mass in the presence of a first-order reaction. With a suitable choice of dimensionless variables, this system leads to the set of coupled partial differential equations

$$\frac{\partial c}{\partial t} = \frac{\partial^2 c}{\partial x^2} - \phi e^{-\beta/T} c$$

$$\frac{\partial T}{\partial t} = \frac{\partial^2 T}{\partial x^2} + \phi^2 \gamma e^{-\beta/T} c$$

where β, γ, and ϕ are the activation energy, Prater number, and Thiele modulus. Assume that you know all of the dimensionless numbers. We want to solve this system of equations using interleaved variables of the form

$$\mathbf{z} = \begin{bmatrix} c_1 \\ T_1 \\ c_2 \\ T_2 \\ \vdots \\ c_n \\ T_n \end{bmatrix}$$

using the method of lines, centered finite differences, and implicit Euler. Determine the non-zero entries in the fourth line of the Jacobian.

7.14 Derive a centered finite difference formula for the first derivative that is accurate to $O(\Delta x)^4$.

Computer Problems

7.15 Write a MATLAB program that uses the method of lines, RK4, and finite differences to solve the diffusion equation

$$\frac{\partial c}{\partial t} = \frac{\partial^2 c}{\partial x^2}$$

subject to an initial condition $c(x,0) = 1$ and boundary conditions $c(0,t) = 1$ and $c(1,t) = 0$. Your program should automatically plot the solution at the times $t = 0.0005$, 0.005, 0.02, 0.05, 0.1, and 0.5.

7.16 Write a MATLAB program that computes the solution to the (dimensionless) unsteady diffusion equation

$$\frac{\partial c}{\partial t} = \frac{\partial^2 c}{\partial x^2}$$

subject to an initial condition $c(x,0) = 0$ and the boundary conditions $c(0,t) = 1$ and $\partial c/\partial x = 0$ at $x = 1$. In other words, this is the solution for the concentration (or temperature profile) in an initially empty (cold) slab if we instantaneously put it in contact with a reservoir at the left that is at the maximum concentration (temperature). Use centered finite differences for the spatial derivative and RK4 for the time integration. You should have 51 nodes and a time step of 0.0001. Make a plot that has the concentration profile at the 11 dimensionless times $t = 0, 0.025, 0.05, \ldots, 0.25$.

7.17 Consider the partial differential equation

$$\frac{\partial c}{\partial t} = \frac{\partial^2 c}{\partial x^2}$$

subject to $c(0, t) = 0$ and

$$c - \frac{\partial c}{\partial x} = 0 \quad \text{at } x = 1$$

We will consider three cases: $c(x, 0) = 1$, $c(x, 0) = x$ and $c(x, 0) = x^2$. Note that the second initial condition is a steady state solution to the PDE.

(a) Use finite differences and the method of lines to recast this problem as a system of ordinary differential equations. Be sure to specify the initial conditions on each equation.

(b) Recast the result in part (a) as a system of algebraic equations to compute $c_i^{(k+1)}$ using implicit Euler.

(c) Write a MATLAB program that integrates the solution to part (b). For $c(x, 0) = x$, make a plot of the difference between the solution at $t = 0.01$ and $t = 0$. Is this what you expect? For the other initial conditions, make a plot of the concentration at $t = 0, 0.0005, 0.01, 0.05, 0.07$, and 0.2. Be sure to include a legend.

7.18 This problem considers different aspects of the solution of the unsteady heat equation.

(a) Use separation of variables to solve the PDE

$$\frac{\partial T}{\partial t} = \frac{\partial^2 T}{\partial x^2}$$

with no flux boundary conditions at $x = 0$ and $x = 1$ and an initial condition $T = 1 - x$.

(b) Write a MATLAB program that makes a plot of $T(x, t)$ versus x for a given value of t and a number n of non-zero terms in the Fourier series. You should use 200 grid points for x. We want to explore the error in the solution as a function of n. Let us define the error as

$$\text{error} = ||T_{n=100} - T_n||$$

where T_n is the temperature field with n terms in the Fourier series. Naturally, the error will go to zero when you have 100 terms, since this definition assumes that $n = 100$ is sufficient.

(c) Have your program automatically generate a log–log plot of the error versus the number of terms in the series for $n = 1$ to $n = 99$ for two different values of the time, $t = 0.000\,05$ and $t = 0.001$. Explain why these two curves are different.

(d) Now investigate the cost of solving the PDE in part (a) using centered finite differences. First write a code that uses RK4 to perform the integration with ten grid points in x. Your program should automatically plot the numerical solution at the times $t = 0, 0.01, 0.05, 0.1$, and 0.5. It should also make a plot of the exact solution at these times using $n = 100$ terms in the Fourier series. The purpose of this part of the problem is for you to confirm that your code works correctly – even for a very coarse discretization the numerical solution is very close to the exact solution.

7.19   In heat transfer the heat transfer from a slab of thickness $2w$ initially at a uniform temperature T_0 can sometimes be modeled by a lumped parameter analysis

$$\rho \hat{C}_p V \frac{dT}{dt} = -2hA(T - T_\infty)$$

where ρ is the density of the slab, $\hat{C}_p$ is its heat capacity, h is the heat transfer coefficient, and T_∞ is the temperature far from the slab. The volume of the slab is $V = 2wA$, where A is the surface area of one face. The factor of 2 appears in the equation because there are two faces. The solution to this problem is

$$\frac{T - T_\infty}{T_0 - T_\infty} = \exp\left(-\frac{h}{\rho \hat{C}_p w} t\right)$$

If we define the dimensionless temperature $\theta \equiv (T - T_\infty)/(T_0 - T_\infty)$ and the dimensionless time $\tau = ht/\rho \hat{C}_p w$, then the lumped parameter solution is

$$\theta = e^{-\tau}$$

Here, we will consider the accuracy of this approximation as a function of the Biot number

$$\text{Bi} \equiv \frac{hw}{k}$$

where k is the thermal conductivity of the slab. The unsteady heat equation is

$$\rho \hat{C}_p \frac{\partial T}{\partial t} = k \frac{\partial^2 T}{\partial x^2}$$

If the initial condition is uniform, $T(x, 0) = T_0$, then the problem is symmetric about $x = 0$ and we can solve it subject to the symmetry boundary condition

$$\left. \frac{\partial T}{\partial x} \right|_{x=0} = 0$$

and the heat transfer boundary condition

$$-k \left. \frac{\partial T}{\partial x} \right|_{x=w} = h[T(x = w, t) - T_\infty]$$

If we use the same dimensionless time and temperature, and introduce a dimensionless position $z = x/w$, then the differential equation becomes

$$\frac{\partial \theta}{\partial \tau} = \left(\frac{1}{\text{Bi}}\right) \frac{\partial^2 \theta}{\partial z^2}$$

with

$$\left. \frac{\partial \theta}{\partial z} \right|_{z=0} = 0$$

and

$$-\left. \frac{\partial \theta}{\partial z} \right|_{z=1} = \text{Bi}\,\theta(z = 1, t)$$

(a) Write a MATLAB program that uses the method of lines and finite differences to solve this equation, where you do the time integration using RK4. Include the system of ODEs that you need to solve at each step. Have your program integrate out to a dimensionless time of $\tau = 0.001$ using a grid spacing $\Delta z = 0.005$ and time step $h = 10^{-4}$. Your program should automatically plot the temperature profile at the final time for Bi = 1000, 50, 20, 10, 5, and 1 as separate figures. Include an explanation for their behavior.

(b) You should notice that the solution in part (a) is not very good. (Indeed, since this linear problem has a solution by separation of variables, our numerical solution so far is a waste of time!) Rewrite your program to solve the problem using implicit Euler as the integration method, and include the details of the integration scheme in your written solution. This time, have your program integrate out to $\tau = 0.5$ for Bi = 0.01, 1, 5, 10, 50, and 1000. Your program should automatically generate a plot of the temperature profile for all of the Bi numbers on a single graph, which should also have the prediction from the lumped parameter analysis too. Include an explanation of what is going on.

7.20   This problem involves the solution of the diffusion equation

$$\frac{\partial^2 c}{\partial x^2} + \frac{\partial^2 c}{\partial y^2} = 0$$

with no flux boundary conditions at $x = 0$ and $x = 1$ and a concentration $c = 1$ at $y = 1$. The end goal is to determine the concentration profile when the boundary condition at $y = 0$ is zero for $x \in [0, 0.25]$ and $x \in [0.5, 0.75]$ and no-flux otherwise. The first two parts are designed to help you debug your code, and the third part is the goal of the problem.

(a) Write a MATLAB program that uses finite differences to solve this problem where the boundary condition at $y = 0$ is $c = 0$. Your program should automatically generate a plot of the concentration as a function of position. A convenient plotting tool is `mesh`, which requires that you have your concentration in the form of a matrix. You can also provide matrices with the corresponding x and y data to make a nicely formatted plot. What is the exact solution to the problem?

(b) Now modify your program to use no-flux at $y = 0$. In addition to plotting the solution, display the values of the concentration to the screen. (This will help with your interpretation of the result.) What are the new boundary equations and the exact result for this problem?

(c) Now modify your program to use $c = 0$ at $y = 0$ for $x \in [0, 0.25]$ and $x \in [0.5, 0.75]$ and no-flux otherwise. You program should produce two plots. The first plot is the same as the previous problems, with the concentration as a function of two-dimensional position. The second plot is the concentration at the lower boundary, $c(x, 0)$, as a function of x. What was your method for implementing the lower boundary condition?

7.21   This problem takes a closer look at flow inside a square duct. Read this entire problem statement first. You should try to construct your programs in parts (a) and (b) in a way that makes it easier for you to solve part (c).

(a)   For flow inside a square duct the dimensionless velocity profile is given by the infinite series solution

$$v(x, y) = \sum_{n=1,n \text{ odd}}^{\infty} \frac{4}{n^3 \pi^3} \left(1 - \frac{\cosh(n\pi y)}{\cosh(n\pi / 2)} \right) \sin(n\pi x)$$

where $x \in [0, 1]$ and $y \in [-1/2, 1/2]$. Make a MATLAB program that plots this solution using mesh for 100 equally spaced nodes in x and y. You should use enough points in the sum so that the result is not changing much as you add more terms.

(b)   The flow in a square duct is the solution to the Poisson equation,

$$\nabla^2 v = -1$$

subject to $v = 0$ on the boundaries. Write a MATLAB program that solves this equation using centered finite differences with 100 nodes per side using a sparse solver. Your program should automatically plot the solution.

(c)   Let's now figure out the accuracy of the numerical solution. Write a MATLAB program, using the programs that you already wrote, that computes

$$\epsilon = ||\mathbf{v} - \tilde{\mathbf{v}}||$$

where $\mathbf{v}$ is the infinite series results (with k terms sufficient to provide accuracy) and $\tilde{\mathbf{v}}$ is the numerical solution (with n nodes on each side). You can think of this as an error in the solution, assuming that you use enough terms in the infinite series. Your program should automatically make a log–log plot of the error versus the number of nodes for a fixed number of terms in the infinite sum.

7.22   In this problem, we will look at the two-dimensional diffusion problem with a reactive surface at one of the boundaries. The steady state PDE in the domain is

$$\frac{\partial^2 c}{\partial^2 x} + \frac{\partial^2 c}{\partial^2 y} = 0$$

Along all of the non-reactive boundaries, we will set the concentration to unity

$$c(x, -1) = c(-1, y) = c(1, y) = 1$$

On the upper boundary at $y = 1$, we will have a first-order reaction that consumes the reactant

$$\frac{\partial c}{\partial y} = -kc$$

(Note that the negative sign comes from the direction of the normal vector to the reactive surface.) In the dimensionless form of this problem, the quantity k is the Dämkohler number.

Write a MATLAB program that solves this problem using 101 grid points in x and y and centered finite differences. Include a list of all of the equations that you need to

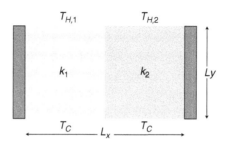

Figure 7.17 Geometry for Problem 7.23.

solve this problem. Even with only 101 grid points in each direction you have a very large matrix ($n^2 = 10201$), so take advantage of the sparse matrix solver. In the first use of your program, make a surface plot of the concentration as a function of position for $k = 1$. If you think a bit about the problem, you should be able to tell if your numerical solution makes sense or not. Include this plot with your written solution. Now have your program make a plot of the concentration at the reactive surface for the following values of the Dämkohler number: $k = 0, 0.1, 0.2, 0.5, 0.75, 1, 2, 3, 5, 10, 20, 100$. Include an explanation of its behavior.

7.23 We would like to solve the steady state heat transfer problem given by the drawing in Fig. 7.17. We have a composite solid material of area $L_x L_y$, with the left half having conductivity k_1 and the right half having conductivity k_2. The lower part of the material is in contact with a cold reservoir maintained at T_c, and the top half is maintained at two different hot temperatures, $T_{H,1}$ and $T_{H,2}$. The side walls of the material are insulating boundary conditions.

(a) Determine the set of dimensionless equations and boundary conditions needed to solve this problem. Your equations should involve a dimensionless temperature, $\theta = (T - T_C)/(T_{H,1} - T_C)$ and length scales $\tilde{x} = x/L_x$ and $\tilde{y} = y/L_x$. These naturally lead to a conductivity ratio $k = k_1/k_2$, a temperature ratio $T_r = (T_{H,2} - T_C)/(T_{H,1} - T_C)$, and an aspect ratio $L = L_y/L_x$.

(b) We now want to write a MATLAB program to compute the temperature profile for given values of the parameters k, T_r, and L and the number of nodes n_x used to discretize the geometry for one of the materials. Although there are many ways to construct the solution, use the grid in Fig. 7.18 (which is the easiest, if not the most efficient). The origin is in the upper left corner. The nodes for material 1 run from $j = 1$ to $j = n_x n_y$, where n_y is the number of grid points needed for a given value of L with constant grid spacing $\Delta x = \Delta y$. The nodes for material 2 run from $j = n_x n_y + 1$ to $j = 2 n_x n_y$. Note that both materials contain nodes at the interface, but they are at the same location. It is possible to reduce the number of nodes by substitution, but let's use this slightly larger grid because it is conceptually simpler.

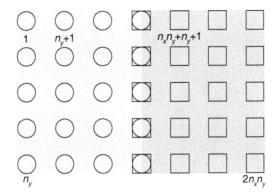

Figure 7.18 Grid for Problem 7.23.

Convert the equations from part (a) into equations on this grid using finite differ-ence. For the no-flux boundaries, use centered-finite differences. For the interface flux conditions, use forward or backward differences.

Write a MATLAB program that solves for the temperature profile leaving the values of k, T_r, L, and n_x as variables that you can easily change. To test your program, use $k = 10$, $T_r = 1.5$, and $L = 0.5$. Set $n_x = 3$, which is the smallest possible grid you can use. If you store the interface temperature conditions in the equations for the interface nodes in material 1 and the interface flux conditions in the equations for the interface nodes in material 2, you should have the following linear system to solve:

$$
\begin{bmatrix}
1 & 0 & 0 & 0 & 0 & 0 & 0 & 0 & 0 & 0 & 0 & 0 & 0 & 0 & 0 & 0 & 0 & 0 \\
1 & -4 & 1 & 0 & 2 & 0 & 0 & 0 & 0 & 0 & 0 & 0 & 0 & 0 & 0 & 0 & 0 & 0 \\
0 & 0 & 1 & 0 & 0 & 0 & 0 & 0 & 0 & 0 & 0 & 0 & 0 & 0 & 0 & 0 & 0 & 0 \\
0 & 0 & 0 & 1 & 0 & 0 & 0 & 0 & 0 & 0 & 0 & 0 & 0 & 0 & 0 & 0 & 0 & 0 \\
0 & 1 & 0 & 1 & -4 & 1 & 0 & 1 & 0 & 0 & 0 & 0 & 0 & 0 & 0 & 0 & 0 & 0 \\
0 & 0 & 0 & 0 & 0 & 1 & 0 & 0 & 0 & 0 & 0 & 0 & 0 & 0 & 0 & 0 & 0 & 0 \\
0 & 0 & 0 & 0 & 0 & 0 & 1 & 0 & 0 & 0 & 0 & 0 & 0 & 0 & 0 & 0 & 0 & 0 \\
0 & 0 & 0 & 0 & 0 & 0 & 0 & 1 & 0 & 0 & -1 & 0 & 0 & 0 & 0 & 0 & 0 & 0 \\
0 & 0 & 0 & 0 & 0 & 0 & 0 & 0 & 1 & 0 & 0 & 0 & 0 & 0 & 0 & 0 & 0 & 0 \\
0 & 0 & 0 & 0 & 0 & 0 & 0 & 0 & 0 & 1 & 0 & 0 & 0 & 0 & 0 & 0 & 0 & 0 \\
0 & 0 & 0 & 0 & -10 & 0 & 0 & 10 & 0 & 0 & 1 & 0 & 0 & -1 & 0 & 0 & 0 & 0 \\
0 & 0 & 0 & 0 & 0 & 0 & 0 & 0 & 0 & 0 & 0 & 1 & 0 & 0 & 0 & 0 & 0 & 0 \\
0 & 0 & 0 & 0 & 0 & 0 & 0 & 0 & 0 & 0 & 0 & 0 & 1 & 0 & 0 & 0 & 0 & 0 \\
0 & 0 & 0 & 0 & 0 & 0 & 0 & 0 & 0 & 0 & 1 & 0 & 1 & -4 & 1 & 0 & 1 & 0 \\
0 & 0 & 0 & 0 & 0 & 0 & 0 & 0 & 0 & 0 & 0 & 0 & 0 & 0 & 1 & 0 & 0 & 0 \\
0 & 0 & 0 & 0 & 0 & 0 & 0 & 0 & 0 & 0 & 0 & 0 & 0 & 0 & 0 & 1 & 0 & 0 \\
0 & 0 & 0 & 0 & 0 & 0 & 0 & 0 & 0 & 0 & 0 & 0 & 0 & 2 & 0 & 1 & -4 & 1 \\
0 & 0 & 0 & 0 & 0 & 0 & 0 & 0 & 0 & 0 & 0 & 0 & 0 & 0 & 0 & 0 & 0 & 1 \\
\end{bmatrix}
$$

$$
\times
\begin{bmatrix}
\theta_1 \\
\theta_2 \\
\theta_3 \\
\theta_4 \\
\theta_5 \\
\theta_6 \\
\theta_7 \\
\theta_8 \\
\theta_9 \\
\theta_{10} \\
\theta_{11} \\
\theta_{12} \\
\theta_{13} \\
\theta_{14} \\
\theta_{15} \\
\theta_{16} \\
\theta_{17} \\
\theta_{18}
\end{bmatrix}
=
\begin{bmatrix}
1 \\
0 \\
0 \\
1 \\
0 \\
0 \\
1.25 \\
0 \\
0 \\
1.25 \\
0 \\
0 \\
1.5 \\
0 \\
0 \\
1.5 \\
0 \\
0
\end{bmatrix}
$$

It is easiest to check your program using the Workspace environment in MATLAB, which will let you open up variables in a format similar to Excel. Once you know the program is working, use $n_x = 32$ to make a high-resolution plot of the temperature profile. You should use the `surf` function in MATLAB for the plot, and the `view` function to have it rotated to the proper orientation and elevation. Since it is a pain to add text to the plots, it is sufficient to get it into the same orientation as the figure in this assignment. Your program should automatically generate this figure.

(c) We now want to compare the numerical solution to the prediction of the resistor models. Recall that the resistor model would predict that the flux coming out of the bottom of each material would be

$$
Q_i = k_i A \frac{\Delta T_i}{L_y}
$$

where A is the cross-sectional area. You can also compute the flux out of the bottom of material $i = 1$ and $i = 2$ by using backwards differences and then integrating across the boundary using Boole's rule. (See Chapter 8.) Modify your program from part (b) to make this calculation for T_r values from 1.0 to 2.0 in increments of 0.1. Report your result as the ratio of Q from the numerical solution to the predicted value of Q for the resistor model for each of the materials. Your program should automatically generate the plot with axis labels, a legend, and a title. What does this plot tell you about the accuracy of the resistor model?

8   Interpolation and Integration

8.1   Introduction

The goal of this chapter is to discuss how to compute integrals of the form

$$I = \int_a^b f(x)dx \tag{8.1.1}$$

for some function $f(x)$. This function can be known in closed form, for example $f(x) = x^2$, or we can have tabulated data for $f(x_i)$ at different values of x_i. The former case arises most often in theoretical work, where in the course of the development of some theory you encounter an integral that cannot be computed in closed form, while the latter is often the case in experiments, for example where you have acquired equilibrium vapor-liquid composition data to design a distillation column.

The basic idea behind numerical integration is shown in Fig. 8.1. Given either tabulated data for $f(x_i)$ at some points x_i or the actual functional form of $f(x)$ itself, we fit an easily integrated function $\tilde{f}(x)$ to the data and then just integrate that function instead. Since $\tilde{f}(x)$ is an approximation to the true function $f(x)$, then the integral of $\tilde{f}(x)$ is an approximation to the integral of $f(x)$. We thus need to begin our discussion with interpolation, which is the approach we will use to determine the function $\tilde{f}(x)$. We then discuss the use of polynomial interpolating functions, which lead to the so-called Newton–Coates formulas for numerical integration. The simplest of these formulas is the trapezoidal rule, which interpolates the function as a line. The corresponding integral is simply the width of the region multiplied by the average height of the function,

$$I \approx (b - a)\left[\frac{f(b) - f(a)}{2}\right] \tag{8.1.2}$$

While the latter result is intuitive, we will see in this chapter how to derive it in a rigorous way that allows us to produce higher-order formulas with correspondingly lower errors. We will also see how repeated use of the rather simple result in Eq. (8.1.2) can be used to produce more accurate approximations to the integrals. If we do this in a systematic way, we can produce integrals to a desired degree of accuracy without any *a priori* knowledge of the function itself. At the end of this chapter, we will also discuss briefly Gauss quadrature. We will use a rather intuitive method to determine two-point Gauss quadrature that illustrates both its power and limitations. A more detailed derivation of Gauss quadrature requires the use of Lagrange polynomials, which we do not cover here.

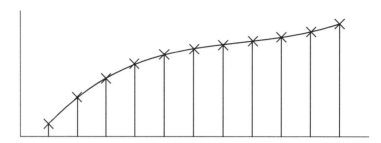

Figure 8.1 Schematic illustration of using an interpolating polynomial for numerical integration.

Finally, we will wrap up the book with an example of using interpolation and numerical integration for the design of a differential distillation column.

8.2   Polynomial Interpolation

In Section 2.14, we discussed the topic of regression to experimental data and its relationship to interpolation. Recall that in regression, we try to find a function $\tilde{f}(x)$ that describes the overall trend in the function $f(x)$ without requiring that this function $\tilde{f}(x)$ actually go through any of the data points. Interpolation is the opposite case. Here, we are trying to find a function that passes through all of the data points. For the case of polynomial interpolation, which is our focus here, interpolation to n data points requires an $(n-1)$th order polynomial.

8.2.1   Linear Interpolation

Let us begin with the simplest type of interpolation, namely linear interpolation between two data points $f(x_0)$ and $f(x_1)$. You may recall doing this type of interpolation from steam table data in mass and energy balances or in your thermodynamics class. For example, you may have data for the specific volume at two different values of the temperature and you want to determine the specific volume at some intermediate temperature.

In linear interpolation, you are trying to find the linear function $\tilde{f}(x)$ that connects the points $f(x_0)$ and $f(x_1)$,

$$\tilde{f}(x) = ax + b \tag{8.2.1}$$

Applying this equation to the points x_0 and x_1, we can set up a system of equations for the unknowns a and b,

$$\begin{bmatrix} x_0 & 1 \\ x_1 & 1 \end{bmatrix} \begin{bmatrix} a \\ b \end{bmatrix} = \begin{bmatrix} f(x_0) \\ f(x_1) \end{bmatrix} \tag{8.2.2}$$

Our derivation will connect more closely to what follows if we note instead that the slope from x_0 to x needs to be the same as the slope from x_0 to x_1 (see Fig. 8.2),

$$\frac{\tilde{f}(x) - f(x_0)}{x - x_0} = \frac{f(x_1) - f(x_0)}{x_1 - x_0} \tag{8.2.3}$$

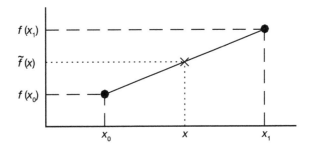

Figure 8.2 Schematic illustration of linear interpolation.

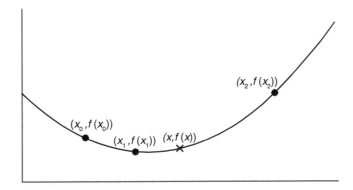

Figure 8.3 Schematic illustration of quadratic interpolation.

If we solve Eq. (8.2.3) for $\tilde{f}(x)$, we get the linear interpolation formula

$$\tilde{f}_1(x) = f(x_0) + \left[\frac{f(x_1) - f(x_0)}{x_1 - x_0}\right](x - x_0) \tag{8.2.4}$$

We have added the subscript "1" to $\tilde{f}(x)$ to denote that this is a linear interpolating function – we will shortly see how to derive the quadratic interpolant $\tilde{f}_2(x)$ and the higher order interpolants $\tilde{f}_n(x)$. The term in the brackets in Eq. (8.2.4) looks familiar from our discussion of boundary value problems in Chapter 6. In the context of interpolation, this quantity is called the first divided difference, and it is an approximation for the first derivative of the function. While it may seem that we ended up with this divided difference by coincidence, this is not the case and the idea of a divided difference will play a key role in what follows.

8.2.2 Quadratic Interpolation

Let us now consider how to fit a quadratic polynomial to three data points, as illustrated in Fig. 8.3. If we took a naive approach we would now say we want to find the coefficients to the function

$$\tilde{f}(x) = ax^2 + bx + c \tag{8.2.5}$$

that interpolates between x_0, x_1, and x_2. The corresponding linear algebraic system we would need to solve is then

$$\begin{bmatrix} x_0^2 & x_0 & 1 \\ x_1^2 & x_1 & 1 \\ x_2^2 & x_2 & 1 \end{bmatrix} \begin{bmatrix} a \\ b \\ c \end{bmatrix} = \begin{bmatrix} f(x_0) \\ f(x_1) \\ f(x_2) \end{bmatrix} \tag{8.2.6}$$

While we can proceed with the latter approach to derive all higher-order formulas, since you probably see how the structure of the matrix evolves with the order of the interpolation formula, we will take a different tack that is even more useful. Let us write the interpolant in the form

$$\tilde{f}_2(x) = b_0 + b_1(x - x_0) + b_2(x - x_0)(x - x_1) \tag{8.2.7}$$

You can readily confirm that Eq. (8.2.7) is equivalent to Eq. (8.2.5). We prefer the form of Eq. (8.2.7) to simplify the algebra, which will be obvious in a moment, and to allow us to see how we would go about deriving the formulas for higher-order interpolants. We thus need to compute the coefficients b_i.

To compute b_0, we evaluate Eq. (8.2.7) at x_0,

$$f(x_0) = b_0 + b_1(x_0 - x_0) + b_2(x_0 - x_0)(x_0 - x_1) \tag{8.2.8}$$

which gives us

$$b_0 = f(x_0) \tag{8.2.9}$$

and reduces Eq. (8.2.7) to

$$\tilde{f}_2(x) = f(x_0) + b_1(x - x_0) + b_2(x - x_0)(x - x_1) \tag{8.2.10}$$

This is the reason for writing the interpolation in the form of Eq. (8.2.7) – at each point x_i, one of the terms will disappear until we get to the last interpolation point.

To compute b_1, we now evaluate Eq. (8.2.10) at x_1,

$$f(x_1) = f(x_0) + b_1(x_1 - x_0) + b_2(x_1 - x_0)(x_1 - x_1) \tag{8.2.11}$$

This gives us

$$b_1 = \frac{f(x_1) - f(x_0)}{x_1 - x_0} \tag{8.2.12}$$

which was the divided difference of the first derivative that we already saw in Eq. (8.2.4). We have now reduced Eq. (8.2.10) down to

$$\tilde{f}_2(x) = f(x_0) + \left[\frac{f(x_1) - f(x_0)}{x_1 - x_0}\right](x - x_0) + b_2(x - x_0)(x - x_1) \tag{8.2.13}$$

To compute b_2, we now evaluate Eq. (8.2.13) at x_2,

$$f(x_2) = f(x_0) + \left[\frac{f(x_1) - f(x_0)}{x_1 - x_0}\right](x_2 - x_0) + b_2(x_2 - x_0)(x_2 - x_1) \tag{8.2.14}$$

Let's do a bit of algebra to get the result into a nice form for later. To do so, we subtract $f(x_0)$ from both sides and divide through by the factor $x_2 - x_0$ to give

$$\frac{f(x_2) - f(x_0)}{x_2 - x_0} = \frac{f(x_1) - f(x_0)}{x_1 - x_0} + b_2(x_2 - x_1) \tag{8.2.15}$$

Solving for b_2 gives

$$b_2 = \left[\frac{f(x_2) - f(x_0)}{x_2 - x_0} - \frac{f(x_1) - f(x_0)}{x_1 - x_0} \right] \frac{1}{x_2 - x_1} \tag{8.2.16}$$

While this is an OK way to write b_2, let us do just a bit more work. (We promise all of this work will be worthwhile in a moment!) Let's rewrite Eq. (8.2.16) with a common denominator,

$$b_2 = \frac{[f(x_2) - f(x_0)](x_1 - x_0) - [f(x_1) - f(x_0)](x_2 - x_0)}{(x_2 - x_1)(x_1 - x_0)(x_2 - x_0)} \tag{8.2.17}$$

The idea is then to write b_2 as divided differences of the first derivative over the intervals $x \in [x_1, x_2]$ and $x \in [x_0, x_1]$. If we note that

$$x_2 - x_0 = (x_2 - x_1) + (x_1 - x_0) \tag{8.2.18}$$

then we can write

$$b_2 = \frac{f(x_2)(x_1 - x_0) + f(x_0)(x_2 - x_1) - f(x_1)[(x_2 - x_1) + (x_1 - x_0)]}{(x_2 - x_1)(x_1 - x_0)(x_2 - x_0)} \tag{8.2.19}$$

Grouping the terms in the numerator by the weighting in the interval gives us the desired result,

$$b_2 = \left[\frac{f(x_2) - f(x_1)}{x_2 - x_1} - \frac{f(x_1) - f(x_0)}{x_1 - x_0} \right] \frac{1}{x_2 - x_0} \tag{8.2.20}$$

The two terms in the brackets are the divided difference of the first derivative over the intervals $x \in [x_1, x_2]$ and $x \in [x_0, x_1]$, respectively, with a normalization factor of the total interval, $x \in [x_0, x_2]$. The result is known as the second divided difference.

8.2.3   Newton's Divided Differences and Polynomial Interpolation

The approach we used to derive the formula for quadratic interpolation can be used to derive formulas for cubic interpolation, quartic interpolation, and so on. These higher-order interpolation formulas follow a pattern. The general form of an nth order interpolant is

$$f_n(x) = b_0 + b_1(x - x_0) + b_2(x - x_0)(x - x_1) + \cdots$$
$$\cdots + b_n(x - x_0)(x - x_1) \cdots (x - x_{n-1}) \tag{8.2.21}$$

where the coefficients are the divided differences,

$$b_0 = f(x_0) \tag{8.2.22}$$
$$b_1 = f[x_1, x_0] \tag{8.2.23}$$

$$b_2 = f[x_2, x_1, x_0] \tag{8.2.24}$$

$$\vdots \qquad \vdots$$

$$b_n = f[x_n, x_{n-1}, \ldots, x_1, x_0] \tag{8.2.25}$$

Here, we have introduced a compact notation for the divided difference. To find a divided difference with two terms, we use

$$f[x_i, x_j] = \frac{f(x_i) - f(x_j)}{x_i - x_j} \tag{8.2.26}$$

which is just the rise over run of the first two terms in the interval. The next divided difference, which has three terms, is defined as a difference of divided differences,

$$f[x_i, x_j, x_k] = \frac{f[x_i, x_j] - f[x_j, x_k]}{x_i - x_k} \tag{8.2.27}$$

To evaluate Eq. (8.2.27), we need to first compute Eq. (8.2.26) for the two sub-intervals. This recursive algorithm allows us to go up to any order, since we define the nth finite divided difference in terms of the two $(n-1)$th divided differences,

$$f[x_n, x_{n-1}, \ldots, x_1, x_0] = \frac{f[x_n, x_{n-1}, \ldots, x_2, x_1] - f[x_{n-1}, \ldots, x_1, x_0]}{x_n - x_0} \tag{8.2.28}$$

Example 8.1   Show that Eq. (8.2.27) reduces to Eq. (8.2.20).

Solution
Let $x_i = x_2$, $x_j = x_1$, and $x_k = x_0$. Then we have

$$f[x_2, x_1, x_0] = \frac{f[x_2, x_1] - f[x_1, x_0]}{x_2 - x_0} \tag{8.2.29}$$

The first divided differences are computed from Eq. (8.2.26),

$$f[x_2, x_1] = \frac{f(x_2) - f(x_1)}{x_2 - x_1} \tag{8.2.30}$$

$$f[x_1, x_0] = \frac{f(x_1) - f(x_0)}{x_1 - x_0} \tag{8.2.31}$$

Substituting the latter into Eq. (8.2.29) gives

$$f[x_2, x_1, x_0] = \left[\frac{f(x_2) - f(x_1)}{x_2 - x_1} - \frac{f(x_1) - f(x_0)}{x_1 - x_0} \right] \frac{1}{x_2 - x_0} \tag{8.2.32}$$

which is indeed Eq. (8.2.20) for b_2.

The latter example should make it clear to you now why we went through so much effort in Section 8.2.2 to write the final result for b_2 in the particular form of Eq. (8.2.20). So that you are confident with the recursive form of the finite differences, let's also work on one higher term.

Example 8.2   Compute the third divided difference using the recursion relationship Eq. (8.2.28).

Solution

The recursion relationship gives us

$$f[x_3, x_2, x_1, x_0] = \frac{f[x_3, x_2, x_1] - f[x_2, x_1, x_0]}{x_3 - x_0} \qquad (8.2.33)$$

We then apply the recursion relationship to each term on the right-hand side,

$$f[x_3, x_2, x_1] = \frac{f[x_3, x_2] - f[x_2, x_1]}{x_3 - x_1} \qquad (8.2.34)$$

$$f[x_2, x_1, x_0] = \frac{f[x_2, x_1] - f[x_1, x_0]}{x_2 - x_0} \qquad (8.2.35)$$

Finally, we need to apply the recursion relationship again to the terms on the right-hand side to get the remaining three divided differences,

$$f[x_3, x_2] = \frac{f(x_3) - f(x_2)}{x_3 - x_2} \qquad (8.2.36)$$

$$f[x_2, x_1] = \frac{f(x_2) - f(x_1)}{x_2 - x_1} \qquad (8.2.37)$$

$$f[x_1, x_0] = \frac{f(x_1) - f(x_0)}{x_1 - x_0} \qquad (8.2.38)$$

Having gotten to the end of the recursions, we now work our way upward to generate the desired result. We first substitute the first divided differences into the second divided differences,

$$f[x_3, x_2, x_1] = \frac{f(x_3) - f(x_2)}{(x_3 - x_2)(x_3 - x_1)} - \frac{f(x_2) - f(x_1)}{(x_2 - x_1)(x_3 - x_1)} \qquad (8.2.39)$$

$$f[x_2, x_1, x_0] = \frac{f(x_2) - f(x_1)}{(x_2 - x_1)(x_2 - x_0)} - \frac{f(x_1) - f(x_0)}{(x_1 - x_0)(x_2 - x_0)} \qquad (8.2.40)$$

The latter results are then substituted into Eq. (8.2.33) to get the final result. In the general case, where the positions x_i are not specified, the end result can be a bit of a mess. Let's hold off on any further algebra here until Example 8.6.

PROGRAM 8.1   *The recursive nature of Eq. (8.2.27) makes it relatively easy to set up a program that will generate the coefficients b_i for $i = 0$ to $i = n - 1$ given n data points.*

```
1  function [n,coeff] = integrate_interpolation_tabulated(x,f)
2  % x = tabulated interpolation points
3  % f = tabulated values of the function
4  n = length(x) - 1; %order of polynomial
5  U = zeros(n); %upper triangular matrix
6            %to hold finite-divided differences
7
8  %compute first-order divided differences
```

```
 9   j = 1;
10   for i = 1:n
11       U(i,1) = (f(i+1) - f(i))/(x(i+j)-x(i));
12   end
13
14   %compute higher-order divided differences
15   for j = 2:n
16       for i = 1:n-j+1
17           U(i,j) = (U(i+1,j-1) - U(i,j-1))/(x(i+j)-x(i));
18       end
19   end
20   coeff = zeros(1,n+1);
21   coeff(1,1) = f(1);
22   coeff(1,2:n+1) = U(1,:);
```

This program takes in two inputs, a vector x of points where we want to do the interpolation and a second vector f that contains the value of f(x) at each of the interpolation points. This program is thus well suited to handle tabulated data in the form of $f(x_i)$. If we have a continuous function, we simply would need to write a program that generates the vector f(x). We will see how to do this shortly in an example. The function returns two arguments, a scalar n that gives the order of the polynomial and a vector coeff that contains the values of the b_i. In the program, we are choosing to determine the order n of the polynomial as the first step in the program on line 4. Note that we could have easily made n an input to the program instead.

*Now let's see how this program operates. The values of all of the different divided differences are stored in an upper triangular matrix called U. In the first **for** loop from lines 10–12, we fill in the first column of U with the first divided differences. Note that these calculations require that we refer to the function evaluations in the vector f, as they correspond to Eq. (8.2.26). The set of nested **for** loops in lines 15–19 computes all the higher-order divided differences. The outer loop counter j is keeping track of the order of the divided differences, with the upper bound being the nth divided difference. The inner loop then constructs each of the divided differences corresponding to the current value of j and stores them in the appropriate location in the upper triangular matrix U.*

For an nth order polynomial, we will have $n + 1$ coefficients since there is a zero term. The first coefficient, b_0, is obtained from the first entry in f. The remaining coefficients are the first row of U.

PROGRAM 8.2   *Once you have the coefficients, you also want to be able to construct the approximation $\tilde{f}_n(x)$. This is accomplished for a single point x with the following short program:*

```
1   function y = integrate_interpolate_f(x,xpts,coeff,n)
2   % x = value of x to interpolate y
3   % xpts = interpolation points
4   % coeffs = interpolation coefficients
5   % n = power of interpolant
6   y = coeff(1); %zero-order interpolant
7   for i = 2:n+1
```

```
8        p = coeff(i); %interpolation coefficient
9        xdiff = 1; %initialize determination of polynomial
10       for j = 1:i-1
11           xdiff = xdiff*(x-xpts(j));
12       end
13       y = y + coeff(i)*xdiff;
14   end
```

This program requires four inputs. The vector coeff *and the scalar* n *are the outputs of the previous program. The vector* xpts *is the values of x_i that we used for the interpolation, and the scalar x is the location where we would like to determine $\tilde{f}_n(x)$. The nested* for *loops in lines 7–14 implement Eq. (8.2.21). The outer loop goes through each term of the equation, and the inner loop computes the product of $(x-x_0)(x-x_1)\cdots$ that multiplies this particular b_i value. Note that the first term, b_0, is handled separately in line 6, which is the reason why the counter in line 7 starts at $i = 2$. If we want to construct a graph, we can simply call this program repeatedly for each value of x on the plot.*

Example 8.3 Determine a cubic interpolation to $\tan(x)$ using the interpolation points $x = 0$, $\pi/3$, $\pi/6$, and $\pi/4$.

Solution

To do this example, we will use Program 8.1. We chose to use a vector x for the program that is not in sequential order to illustrate that the interpolation formulas are independent of the ordering of the interpolation points x_i (even though you will normally have the data in ascending or descending order). To generate the function evaluations, we can write a short program that takes in the interpolation points and returns the function at those points:

```
1   function f = getf(x)
2   n = length(x);
3   f = zeros(n,1);
4   for i = 1:n
5       f(i) = tan(x(i));
6   end
```

The corresponding values are f = [0, 1.7321, 0.5774, 1.0000].

We can then feed this vector f and the vector x into Program 8.1 to generate the b_i values. While the output of that program is our desired result, it is worthwhile to look at the vector U as the program progresses to see each of the divided differences and how the higher order ones are constructed from the lower order ones. When we reach line 13 of the program, we have

$$\mathbf{U} = \begin{bmatrix} 1.6540 & 0 & 0 \\ 2.2053 & 0 & 0 \\ 1.6144 & 0 & 0 \end{bmatrix} \tag{8.2.41}$$

The first column of **U** corresponds to the three first divided differences between the four points. When we execute the first iteration of the loop in lines 15–19, we have two new values in the second column corresponding to the second divided differences,

$$\mathbf{U} = \begin{bmatrix} 1.6540 & 1.0530 & 0 \\ 2.2053 & 2.2571 & 0 \\ 1.6144 & 0 & 0 \end{bmatrix} \qquad (8.2.42)$$

There is one more iteration of the loop in lines 15–19 that computes the third divided difference for this problem, giving

$$\mathbf{U} = \begin{bmatrix} 1.6540 & 1.0530 & 1.5332 \\ 2.2053 & 2.2571 & 0 \\ 1.6144 & 0 & 0 \end{bmatrix} \qquad (8.2.43)$$

The coefficients are thus the first line of this matrix along with the value of $f(x_0)$, meaning

$$b_0 = 0 \qquad (8.2.44)$$
$$b_1 = 1.6540 \qquad (8.2.45)$$
$$b_2 = 1.0530 \qquad (8.2.46)$$
$$b_3 = 1.5332 \qquad (8.2.47)$$

Once we have the values of b_i, we can use Eq. (8.2.21) to write down the interpolation for this function, noting that $x_0 = 0$, $x_1 = \pi/3$, $x_2 = \pi/6$, and $x_3 = \pi/4$. The complete result is

$$\tilde{f}_3(x) = 1.6540x + 1.0530x\left(x - \frac{\pi}{3}\right) + 1.5332x\left(x - \frac{\pi}{3}\right)\left(x - \frac{\pi}{6}\right) \qquad (8.2.48)$$

If you plug $x = \pi/4$ into this equation, you will get back $\tilde{f}_3(x) = 1.0000$, which must be the case since $\pi/4$ was one of the interpolation points.

Figure 8.4 shows that Eq. (8.2.48) is a good approximation to the tangent function over the range of the interpolation $x \in [0, \pi/3]$. However, we should be very wary about

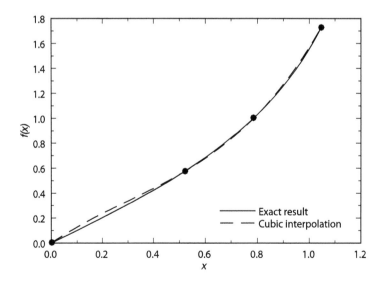

Figure 8.4 Comparison of a cubic interpolation to tan(x) at the points $x_0 = 0$, $x_1 = \pi/3$, $x_2 = \pi/6$, and $x_3 = \pi/4$ with the exact result. Note that the interpolation is equal to the function at the interpolation points.

using interpolation formulas outside the range of the interpolation. For example, if we substitute $\pi/2$ into Eq. (8.2.48), we have $\tilde{f}_3(x) = 4.7846$. This is *very* different than the value of the tangent function at $\pi/2$, which diverges. Constructing accurate interpolation formulas near singular points in a function is numerically challenging, requiring very high-order interpolants.

8.2.4   Equally Spaced Data

So far, we have not placed any restriction on the data points x_i. In the divided differences formula in Eq. (8.2.28), there can be arbitrary spacing between the neighboring values of x_i. Indeed, the values of x_i do not even need to be in ascending or descending order, although that is often the case in tabulated data.

Consider the special case where the data in x_i are in ascending order with a uniform spacing h between data points. In other words, we require that

$$
\begin{aligned}
x_1 &= x_0 + h \\
x_2 &= x_1 + h = x_0 + 2h \\
&\vdots \quad \vdots \\
x_n &= x_0 + nh
\end{aligned}
\tag{8.2.49}
$$

The divided difference formulas in this case reduce to so-called "Newton–Gregory" polynomials. Let us see what happens for the first few divided difference formulas with evenly spaced data.

Example 8.4 Simplify Eq. (8.2.26) for equally spaced data.

Solution
With $x_i = x_0 + h$ and $x_j = x_0$, Eq. (8.2.26) becomes

$$
f[x_1, x_0] = \frac{f(x_1) - f(x_0)}{h}
\tag{8.2.50}
$$

Example 8.5 Simplify Eq. (8.2.32) for evenly spaced data.

Solution
With $x_2 = x_0 + 2h$ and $x_1 = x_0 + h$, we have

$$
f[x_2, x_1, x_0] = \left[\frac{f(x_2) - f(x_1)}{h} - \frac{f(x_1) - f(x_0)}{h} \right] \frac{1}{2h}
\tag{8.2.51}
$$

which reduces to

$$
f[x_2, x_1, x_0] = \frac{f(x_2) - 2f(x_1) + f(x_0)}{2h^2}
\tag{8.2.52}
$$

Does this look familiar? We will return to this question shortly.

Example 8.6   Simply Eq. (8.2.33) for evenly spaced data.

Solution

We can begin by applying Eq. (8.2.52) to get convenient expressions for the two second divided differences,

$$f[x_3, x_2, x_1] = \frac{f(x_3) - 2f(x_2) + f(x_1)}{2h^2} \qquad (8.2.53)$$

$$f[x_2, x_1, x_0] = \frac{f(x_2) - 2f(x_1) + f(x_0)}{2h^2} \qquad (8.2.54)$$

Substituting the latter into Eq. (8.2.33) gives

$$f[x_3, x_2, x_1, x_0] = \frac{\left[f(x_3) - 2f(x_2) + f(x_1)\right] - \left[f(x_2) - 2f(x_1) + f(x_0)\right]}{6h^3} \qquad (8.2.55)$$

which we can simplify to

$$f[x_3, x_2, x_1, x_0] = \frac{f(x_3) - 3f(x_2) + 3f(x_1) - f(x_0)}{3!h^3} \qquad (8.2.56)$$

In the latter, note the factorial in the denominator. For evenly spaced data, the higher-order formulas are reasonably compact. For arbitrarily spaced data, which was the case in Example 8.2, the general formulas get to be very cumbersome.

PROGRAM 8.3   *There is no reason that we cannot use Program 8.1 for equally spaced data, since we just need to remember to feed it such data. However, it is worth taking a moment to modify the program to see how it can be simplified for equally spaced data:*

```
1  function [x,coeff] = integrate_interpolation_equal(xmin,xmax,n)
2  %xmin = lowest interpolation point
3  %xmax = highest interpolation point
4  %n = polynomial power (number of points - 1)
5
6  U = zeros(n); %upper triangular matrix
7                %to hold finite-divided differences
8  h = (xmax - xmin)/n; %step size
9  x = linspace(xmin,xmax,n+1); %interpolation points
10
11 %compute first-order divided differnces
12 for i = 1:n
13     U(i,1) = (getf(x(i+1)) - getf(x(i)))/h;
14 end
15
16 %compute higher-order divided differences
17 for j = 2:n
18     for i = 1:n-j+1
19         U(i,j) = (U(i+1,j-1) - U(i,j-1))/(j*h);
20     end
21 end
22 coeff = zeros(1,n+1);
23 coeff(1,1) = getf(xmin);
24 coeff(1,2:n+1) = U(1,:);
```

In this program, we give the bounds for $x \in [x_{\min}, x_{\max}]$ and the polynomial order that we want to use. Line 8 of the program determines the corresponding step size and line 9 sets up a vector x with all of the interpolation points. The remainder of the program is quite similar to Program 8.1. The key differences are (i) in line 13, we call a subfunction getf *to get the value of the function at the interpolation points, and (ii) both line 13 and line 19 have the distance between the interpolation points coded in terms of the step size h. The output of this program is slightly different than the one we used for arbitrary tabulated data, returning the location of the interpolation points and the coefficients. We do not need to return the order of the interpolating polynomial, since that was an input this time. The present program is readily combined with Program 8.2 to generate plots of the interpolating polynomial.*

Often, the divided differences for equally spaced data are written in a short-hand form as

$$f[x_0, x_1, \ldots, x_n] = \frac{1}{n!} \left(\frac{\Delta^n f(x_0)}{h^n} \right) \tag{8.2.57}$$

The quantity $\Delta^n f(x_0)$ is often called the nth order finite difference equation. The first finite difference equations are

$$\Delta f(x_0) = f(x_1) - f(x_0) \tag{8.2.58}$$
$$\Delta^2 f(x_0) = f(x_2) - 2f(x_1) + f(x_0) \tag{8.2.59}$$
$$\Delta^3 f(x_0) = f(x_3) - 3f(x_2) + 3f(x_1) - f(x_0) \tag{8.2.60}$$

We wrote Eq. (8.2.57) in a suggestive form. The term in parenthesis is exactly what we derived in Section 6.2.2 for centered finite difference approximations of the derivative. For example, if we use Eq. (8.2.52) in Eq. (8.2.57), we see that

$$\frac{f(x_2) - 2f(x_1) + f(x_0)}{2h^2} = \frac{1}{2!} \left(\frac{\Delta^2 f(x_0)}{h^2} \right) \tag{8.2.61}$$

which means that

$$\frac{\Delta^2 f(x_0)}{h^2} = \frac{f(x_2) - 2f(x_1) + f(x_0)}{h^2} \tag{8.2.62}$$

The latter result is exactly the same as the centered finite differences approximation for the second derivative in Eq. (6.2.13) about the point x_1. Indeed, the divided differences for equally spaced data are an alternate approach to derive centered finite difference formulas.

8.2.5 Error of Newton's Interpolating Polynomials

The form of Eq. (8.2.57) also suggests an analogy between polynomial interpolation and a Taylor series. If we recall the terms in the interpolating polynomial equation (8.2.21) and use Eq. (8.2.57), we could rewrite the interpolation formula for equally spaced data as

$$f_n(x) = f(x_0) + \left(\frac{\Delta f(x_0)}{h} \right)(x - x_0) + \frac{1}{2} \left(\frac{\Delta^2 f(x_0)}{h^2} \right)(x - x_0)(x - x_1) + \cdots$$

$$\cdots + \frac{1}{n!}\left(\frac{\Delta^n f(x_0)}{h^n}\right)(x-x_0)(x-x_1)\cdots(x-x_{n-1}) \tag{8.2.63}$$

This looks remarkably like a Taylor series. In general, the form for the error in a Taylor series up to n terms is

$$R_n = \frac{f^{(n+1)}(\xi)}{(n+1)!}(x-x_0)^{n+1}, \quad \xi \in [x_0, x] \tag{8.2.64}$$

The error in an interpolating polynomial has a very similar form,

$$R_n = \frac{f^{(n+1)}(\xi)}{(n+1)!}(x-x_0)(x-x_1)\cdots(x-x_n) \tag{8.2.65}$$

where ξ is on the interval containing the known data points x_i and x.

This result is fine when you know the functional form of $f(x)$, since you can compute the $(n+1)$th derivative of the function. In cases where you only have tabulated data and thus cannot compute the derivatives of $f(x)$, which is the more general case, then you can estimate the error from a divided difference at the point x,

$$R_n = f[x, x_n, x_n - 1, \ldots, x_0](x-x_0)(x-x_1)\cdots(x-x_n) \tag{8.2.66}$$

The error in the interpolation for evenly spaced data can be written in a more convenient form than Eq. (8.2.65), the latter being valid for any spacing of the data. To do so, let us define a new variable

$$\alpha = \frac{x - x_0}{h} \tag{8.2.67}$$

which is just the number of steps of size h to get from the start of the interval at x_0 to some point x. This allows us to rewrite the interpolating polynomial in Eq. (8.2.21) as

$$f_n(x) = f(x_0) + \Delta f(x_0)\alpha + \frac{\Delta^2 f(x_0)}{2!}\alpha(\alpha-1) + \cdots$$
$$\cdots + \frac{\Delta^n f(x_0)}{n!}\alpha(\alpha-1)\cdots(\alpha-n+1) \tag{8.2.68}$$

This provides a nice result for the error,

$$R_n = \frac{f^{(n+1)}(\xi)}{(n+1)!}h^{n+1}\alpha(\alpha-1)(\alpha-2)\cdots(\alpha-n) \tag{8.2.69}$$

The definition of α, along with this error estimate, will be very convenient in our discussion of Newton–Coates integration in the next section.

8.3   Newton–Coates Integration

The idea in Newton–Coates formulas is to replace $f(x)$ in the integral

$$I = \int_a^b f(x)\,dx \tag{8.3.1}$$

with an nth order polynomial $\tilde{f}_n(x)$

$$I \approx \int_a^b \tilde{f}_n(x)\,dx \tag{8.3.2}$$

which can be integrated easily. In the Newton–Coates methods, we use Eq. (8.2.68) as the polynomial interpolation of evenly spaced data and Eq. (8.2.69) as the estimate of the error. We will work out the first few Newton–Coates formulas in detail, and then cite the higher-order results without proof since their derivation follows the same approach but with more tedious algebra.

8.3.1 Trapezoidal Rule ($n = 1$)

The trapezoidal rule corresponds to linear interpolation of the function, where we simply connect $f(a)$ and $f(b)$ to get a straight line. In terms of the notation above, this is a polynomial interpolation with $n = 1$ where

$$\tilde{f}_1(x) = f(x_0) + \Delta f(x_0)\alpha + \frac{f''(\xi)}{2}h^2\alpha(\alpha - 1) \tag{8.3.3}$$

For the interval $x \in [a, b]$, we then have $x_0 = a$ and $x_1 = b$. Since we span with a single step, we also have $h = b - a$. To approximate the integral we use

$$I = \int_a^b \left[f(a) + \Delta f(a)\alpha + \frac{f''(\xi)}{2}h^2\alpha(\alpha - 1) \right] dx \tag{8.3.4}$$

If we note from Eq. (8.2.67) that

$$d\alpha = \frac{dx}{h} \tag{8.3.5}$$

the approximation to the integral becomes

$$I = h \int_0^1 \left[f(a) + \Delta f(a)\alpha + \frac{f''(\xi)}{2}h^2\alpha(\alpha - 1) \right] d\alpha \tag{8.3.6}$$

Computing the polynomial integrals gives

$$I = \frac{h}{2}\left[f(b) + f(a) \right] - \frac{1}{12}f''(\xi)h^3 \tag{8.3.7}$$

The first term is the formula for the trapezoidal rule given in Eq. (8.1.2). The second term is the truncation error. One common source of confusion is the unknown value of ξ, which leads students to wonder how to compute the integral. Note that you do not actually evaluate the truncation error; if you want to evaluate an integral with trapezoidal rule you just compute

$$I = (b - a)\left[\frac{f(b) + f(a)}{2} \right] \tag{8.3.8}$$

The truncation error gives you an idea of how the error will depend on the step size h and the smoothness of the function.

8.3.2   Simpson's 1/3 Rule ($n = 2$)

The derivation proceeds in basically the same way as the trapezoidal rule, but there is an interesting twist for even values of n in the Newton–Coates formulas that makes a detailed derivation worthwhile. The integral we want to calculate here is

$$I = \int_{x_0}^{x_2} \left[f(x_0) + \Delta f(x_0)\alpha + \frac{\Delta^2 f(x_0)}{2}\alpha(\alpha - 1) + \frac{\Delta^3 f(x_0)}{6}\alpha(\alpha - 1)(\alpha - 2) \right.$$
$$\left. + \frac{f^{(4)}(\xi)}{24}\alpha(\alpha - 1)(\alpha - 2)(\alpha - 3)h^4 \right] dx \tag{8.3.9}$$

You may be wondering why we used terms up to $n = 3$ in the approximation – this will be clear in a moment. Since we have used two steps to span the interval,

$$h = \frac{b - a}{2} \tag{8.3.10}$$

where the points in the interval are

$$x_0 = a \tag{8.3.11}$$
$$x_1 = a + h = \frac{a + b}{2} \tag{8.3.12}$$
$$x_2 = a + 2h = b \tag{8.3.13}$$

We use the same change of variable as Eq. (8.3.5) to get

$$I = h \int_0^2 \left[f(x_0) + \Delta f(x_0)\alpha + \frac{\Delta^2 f(x_0)}{2}\alpha(\alpha - 1) + \frac{\Delta^3 f(x_0)}{6}\alpha(\alpha - 1)(\alpha - 2) \right.$$
$$\left. + \frac{f^{(4)}(\xi)}{24}\alpha(\alpha - 1)(\alpha - 2)(\alpha - 3)h^4 \right] d\alpha \tag{8.3.14}$$

The interesting term to look at is the cubic one

$$\int_0^2 \alpha(\alpha - 1)(\alpha - 2)d\alpha = \left[\frac{\alpha^4}{4} - \alpha^3 + \alpha^2 \right]_0^2 = 0 \tag{8.3.15}$$

The non-zero integrals give

$$I = h \left[2f(x_0) + 2\Delta f(x_0) + \frac{\Delta^2 f(x_0)}{3} - \frac{1}{90}f^{(4)}(\xi)h^4 \right] \tag{8.3.16}$$

Using the formulas for the finite differences in Eqs. (8.2.58) and (8.2.59) gives

$$I = \frac{h}{3} \left[f(x_0) + 4f(x_1) + f(x_2) \right] - \frac{1}{90}f^{(4)}(\xi)h^5 \tag{8.3.17}$$

The latter equation is the reason why this is called Simpson's 1/3 rule – the prefactor is $h/3$ when you write the result in terms of the Newton–Gregory formulas. This is often confusing if you look up the formula in a table because the normal way to write the Simpson's 1/3 rule is to substitute for the x_i values and h. The resulting formula

$$I = \frac{b-a}{6}\left[f(a) + 4f\left(\frac{a+b}{2}\right) + f(b)\right] - \frac{1}{90}f^{(4)}(\xi)h^5 \qquad (8.3.18)$$

looks like a 1/6th rule!

The really amazing thing about Simpson's 1/3 rule is that you get an exact result for cubic equations using only a parabola for interpolation owing to Eq. (8.3.15). This additional level of accuracy occurs for all even values of n.

8.3.3   Higher-Order Formulas

You can proceed with the same ideas to get higher-order formulas. We will just state the results here. For $n = 3$, you have Simpson's 3/8 rule,

$$I = \frac{b-a}{8}\left[f(x_0) + 3f(x_1) + 3f(x_2) + f(x_3)\right] - \frac{3}{80}f^{(4)}(\xi)h^5 \qquad (8.3.19)$$

Note that the error in this method is of the same order (h^5) as Simpson's 1/3 rule. For $n = 4$, you have Boole's rule,

$$I = \frac{b-a}{90}\left[7f(x_0) + 32f(x_1) + 12f(x_2) + 32f(x_3) + 7f(x_4)\right] - \frac{8}{945}f^{(6)}(\xi)h^7 \quad (8.3.20)$$

and for $n = 5$ you have

$$I = \frac{b-a}{288}\left[19f(x_0) + 75f(x_1) + 50f(x_2) + 50f(x_3) + 75f(x_4) + 19f(x_5)\right]$$
$$- \frac{275}{12096}f^{(6)}(\xi)h^7 \qquad (8.3.21)$$

Again, you have a zero integral at $n = 4$ so that the error of $O(h^7)$ is the same as for $n = 5$.

PROGRAM 8.4   *We can write a nice compact program that allows us to integrate some function $f(x)$ using a particular order of the Newton–Coates formula by taking advantage of the* **switch** *operator, which is a useful alternative to implementing many* **if-else** *statements.*

```
1  function out = integrate_Newton_Coates(a,b,n)
2  % a = lower bound of integral
3  % b = upper bound of integral
4  % n = order of formula
5  x = linspace(a,b,n+1);
6  switch n
7      case 1
8          w = [1,1]/2;
9      case 2
10         w = [1,4,1]/6;
11     case 3
12         w = [1,3,3,1]/8;
13     case 4
14         w = [7,32,12,32,7]/90;
15     case 5
16         w = [19,75,50,50,75,19]/288;
```

```
17      otherwise
18          fprintf('Use a value of n between 1 and 5\n')
19          return
20  end
21
22  I = 0;
23  for i = 1:n+1
24      I = I + w(i)*getf(x(i));
25  end
26  out = I*(b-a);
```

The function takes as its inputs the upper and lower bounds of the integral along with the value n for the Newton–Coates formula. Line 5 sets up the locations of the function evaluations according to the value of n, remembering that the number of function evaluations is one higher than the order of the polynomial. Lines 6–20 implement a **switch** *operator to select the correct set of weights to use for the various Newton–Coates formulas. The word* **switch** *is followed by the variable that operates the switch, which in our case is the order of the polynomial. We then go through a sequence of* **case** *commands that are operated if the value of n is equal to the value of the case. While we are using numbers here, it is worth noting that the* **switch** *operator is also convenient for evaluating text strings. When we reach the case with the proper value of n, we create a vector of weights using the Newton–Coates formulas. If we reach the end of the* **switch** *and did not set weights, we write an error message to the screen and exit the program.*

Once we know the weights, lines 22–26 compute the value of the integral. Note that the counter in line 23 has to go to n+1 *because we make one more function evaluation than the order of the polynomial.*

Example 8.7　To test our claim about the accuracy of these methods, use Program 8.4 to calculate the integral

$$I = (m+1) \int_0^1 x^m dx \tag{8.3.22}$$

which we know has the value $I = 1$ for any positive integer m.

Solution

Table 8.1 shows the results up to $m = 6$. There are two important things to note here. First, as we move down a given column in the table, once we exceed the order of the polynomial used for the interpolation in a particular method, then the error increases monotonically with m. This occurs because the higher-order polynomials are increasingly curved, so the values of their derivatives, which appear in the equation for R_n in Eq. (8.2.69), are increasing. Second, note that the error for $n = 2$ and $n = 3$ begins at $m = 4$, and the error for $n = 4$ and $n = 5$ begins at $m = 6$. This is the manifestation of the zero value of the odd integral in Eq. (8.3.15) for $n = 2$ and a similar integral at the fifth power for $n = 4$.

Table 8.1 Values of the integral in Eq. (8.3.22) using Newton–Coates formulas of order n.

m	$n=1$	$n=2$	$n=3$	$n=4$	$n=5$
1	1.000 000	1.000 000	1.000 000	1.000 000	1.000 000
2	1.500 000	1.000 000	1.000 000	1.000 000	1.000 000
3	2.000 000	1.000 000	1.000 000	1.000 000	1.000 000
4	2.500 000	1.041 667	1.018 519	1.000 000	1.000 000
5	3.000 000	1.125 000	1.055 556	1.000 000	1.000 000
6	3.500 000	1.239 583	1.109 053	1.002 604	1.001 467

8.4 Applications of Trapezoidal Rule

The trapezoidal rule in Eq. (8.3.7) is the simplest of all of the Newton–Coates formulas. Despite its simplicity, the trapezoidal rule turns out to be very useful for constructing more advanced algorithms for numerical integration. We will consider three such algorithms here.

8.4.1 Multiple Trapezoidal Rule

All of the Newton–Coates formulas use a single interpolant over the entire range $x \in [a,b]$. However, if you remember that the integral is just the area under the curve, then you can see that you could also break up the integral over $x \in [a,b]$ into a sum of integrals,

$$I = \int_{x_0}^{x_1} f(x)\,dx + \int_{x_1}^{x_2} f(x)\,dx + \cdots + \int_{x_{n-1}}^{x_n} f(x)\,dx \tag{8.4.1}$$

Figure 8.5 explains the utility of breaking up the integral. In some cases, you can have a reasonably high degree of curvature to the function or large variations in the curvature. In these cases, you would need to use a very high-order polynomial to get a good interpolation for the function. While you can certainly continue in the manner of Section 8.3 to derive the weights for a higher-order function, it is much simpler to break up the integral into a sufficiently small number of trapezoids. The computational cost is essentially the same because you have to make n function evaluations in either case.

The multiple trapezoidal rule has a convenient form and a nice geometrical interpretation. Applying the trapezoidal rule on each integral in Eq. (8.4.1), we get

$$I = h\frac{f(x_0)+f(x_1)}{2} + h\frac{f(x_1)+f(x_2)}{2} + \cdots + h\frac{f(x_{n-1})+f(x_n)}{2} \tag{8.4.2}$$

Every function evaluation appears twice except for the first and last terms. As a result, we can write the latter equation in the compact form

$$I = \frac{h}{2}\left[f(x_0) + \left\{ 2\sum_{i=1}^{n-1} f(x_i) \right\} + f(x_n) \right] \tag{8.4.3}$$

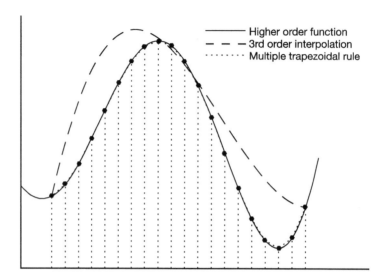

Figure 8.5 Utility of multiple trapezoidal rule to handle functions with significant curvature.

The geometrical interpretation becomes more apparent if we rewrite this equation in the form

$$I = (b - a) \left[\frac{f(x_0) + 2 \sum_{i=1}^{n-1} f(x_i) + f(x_n)}{2n} \right] \tag{8.4.4}$$

The term $b - a$ is the width of the function and the terms in the brackets is the average height.

Since we broke up the integral into a bunch of pieces, the error estimate is the sum of the errors for each segment,

$$E = -\frac{(b - a)^3}{12n^3} \sum_{i=1}^{n} f''(\xi) \tag{8.4.5}$$

Since we do not actually know the point ξ at which the error is evaluated, it is convenient to simply define the average value of the second derivative, $\bar{f}''$, which allows us to write an estimate of the error in the form

$$E = -\frac{(b - a)^3}{12n^2} \bar{f}'' \tag{8.4.6}$$

Note that we used one factor of n to convert the sum in Eq. (8.4.5) into an average value in Eq. (8.4.6). If you want to write this in terms of the step size for each of the intervals,

$$E = -\frac{(b - a)}{12} h^2 \bar{f}'' \tag{8.4.7}$$

PROGRAM 8.5   *The program for implementing multiple trapezoidal rule follows naturally from the general program we wrote for Newton–Coates in Program 8.4.*

```
1  function out = integrate_multiple_trapezoidal(xmin,xmax,n)
2  x = linspace(xmin,xmax,n+1);
3  I = getf(xmin) + getf(xmax);
4  if n > 1
5      for i = 2:n
6          I = I + 2*getf(x(i));
7      end
8  end
9  out = (xmax-xmin)*I/(2*n);
```

The program takes as its input the bounds of the integral and the number of trapezoids. Line 2 sets up the locations for the function evaluations at evenly spaced points along the interval. Line 3 treats the endpoints, and any interior points are handled by lines 4–8 to build up the sum of the values of the function. Line 9 then implements multiple trapezoidal rule in the form of Eq. (8.4.4).

Example 8.8   Use the multiple trapezoidal rule to determine how the error in the integral

$$I = \int_0^{1/2} \cos \pi x \, dx \qquad (8.4.8)$$

depends on the number of trapezoids.

Solution

We know that the integral is $I = \pi^{-1}$. Figure 8.6 shows the value of the error as a function of the number of trapezoids. For the smallest values of n, we can get a

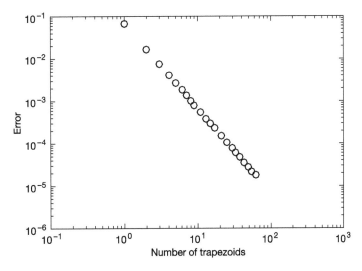

Figure 8.6 Error in the multiple trapezoidal rule integration of Eq. (8.4.8) as a function of the number of trapezoids.

non-monotonic dependence of the error on n because we use a small number of line segments to represent a continuously varying function. Once we reach a modest number of trapezoids, the error is already quite small. For example, we are underestimating the integral by only 0.03. The plot corresponds to 100 trapezoids, which produces an underestimate of only 8.5×10^{-5}.

8.4.2   Richardson Extrapolation

The idea behind Richardson extrapolation is to use two separate trapezoidal integrations to get an even more accurate answer that you could get from either trapezoidal rule itself. To see how you can seemingly get something for nothing, let's consider what happens if we use a step size

$$h_1 = \frac{b - a}{n_1} \tag{8.4.9}$$

where n_1 is the number of intervals in the multiple trapezoidal rule. The integral is then

$$I_1 = \frac{h_1}{2} \left[f(a) + 2 \sum_{i=1}^{n_1 - 1} f(x_i) + f(b) \right] \tag{8.4.10}$$

and the error of this integral is

$$E_1 = -\frac{b - a}{12} h_1^2 \bar{f}'' \tag{8.4.11}$$

We now perform a second integration with a smaller step size

$$h_2 = \frac{b - a}{n_2} \tag{8.4.12}$$

where n_2 is the number of intervals in the multiple trapezoidal rule. The integral is then

$$I_2 = \frac{h_2}{2} \left[f(a) + 2 \sum_{i=1}^{n_2 - 1} f(x_i) + f(b) \right] \tag{8.4.13}$$

and the error of this integral is

$$E_2 = -\frac{b - a}{12} h_2^2 \bar{f}'' \tag{8.4.14}$$

The clever idea behind Richardson extrapolation is to use both estimates of the integral to reduce the error. If the actual value of the integral is I, then we estimate that

$$I \approx I_1 + E_1 \approx I_2 + E_2 \tag{8.4.15}$$

If we further assume that $\bar{f}''$ is independent of the step size, then Eqs. (8.4.11) and (8.4.14) give

$$\frac{E_1}{E_2} \approx \frac{h_1^2}{h_2^2} \tag{8.4.16}$$

If we use Eq. (8.4.16) in our estimate of the integral (8.4.15), we have

$$I_1 + E_2 \left(\frac{h_1}{h_2}\right)^2 \approx I_2 + E_2 \tag{8.4.17}$$

If we solve this equation for E_2, we have

$$E_2 = \frac{I_1 - I_2}{1 - \left(\dfrac{h_1}{h_2}\right)^2} \tag{8.4.18}$$

We now have an estimate for the error! We then get a better result for the actual value of the integral,

$$I = I_2 + E_2 \approx I_2 + \frac{I_1 - I_2}{1 - \left(\dfrac{h_1}{h_2}\right)^2} \tag{8.4.19}$$

which simplifies to

$$I \approx \frac{I_1 - (h_1/h_2)^2 I_2}{1 - (h_1/h_2)^2} \tag{8.4.20}$$

The error of Eq. (8.4.20) is now $\mathcal{O}(h^4)$ even though each of the integrals we computed has an error of $\mathcal{O}(h^2)$.

Since we know that the number of trapezoids n is inversely related to the step size h, Eq. (8.4.20) can also be written in the form

$$I = \frac{I_1 - (n_2/n_1)^2 I_2}{1 - (n_2/n_1)^2} \tag{8.4.21}$$

The latter form is somewhat easier to implement, since we require that the number of trapezoids be an integer value. Directly using Eq. (8.4.20) requires some care, since you need to be sure to pick the step sizes h_1 and h_2 such that you get integer numbers of trapezoids.

PROGRAM 8.6   *The program to implement Richardson extrapolation is relatively easy to create once we already have Program 8.5 for multiple trapezoidal rule.*

```
1  function I = integrate_Richardson(xmin,xmax,n1,n2)
2  % xmin = lower bound for integration
3  % xmax = upper bound for integration
4  % n1 = number of trapezoids for I1
5  % n2 = number of trapezoids for I2
6  I1 = integrate_multiple_trapezoidal(xmin,xmax,n1);
7  I2 = integrate_multiple_trapezoidal(xmin,xmax,n2);
8  I = (I1 - (n2/n1)^2*I2)/(1- (n2/n1)^2);
```

This program simply makes two estimates of the integral, I_1 and I_2, using the multiple trapezoidal rule program. Line 8 then combines the integrals according to Eq. (8.4.21) to produce the better estimate of the integral.

8.4.3   Romberg Integration

The idea of Romberg integration builds upon Richardson extrapolation for the case where $h_1 = 2h_2$. In this case, Richardson extrapolation (8.4.20) becomes

$$I = \frac{4}{3}I_2 - \frac{1}{3}I_1 \qquad (8.4.22)$$

Romberg integration uses the general formula for combining integrals of order $h_i = 2h_{i+1}$ to get a better estimate,

$$I_{j,k} \approx \frac{4^{k-1}I_{j+1,k-1} - I_{j,k-1}}{4^{k-1} - 1} \qquad (8.4.23)$$

In the latter, $I_{j,k}$ is the improved estimate of the integral, $I_{j+1,k-1}$ is the better integral from the previous step, and $I_{j,k-1}$ is the less accurate integral from the previous step. We then recognize that Eq. (8.4.22) corresponds to Eq. (8.4.23) for the case $j = 1$ and $k = 2$.

The idea behind the algorithm is illustrated in Fig. 8.7.

(1) Make a first pair of estimates $I_{1,1}$ and $I_{2,1}$.
(2) Combine these two estimates to get a new value $I_{1,2}$. This is what we did in Eq. (8.4.22).
(3) Compute $I_{3,1}$ using a step size half of $I_{2,1}$.
(4) Combine $I_{3,1}$ and $I_{2,1}$ to get $I_{2,2}$.
(5) Combine $I_{1,2}$ and $I_{2,2}$ to get $I_{1,3}$.

We can continue this algorithm to higher order by computing $I_{4,1}$, which is then combined with previous integrals to get $I_{3,2}$, $I_{2,3}$, and finally $I_{1,4}$, which is our new best estimate of the integral.

Each time the index k is increased, we get a better estimate of the integral as the term $I_{1,k}$ by combining all of the previous estimates at the level $I_{j,1}$ for $j = 1, 2, \ldots, k$. Romberg integration is very well suited for a computer algorithm and it is more efficient that the trapezoidal rule or Simpson's rule. The best part about Romberg integration is that you do not need to know anything about the function in advance to set the accuracy of the integration. You simply continue to a sufficiently high value of k to get the desired accuracy. As is usually the case, you should set a cut-off in the number of integrals you are willing to compute to avoid getting stuck in an infinite `while` loop.

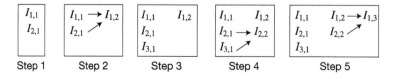

Figure 8.7 Schematic illustration of the Romberg integration algorithm.

PROGRAM 8.7   *We can build on Program 8.5 to do Romberg integration.*

```
1   function out = integrate_Romberg(xmin,xmax,n,tol)
2   % xmin = lower bound for integral
3   % xmax = upper bound for integral
4   % n = number of trapezoids for first estimate
5   % tol = desired accuracy
6   I(1,1) = integrate_multiple_trapezoidal(xmin,xmax,n);
7   I(2,1) = integrate_multiple_trapezoidal(xmin,xmax,2*n);
8   I(1,2) = (4*I(2,1) - I(1,1))/3;
9   m = 2;
10  err = abs(I(1,2)-I(1,1));
11  while err > tol
12      m = m + 1;
13      I(m,1) = integrate_multiple_trapezoidal(xmin,xmax,2^(m-1)*n);
14      for k = 2:m
15          j = m - k + 1;
16          I(j,k) = (4^(k-1)*I(j+1,k-1) - I(j,k-1))/(4^(k-1)-1);
17      end
18      err = abs(I(1,m) - I(1,m-1));
19      if m >= 50
20          fprintf('Did not converge to %6.4e.\n',tol)
21          break
22      end
23  end
24  out = I(1,m);
```

The program takes as inputs the upper and lower bound for the integral, the number of trapezoids to use or the first integral, and the tolerance for convergence of the estimate of the integral. Lines 6–8 implement Richardson extrapolation with half the step size in Eq. (8.4.22) and stores the results in the form of Fig. 8.7. Lines 9 and 10 then set up the counter and error for the **while** *loop, where m is the current value for the most accurate integral. We then implement successive rounds of Romberg integration by incrementing the counter and then making a new estimate by multiple trapezoidal rule in line 13. Note that the number of trapezoids is $2^{(m-1)}n$ as we improve the estimate, since h is the inverse of n. The* **for** *loop in lines 14–17 uses this new estimate and combines it with all of the previous estimates via Eq. (8.4.23) to get a new best estimate, again following the logic outlined in Fig. 8.7. We then check the error and see how we are doing, along with a check on the number of iterations in lines 19–22.*

It is worthwhile to point out that this program is somewhat inefficient, since it directly applies our multiple trapezoidal rule program for each value of n. As we see in Fig. 8.8, we can reduce the number of function evaluations by noting that, when we go from j

Figure 8.8 Schematic illustration of the computational advantage of increasing the number of trapezoids by a factor of 2 rather than some arbitrary number. At the second step, only those points indicated by an O required a function evaluation.

trapezoids for $I_{j,1}$ to $2j$ trapezoids for $I_{j+1,1}$, the integral corresponding to j trapezoids already contains half of the function evaluations required to compute the integral with $2j$ trapezoids. If we want to be more computationally efficient, we should compute the first estimate $I_{1,1}$ using our multiple trapezoidal rule program and then compute all further integrals $I_{j,1}$ by using the result from $I_{j-1,1}$ and only making function evaluations at the new intermediate points illustrated in Fig. 8.8.

Example 8.9   Use Romberg integration to compute the integral from Example 8.8.

Solution

Let's see how this solution works if we start with $n = 4$ trapezoids for $I_{1,1}$. This means that we need to use $n = 8$ trapezoids for $I_{2,1}$. We already know from Fig. 8.6 that neither of these multiple trapezoidal rule integrals is very good. At the start of the `while` loop, the matrix $\mathbf{I}$ in the program has the value

$$\mathbf{I} = \begin{bmatrix} 0.314\,209 & 0.318\,313 \\ 0.317\,287 & \end{bmatrix} \tag{8.4.24}$$

The entry $I_{1,1} = 0.314\,209$ is the result for $n = 4$ trapezoids and the entry $I_{2,1} = 0.317\,287$ is the result for $n = 8$ trapezoids. The difference between the estimates of the integral in the first row of $\mathbf{I}$ is relatively large. As a result, we should do at least one iteration of the `while` loop. This requires computing an integral with $n = 16$ trapezoids and leads to

$$\mathbf{I} = \begin{bmatrix} 0.314\,209 & 0.318\,313 & 0.318\,310 \\ 0.317\,287 & 0.318\,310 & \\ \mathbf{0.318\,054} & & \end{bmatrix} \tag{8.4.25}$$

where the entry in bold is the multiple trapezoidal rule value for $n = 16$. We are already doing much better – the difference between $I_{1,2}$ and $I_{1,3}$ is already in the sixth decimal place. If we go another round of integration with $n = 32$ trapezoids we have

$$\mathbf{I} = \begin{bmatrix} 0.314\,209 & 0.318\,313 & 0.318\,310 & 0.318\,310 \\ 0.317\,287 & 0.318\,310 & 0.318\,310 & \\ 0.318\,054 & 0.318\,310 & & \\ \mathbf{0.318\,246} & & & \end{bmatrix} \tag{8.4.26}$$

where the entry in bold is the multiple trapezoidal rule value for $n = 32$. At this point we have no change in the value of the integral out to six decimal places when we compare $I_{1,3}$ to $I_{1,4}$, which seems pretty good! If we run the loop just one more time with $n = 64$ trapezoids, the estimate of the integral becomes

$$I_{1,5} = 0.318\,309\,886\,183\,79 \tag{8.4.27}$$

The difference between the latter estimate and the exact answer $1/\pi$ is 1.1102×10^{-16}, which is almost at machine precision. The comparison between Romberg integration and the results by simply applying multiple trapezoidal rule in Example 8.8 is striking. Each of the individual estimates of the integral from multiple trapezoidal rule that we used here, corresponding to the values $I_{j,1}$, are not especially accurate. Indeed, even the first round of Romberg integration using $n = 4$ and $n = 8$ trapezoids, which produced the result $I_{1,2} = 0.318\,313$, leads to a better estimate of the integral than the $I_{4,1}$ entry, which corresponds to $n = 64$ trapezoids. Simply by combining these estimate via Eq. (8.4.23) we are able to almost reach the limits of machine precision. Moreover, we did not need to know *a priori* how many trapezoids to use to get the desired error. We simply continued to compute new values of $I_{1,k}$ until the solution was satisfactory. These two aspects of Romberg integration are especially attractive for computing integrals numerically.

8.5   Gauss Quadrature

In Newton–Coates formulas, we fixed the points on the interval and then used interpolation to arrive at an estimate for the integral. The evenly spaced points, which are useful for constructing a simple polynomial interpolant using finite differences, may not be the best locations for numerical accuracy. In Gauss quadrature, we pick both the points x_i and their weights to provide an optimal estimate of the integral. In doing so, we can try to offset the cases where the integral is overestimated or underestimated, as illustrated in Fig. 8.9. The particular method we will discuss here is Gauss–Legendre quadrature.

We will show how to derive the formula for two-point Gauss quadrature. Let's approximate the integral in the form

$$I \approx c_0 f(x_0) + c_1 f(x_1) \tag{8.5.1}$$

There are four parameters in this equation, the location of the function evaluations (x_0 and x_1), which are known as Gauss points, and the weights (c_0 and c_1). We will use these four degrees of freedom to ensure that we can capture a polynomial up to a cubic equation exactly on the interval $x \in [-1, 1]$.

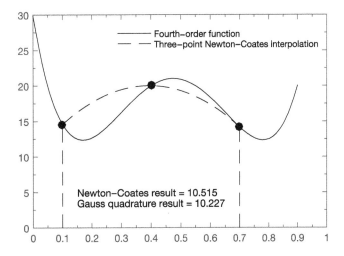

Figure 8.9 Uniformly spaced interpolation points may lead to overestimates or underestimates of the area under the curve.

Restricting the bounds of the integral simply makes less work for the algebra, as we can always make a change of variable to put an integral on $\tilde{x} \in [a, b]$,

$$I = \int_a^b f(\tilde{x}) d\tilde{x} \qquad (8.5.2)$$

on the interval $x \in [-1, 1]$ via

$$\tilde{x} = \frac{(b + a) + (b - a)x}{2} \qquad (8.5.3)$$

This change of variable means that

$$d\tilde{x} = \left(\frac{b - a}{2}\right) dx \qquad (8.5.4)$$

whereupon our integral in Eq. (8.5.2) becomes

$$I = \frac{b - a}{2} \int_{-1}^1 f(x) dx \qquad (8.5.5)$$

Let's see how to find the weights and Gauss points by requiring that Eq. (8.5.1) be exact for a constant function $f(x) = 1$,

$$c_0 + c_1 = \int_{-1}^1 dx = 2 \qquad (8.5.6)$$

a linear function $f(x) = x$,

$$c_0 x_0 + c_1 x_1 = \int_{-1}^1 x \, dx = 0 \qquad (8.5.7)$$

a quadratic function $f(x) = x^2$,

$$c_0 x_0^2 + c_1 x_1^2 = \int_{-1}^1 x^2 \, dx = \frac{2}{3} \qquad (8.5.8)$$

and a cubic function $f(x) = x^3$,

$$c_0 x_0^3 + c_1 x_1^3 = \int_{-1}^1 x^3 \, dx = 0 \qquad (8.5.9)$$

Equations (8.5.6)–(8.5.9) are a nonlinear system of algebraic equations. For the two-point Gauss quadrature, these can be solved quite readily. We first use (8.5.7) to get

$$c_1 = -\frac{c_0 x_0}{x_1} \qquad (8.5.10)$$

If we substitute this result in (8.5.9),

$$c_0 x_0^3 = \left(\frac{c_0 x_0}{x_1}\right) x_1^3 \qquad (8.5.11)$$

The result is

$$x_0^2 = x_1^2 \qquad (8.5.12)$$

Table 8.2 Gauss points and weight for three-point, four-point, and five-point Gauss quadrature on the interval $x \in [-1, 1]$.

Number of points	Gauss points	Weights
2	$\pm 0.577\,350$	1
3	0	0.888\,889
	$\pm 0.774\,597$	0.555\,556
4	$\pm 0.339\,981$	0.652\,145
	$\pm 0.861\,136$	0.347\,855
5	0	0.568\,889
	$\pm 0.538\,469$	0.478\,628
	$\pm 0.906\,180$	0.23\,6927

While one possibility is to use the same point twice, the much more useful result is to pick two points of the same magnitude but opposite sign,

$$x_0 = -x_1 \qquad (8.5.13)$$

The latter result in Eq. (8.5.10) with the normalization given by Eq. (8.5.6) gives

$$c_0 = c_1 = 1 \qquad (8.5.14)$$

If we now use the remaining constraint (8.5.8), we have

$$2x_0^2 = \frac{2}{3} \qquad (8.5.15)$$

which gives us the Gauss points,

$$x_0 = \frac{-1}{\sqrt{3}}, \quad x_1 = \frac{1}{\sqrt{3}} \qquad (8.5.16)$$

Two-point Gauss quadrature is thus

$$I = f\left(\frac{-1}{\sqrt{3}}\right) + f\left(\frac{1}{\sqrt{3}}\right) \qquad (8.5.17)$$

Gauss quadrature is remarkable because you use n function evaluations to get an exact result for polynomials up to order $2n - 1$. For smooth functions, Gauss quadrature gives very high accuracy. However, Gauss quadrature requires that you have information about the function at the Gauss points. When you know the function that you are trying to integrate, this is not an issue. However, if you have tabulated data, then it would be rather remarkable that you would have the data at the Gauss points. In the latter case, it is often preferable to use a Newton–Coates formula for the integral.

The derivation of two-point Gauss quadrature is relatively straightforward, since it is simple to compute the location of the Gauss points and their weights. When you move to higher-order Gauss quadrature, it is not so easy to solve the nonlinear algebraic equations. Fortunately, there is deeper mathematics to Gauss quadrature based on Legendre polynomials that allows you to determine the weights. Table 8.2 has the weights and

Gauss points for some other Gauss quadratures. Note that five-point Gauss quadrature can capture a ninth-order polynomial exactly.

PROGRAM 8.8   *Let's write a program similar to Program 8.4 that instead does Gauss quadrature.*

```
1   function I = integrate_gauss(xmin,xmax,n)
2   % xmin = lower bound for integral
3   % xmax = upper bound for integral
4   % n = order of Gauss quadrature
5
6   switch n
7       case 2
8           w = [1, 1];
9           x = [-1/sqrt(3), 1/sqrt(3)];
10      case 3
11          w = [5/9, 8/9, 5/9];
12          x = [-sqrt(3/5), 0, sqrt(3/5)];
13      case 4
14          a = sqrt(30)/36;
15          b = (2/7)*sqrt(6/5);
16          c = sqrt(3/7 - b);
17          d = sqrt(3/7 + b);
18          w = [0.5 + a, 0.5 + a, 0.5 - a, 0.5 - a];
19          x = [c, -c, d, -d];
20      case 5
21          a = 322/900;
22          b = 13*sqrt(70)/900;
23          c = 2*sqrt(10/7);
24          d = sqrt(5 - c)/3;
25          e = sqrt(5 + c)/3;
26          w = [128/225, a + b, a + b, a - b, a - b];
27          x = [0, d, -d, e, -e];
28      otherwise
29          fprintf('Value of n is not valid.\n')
30          x = 0;
31          w = 0;
32  end
33
34  for i = 1:n
35      y(i) = ((xmax + xmin) + (xmax - xmin)*x(i))/2;
36  end
37
38  I = 0;
39  for i = 1:n
40      I = I + w(i)*getf(y(i));
41  end
42  I = (xmax-xmin)*I/2;
```

Like we did for Newton–Coates integration, the first part of the program from lines 6–32 is a **switch** *operator that determines the order of the integration. The difference here is that we need to specify both the weights and the Gauss points, since they are not evenly spaced on the interval $x \in [-1, 1]$. Notice also that we used the exact results for the Gauss points and weights, rather than the numerical values in Table 8.2. (Indeed, this*

Table 8.3 Values of the integral in Eq. (8.3.22) using Gauss quadrature formulas of order n.

m	2	3	4	5
1	1.000 000	1.000 000	1.000 000	1.000 000
2	1.000 000	1.000 000	1.000 000	1.000 000
3	1.000 000	1.000 000	1.000 000	1.000 000
4	0.972 222	1.000 000	1.000 000	1.000 000
5	0.916 667	1.000 000	1.000 000	1.000 000
6	0.842 593	0.997 500	1.000 000	1.000 000
7	0.759 259	0.990 000	1.000 000	1.000 000
8	0.673 611	0.976 125	0.999 796	1.000 000
9	0.590 278	0.955 625	0.998 980	1.000 000
10	0.512 088	0.929 019	0.997 055	0.999 984

is how we generated the table.) If the integral that we want to compute is not in this interval, Lines 34–36 do the mapping in Eq. (8.5.3). lines 38–41 sum up the terms in the estimate for the integral, and line 42 makes the correction in Eq. (8.5.5) if the interval is not on $x \in [-1, 1]$.

Example 8.10 Repeat Example 8.7 for Gauss quadrature using Program 8.8.

Solution

For Gauss quadrature, we expect the integral to be exact up to a polynomial of order $m = 2n - 1$. Table 8.3 indicates that this is indeed the case – the integral value is $I = 1$ up to a polynomial of the proper order. You should also note that the error in the integral estimate increases with the order of the polynomial, as was the case with Newton–Coates integration.

8.6 Case Study: Concentrated Differential Distillation

8.6.1 Problem Statement

In this final case study, we will consider the case of differential distillation of benzene and toluene from our colleague Ed Cussler's book *Diffusion*, which is illustrated schematically in Fig. 8.10. In this system, we have a saturated liquid feed of 3500 mol/h of 40% benzene that is to be distilled into a top product of $D = 1400$ mol/h that is $x_D = 98\%$ benzene and a bottom product of 2100 mol/h that is $x_B = 2\%$ benzene. The flow of gas in the (upper) rectifying part of the column is $G = 4730$ mol/h and the flow of liquid in the rectifying part of the column is $L = 3330$ mol/h. The balance on toluene on the upper part of the column gives the operating line for rectification

$$y_r = \frac{Dx_D}{G} + \frac{L}{G}x = 0.287 + 0.704x \qquad (8.6.1)$$

Table 8.4 Vapor–liquid equilibrium data for the mole fraction of toluene.

x	0.01	0.08	0.17	0.35	0.40	0.52	0.58	0.62	0.75	0.90	0.97
y^*	0.03	0.17	0.33	0.57	0.63	0.73	0.78	0.81	0.88	0.96	0.99

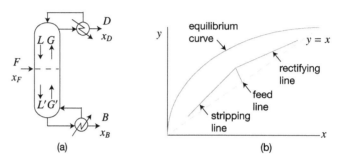

(a)   (b)

Figure 8.10 Concentrated differential distillation. (a) Schematic illustration of the distillation column with the various flow rates and compositions. (b) Sketch of the operating lines (stripping, rectifying, and feed) on a vapor–liquid x–y equilibrium diagram.

where x is the mole fraction of toluene in the liquid flowing downward at some point in the column and y is the corresponding mole fraction of toluene in the vapor flowing upward at that point in the column. Since the feed is saturated liquid, it all goes down into the stripping part of the column such that the liquid flow rate is $L' = 6830$ mol/h and the gas flow rate is $G' = 4730$ mol/h to satisfy the mass and energy balances at the feed location. The balance on toluene in the lower part of the column gives the second operating line for stripping

$$y_s = -\frac{Bx_B}{G'} + \frac{L'}{G'}x = -0.009 + 1.444x \tag{8.6.2}$$

These two lines intersect at the feed line, which for a saturated liquid at this concentration is $x = 0.4$.

The goal of the problem is to determine the number of theoretical units (NTU), which is a measure of the difficulty of the separation and an intermediate step to compute the height of the column (which also requires knowing the height of a theoretical unit, or HTU). The analysis of the mass transfer inside the column leads to

$$\text{NTU} = \int_{x_D}^{y_l(x_B)} \frac{dy}{y - y^*} \tag{8.6.3}$$

where $y_l(x_B)$ is the composition of the vapor phase that would be in equilibrium with the bottoms product at x_B. For the toluene–benzene system with $x_B = 0.02$, the equilibrium phase data gives $y_l(x_B) = 0.085$. The quantity y^* is the vapor-phase equilibrium composition, which is given in tabulated form in Table 8.4.

The goal of this problem is to use the methods in this chapter to compute the NTU for this process.

8.6.2   Solution

The first item of business is to determine a relationship between y^* and x so that we can use it in our integration. The simplest approach is to create an interpolating function to the equilibrium data, which we can then use directly in our integral. We can readily accomplish this task using Program 8.1 to create a tenth-order polynomial for $y^*(x)$. We can then break up the integral in Eq. (8.6.3) into two pieces,

$$\text{NTU} = \int_{x_D}^{y_F} \frac{dy_r}{y_s - y^*} + \int_{y_F}^{y_l(x_B)} \frac{dy_s}{y_s - y^*} \tag{8.6.4}$$

where $y_F = 0.5686$ is the point at which the two operating lines meet, i.e. the intersection of Eq. (8.6.1) and Eq. (8.6.2). We can then use any of our methods to evaluate the integrals. Let's use five-point Gauss quadrature, since it is quite accurate and this is a relatively smooth integrand. To do so, we first identify the Gauss points in y_r or y_s over their range of integration. At each Gauss point y_i, we can compute the corresponding value of x_i from either Eq. (8.6.1) or Eq. (8.6.2). We then use Program 8.2 to compute the corresponding value of y^* at the location x_i. We then are able to do the function evaluations for Gauss quadrature and get the result.

Let's see how this strategy is implemented. The main challenge is implementing the function evaluations, which are handled by the following program:

```
1  function f = getf(y,xpts,coeff)
2  n = length(coeff)-1;
3  if y >= 0.5686
4      x = (y - 0.287)/0.704;
5  else
6      x = (y + 0.009)/1.444;
7  end
8  ys = integrate_interpolate_f(x,xpts,coeff,n);
9  f = 1/(y - ys);
```

The **if-else** block selects the appropriate operating line, and we then evaluate the integrand at that value of y. Note that line 8 calls Program 8.2.

Actually computing the integrals is handled by a master program:

```
1  function integration_distillation
2  clc
3  xD = 0.98;
4  yF = 0.5868;
5  yl = 0.085;
6  %equilibrium data
7  x = [0.01,0.08,0.17,0.35,0.40,0.52,0.58,0.62,0.75,0.90,0.97];
8  ys = [0.03,0.17,0.33,0.57,0.63,0.73,0.78,0.81,0.88,0.96,0.99];
9  %interpolate ys(x)
10 [n,coeff] = integrate_interpolation_tabulated(x,ys);
11 %integrate lower
12 I1 = integrate_gauss(xD,yF,5,x,coeff);
13 I2 = integrate_gauss(yF,yl,5,x,coeff);
14 I = I1 + I2;
15 fprintf('NTU = %6.2f \n',I)
```

This function calls a modified version of Program 8.8 for Gauss quadrature that also passes the interpolation parameters, which are required by our function `getf.m`

```
1   function I = integrate_gauss(xmin,xmax,n,xpts,coeff)
2   % xmin = lower bound for integral
3   % xmax = upper bound for integral
4   % n = order of Gauss quadrature
5   % xpts, coeff  = parameters needed for getf
6
7   switch n
8       case 2
9           w = [1, 1];
10          x = [-1/sqrt(3), 1/sqrt(3)];
11      case 3
12          w = [5/9, 8/9, 5/9];
13          x = [-sqrt(3/5), 0, sqrt(3/5)];
14      case 4
15          a = sqrt(30)/36;
16          b = (2/7)*sqrt(6/5);
17          c = sqrt(3/7 - b);
18          d = sqrt(3/7 + b);
19          w = [0.5 + a, 0.5 + a, 0.5 - a, 0.5 - a];
20          x = [c, -c, d, -d];
21      case 5
22          a = 322/900;
23          b = 13*sqrt(70)/900;
24          c = 2*sqrt(10/7);
25          d = sqrt(5 - c)/3;
26          e = sqrt(5 + c)/3;
27          w = [128/225, a + b, a + b, a - b, a - b];
28          x = [0, d, -d, e, -e];
29      otherwise
30          fprintf('Value of n is not valid.\n')
31          x = 0;
32          w = 0;
33  end
34
35  for i = 1:n
36      y(i) = ((xmax + xmin) + (xmax - xmin)*x(i))/2;
37  end
38
39  I = 0;
40  for i = 1:n
41      I = I + w(i)*getf(y(i),xpts,coeff);
42  end
43  I = (xmax-xmin)*I/2;
```

The result of this program is NTU = 8.42. If you used Professor Cussler's book for mass transfer, you may recall that he got an answer of NTU = 13.9 for the same distillation specifications. These are substantially different answers, since the column required from Professor Cussler's calculation is 65% larger than our result. The difference lies in the thermodynamic data. The data you are provided with here are quite similar to those used by Professor Cussler in his book, but we have added on 5% random noise to each of his data points. If we run the same program but use Professor Cussler's data points, we indeed get the same answer, NTU = 13.9.

The sensitivity of the height of the distillation column to small fluctuations in the data is worth keeping in mind in other courses. It would seem that 5% uncertainty in the data is relatively small, but it is amplified in the final result via the interpolation and then the integration. In many instances, you may have an estimate for the error in the data. One very nice feature of understanding how numerical integration works (and, indeed, all of the other numerical methods in this book) is that you can readily do propagation of error through the numerical method to figure out exactly how the uncertainty in the data leads to uncertainty in the NTU. Since you also know something about the inherent error in the numerical methods, you can also make an informed conclusion about which factor is more important – error in your data or error in your calculation.

8.7   Implementation in MATLAB

The standard method in MATLAB to numerical integrate a function is called `integral`. If you are familiar with older versions of MATLAB, note that `integral` is the replacement for the function `quad`, which implemented a version of Simpson's rule.

Example 8.11   Use `integral` to evaluate the integral

$$I = \int_0^1 e^{-x^2} dx \qquad (8.7.1)$$

Solution
A MATLAB script to solve this problem is:

```
1  function matlab_integral
2  I = integral(@(x) getf(x),0,1)
3
4  function f = getf(x)
5  f = exp(-x.^2);
```

As we've seen previously, using `integral` requires an anonymous function call as the first entry. The second entry is the lower bound of the integral, and the third entry is the upper bound of the integral. There is one very important point to notice in line 5, where we provide the form of the integrand. When evaluating the integral, MATLAB needs to pass a vector of values of x_i points to evaluate the integral. Since we want the value of x_i^2, we need to use element-by-element multiplication, `x.^2`. This is different than writing `x^2`, which is multiplication of two vectors in MATLAB. If you write the latter, you will get a sea of red error messages!

MATLAB also has a function to implement multiple trapezoidal rule from tabulated data. If your data are spaced by $\Delta x = 1$, then you can just call `trapz` on the vector of

Table 8.5 Tabulated data for Example 8.12.

x	0.0000	0.2500	0.5000	0.7500	1.0000
y	1.0000	0.9394	0.7788	0.5698	0.3679

data. You need to make a rescaling if the data are not spaced by $\Delta x = 1$, which you can see from the following example.

Example 8.12   Use `trap` to find the integral using the tabulated data in Table 8.5.

Solution
A MATLAB script to solve this problem is:

```
1  function matlab_trapz
2  y = [1.0000, 0.9394, 0.7788, 0.5698, 0.3679];
3  I = trapz(y)/4
```

Our data have a spacing $\Delta x = 1/4$, so we have to multiply the result from `trapz` by the spacing Δx. The data in Table 8.5 are simply the values of $y = e^{-x^2}$, so this problem is equivalent to solving Eq. (8.7.1). The result of the five-point multiple trapezoidal rule, $I = 0.7430$, is not so far from the result from Example 8.7.1, $I = 0.7468$.

8.8   Further Reading

Numerical integration and interpolation are discussed at length in most standard numerical methods textbooks. A good reference that expands on some of the topics we covered here is

• S. C. Chapra and R. P. Canale, *Numerical Methods for Engineers*, McGraw Hill, 2010.

The content on concentrated distillation used in our case study is discussed at length in

• E. L. Cussler, *Diffusion: Mass Transfer in Fluid Systems*, Cambridge University Press, 2009.

Problems

8.1   How many points do you need to construct a cubic interpolation?

8.2   Derive the fourth-order finite difference $\Delta^4 f$.

8.3   Derive the truncation error for a sixth-order Newton–Coates integration formula.

8.4   Use Newton's divided difference interpolation to find erf(0.5) based on the data provided in the following table. You need to go up to the fourth divided difference. Compare your result with that obtained from typing `erf(0.5)` at the MATLAB command prompt.

x	erf(x)
0.2	0.222 70
0.358	0.387 35
0.41	0.437 97
0.537	0.552 41
0.6	0.603 86

8.5 If I want to compute the integral

$$\int_0^1 x^{\cos \pi x} dx$$

using two point Gauss quadrature, at what values of x do you evaluate the function $x^{\cos \pi x}$?

8.6 Answer the following questions about this code:

```
1   function problem8_6
2   clc
3   n = 3;
4   a = 0; b = 1;
5   j = getI(a,b,n);
6   k = getI(a,b,2*n);
7   m = getI(a,b,4*n);
8   o = (4*k - j)/3;
9   p = (4*m - k)/3;
10  I = (16*p - o)/15;
11  fprintf('%14.12f \n',I)
12
13  function out = getI(a,b,n)
14  h = (b-a)/n;
15  I = 0;
16  I = I + feval(a)/2 + feval(b)/2;
17  for j = 1:n-1
18      I = I + feval(a + j*h);
19  end
20  out = I*h;
21
22  function out = feval(x)
23  out = sin(pi*x^2)*exp(-x);
```

(a) What *mathematical* problem is being solved?
(b) What *numerical method* is used to solve the problem?
(c) What is the significance of the value of n on line 3?
(d) What numerical method is implemented by the function getI?
(e) Explain why there are two factors of 1/2 in line 16.
(f) There are six different estimates of the answer computed in this code. Which one is the least accurate?

Computer Problems

8.7 Compute the first four polynomial interpolations (linear, quadratic, etc.) for the function $f(x) = 1 - \exp(-x)$ with evenly spaced data on the interval $x \in [0, 3]$. Make a plot of the interpolants and the function itself using MATLAB. Also make a plot of the difference between the function and the interpolant. You can either work it out by hand or by writing a program. If you work it out by hand, include the values of all the coefficients for each interpolation formula.

8.8 Assume you have $n+1$ data points (x_i, y_i), $i = 0, \ldots, n$, where x is the independent variable and y is the dependent variable. You suspect that these data come from some unknown function of the form $y = f(x)$. Since the function is unknown, you choose to construct an approximation to the function based on interpolating the data with polynomials. Typically the x_i values are equispaced. In this problem, you will examine how well a polynomial interpolant based on equispaced data approximates a function. To do this, one way is to evaluate a known function at $n + 1$ points to get (x_i, y_i), interpolate through the points, and finally evaluate the interpolant as well as the original function at various points to check the error.

Consider the function

$$f(x) = \frac{1}{1 + 25x^2}, \qquad x \in [-1, 1].$$

Write a MATLAB program to do the following.

(a) Generate 500 equispaced points from $[-1, 1]$ using the `linspace` command and store it in a vector $\tilde{x}$. Determine $\tilde{y}$, where $\tilde{y}_i = f(\tilde{x}_i)$.

(b) For $n = 5$, 10, and 15:

    (1) Generate $n + 1$ equispaced points from $[-1, 1]$ and store it in a vector x. Evaluate y, where $y_i = f(x_i)$. These are the points based on which you will interpolate. Your interpolant will be an nth degree polynomial.

    (2) Determine the nth degree polynomial which interpolates through all pairs (x_i, y_i) using the Newton–Gregory method and evaluate it at all $\tilde{x}_i$, $i = 1, \ldots, 500$. Store the evaluated results in a vector $\hat{y}$. What is important is the value of the polynomial at different points, so you do not need to explicitly state or print this polynomial.

    (3) On the same graph, plot $\tilde{y}$ versus $\tilde{x}$, y versus x, and $\hat{y}$ versus $\tilde{x}$. The original function is $\tilde{y}$, its values at the interpolating points are given by y, and the values from your approximation to the function are given by $\hat{y}$.

Your program should not print anything to the screen. It should only generate the plots with appropriate titles. Here is one way to generate the plots:

```
1    function out = blahblahblah
2        for n = 5:5:15
3            Compute x_tilde, y_tilde, x, y, and y_hat ...
4            figure()
5            plot(x_tilde, y_tilde, '-k', x, y, 'sk', x_tilde, y_hat)
```

```
6              legend('ytilde', 'y', 'yhat')
7              title(['Plot for n = ', num2str(n)])
8         end
9      end
```

Do you observe any problem as you increase n? If you think there is a problem, how can this be fixed? Answer in words only.

8.9 Write a program that uses multiple trapezoidal rule to compute the integral

$$I = \int_0^3 1 - \exp(-x)dx$$

Your program should automatically generate a plot of I versus the number n of trapezoids used to compute the answer up to a reasonable accuracy. Include the exact result for the integral as a dashed line on the plot.

8.10 Write a MATLAB program that computes the integral

$$\int_{-1}^{1} \ln(\ln(x+3))dx$$

using the following methods:

(a) trapezoidal rule
(b) Simpson's 1/3 rule
(c) Simpson's 3/8 rule
(d) Boole's rule
(e) fifth-order Newton–Coates
(f) multiple trapezoidal rule (ten times)
(g) multiple trapezoidal rule (100 times)
(h) Gauss quadrature (two-point)
(i) Richardson extrapolation (using four and eight trapezoidal rule evaluations)
(j) Richardson extrapolation (using ten and 100 trapezoidal rule evaluations).

You program should print the results to the screen along with some text that explains which result is being printed using the fprintf command.

8.11 Write a MATLAB program that computes the integral

$$\int_{-1}^{1} e^{-3x} \ln[2 + \sin(\pi x)]\, dx$$

using the following methods:

(a) trapezoidal rule
(b) Simpson's 1/3 rule
(c) Simpson's 3/8 rule
(d) Boole's rule
(e) fifth-order Newton–Coates
(f) multiple trapezoidal rule (ten times)
(g) multiple trapezoidal rule (100 times)

Table 8.6 Rate data for Problem 8.12.

f_A	$-r_A$ (M/s)
0	3.94
0.1	2.91
0.2	2.33
0.3	1.95
0.4	1.47
0.5	1.08
0.6	0.82
0.7	0.49
0.8	0.34

(h) Gauss quadrature
(i) Richardson extrapolation (using four and eight trapezoidal rule evaluations)
(j) Richardson extrapolation (using ten and 100 trapezoidal rule evaluations).
You program should print the results to the screen along with some text that explains which result is being printed using the `fprintf` command.

8.12 An important time parameter in reaction engineering is the reactor space time τ, the time needed to process one reactor volume of fluid. For an ideal plug-flow reactor (PFR), a mass balance gives a design equation of

$$\tau = C_{A0} \int_{f_{Ain}}^{f_{Aout}} \frac{df_A}{-r_A}$$

where C_{A0} is the initial concentration of reactant A, f_A is the fractional conversion, and r_A is the rate of reaction. We can see that by plotting fractional conversion data against $-1/r_A$, space time can be calculated by integration. Rate data were collected straight from the reactor with the innovative rate meter. Given $C_{A0} = 5M$ and the data in Table 8.6, write a MATLAB program to calculate the space time of an ideal PFR via interpolation and multiple trapezoid integration to a reasonable accuracy. Let $f_{Ain} = 0.05$ and $f_{Aout} = 0.75$.

8.13 In reaction engineering, the average residence time $\bar{t}$ is the amount of time an element of fluid spends in a reactor and is related to the amount of substance present in the system. The easiest method to determine $\bar{t}$ is via a pulse stimulus, where a small amount of a tracer is put into a reactor operating under steady state and the effluent concentration measured over time. The average residence time is calculated as

$$\bar{t} = \frac{\int_0^\infty tC dt}{\int_0^\infty C dt}$$

Given the data in Table 8.7, write a MATLAB program to interpolate the data to the highest order, display the coefficients, and plot the polynomial. Use multiple trapezoid integration to calculate $\bar{t}$ to a reasonable accuracy.

Table 8.7 Concentration data of effluent in Problem 8.13.

t (s)	C (ppm)
0	0
100	20
200	20
300	16
400	10
500	7
600	5
700	3
800	1
900	0

8.14 Breakthrough curves describe the concentration of the solution eluted from an adsorption column over time. The efficiency of a column is calculated from breakthrough curves as

$$\theta = \frac{\int_0^{t_b} Q(c_0 - c)dt}{\int_0^\infty Q(c_0 - c)dt}$$

where Q is the volumetric flow rate, c_0 is the original concentration, and t_b is the breakthrough time, the time when the eluent concentration is no longer zero. Note that for an ideal adsorption column, the eluent concentration would be equal to the original at t_b and would give an efficiency of 100%. In the case of a real column, let t_b be the time when the concentration is 1% of the original, or $c/c_0 = 0.01$.

Given the data in Table 8.8, write a MATLAB program to interpolate the data to the highest order, display the coefficients, and plot the data and polynomial together. Use the same program to use multiple trapezoid integration to calculate the efficiency to a reasonable accuracy. By hand, compute the efficiency by replacing the integration with right-hand sums and compare to your program output.

Hint: It would be best to split the integrals into two sections (before and after $t = 12\,\text{h}$) for a better interpolation. The eluent concentration before $t = 12\,\text{h}$ is zero.

8.15 The kinetic theory of adsorption of polymers under shear flow near a plane wall leads to the following expression for the steady state concentration $C(z)$ in the wall-normal direction (z-axis)

$$C(z) = \exp\left[-\int_z^\infty \frac{L_d}{q^2}\,dq + 2\varepsilon_s\left\{\left(\frac{\sigma_s}{z}\right)^6 - \left(\frac{\sigma_s}{z}\right)^{12}\right\}\right],$$

where L_d is a characteristic length scale of polymer migration under shear flow, ε_s is the strength of polymer–wall attraction, and σ_s is a constant. L_d is some function of z, but it is only known numerically. In the far-field limit, you can use treat L_d as a constant, say

Table 8.8 Concentration data for breakthrough curve in Problem 8.14.

t (h)	$c(ppm)$
12	0
12.5	20
13	50
13.5	250
14	550
14.5	810
15	980
15.5	1220
16	1430
16.5	1650
17	1870
17.5	1930
18	1970
18.5	1990
19	2000

L. Then the integral inside the square brackets can be calculated analytically, and you obtain the much simpler expression

$$C(z) = \exp\left[-\frac{L}{z} + 2\varepsilon_s\left\{\left(\frac{\sigma_s}{z}\right)^6 - \left(\frac{\sigma_s}{z}\right)^{12}\right\}\right]$$

We want to calculate the amount of adsorbed Γ using the relation

$$\Gamma = \int_{z_i}^{z_c}\left[C(z) - 1\right]dz,$$

where z_i is the distance from the wall set by the polymer size and z_c is some distance from the wall.

Write a MATLAB program to calculate an approximation to the integral in Eq (8.4.15) using the multiple trapezoidal rule, with $\varepsilon_s = 4.0$, $\sigma_s = 2.0$, $L = 0.5$, $z_i = 0.5$, and $z_c = 10.0$. Plot the value you obtain with the number of segments, $n = 1, \ldots, 60$.

Your program should automatically generate the plot. If your program is right, your plot will show oscillatory behavior. Answer in words the possible cause for this. Usually, the trapezoidal rule is not accurate for the above type of function. Explain in words why it is so.

8.16 In thermodynamics, you learned that the saturation pressure for a one-component vapor–liquid equilibrium can be computed using an integral construction from the phase diagram, where the area "under" or "over" the curve needs to be equal. Using a departure function method, the saturation pressure at 395 K for the van der Waals fluid in Problem 5 was 1.404 886 MPa. Write a MATLAB program that uses Gauss quadrature to see if this result also agrees with the integral construction.

8.17   Derive the equations for the coefficients and function evaluation points that you would need for a n-point Gauss quadrature that is exact for $2n - 1$-order polynomials. Write a MATLAB program that uses Newton–Raphson to determine the coefficients and evaluation points for an arbitrary value of n. In your written solution, include the entries in the Jacobian. Use your program to solve for the coefficients and evaluation points for $n = 5$.

Appendix A **MATLAB "Tutorial"**

Let us to begin by saying clearly that this appendix is not a general tutorial on how to use MATLAB. There are many excellent resources on MATLAB in the suggested reading for this appendix, most notably the native documentation for MATLAB. Our goal here is to simply provide you with a condensed, collated list of the pieces of MATLAB that we use in the programs in this book. As such, it is a tutorial for understanding this book, rather than a tutorial in general.

All of our programs in the main text are saved as m-files, which are scripts that execute line-by-line. For the start of this tutorial, we will use the command line interface in MATLAB so that you can see the output of each line. We will then switch to scripts for multi-line commands, and finally end with functions when we summarize how they work.

A.1   Data Structures

Our programs use four types of data: scalars, vectors, matrices, and strings.

To define a scalar quantity, we type

```
>> x = 3.2
```

which assigns the number 3.2 to the memory space for the variable x. If we just want the unsigned value of a scalar variable, we can write abs(x). If we want to find the remainder from division of two scalars, we use the modulo operator,

```
>> mod(4,x)

ans =

    0.8000
```

In this example, MATLAB finds the remainder of 4 divided by x, which has a value of 3.2.

While it is possible to declare your variables as particular types in MATLAB, we take advantage of MATLAB's ability to determine the variable type on the fly. This usually works fine, although you can sometimes run into problems when your data are a floating point and you need a scalar, for example when determining the number of steps for numerical integration. So you need to be careful. If you do not declare a variable as an integer but need it to be an integer, one shortcut is to round off the number, e.g.

```
>> x = round(x)
```

There are options for how you round; round(x) goes to the nearest integer, floor(x) rounds down; and ceil(x) rounds up.

While we normally expect scalars to be actual numbers, there are two types of problems that can happen in MATLAB when the numbers get too larger or too small. MATLAB will either call them infinite, which is denoted as Inf, or "not a number," which is denoted NaN.

To create a vector, you need to enter its entries separated by a space or comma (to make a column vector) or by semicolons (to create a row vector). For example, the input

```
>> y = [3 2 1]

y =

     3     2     1
```

makes a 1×3 column vector for y. If you wanted a column vector, you would enter

```
>> y = [3; 2; 1]

y =

     3
     2
     1
```

Often, we would like to make vectors that have linearly spaced data between two values, for example when we set up a grid for a boundary value problem. MATLAB offers a convenient function, linspace, that implements this operation. If we write

```
>> y = linspace(0,1,5)

y =

     0    0.2500    0.5000    0.7500    1.0000
```

the vector y contains five evenly spaced points between 0 and 1. If you need logarithmically spaced data, there is a function logspace that acts similarly. An alternate approach is to specify the bounds and the increment,

```
>> y = 0:0.25:1

y =

     0    0.2500    0.5000    0.7500    1.0000
```

The entry here asks for numbers from 0 to 1 in increments of 0.25. If your increment overruns the upper bound, the last entry is just ignored. For example:

```
>> y = 0:0.25:1.1

y =

        0    0.2500    0.5000    0.7500    1.0000
```

gives the same result since the next entry, 1.25, is above the upper bound of 1.1.

If you need to address an entry in a vector, you simply refer to its index. For example, if we wanted to change the second entry in y from 0.25 to 0.5, we would write

```
>> y(2) = 0.5

y =

        0    0.5000    0.5000    0.7500    1.0000
```

If you need to know the size of a vector, you can use the function `length`,

```
>> length(y)

ans =

     5
```

To create a matrix, separate the entries on a given row by commas and then separating the rows by semicolons,

```
>> A = [1, 3, 2; 2, 1 3; 1, 1, 1]

A =

     1     3     2
     2     1     3
     1     1     1
```

Similar to vectors, we can address individual entries in a matrix by their row and column number. For example, we can change the $A_{3,2} = 1$ to $A_{3,2} = 4$ by writing

```
>> A(3,2) = 4

A =

     1     3     2
     2     1     3
     1     4     1
```

If we need to know the size of a matrix, we can use the command `size`

```
>> size(A)

ans =

      3      3
```

This function returns a vector where the first entry is the number of rows and the second entry is the number of columns. For a square matrix, as was the case here, these two numbers are equal.

For the programs in our book, one of the most important MATLAB functions for vectors and matrices is `zeros`. This function makes a vector or matrix full of zeros. If we want a vector, we can write

```
>> z = zeros(1,3)

z =

      0      0      0
```

which makes a 1×3 vector of zeros. If we want a matrix, we can write

```
>> A = zeros(3,2)

A =

      0      0
      0      0
      0      0
```

If the matrix is square, then we just need to enter one value,

```
>> A = zeros(3)

A =

      0      0      0
      0      0      0
      0      0      0
```

Having made space in the memory for the vector or matrix, we can now enter each entry individually. For example, we can put $A_{1,1} = 2$ by writing

```
>> A(1,1) = 2

A =

      2      0      0
      0      0      0
      0      0      0
```

When matrices have many zeros in them, it is convenient to store them in sparse form. For the previous case, we would get

```
>> A = sparse(A)

A =

   (1,1)        2
```

where the result is a listing of the non-zero entries in the matrix and their values.

The `zeros` function is very important for setting up space in the memory when we have very large vectors or matrices, which happens frequently in our solutions to differential equations. If we do not make the space first and then fill it in later, when we add a new dimension to the matrix, MATLAB will copy all the old data to the new matrix and then add the new entry. You will not notice the slowdown if you have a small problem, but it becomes very apparent if you are copying large amounts of information.

Sometimes we know that many of the entries in a matrix or vector are going to be 1 instead of 0, for example as the initial condition to a boundary value problem. In this case, we can save time by initializing the matrix as all 1 by using `ones` instead of `zeros`. Naturally, this provides us with a way to set all the entries of the matrix to the same value by multiplying the result from `ones` by some scalar.

Strings are series of text and are entered using single quotation marks,

```
>> s = 'This is a string'

s =

This is a string
```

In MATLAB, the string will be purple-colored to make it easy to see. String manipulation is a major part of many programs, but it is less frequently used in scientific computation. For the most part, we will simply use strings to set output files and formatting commands.

If you want to clear out the memory, simply type

```
>> clear all
```

If you want to remove specific items from the memory, type them after the word `clear`. For example,

```
>> clear x
```

would delete x from the memory.

A.2   Matrix and Vector Manipulation

MATLAB has some very powerful tools for handling matrices and vectors. For the most part, we do not use these tools in our codes. However, there are a few tools that we do use at different points in the book.

One useful tool is to transpose a matrix or vector. For example, imagine we typed in a vector in the form

```
>> x = [1 3 2]

x =

     1     3     2
```

but realized that we wanted it as a row vector. We can do the transpose by typing

```
>> x = x.'

x =

     1
     3
     2
```

Naturally, we can also use transpose on matrices.

A second useful tool is the `norm` function, which we use ubiquitously to check for convergence of iterative methods. This function is implemented in a straightforward way,

```
>> norm(x)

ans =

    3.7417
```

This is the Euclidean norm of **x**. For a matrix, the `norm` function returns the spectral norm.

We can also add up the entries in a matrix or vector with the `sum` function. If we use `sum` for a vector, we just get back the sum of the entries,

```
>> sum([1 3 2])

ans =

     6
```

If we use `sum` for a matrix, we get back a vector with the sum of the columns,

```
>> sum([1,2;3 4])

ans =

     4     6
```

If we needed to find the sum of all the entries in a matrix, we could nest two `sum` functions,

```
>> sum(sum([1,2;3,4]))

ans =

    10
```

There are built-in functions to also find the minimum (`min`) and maximum (`max`) of a vector or matrix. For a vector, the implementation looks like

```
>> min([3,1,4])

ans =

    1
```

Note that you can use this function without creating a vector, for example by writing

```
>> max(3,2)

ans =

    3
```

While this seems silly for comparing two numbers, it is a useful function for comparing the size of two variables, especially if they are changing throughout the calculation.

For a matrix, the result of `max` or `min` is a vector,

```
>> max([1 2; 3 4])

ans =

    3    4
```

where the entries are the maximum values in each column.

It is also relatively easy to move around sections of vectors and matrices. If you want to build up a matrix from vectors, you first need to make the matrix

```
>> A = zeros(3)

A =

    0    0    0
    0    0    0
    0    0    0
```

and then you can put the vector somewhere inside that matrix

```
>> A(:,1) = x

A =

    1    0    0
```

```
         3      0      0
         2      0      0
```

In this case, we are putting the vector x from the start of this subsection as "All rows, Column 1" of A. If the matrix is bigger than the vector, we need to say where to put the vector. For example, if we write

```
>> A = zeros(4)

A =

       0      0      0      0
       0      0      0      0
       0      0      0      0
       0      0      0      0
```

to make a 4 × 4 matrix, we could then write

```
>> A(1:3,1) = x

A =

       1      0      0      0
       3      0      0      0
       2      0      0      0
       0      0      0      0
```

to put x in "Rows 1–3, Column 1" of A. If you want to extract parts out of matrix, you could write something like

```
>> A(3,1:2)

ans =

       2      0
```

which pulls out the third row, columns 1 and 2 of A as a new vector.

One other item to know about MATLAB is that there are two different types of multiplication. For example, let's consider a pair of vectors,

```
>> x = [1 2 3], y = [3; 2; 1]

x =

       1      2      3


y =

       3
       2
       1
```

Note that we were able to enter two commands on a single line by separating them with a comma. If you wanted to do normal vector multiplication, you just use the multiplication symbol,

```
>> x*y

ans =

     10
```

MATLAB also has a facility to do element-by-element multiplication, which has a different notation:

```
>> x.*y'

ans =

     3     4     3
```

This produces a vector output where each entry is $x_i y_i$. Notice the transpose operator in this last example, which is required so that x and y have the same size.

Both types of multiplication could have been used in our programs in the main text to increase the speed of the programs. However, in many instances we choose to work with **for** loops so that you can more easily see the algorithm.

A.3   Order of Operations

When you write statements to do calculations, it makes sense to put them together into more complicated statements. For example, you could write

```
>> x = 5 + 2

x =

     7

>> x = x^2

x =

     49
```

but it would be simpler just to write

```
>> x = (5+2)^2

x =

     49
```

In order to be sure you get the answer you want, you need to keep in mind the order of precedence for different operators in MATLAB:

(1) Parentheses, ( )
(2) Exponentiation, ^
(3) Negation, –
(4) Multiplication and division, * or /
(5) Addition and subtraction, + or –

Note that you can compute square roots either by writing x^(0.5) or sqrt(x).

A.4 Displaying Data on the Screen

You will notice in the previous examples that every time we entered a command, it was echoed back to screen. If you want to suppress this output, you add a semicolon at the end of the line. For example,

```
>> s = 'This is a string';
```

makes a string s in the memory but nothing is returned to the screen. This is *very* important when you are writing programs that handle large amounts of data or iterate many times. If you display the result of each line to the screen, you will be spending a large amount of time waiting for those data to be printed. This can lead to a substantial slowdown in your program.

If you want to display a variable, there are a number of options. In MATLAB, the simplest option is to use the disp command. This is an unformatted display. For example, the last time we defined A we set it to be a 3 × 3 matrix of zeros. If we now use disp(A) we get

```
>> disp(A)
     0     0     0
     0     0     0
     0     0     0
```

which are all of the entries in A.

A more robust printing method is to use the fprintf command, which is from the C programming language. Any reference on C will have extensive documentation on using fprintf, which is rather powerful. We will only use the most basic functionality here. If you want to display text only, it is very simple:

```
>> fprintf('This is \t text \n')
This is    text
```

In making this example, we included the two most important text formatting commands. If you type \t, this creates a tab space, which is useful for aligning multiple rows of text, for example when printing a table. The formatting command \n makes a carriage return.

If you want to print a variable to the screen, you need to know how much space you want and the type of variable. For example, imagine we previously set x = 3. If we type

```
>> fprintf('x = %3d \n', x)
x =    3
```

we decided to make an integer (d) with three spaces. The % symbol is a placeholder, which then refers to the argument after the format string. If instead we wanted to print a floating point, we change the formatting symbol to f and provide a placeholder with the number of "spaces.number of decimal points",

```
>> fprintf('x = %6.4f \n', x)
x = 3.0000
```

Here, we asked MATLAB to write x as a floating point (f) number with four decimal places, holding six spaces for the entire number. While the first number in the formatting command needs to be large enough to fit the entire number in other programming languages, MATLAB will correct it if you do not provide enough space. The other type of output is in exponential notation,

```
>> fprintf('x = %5.3e \n', x)
x = 3.000e+00
```

In this case, we asked for an exponential notation (e) with three decimal places and five total characters to hold the variable.

While the examples above only have a single variable, you can put multiple variables in the same line by adding additional placeholders. For example, the entry

```
>> fprintf('y1 = %4.2f \t y2 = %4.2f \n',y(1),y(2))
y1 = 0.00    y2 = 0.50
```

writes the first two entries in our vector y to screen with a tab between each entry.

If you want to clear the screen, type

```
>> clc
```

A.5   Logical Flow

The **if**, **if-else**, and **if-elseif** commands are the standard ways to control logical information and discussed in Section 1.4.4. Here we simply want to summarize the syntax of these functions. Since we are now getting a bit more complicated, let's start writing the examples as m-files to make the multi-line commands easier to see.

Table A.1 Logical operators in MATLAB. Note that the function `strcmp(s1,s2)` is used to see if two strings are the same, rather than `s1 == s2`.

Is equal	Is not equal	Greater than	Less than	Greater or equal	Less or equal
==	~=	>	<	>=	<=

For an **if** statement, we have a single block:

```
1  if x == 3
2      y = 2
3  end
```

This command checks the value of x. If the value is 3, then it sets the value of y to be 2. If x is not equal to 3, then the loop does nothing. There are six logical operators listed in Table A.1. The most important ones to note are the "equals" operator, which requires two equals signs since a single equal sign is used to assign a value, and the "not equals" operator. Note that the output of these arguments is a logical variable, either true or false, which is then the input to the **if** operation.

While our example has only a single logical argument, we can provide multiple arguments in the same **if** block. If we want to have "and" logic, we could write **if x < 2 && x > -2** which would be true if $-2 < x < 2$ and false otherwise. If we want to implement "or" logic, we could write **if x > 2 || x < -2**, which would be true if either $x < -2$ or $x > 2$.

If we want to have MATLAB make a decision between two options, we use an **if-else** block:

```
1  if x == 3
2      y = 2
3  else
4      y = 1
5  end
```

In this example, we set `y = 2` if `x = 3`. If this is not true, then we set `y = 1`.

If you need multiple options, then you can use an **if-elseif** structure. Continuing with our previous example, we can add a third option by writing:

```
1  if x == 3
2      y = 2
3  elseif x < 4
4      y = 1
5  else
6      y = 0
7  end
```

This block will set `y = 2` if `x = 3` and exit. If the first argument is false, then the next step in the block checks if `x < 4`. If this is true, then `y = 1` *unless* we already set `y = 2` from the first block. If the second statement is also false, then we end up at the **else** block and set `y = 0`.

You can construct extremely complicated logical operations by nesting various **if** statements. Eventually, this can start to look like spaghetti programming. If you have lots of options, it is worthwhile to implement a **switch-case** structure. An example is

```
1  switch x
2      case x == 1
3          y = 2
4      case x == 2
5          y = 3
6      case x > 4
7          y = 5
8      otherwise
9          y = 0
10  end
```

The variable after **switch** is the one we want to use for the different cases below. We then see if the cases are true. If so, then the block is executed. If all of the cases are false, then we execute the **otherwise** block. The **switch-case** structure is very easy to read and often preferable to a very complicated set of **if** statements.

A.6   Loops

We use two different types of loops in the book, **for** loops (discussed in Section 1.4.2) and **while** loops (discussed in Section 1.4.3).

The syntax of a **for** loop is illustrated in the following example:

```
1  for i = 1:10
2      x = x + i
3  end
```

The command after the word **for** sets the range of an integer counter, in this case i, from $1, 2, \ldots, 10$. The commands between **for** and **end** execute each time the loop runs. For this problem, if we started with x = 0, we would end up with x = 55 at the end of the loop.

While most times you will count forward in increments of 1 for a **for** loop, it can be quite useful to count in other ways. For example, if you want to count backwards, you could write

```
1  for i = 10:-1:1
2      x = x + i
3  end
```

This script does the same thing as the previous one, except that it starts with i = 10 and counts backwards by 1 until i = 1.

The structure of a **while** loop is slightly different than a **for** loop,

```
1  while x < 10
2      x = x + 1
3  end
```

The line with the word **while** is followed by a logical statement, and the entries in the block are executed so long as the statement is true. There are two important things to remember about **while** loops that are common bugs in student codes. First, if the conditional statement is false before the start of the **while** loop, then the loop will not execute. Second, if you do not update the variable in the conditional statement, then the **while** loop will execute forever.

The standard way to avoid infinite **while** loops is to keep track of the number of times the loop executes and quit if the loop goes too long.

```
1   k = 0;
2   while x < 10
3       x = x + 1;
4       k = k + 1;
5       if k > 20
6           break
7       end
8   end
```

In this example, we use the variable k to keep track of the number of iterations and exit if we exceed 20 iterations. The word **break** will break out of a loop. Of course, you need to remember to update k inside the **while** loop, otherwise the **if** statement does not help!

A.7   Plotting

One of the best parts of MATLAB is the ability to generate plots. This feature, along with the easy debugging of codes, is a major reason why we like to use MATLAB for teaching numerical methods. The plotting capabilities of MATLAB are rather extensive, and we used MATLAB to generate many of the figures appearing in this book. Making nice figures in MATLAB takes some skill and care. Making simple figures is very quick and, at least for our classes, more than sufficient for solving homework problems. Let's review the main ways to make plots here.

Let's imagine we have some vector of data y as a function of x. The following script makes the reasonably nice plot of the data in Fig. A.1:

```
1   h = figure;
2   plot(x,y,'-ok')
3   xlabel('x','FontSize',14)
4   ylabel('y','FontSize',14)
5   legend('y(x)','Location','SouthEast')
6   legend boxoff
7   saveas(h,'fig_appendix_plot1.eps')
```

The first line of the script creates a figure window and gives it a handle called h, and the next line plots the data in that figure window. Note that the first line is not required; if you call plot, it will plot the data in the most recent figure window. If there is no figure window, then MATLAB will make one. The problem with this lazy approach is that it will overwrite anything in the figure window, which is troublesome when we want to

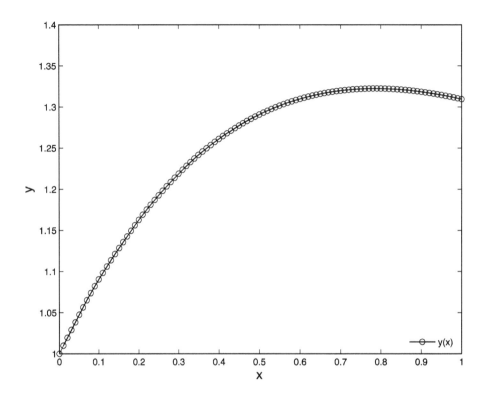

Figure A.1 Example of MATLAB plotting a single function.

make a number of figures. The string in line 2 provides a formatting command to make black circles connected with a solid line. You can find the list of default formatting commands if you type >> help plot at the command line. Lines 3 and 4 create the axis labels. We find that the default font sizes for the axis labels are a bit small, so we increase them to 14 pt. Lines 5 and 6 create a legend and put it at the lower-right corner of the plot. Line 7 saves the plot as an eps figure, which is our preferred format. If you wanted to save in a different format, like jpg, you just replace the ".eps" extension with ".jpg".

If we have a second function $z(x)$ that we want to plot from data in a vector z at the same values of x, we can create the plot (see Fig. A.2) using the following script:

```
1  g = figure;
2  plot(x,y,'-ok',x,z,'-xk')
3  xlabel('x','FontSize',14)
4  ylabel('y,z','FontSize',14)
5  legend('y(x)','z(x)','Location','SouthEast')
6  legend boxoff
7  saveas(g,'fig_appendix_plot2.eps')
```

We had to make a few changes from the previous figure. First, note that line 1 changes the figure handle to g since we already have a figure window open with handle h. The

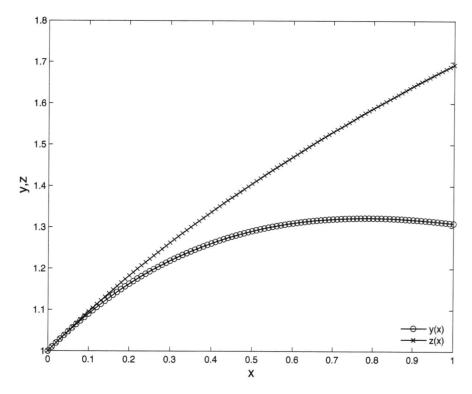

Figure A.2 Example of MATLAB plotting two functions.

second line has been modified to plot z versus x with a solid line with black xs. In line 4, we changed the axis label, and in line 5 we changed the legend to include the second line.

Both of the plots we made here use linear axes. If you want to make a semi-logarithmic plot, you replace `plot` with either `semilogx` or `semilogy`, depending on which axis should be logarithmic. If you want to make a log–log plot, you replace `plot` with `loglog`.

There are several nice functions in MATLAB, such as `mesh` and `surf` for making two-dimensional plots such as the ones in Chapter 7. The syntax and operation of these functions are a bit complicated, so we suggest you take a look at the MATLAB documentation if you need to make these types of plots.

If you make lots of plots by creating new figure windows each time, you can get quite a pile on the screen. More important than the aesthetics is the memory taken up by the plots. So it can be convenient to close them all by entering the command `close all`.

A.8 Functions

The programs in this book are exclusively created by functions. Function files are very similar in format to the scripts that we wrote above (all the boxes with numbered lines).

In both cases, MATLAB executes the entries on each line of the file exactly the same as if those lines were entered on the screen. Likewise, both files are saved with the extension ".m" and called m-files. However, there is a very important difference between scripts and functions in the way that they interact with the memory. In a script, you have access to the working memory (called the Workspace in the current version of MATLAB), and anything you do in the script will affect the local memory. In this sense, the script is literally the same as typing at the command line. In contrast, all variables inside a function are defined locally. They are created when the function starts, and deleted when the function ends. This compartmentalization of information is a tremendous advantage in constructing modular programs, since each function can operate independently of other functions.

As you can see in the numerous examples inside the text, function files start with a very specific format on the first line:

```
function [out1, out2] = function_name(input1, input2, input3)
```

This line must appear as the first line of your function, and `function_name` must be the name of your m-file. The items `input1` and so forth are the inputs to the function. They can be any type of data structure. The function file we wrote here returns two outputs, `out1` and `out2`, in a single vector. Functions can only return one argument, but using vectors in MATLAB allows you to easily return many variables as output of the function.

An important concept in functions (and scripts too, if you write long ones) is the use of comments. A comment starts with the % symbol. All text following the % symbol is not read by the program. This provides a convenient way to explain what your code does, a procedure unsurprisingly called commenting the code. You will see that most of the functions in this book begin with comment lines that explain the inputs to the function.

It is straightforward to combine many functions together to create a bigger program. Indeed, this is the approach that we use ubiquitously in this book. There are two ways to accomplish this task. The most straightforward approach is to cut-and-paste all of the functions into a single m-file. The first function in the file executes first, and the name of that first function should also be the name of the m-file. The remaining functions fit inside a hierarchy, where the use of each function can eventually be traced back to the original file. As an example, Fig. A.3 shows the hierarchy for the liquid–liquid phase equilibrium case study in Section 3.10.

The other option you have is to put the functions in the MATLAB path as separate files. This approach is what you are doing when you call built-in MATLAB functions. MATLAB has a specific order in which it searches through the directories that are in its path. The important point to remember is that the "Current Folder" is the first place MATLAB searches in its path. So if you do not create a single m-file that has all of the functions in the hierarchy, the next easiest approach is to put all of the individual function files in the same directory. You then run the file at the top of the hierarchy, and it calls all of the subsequent functions.

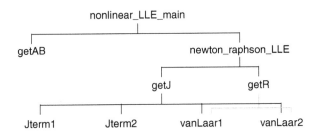

Figure A.3 Example of a hierarchy of functions for the case study in Section 3.10.

It is also important to keep in mind how a function ends. The word **return** inside a function forces the function to quit. In some languages, this word is required at the end of a function to force it to end. This is not the case in MATLAB, but **return** can be a convenient tool to use deep inside your function if you want it to terminate because it is not working well. MATLAB does offer the option of having the word **end** at the end of the function. Note that if one of your functions uses **end**, then all of the functions need to use **end**.

Appendix B **Determinant of a Matrix**

A system of linear equations will have a unique solution if the determinant of the matrix, $\mathbf{A}$, is not equal to zero. The standard approach to compute the determinant of a matrix is to use co-factor expansions. A co-factor expansion along an arbitrary row i is

$$\det \mathbf{A} \equiv |\mathbf{A}| \equiv a_{i,1} A_{i,1} + a_{i,2} A_{i,2} + \cdots + a_{i,n} A_{i,n} \tag{B.1}$$

The entries a_{ij} are from the original matrix in Eq. (2.1.3). The quantities A_{ij} are the co-factors defined as

$$A_{ij} = (-1)^{i+j} \det \mathbf{M}_{ij} \tag{B.2}$$

where $\mathbf{M}_{ij}$ is a matrix formed by deleting the ith row and jth column of $\mathbf{A}$. Co-factor expansion thus reduces the problem of finding the determinant of an $n \times n$ matrix to the problem of finding the determinant of n matrices of size $(n-1) \times (n-1)$. Since the determinant of a scalar is just the scalar itself, we can keep reducing the size of the matrix until the matrices have size 1×1.

Co-factor expansion allows us to develop a very nice formula for the determinant of a 2×2 matrix. If we do co-factor expansion on the first row we have

$$\det \mathbf{A} = a_{1,1} A_{1,1} + a_{1,2} A_{1,2} \tag{B.3}$$

Substituting the definition of the A_{ij} terms gives

$$\det \mathbf{A} = a_{1,1}(-1)^{1+1} \det a_{2,2} + a_{1,2}(-1)^{1+2} \det a_{2,1} \tag{B.4}$$

which gives the standard formula,

$$\det \mathbf{A} = a_{1,1} a_{2,2} - a_{1,2} a_{2,1} \tag{B.5}$$

We can still do a reasonable job of computing the determinant of a 3×3 matrix. The first step reduces the problem to adding three determinants of 2×2 matrices,

$$\det \mathbf{A} = a_{1,1} \det \begin{bmatrix} a_{2,2} & a_{2,3} \\ a_{3,2} & a_{3,3} \end{bmatrix} - a_{1,2} \det \begin{bmatrix} a_{2,1} & a_{2,3} \\ a_{3,1} & a_{3,3} \end{bmatrix}$$
$$+ a_{1,3} \det \begin{bmatrix} a_{2,1} & a_{2,2} \\ a_{3,1} & a_{3,2} \end{bmatrix} \tag{B.6}$$

If we apply Eq. (B.5) to each of the 2×2 arrays we get

$$\det \mathbf{A} = a_{1,1} a_{2,2} a_{3,3} + a_{1,2} a_{2,3} a_{3,1} + a_{1,3} a_{2,1} a_{3,2}$$
$$- a_{1,3} a_{2,2} a_{3,1} - a_{1,1} a_{2,3} a_{3,2} - a_{1,2} a_{2,1} a_{3,3} \tag{B.7}$$

This is also a standard formula that we can remember by writing the original matrix in the form

$$
\begin{bmatrix}
a_{1,1} & a_{1,2} & a_{1,3} \\
a_{2,1} & a_{2,2} & a_{2,3} \\
a_{3,1} & a_{3,2} & a_{3,3}
\end{bmatrix}
\begin{matrix}
a_{1,1} & a_{1,2} \\
a_{2,1} & a_{2,2} \\
a_{3,1} & a_{3,2}
\end{matrix}
\tag{B.8}
$$

The positive terms in Eq. (B.7) are the products of the three right downward diagonals and the negative terms in Eq. (B.7) are the products of the three left downward diagonals.

There is no comparable simple mnemonic aid for 4×4 systems and larger ones. Thus, we generally use co-factor expansion to reduce the original matrix down to a sum of 3×3 matrices and then apply Eq. (B.7) to each of these matrices. While this approach works, there are much faster ways to compute the determinant of a matrix!

There are many useful properties of determinants that are discussed and proved in standard textbooks on linear algebra. As this is a text intended for engineers, let's go through some of the most important properties and show that they are correct for a small system.

Example B.1   In this set of examples, we will use the following matrices

$$
\mathbf{A} = \begin{bmatrix} 1 & 2 \\ 3 & 4 \end{bmatrix} \quad
\mathbf{B} = \begin{bmatrix} 5 & 6 \\ 7 & 8 \end{bmatrix} \quad
\mathbf{U} = \begin{bmatrix} 3 & 2 & 4 \\ 0 & 1 & 6 \\ 0 & 0 & 2 \end{bmatrix}
\tag{B.9}
$$

to demonstrate some properties of determinants. The first one is $\det \mathbf{A}^\dagger = \det \mathbf{A}$.

Solution
Let

$$
\mathbf{A} = \begin{bmatrix} 1 & 2 \\ 3 & 4 \end{bmatrix}
\tag{B.10}
$$

Taking the transpose gives

$$
\det \begin{bmatrix} 1 & 3 \\ 2 & 4 \end{bmatrix} = \det \begin{bmatrix} 1 & 2 \\ 3 & 4 \end{bmatrix}
\tag{B.11}
$$

Applying the rule for a 2×2 determinant gives

$$
4 - 6 = 4 - 6
\tag{B.12}
$$

which is true.

Example B.2   If two rows of a matrix are identical, $\det \mathbf{A} = 0$.

Solution
As an example, use the first row of $\mathbf{A}$ from Example B.1:

$$
\det \begin{bmatrix} 1 & 2 \\ 1 & 2 \end{bmatrix} = 2 - 2 = 0.
\tag{B.13}
$$

Example B.3 Adding a multiple of one row to another leaves the determinant unchanged.

Solution

For example, add the rows of **A** from Example B.1 together to get

$$\mathbf{A}^* = \begin{bmatrix} 1 & 2 \\ 4 & 6 \end{bmatrix} \tag{B.14}$$

The determinant is

$$\det \mathbf{A}^* = 6 - 8 = -2 \tag{B.15}$$

which is the same as what we got for $\det \mathbf{A}$.

Example B.4 The determinant changes sign when two rows are interchanged.

Solution

Reverse the rows of **A** from Example B.1 to get

$$\det \begin{bmatrix} 3 & 4 \\ 1 & 2 \end{bmatrix} = 6 - 4 = 2 \tag{B.16}$$

which is the alternate sign to $\det \mathbf{A}$.

Example B.5 The determinant of an upper triangular matrix is the product of the entries on the diagonal.

Solution

For example, let's use

$$\mathbf{U} = \begin{bmatrix} 3 & 2 & 4 \\ 0 & 1 & 6 \\ 0 & 0 & 2 \end{bmatrix} \tag{B.17}$$

If we apply the rule for 3×3 determinants:

$$\det \mathbf{U} = [(3)(1)(2) + (2)(6)(0) + (4)(0)(0)]$$
$$-[(4)(1)(0) + (3)(6)(0) + (2)(0)(2)]$$
$$= (3)(1)(2) = 6 \tag{B.18}$$

The only entry without a zero is the diagonal. This identity is required to compute the determinant of a matrix by Gauss elimination.

Example B.6 One of the most interesting properties is

$$\det(\mathbf{AB}) = \det(\mathbf{A})\det(\mathbf{B}) \tag{B.19}$$

Solution

Let's try this with $\mathbf{A}$ from Example B.1 and

$$\mathbf{B} = \begin{bmatrix} 5 & 6 \\ 7 & 8 \end{bmatrix} \tag{B.20}$$

To test this out, we first need to do the matrix multiplication:

$$\mathbf{AB} = \begin{bmatrix} 1 & 2 \\ 3 & 4 \end{bmatrix} \begin{bmatrix} 5 & 6 \\ 7 & 8 \end{bmatrix} = \begin{bmatrix} 19 & 22 \\ 43 & 50 \end{bmatrix} \tag{B.21}$$

The determinant is

$$\det(\mathbf{AB}) = (19)(50) - (43)(22) = 950 - 946 = 4 \tag{B.22}$$

We have not yet computed

$$\det \mathbf{B} = \det \begin{bmatrix} 5 & 6 \\ 7 & 8 \end{bmatrix} = 40 - 42 = -2 \tag{B.23}$$

So we also have

$$(\det \mathbf{A})(\det \mathbf{B}) = 4 \tag{B.24}$$

There are two important corollaries to Example B.6. Consider first the case where $\mathbf{B} = \mathbf{A}^{-1}$. Then we have

$$\det(\mathbf{AB}) = \det(\mathbf{AA}^{-1})$$
$$= \det \mathbf{I}$$
$$= 1 \tag{B.25}$$

If we use Eq. (B.19), then we have

$$\det \mathbf{A} = \frac{1}{\det \mathbf{A}^{-1}} \tag{B.26}$$

The other corollary is that if $\det \mathbf{A} = 0$, then $\mathbf{A}$ cannot be inverted, which follows from Eq. (B.26). This is exactly the property we stated at the outset that allows us to use determinants to determine whether a problem can be solved.

Index

CPSIA information can be obtained
at www.ICGtesting.com
Printed in the USA
LVHW020413201218
601087LV00019B/536/P

9 781107 135116